Applied Strength of Materials

Third Edition

Robert L. Mott, P.E.
University of Dayton

Prentice Hall
Upper Saddle River, New Jersey 07458

Library of Congress Cataloging-in-Publication Data

Mott, Robert L.
    Applied strength of materials / Robert L. Mott.—3rd ed.
      p. cm.
    Includes bibliographical references and index.
    ISBN 0-13-376278-5
    1. Strength of materials.   I. Title.
TA405.M883   1996
620.1′12—dc20
                                                95-19852
                                                  CIP

Editor: Stephen Helba
Production Editor: Patricia A. Skidmore
Production Coordination: Deborah S. Powers, Beacon Graphics Corp.
Text Designer: Rebecca M. Bobb
Cover Designer: Retter/Patton & Associates, Inc.
Production Manager: Laura Messerly
Marketing Manager: Debbie Yarnell

This book was set in Times Roman by Beacon Graphics Corporation and was printed and bound by R. R. Donnelley & Sons Company. The cover was printed by Phoenix Color Corp.

©1996, 1990, 1978 Prentice Hall, Inc.
Simon & Schuster Company / A Viacom Company
Upper Saddle River, New Jersey 07458

Printed in the United States of America

10 9 8 7 6

ISBN: 0-13-376278-5

Prentice-Hall International (UK) Limited, *London*
Prentice-Hall of Australia Pty. Limited, *Sydney*
Prentice-Hall Canada Inc., *Toronto*
Prentice-Hall Hispanoamericana, S.A., *Mexico*
Prentice-Hall of India Private Limited, *New Delhi*
Prentice-Hall of Japan, Inc., *Tokyo*
Simon & Schuster Asia Pte. Ltd., *Singapore*
Editora Prentice-Hall do Brasil, Ltda., *Rio de Janeiro*

To my wife **Marge**
To our children **Lynné, Robert, Jr.,** and **Stephen**
And to my **Mother** and **Father**

Preface

OBJECTIVES OF THE BOOK

Applied Strength of Materials provides a comprehensive coverage of the important topics in strength of materials with an emphasis on applications, problem solving, and design of structural members, mechanical devices, and systems. The book is written for the student in a course called "Strength of Materials" in an engineering technology program at the associate degree or baccalaureate degree level, or in an applied engineering course. It is the intent of this book to provide good readability for the student, appropriate coverage of the principles of strength of materials for the faculty member teaching the subject, and a problem solving and design approach that is useful for the practicing designer of mechanical devices, structures, and systems. Educational programs in the mechanical, civil, construction, and manufacturing fields should find the book to be suitable for an introductory course in strength of materials.

STYLE

There is a heavy emphasis on the *applications* of the principles of strength of materials to mechanical, structural, and construction problems while providing a firm foundation of understanding of those principles. At the same time, the *limitations* on the use of analysis techniques are emphasized to ensure that they are applied properly.

Units used are a mixture of SI Metric and U.S. Customary units, in keeping with the dual unit usage evident in U.S. industry and construction.

PREREQUISITES

Students are expected to be able to apply the principles of statics prior to using this book. While not essential, it is recommended that students have completed an introductory course in calculus prior to studying this course. As recommended by accrediting agencies, calculus is used to develop the key principles and formulas used in this book. But the application of the formulas and the problem solving and design techniques can be accomplished without the use of the calculus.

FEATURES OF THE BOOK

The very large number of completely solved example problems are presented in a uniform, logical, and attractive format. Example Problems are set off with a distinctive graphic design and type font. Readers are guided in the process of formulating an approach to problem solving that includes:

a. Statement of the objective of the problem

b. Summary of the given information

c. Definition of the analysis technique to be used

d. Detailed development of the results with all of the equations used and unit manipulations

e. Comments on the solution to remind the reader of the important concepts involved and to judge the appropriateness of the solution. The comments often present alternate approaches or improvements to the machine element or structural member being analyzed. Thus the reader's thought process is carried beyond the requested answer into a critical review of the result. With this process, designers gain good habits of organization when solving their own problems.

The book has a strong presentation and use of *design approaches* in addition to *analysis*. There is much more information about guidelines for design of mechanical devices and structural members than in most books on this subject. The approaches are based on another book of mine, *Machine Elements in Mechanical Design*. To complement the emphasis on design, a large amount of design data is given in the Appendix and in selected chapters throughout the book.

Chapter 2, "Design Properties of Materials," includes much information and discussion on the selection and proper application of engineering materials of many types, both metallic and nonmetallic. There is an extensive introduction to composite materials given along with commentary throughout the book on the application of composites to various kinds of load-carrying members. Readers are given information about the advantages of composites relative to traditional structural materials such as metals, wood, concrete, and plastics. The reader is encouraged to seek more education and experience to learn the unique analysis and design techniques required for the proper application of composite materials. It is logical that such study would *follow* an introductory course in strength of materials, for which this book is designed.

ELECTRONIC AIDS TO PROBLEM SOLVING AND DESIGN

Most chapters include computer programming assignments along with suggestions for the use of spreadsheets, equation-solving software such as MATLAB, and graphing calculators pertinent to strength of materials. Such electronic aids, when used to supplement the basic understanding of the principles of strength of materials, lead to a deeper appreciation of those principles and their application to more complex situations.

ACKNOWLEDGMENTS

I appreciate the helpful suggestions of William Wood, Youngstown State University, and Jerome Lange, Milwaukee Area Technical College, who reviewed this text.

Robert L. Mott

Contents

CHAPTER 8   STRESS DUE TO BENDING   274

CHAPTER 9   SHEARING STRESSES IN BEAMS   326

CHAPTER 10   THE GENERAL CASE OF COMBINED
STRESS AND MOHR'S CIRCLE   361

1

Basic Concepts in Strength of Materials

1–1 OBJECTIVES OF THIS BOOK

It is essential that any product, machine, or structure be safe and stable under the loads exerted on it during any foreseeable use. The analysis and design of such devices or structures to ensure safety is the primary objective of this book.

Failure of a component of a structure can occur in several ways:

1. The material of the component could fracture completely.
2. The material may deform excessively under load so that the component is no longer suitable for its purpose.
3. The structure could become unstable and buckle, and thus be unable to carry the intended loads.

Examples of these failure modes should help you to understand the importance of your learning the principles of applied strength of materials as presented in this book.

Preventing failure by fracture. Figure 1–1 shows two rods carrying a heavy casting. Imagine that you are the person responsible for designing the rods. Certainly you would want to ensure that the rods were sufficiently strong so they will not break and allow the casting to fall, possibly causing great damage or injury to people. As the

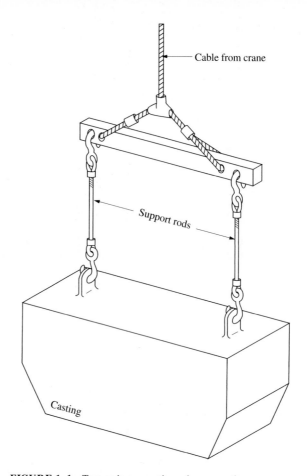

FIGURE 1–1 Two rods supporting a heavy casting.

designer of the rods, what information would you need? What design decisions must you make? Here is a partial list.

1. What is the weight and physical size of the casting?
2. Where is its center of gravity? This is important so you can decide where to place the points of attachment of the rods to the casting.
3. How will the rods be attached to the casting and to the support system at the top?
4. Of what material will the rods be made? What is its strength?
5. What will be the shape and size of the cross section of the rods?
6. How will the load of the casting be initially applied to the rods: slowly, with shock or impact, or with a jerking motion?
7. Will the rods be used for many cycles of loading during their expected life?

Knowing these factors will permit you to design the rods to be safe; that is, so they will not break under the anticipated service conditions. This will be discussed in more detail in Chapters 1 and 3.

Preventing excessive deformation.   Gears are used in mechanical drives to transmit power in such familiar systems as the transmission of a truck, the drive for a conveyor belt, or the spindle of a machine tool. For proper operation of the gears, it is essential that they are properly aligned so the teeth of the driving gear will mesh smoothly with those of the mating gear. Figure 1–2 shows two shafts carrying gears in mesh. The shafts are supported in bearings that are in turn mounted rigidly in the housing of the transmission. When the gears are transmitting power, forces are developed that tend to push them apart. These forces are resisted by the shafts so that they are loaded as shown in Figure 1–3. The action of the forces perpendicular to the shafts tends to cause them to bend which would cause the gear teeth to become misaligned. Therefore, the shafts must be designed to keep deflections at the gears to a small, acceptable level. Of course, the shafts must also be designed to be safe under the applied loads. In this type of loading, the shafts are considered to be *beams*. Chapters 8 and 12 discuss the principles of the design of beams for strength and deflection.

Stability and buckling.   A structure may collapse because one of its critical support members is unable to hold its shape under applied loads, even though the material does not fail by fracturing. An example is a long, slender post or column subjected to a downward, compressive load. At a certain critical load, the column will *buckle*. That is, it will suddenly bend or bow out, losing its preferred straight shape. When this happens, if the load remains applied, the column will totally collapse. Figure 1–4 shows a sketch of such a column that is relatively long with a thin rectangular cross section. You can demonstrate the buckling of this type of column using a simple ruler or meter stick. To prevent buckling, you must be able to specify an appropriate material, shape, and size for the cross section of a compression member of a given length so it will remain straight under the expected loads. Chapter 14 presents the analysis and design of columns.

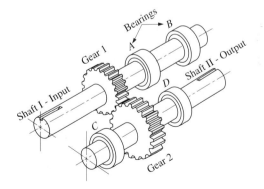

FIGURE 1–2   Two shafts carrying gears in mesh.

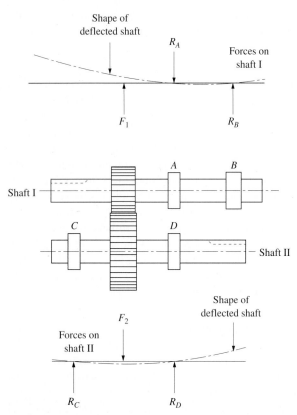

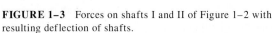

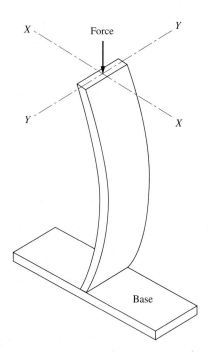

FIGURE 1–4 A slender column in compression, illustrating elastic instability or buckling.

FIGURE 1–3 Forces on shafts I and II of Figure 1–2 with resulting deflection of shafts.

In summary, design and analysis, using the principles of strength of materials, are required to ensure that a component is safe with regard to *strength, rigidity,* and *stability*. It is the objective of this book to help you gain the ability to design and analyze load-carrying components of structures and machines that will be safe and suitable for their intended functions.

1–2 OBJECTIVES OF THIS CHAPTER

In this chapter we present the basic concepts in strength of materials that will be expanded on in later chapters. After completing this chapter, you should be able to:

1. Use correct units for the quantities encountered in the study of strength of materials in both the SI metric unit system and the U.S. Customary unit system.
2. Use the terms *mass* and *weight* correctly and be able to compute the value of one when the value of the other is given.
3. Define *stress*.

4. Define *direct normal stress* and compute the value of this type of stress for both tension and compression loading.

5. Define *direct shear stress* and compute its value.

6. Identify conditions in which a load-carrying member is in *single shear* or *double shear.*

7. Draw *stress elements* showing the normal and shear stresses acting at a point in a load-carrying member.

8. Define *bearing stress* and compute its value.

9. Define *normal strain* and *shearing strain.*

10. Define *Poisson's ratio* and give its value for typical materials used in mechanical and structural design.

11. Recognize standard structural shapes and standard screw threads and use data concerning them.

12. Define *modulus of elasticity in tension.*

13. Define *modulus of elasticity in shear.*

14. Understand the responsibilities of designers.

1–3 BASIC UNIT SYSTEMS

Computations required in the application of strength of materials involve the manipulation of several sets of units in equations. For numerical accuracy it is of great importance to ensure that consistent units are used in the equations. Throughout this book units will be carried with the applicable numbers.

Because of the present transition in the United States from the traditional U.S. Customary units to metric units, both are used in this book. It is expected that persons entering or continuing an industrial career within the next several years will be faced with having to be familiar with both systems. On the one hand, many new products such as automobiles and business machines are being built using metric dimensions. Thus components and manufacturing equipment will be specified in those units. However, the transition is not being made uniformly in all areas. Designers will continue to deal with such items as structural steel, aluminum, and wood whose properties and dimensions are given in English units in standard references. Also, designers, sales and service people, and those in manufacturing must work with equipment that has already been installed and that was made to U.S. Customary system dimensions. Therefore, it seems logical that persons now working in industry should be capable of working and thinking in both systems.

The formal name for the U.S. Customary unit system is the English Gravitational Unit System (EGU). The metric system, which has been adopted internationally, is called by the French name Systéme International d'Unités, the International System of Units, abbreviated SI in this book.

In most cases, the problems in this book are worked either in the U.S. Customary unit system or the SI system rather than mixing units. In problems for which data are given in both unit systems, it is most desirable to change all data to the same system before completing the problem solution. Appendix A–25 gives conversion factors for use in performing conversions.

The basic quantities for any unit system are length, time, force, mass, temperature, and angle. Table 1–1 lists the units for these quantities in the SI unit system, and Table 1–2 lists the quantities in the U.S. Customary unit system.

Prefixes for SI units.   In the SI system, prefixes should be used to indicate orders of magnitude, thus eliminating digits and providing a convenient substitute for writing powers of 10, as generally preferred for computation. Prefixes representing steps of 1000 are recommended. Those usually encountered in strength of materials are listed in Table 1–3. Table 1–4 shows how computed results should be converted to the use of the standard prefixes for units.

TABLE 1–1   Basic quantities in the SI metric unit system

Quantity	SI unit	Other metric units
Length	meter (m)	millimeter (mm)
Time	second (s)	minute (min), hour (h)
Force	newton (N)	$kg \cdot m/s^2$
Mass	kilogram (kg)	$N \cdot s^2/m$
Temperature	kelvin (K)	degrees Celsius (°C)
Angle	radian	degree

TABLE 1–2   Basic quantities in the U.S. Customary unit system

Quantity	U.S. Customary unit	Other U.S. units
Length	foot (ft)	inch (in)
Time	second (s)	minute (min), hour (h)
Force	pound (lb)	kip*
Mass	slug	$lb \cdot s^2/ft$
Temperature	°F	
Angle	degree (deg)	radian (rad)

*1.0 kip = 1000 pounds. The name is derived from the term *kilopound*.

TABLE 1–3   Prefixes for SI units

Prefix	SI symbol	Factor
giga	G	$10^9 = 1\,000\,000\,000$
mega	M	$10^6 = 1\,000\,000$
kilo	k	$10^3 = 1\,000$
milli	m	$10^{-3} = 0.001$
micro	μ	$10^{-6} = 0.000\,001$

TABLE 1–4 Proper method of reporting computed quantities

Computed result	Reported result
0.005 48 m	5.48×10^{-3} m, or 5.48 mm
12 750 N	12.75×10^{3} N, or 12.75 kN
34 500 kg	34.5×10^{3} kg, or 34.5 Mg (megagrams)

1–4 RELATIONSHIP AMONG MASS, FORCE, AND WEIGHT

Force and mass are separate and distinct quantities. Weight is a special kind of force.

Mass refers to the amount of the substance in a body.

Force is a push or pull effort exerted on a body either by an external source or by gravity.

Weight is the force of gravitational pull on a body.

Mass, force, and weight are related by Newton's law:

$$\text{force} = \text{mass} \times \text{acceleration}$$

We often use the symbols F for force, m for mass, and a for acceleration. Then,

$$F = m \times a \quad \text{or} \quad m = F/a$$

When the pull of gravity is involved in the calculation of the weight of a mass, a takes the value of g, the acceleration due to gravity. Then, using W for weight,

➪ **Weight–Mass Relationship**

$$W = m \times g \quad \text{or} \quad m = W/g \tag{1–1}$$

We will use the following values for g:

$$\text{SI units: } g = 9.81 \text{ m/s}^2 \qquad \text{U.S. units: } g = 32.2 \text{ ft/s}^2$$

Units for mass, force, and weight. Tables 1–1 and 1–2 show the preferred units and some other convenient units for mass and force in both the SI and U.S. unit systems. The units for force are also used as the units for weight.

The newton (N) in the SI unit system is named in honor of Sir Isaac Newton and it represents the amount of force required to give a mass of 1.0 kg an acceleration of 1.0 m/s². Equivalent units for the newton can be derived as follows using Newton's law with units only:

$$F = m \times a = \text{kg·m/s}^2 = \text{newton}$$

In the U.S. unit system, the unit for force is defined to be the pound, while the unit of mass (slug) is derived from Newton's law as follows:

$$m = \frac{F}{a} = \frac{lb}{ft/s^2} = \frac{lb \cdot s^2}{ft} = slug$$

The conversion of weight and mass is illustrated in the following example problems.

Example Problem 1–1 (SI system)

A hoist lifts 425 kg of concrete. Compute the weight of the concrete that is the force exerted on the hoist by the concrete.

Solution

Objective Compute the weight of a mass of concrete.

Given $m = 425$ kg

Analysis $W = m \times g$; $g = 9.81$ m/s^2

Results $W = 425$ kg $\times$ 9.81 m/s^2 = 4170 kg·m/s^2 = 4170 N

Comment Thus 425 kg of concrete weighs 4170 N.

Example Problem 1–2 (U.S. system)

A hopper of coal weighs 8500 lb. Determine its mass.

Solution

Objective Compute the mass of a hopper of coal.

Given $W = 8500$ lb

Analysis $m = W/g$; $g = 32.2$ ft/s^2

Results $m = 8500$ lb/32.2 ft/s^2 = 264 lb·s^2/ft = 264 slugs

Comment Thus 8500 lb of coal has a mass of 264 slugs.

Chapter 1 ▪ Basic Concepts in Strength of Materials

Density and specific weight. To characterize the mass or weight of a material relative to its volume, we use the terms *density* and *specific weight,* defined as follows:

Density is the amount of mass per unit volume of a material.

Specific weight is the amount of weight per unit volume of a material.

We will use the Greek letter ρ (rho) as the symbol for density. For specific weight we will use γ (gamma).

The units for density and specific weight are summarized below.

	U.S. Customary	SI Metric
Density	slugs/ft^3	kg/m^3
Specific weight	lb/ft^3	N/m^3

Other conventions are sometimes used, often leading to confusion. For example, in the U.S., density is occasionally expressed in lb/ft^3 or lb/in^3. Two interpretations are used for this. One is that the term implies *weight density* with the same meaning as specific weight. Another is that the quantity *lb* is meant to be *pound mass* rather than *pound weight,* and the two have equal numerical values.

1–5   THE CONCEPT OF STRESS

The objective of any strength analysis is to ensure safety. Accomplishing this requires that the *stress* produced in the material of the member being analyzed is below a certain safe level that will be described in Chapter 3. Understanding the meaning of *stress* in a load-carrying member, as given below, is of utmost importance in studying strength of materials.

Stress is the internal resistance offered by a unit area of the material from which a member is made to an externally applied load.

We are concerned with what happens *inside* a load-carrying member. We must determine the magnitude of *force* exerted on each *unit area* of the material. The concept of stress can be expressed mathematically as

⇨ **Definition of Stress**

$$\text{stress} = \frac{\text{force}}{\text{area}} = \frac{F}{A} \tag{1–2}$$

In some cases, as described in the following section on *Direct Normal Stress,* the applied force is shared uniformly by the entire cross section of the member. In such cases stress can be computed by simply dividing the total force by the area of the part that resists the force. Then the level of stress will be the same at any point in any cross section.

In other cases, such as for *stress due to bending* presented in Chapter 8, the stress will vary at different positions within the cross section. Then, it is essential that you consider the level of stress *at a point*. Typically, your objective is to determine at which point the maximum stress occurs and what is its magnitude.

In the U.S. Customary unit system, the typical unit for force is the pound and the most convenient unit of area is the square inch. Thus stress is indicated in lb/in², abbreviated psi. Typically encountered stress levels in machine design and structural analysis are several thousand psi. For this reason, the unit of kip/in² is often used, abbreviated ksi. For example, if a stress is computed to be 26 500 psi, it might be reported as

$$\text{stress} = \frac{26\,500 \text{ lb}}{\text{in}^2} \times \frac{1 \text{ kip}}{1000 \text{ lb}} = \frac{26.5 \text{ kip}}{\text{in}^2} = 26.5 \text{ ksi}$$

In the SI unit system, the standard unit for force is the newton and area is in square meters. Thus the standard unit for stress is the N/m², given the name *pascal* and abbreviated Pa. Typical levels of stress are several million pascals, so the most convenient unit for stress is the megapascal or MPa. This is convenient for another reason. In calculating the cross-sectional area of load-carrying members, measurements of dimensions in mm are usually used. Then the stress would be in N/mm² and it can be shown that this is numerically equal to the unit of MPa. For example, assume that a force of 15 000 N is exerted over a square area 50 mm on a side. The resisting area would be 2500 mm² and the resulting stress would be

$$\text{stress} = \frac{\text{force}}{\text{area}} = \frac{15\,000 \text{ N}}{2500 \text{ mm}^2} = \frac{6.0 \text{ N}}{\text{mm}^2}$$

Converting this to pascals would produce

$$\text{stress} = \frac{6.0 \text{ N}}{\text{mm}^2} \times \frac{(1000)^2 \text{ mm}^2}{\text{m}^2} = 6.0 \times 10^6 \text{ N/m}^2 = 6.0 \text{ MPa}$$

This demonstrates that the unit of N/mm² is identical to the MPa, an observation that will be used throughout this book.

1–6 DIRECT NORMAL STRESS

One of the most fundamental types of stress that exists is the *normal stress,* indicated by the lowercase Greek letter σ (sigma), in which the stress acts perpendicular, or normal, to the cross section of the load-carrying member. If the stress is also uniform across the resisting area, the stress is called a *direct normal stress.*

Normal stresses can be either compressive or tensile. A compressive stress is one that tends to crush the material of the load-carrying member and to shorten the member itself. A tensile stress is one that tends to stretch the member and pull the material apart.

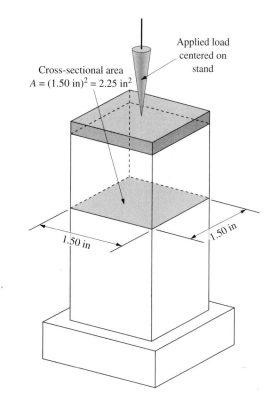

FIGURE 1–5 Example of direct compressive stress.

An example of a member subjected to compressive stress is shown in Figure 1–5. The support stand is designed to be placed under heavy equipment during assembly and the weight of the equipment tends to crush the square shaft of the support, placing it in compression.

Example Problem 1–3

Figure 1–5 shows a support stand designed to carry downward loads. Compute the stress in the square shaft at the upper part of the stand for a load of 27 500 lb. The line of action of the applied load is centered on the axis on the shaft and the load is applied through a thick plate that distributes the force to the entire cross section of the stand.

Solution

Objective Compute the stress in the upper part of the stand.

Given Load $= F = 27\,500$ lb; load is centered on the stand.
Cross section is square; dimension of each side is 1.50 in.

Analysis On any cross section of the stand, there must be an internal resisting force that acts upward to balance the downward applied load. The internal force is distributed over the cross-sectional area, as shown in Figure 1–6. Each small unit area of the cross section would support the same part of the total load. The stress produced in the square shaft tends to crush the material

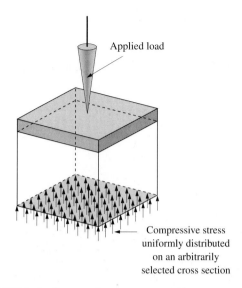

Applied load

Compressive stress
uniformly distributed
on an arbitrarily
selected cross section

FIGURE 1–6 Compressive stress on an arbitrary cross section of support stand shaft.

and it is therefore a compressive stress. Equation (1–2) can be used to compute the magnitude of the stress.

Results   stress $= \sigma =$ force/area $= F/A$ (compressive)

$A = (1.50 \text{ in})^2 = 2.25 \text{ in}^2$

$\sigma = F/A = 27\,500 \text{ lb}/2.25 \text{ in}^2 = 12\,222 \text{ lb/in}^2 = 12\,222 \text{ psi}$

Comment   This level of stress would be present on any cross section of the square shaft between its ends.

An example of a member subjected to a tensile load is shown in Figure 1–1. The two rods suspend a heavy casting from a crane and the weight of the casting tends to stretch the rods and pull them apart.

Example Problem 1–4   Figure 1–1 shows two circular rods carrying a casting weighing 11.2 kN. If each rod is 12.0 mm in diameter and the two rods share the load equally, compute the stress in the rods.

Solution   **Objective**   Compute the stress in the support rods.

Given   Casting weighs 11.2 kN. Each rod carries half the load.
Rod diameter $= D = 12.0$ mm.

Analysis Direct tensile stress produced in each rod. Use Equation (1–2).

Results $F = 11.2$ kN/2 = 5.60 kN   or   5600 N on each rod
Area = $A = \pi D^2/4 = \pi(12.0$ mm$)^2/4 = 113$ mm^2
$\sigma = F/A = 5600$ N/113 mm$^2 = 49.5$ N/mm$^2 = 49.5$ MPa

Comment Figure 1–7 shows an arbitrarily selected part of the rod with the applied load on the bottom and the internal tensile stress distributed uniformly over the cut section.

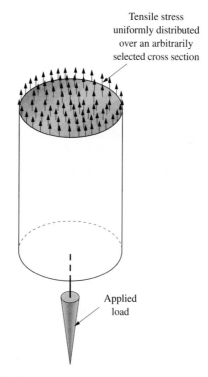

Tensile stress
uniformly distributed
over an arbitrarily
selected cross section

Applied
load

FIGURE 1–7 Tensile stress on an arbitrary cross section of a circular rod.

1–7 STRESS ELEMENTS FOR DIRECT NORMAL STRESSES

The illustrations of stresses in Figures 1–6 and 1–7 are useful for visualizing the nature of the internal resistance to the externally applied force, particularly for these cases in which the stresses are uniform across the entire cross section. In other cases it is more convenient to visualize the stress condition on a small (infinitesimal) element. Consider a small cube of material anywhere inside the square shaft of the support stand shown in Figure 1–5. There must be a net compressive force acting on the top and bottom faces of the cube, as shown in Figure 1–8(a). If the faces are considered to be unit areas,

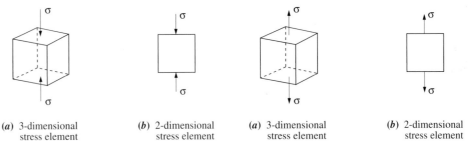

(*a*) 3-dimensional
stress element

(*b*) 2-dimensional
stress element

(*a*) 3-dimensional
stress element

(*b*) 2-dimensional
stress element

FIGURE 1–8 Stress element for compressive stresses.

FIGURE 1–9 Stress element for tensile stresses.

these forces can be considered to be the stresses acting on the faces of the cube. Such a cube is called a *stress element.*

Because the element is taken from a body in equilibrium, the element itself is also in equilibrium and the stresses on the top and bottom faces are the same. A simple stress element like this one is often shown as a two-dimensional square element rather than the three-dimensional cube, as shown in Figure 1–8(b).

Similarly, the tensile stress on any element of the rod in Figure 1–1 can be shown as in Figure 1–9, with the stress vector acting outward from the element. Note that either compressive or tensile stresses are shown acting perpendicular (normal) to the surface of the element.

1–8 DIRECT SHEAR STRESS

Shear refers to a cutting-like action. When you use common household scissors, often called *shears,* you cause one blade of the pair to slide over the other to cut (shear) paper, cloth, or other material. A sheet metal fabricator uses a similar shearing action when cutting metal for ductwork. In these examples, the shearing action progresses along the length of the line to be cut so only a small part of the total cut is being made at any given time. And, of course, the objective of the action is to actually cut the material. That is, you *want* the material to fail.

The examples described in this section along with their accompanying figures illustrate several cases where *direct shear* is produced. That is, the applied shearing force is resisted uniformly by the area of the part in shear, producing a uniform level of shearing force across the area. The symbol used for shear stress is τ, the lower case Greek letter *tau.* Then the direct shear stress can be computed from

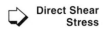
Direct Shear Stress

$$\text{shear stress} = \tau = \frac{\text{applied force}}{\text{shear area}} = \frac{F}{A_s} \qquad (1\text{–}3)$$

Figure 1–10 shows a punching operation where the objective is to actually cut one part of the material from the other. The punching action produces a slot in the flat sheet metal. The part removed in the operation is sometimes called a slug. Many different shapes can be produced by punching with either the slug or the sheet having the

Chapter 1 ▪ Basic Concepts in Strength of Materials

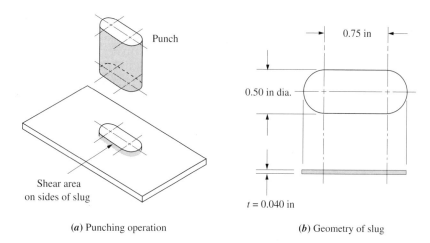

Punch

0.75 in

0.50 in dia.

Shear area
on sides of slug

$t = 0.040$ in

(a) Punching operation

(b) Geometry of slug

FIGURE 1–10   Illustration of direct shear stress in a punching operation.

hole being the desired part. Normally, the punching operation is designed so the entire shape is punched out at the same time. Therefore, the shearing action occurs along the *sides* of the slug. The area in shear for this case is computed by multiplying the length of the perimeter of the cut shape by the thickness of the sheet. That is, for a punching operation,

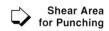

Shear Area for Punching

$$A_s = \text{perimeter} \times \text{thickness} = p \times t \qquad (1\text{–}4)$$

Example Problem 1–5   For the punching operation shown in Figure 1–10, compute the shear stress in the material if a force of 1250 lb is applied through the punch. The thickness of the material is 0.040 in.

Solution   **Objective**   Compute the shear stress in the material.

Given $F = 1250$ lb; shape to be punched shown in Fig. 1–10; $t = 0.040$ in.

Analysis   The sides of slug are placed in direct shear resisting the applied force. Use Equations (1–3) and (1–4).

Results   The perimeter, p, is

$$p = 2(0.75 \text{ in}) + \pi(0.50 \text{ in}) = 3.07 \text{ in}$$

The shear area,

$$A_s = p \times t = (3.07 \text{ in})(0.040 \text{ in}) = 0.1228 \text{ in}^2$$

Then the shear stress is

$$\tau = F/A_s = 1250 \text{ lb}/0.1228 \text{ in}^2 = 10\,176 \text{ psi}$$

Comment   At this time, we do not know whether or not this level of stress will cause the slug to be punched out; it depends on the shear strength of the material which is discussed in Chapters 2 and 3.

Single shear.   A pin or a rivet is often inserted into a cylindrical hole through mating parts to connect them, as shown in Figure 1–11. When forces are applied perpendicular to the axis of the pin, there is the tendency to cut the pin across its cross section, producing a shear stress. This action is often called *single shear* because a single cross section of the pin resists the applied shearing force. In this case, the pin is usually designed so the shear stress is below the level that would cause the pin to fail. More will be said in Chapter 3 about allowable stress levels.

Double shear.   When a pin connection is designed as shown in Figure 1–12, there are two cross sections to resist the applied force. In this arrangement the pin is said to be in *double shear.*

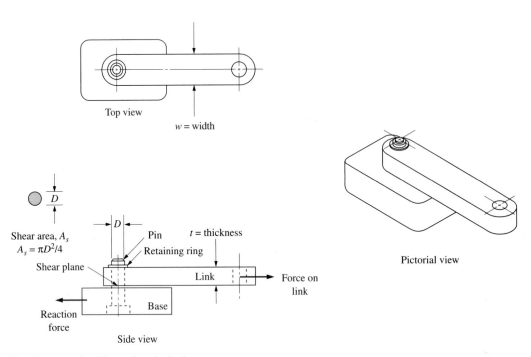

FIGURE 1–11   Pin connection illustrating single shear.

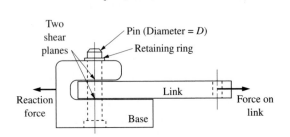

FIGURE 1–12   Pin connection illustrating double shear.

Example Problem
1–6

The force on the link in the simple pin joint shown in Figure 1–11 is 3550 N. If the pin has a diameter of 10.0 mm, compute the shear stress in the pin.

Solution   **Objective**   Compute the shear stress in the pin.

Given $F = 3550$ N; $D = 10.0$ mm

Analysis   Pin is in direct shear with one cross section of the pin resisting all of the applied force (single shear). Use Equation (1–3).

Results   The shear area, A_s, is

$$A_s = \frac{\pi D^2}{4} = \frac{\pi (10.0 \text{ mm})^2}{4} = 78.5 \text{ mm}^2$$

Then the shear stress is

$$\tau = \frac{F}{A_s} = \frac{3550 \text{ N}}{78.5 \text{ mm}^2} = 45.2 \text{ N/mm}^2 = 45.2 \text{ MPa}$$

Comment   This stress is shown in Figure 1–13 on a cut section of the pin.

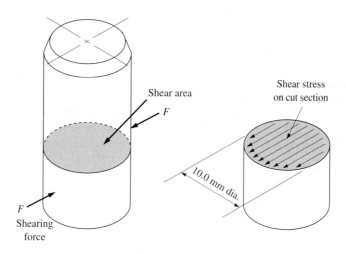

FIGURE 1–13   Direct shear stress in a pin.

Example Problem
1–7

If the pin joint just analyzed was designed as shown in Figure 1–12, compute the shear stress in the pin.

Solution   **Objective**   Compute the shear stress in the pin.

Given $F = 3550$ N; $D = 10.0$ mm (same as in Ex. Prob. 1–6).

Analysis Pin is in direct shear with two cross sections of the pin resisting the applied force (double shear). Use Equation (1–3).

Results The shear area, A_s, is

$$A_s = 2\left(\frac{\pi D^2}{4}\right) = 2\left[\frac{\pi(10.0 \text{ mm})^2}{4}\right] = 157 \text{ mm}^2$$

The shear stress in the pin is

$$\tau = \frac{F}{A_s} = \frac{3550 \text{ N}}{157 \text{ mm}^2} = 22.6 \text{ N/mm}^2 = 22.6 \text{ MPa}$$

Comment The resulting shear stress is 1/2 of the value found for single shear.

Keys. Figure 1–14 shows an important application of shear in mechanical drives. When a power transmitting element, such as a gear, chain sprocket, or belt pulley, is placed on a shaft, a key is often used to connect the two and permit the transmission of torque from one to the other. The torque produces a tangential force at the interface between the shaft and the inside of the hub of the mating element. The torque is reacted by the moment of the force on the key times the radius of the shaft. That is, $T = F(D/2)$. Then the force is $F = 2T/D$. In Figure 1–14, we have shown the force F_1 exerted by the shaft on the left side of the key. On the right side, an equal force F_2 is the reaction exerted by the hub on the key. This pair of forces tends to cut the key, producing a shear stress. Note that the shear area, A_s, is a rectangle with dimensions $b \times L$. The following example problem illustrates the computation of direct shear stress in a key.

Example Problem 1–8 Figure 1–14 shows a key inserted between a shaft and the mating hub of a gear. If a torque of 1500 lb·in is transmitted from the shaft to the hub, compute the shear stress in the key. For the dimensions of the key, use $L = 0.75$ in; $h = b = 0.25$ in. The diameter of the shaft is 1.25 in.

Solution

Objective Compute the shear stress in the key.

Given $T = 1500$ lb·in; $D = 1.25$ in; $L = 0.75$ in; $h = b = 0.25$ in

Analysis The key is in direct shear. Use Equation (1–3).

Results Shear area: $A_s = b \times L = (0.25 \text{ in})(0.75 \text{ in}) = 0.1875 \text{ in}^2$. The force on the key is produced by the action of the applied torque. The torque is reacted by the moment of the force on the key times the radius of the shaft. That is, $T = F(D/2)$. Then the force is

$$F = 2T/D = (2)(1500 \text{ lb·in})/(1.25 \text{ in}) = 2400 \text{ lb}$$

Then the shear stress is

$$\tau = F/A_s = 2400 \text{ lb}/0.1875 \text{ in}^2 = 12\,800 \text{ psi}$$

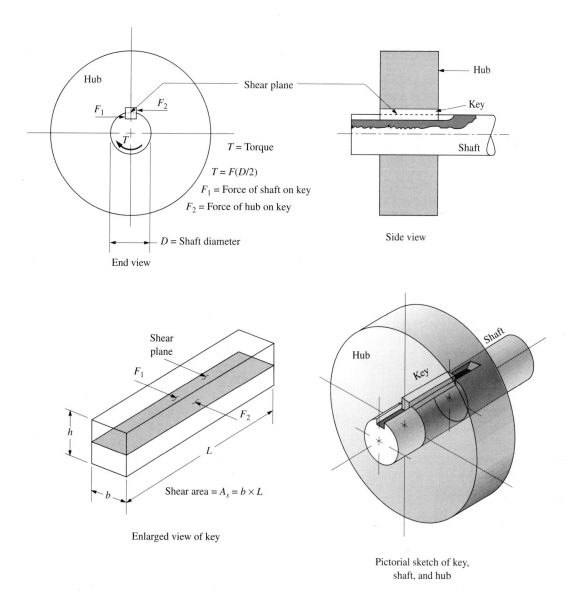

FIGURE 1–14 Direct shearing action on a key between a shaft and the hub of a gear, pulley, or sprocket in a mechanical drive system.

1–9 STRESS ELEMENTS FOR SHEAR STRESSES

An infinitesimally small cubic element of the material from the shear plane of any of the examples shown in Section 1–8 would appear as shown in Figure 1–15 with the shear stresses acting parallel to the surfaces of the cube. For example, an element taken from the shear plane of the key in Figure 1–14 would have a shear stress acting toward the left on its top surface. For equilibrium of the element with regard to

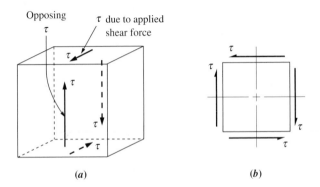

FIGURE 1–15 Stress element showing shear stress.
(a) Three-dimensional stress element. (b) Two-dimensional
stress element.

horizontal forces, there must be an equal stress acting toward the right on the bottom surface. This is the cutting action characteristic of shear.

But the two stress vectors on the top and bottom surfaces cannot exist alone because the element would tend to rotate under the influence of the couple formed by the two shearing forces acting in opposite directions. To balance that couple, a pair of equal shear stresses is developed on the vertical sides of the stress element, as shown in Figure 1–15(a).

The stress element is often drawn in the two-dimensional form shown in Figure 1–15(b). Note how the stress vectors on adjacent faces tend to meet at the corners. Such stress elements are useful in the visualization of stresses acting at a point within a material subjected to shear.

1–10 BEARING STRESS

When one solid body rests on another and transfers a load to it, the form of stress called *bearing stress* is developed at the surfaces in contact. Similar to direct compressive stress, the bearing stress is a measure of the tendency for the applied force to crush the supporting member.

Bearing stress is computed in a manner similar to direct normal stresses:

 **Bearing Stress**

$$\sigma_b = \frac{\text{applied load}}{\text{bearing area}} = \frac{F}{A_b} \tag{1–5}$$

For flat surfaces in contact, the bearing area is simply the area over which the load is transferred from one member to the other. If the two parts have different areas, the smaller area is used. Another requirement is that the materials transmitting the loads must remain nearly rigid and flat in order to maintain their ability to carry the loads. Excessive deflection will reduce the effective bearing area.

Figure 1–16 shows an example from building construction in which bearing stress is important. A hollow steel column 4.00 in square rests on a thick steel plate 6.00 in square. The plate rests on a concrete pier, which in turn rests on a base of gravel.

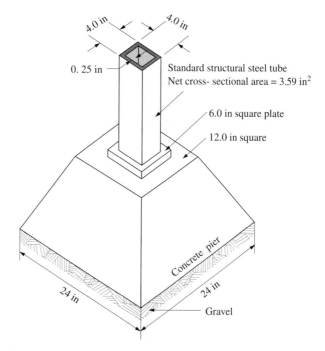

4.0 in   4.0 in

0. 25 in

Standard structural steel tube
Net cross- sectional area = 3.59 in^2

6.0 in square plate

12.0 in square

Concrete pier

24 in

24 in

Gravel

FIGURE 1–16 Bearing stress example.

These successively larger areas are necessary to limit bearing stresses to reasonable levels for the materials involved.

Example Problem 1–9 Refer to Figure 1–16. The square steel tube carries 30 000 lb of axial compressive force. Compute the compressive stress in the tube and the bearing stress between each mating surface. Consider the weight of the concrete pier to be 338 lb.

Solution **Objectives** Compute the compressive stress in the tube. Compute the bearing stress at each surface.

Given Load F = 30 000 lb compression. Weight of pier = 338 lb. Geometry of members shown in Figure 1–16.

Analysis Tube: Use direct compressive stress formula, Equation (1–3). Bearing stresses: Use Equation (1–5) for each pair of mating surfaces.

Results *Compressive stress in the tube:* (area = A = 3.59 in^2)

$$\sigma = F/A = 30\,000 \text{ lb}/3.59 \text{ in}^2 = 8356 \text{ psi}$$

Bearing stress between tube and square plate: This will be equal in magnitude to the compressive stress in the tube because the cross-sectional area of the tube is the smallest area in contact with the plate. Then,

$$\sigma_b = 8356 \text{ psi}$$

Bearing stress between plate and top of concrete pier: The bearing area is that of the square plate because it is the smallest area at the surface.

$$\sigma_b = F/A_b = 30\,000 \text{ lb}/(6.00 \text{ in})^2 = 833 \text{ psi}$$

Bearing stress between the pier and the gravel: The bearing area is that of a square, 24 in on a side. Add 338 lb for the weight of the pier.

$$\sigma_b = F/A_b = 30\,338 \text{ lb}/(24.00 \text{ in})^2 = 52.7 \text{ psi}$$

Comment Chapter 3 presents some data on allowable bearing stresses.

Bearing stresses in pin joints. Frequently in mechanical and structural design, cylindrical pins are used to connect components together. One design for such a connection is shown in Figure 1–11. When transferring a load across the pin, the bearing stress between the pin and each mating member should be computed.

The effective bearing area for the cylindrical pin in a close-fitting hole requires that the *projected area* be used, computed as the product of the diameter of the pin and the length of the surface in contact.

Example Problem 1–10 Refer to Figure 1–11. Compute the bearing stress between the 10.0 mm-diameter pin and the mating hole in the link. The force applied to the link is 3550 N. The thickness of the link is 15.0 mm and its width is 25.0 mm.

Solution **Objective** Compute the bearing stress between the mating surfaces of the pin and the inside of the hole in the link.

Given Load = F = 3550 N. t = 15.0 mm; w = 25.0 mm; D = 10.0 mm
Geometry of members shown in Figure 1–11.

Analysis Bearing stresses: Use Equation (1–5) for each pair of mating surfaces. Use the projected area of the hole for the bearing area.

Results *Between the pin and the link:*

$$A_b = D \times t = (10.0 \text{ mm})(15.0 \text{ mm}) = 150 \text{ mm}^2$$

Then the bearing stress is

$$\sigma_b = \frac{3550 \text{ N}}{150 \text{ mm}^2} = 23.7 \text{ N/mm}^2 = 23.7 \text{ MPa}$$

Contact stress. Cases of bearing stress considered earlier in this section are those in which *surfaces* are in contact and the applied force is distributed over a relatively large area. When the load is applied over a very small area, the concept of *contact stress* must be used.

Examples of contact stress situations are shown in Figure 1–17:

- cylindrical roller on a flat plate such as a steel railroad wheel on a rail
- spherical ball on a flat plate such as a ball transfer device to move heavy loads

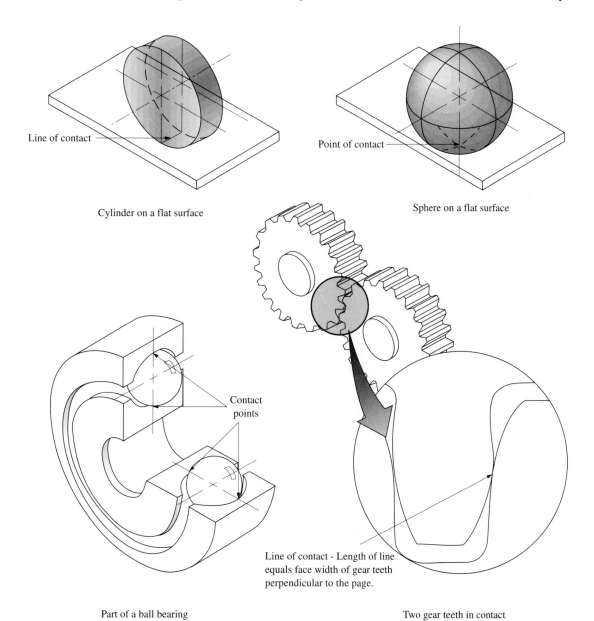

Line of contact

Cylinder on a flat surface

Point of contact

Sphere on a flat surface

Contact points

Line of contact - Length of line equals face width of gear teeth perpendicular to the page.

Part of a ball bearing

Two gear teeth in contact

FIGURE 1–17 Examples of load-carrying members subjected to contact stress.

- spherical ball on curved plate such as a ball bearing rolling in its outer race
- two convex curved surfaces such as gear teeth in contact.

The detailed analyses of contact stresses, sometimes called *Hertz* stresses, are not developed in this book. But it is important to recognize that the magnitude of contact stresses can be very high. Consider the case of a spherical ball on a flat plate carrying a downward load. A perfectly spherical surface will contact a flat plane at only a single, infinitesimally small point. Then, when applying the bearing stress relationship, $\sigma_b = F/A_b$, the magnitude of the area approaches *zero*. Then the stress approaches *infinity*. Actually, because of the elasticity of the materials in contact, there will be some deformation and the contact area becomes a finite, but small, circular area. But the local stress will still be very large. For this reason, load-carrying members subjected to contact stresses are typically made from very hard, high-strength materials.

Similarly, when a cylindrical roller contacts a flat plate, the contact is theoretically a line having a zero width. Therefore, the bearing area is theoretically zero. The elasticity of the materials will produce an actual bearing area that is a narrow rectangle, again resulting in a finite, but large, contact stress. More is said about the special case of steel rollers on steel plates in Chapter 3. See References 6 and 7 for more detailed analyses.

1–11 THE CONCEPT OF STRAIN

Any load-carrying member deforms under the influence of the load applied. The square shaft of the support stand in Figure 1–5 gets shorter as the heavy equipment is placed on the stand. The rods supporting the casting in Figure 1–1 get longer as the casting is hung onto them.

The total deformation of a load-carrying member can, of course, be measured. It will also be shown later how the deformation can be calculated.

Figure 1–18 shows an axial tensile force of 10 000 lb applied to an aluminum bar that has a diameter of 0.75 in. Before the load was applied, the length of the bar was 10.000 in. After the load is applied, the length is 10.023 in. Thus the total deformation is 0.023 in.

Strain, also called *unit deformation,* is found by dividing the total deformation by the original length of the bar. The lowercase Greek letter epsilon (ϵ) is used to denote strain:

▷ **Definition of Strain**

$$\text{strain} = \epsilon = \frac{\text{total deformation}}{\text{original length}} \tag{1–6}$$

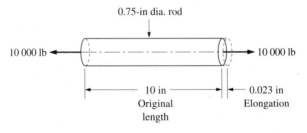

FIGURE 1–18 Elongation of a bar in tension.

For the case shown in Figure 1–18,

$$\epsilon = \frac{0.023 \text{ in}}{10.000 \text{ in}} = 0.0023 \text{ in/in}$$

Strain could be said to be dimensionless because the units in the numerator and denominator could be canceled. However, it is better to report the units as in/in or mm/mm to maintain the definition of deformation per unit length of the member. More will be said about strain and deformation in later chapters.

1–12 POISSON'S RATIO

A more complete understanding of the deformation of a member subjected to normal stress can be had by referring to Figure 1–19. The element shown is taken from the bar in Figure 1–18. The tensile force on the bar causes an elongation in the direction of the applied force, as would be expected. But at the same time, the width of the bar is being shortened. Thus a simultaneous elongation and contraction occurs in the stress element. From the elongation, the axial strain can be determined, and from the contraction, the lateral strain can be determined.

The ratio of the lateral strain on the element to the axial strain is called **Poisson's ratio** *and is a property of the material from which the load-carrying member is made.*

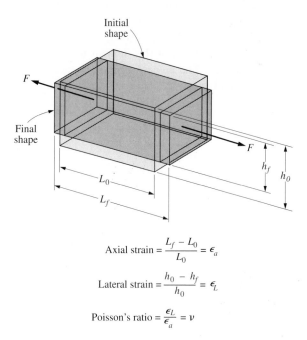

$$\text{Axial strain} = \frac{L_f - L_0}{L_0} = \epsilon_a$$

$$\text{Lateral strain} = \frac{h_0 - h_f}{h_0} = \epsilon_L$$

$$\text{Poisson's ratio} = \frac{\epsilon_L}{\epsilon_a} = \nu$$

FIGURE 1–19 Illustration of Poisson's ratio for an element in tension.

TABLE 1–5 Approximate values of
Poisson's ratio

Material	Poisson's ratio, ν
Aluminum (most alloys)	0.33
Brass	0.33
Cast iron	0.27
Concrete	0.10–0.25
Copper	0.33
Phosphor bronze	0.35
Carbon and alloy steel	0.29
Stainless steel (18–8)	0.30
Titanium	0.30

In this book the lowercase Greek letter nu (ν) is used to denote Poisson's ratio. Note that some references use mu (μ).

Most commonly used metallic materials have a Poisson's ratio value between 0.25 and 0.35. For concrete, ν varies widely depending on the grade and on the applied stress, but it usually falls between 0.1 and 0.25. Elastomers and rubber may have Poisson's ratio as high as 0.50. Approximate values for Poisson's ratio are listed in Table 1–5.

1–13 SHEARING STRAIN

Earlier discussions of strain described normal strain because it is caused by the normal tensile or compressive stress developed in a load-carrying member. Under the influence of a shear stress, shearing strain would be produced.

Figure 1–20 shows a stress element subjected to shear. The shearing action on parallel faces of the element tends to deform it angularly, as shown to an exaggerated degree. The angle γ (gamma), measured in radians, is the *shearing strain*. Only very small values of shearing strain are encountered in practical problems, and thus the dimensions of the element are only slightly changed.

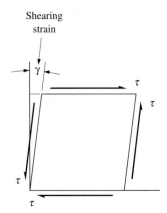

FIGURE 1–20 Shearing strain shown on a stress element.

1–14 MODULUS OF ELASTICITY

A measure of the stiffness of a material is found by computing the ratio of the normal stress on an element to the corresponding strain on the element. This ratio is called the **modulus of elasticity** *and is denoted by* **E.**

That is,

**Modulus of
Elasticity**

$$\text{modulus of elasticity} = \frac{\text{normal stress}}{\text{normal strain}}$$

$$E = \frac{\sigma}{\epsilon} \qquad (1\text{–}7)$$

A material with a higher value of E will deform less for a given stress than one with a lower value of E. A more complete term for E would be modulus of elasticity in tension or compression because it is defined in terms of normal stress. However, the term "modulus of elasticity" without a modifier is usually assumed to be the tensile modulus. More will be said about modulus of elasticity in Chapter 2 and typical values are identified there.

1–15 MODULUS OF ELASTICITY IN SHEAR

The ratio of the shearing stress to the shearing strain is called the **modulus of elasticity in shear** *or the* **modulus of rigidity,** *and is denoted by* **G.**

That is,

**Shear Modulus
of Elasticity**

$$G = \frac{\text{shearing stress}}{\text{shearing strain}} = \frac{\tau}{\gamma} \qquad (1\text{–}8)$$

G is a property of the material and is related to the tensile modulus and Poisson's ratio by

**Relation Between
G and Poisson's
Ratio**

$$G = \frac{E}{2(1 + \nu)} \qquad (1\text{–}9)$$

1–16 PREFERRED SIZES AND STANDARD SHAPES

One responsibility of a designer is to specify the final dimensions for load-carrying members. After completing the analyses for stress and deformation (strain), minimum acceptable values for dimensions are known that will ensure that the member will meet performance requirements. The designer then typically specifies the final dimensions to be standard or convenient values that will facilitate the purchase of materials and the manufacture of the parts. This section presents some guides to aid in these decisions.

Preferred basic sizes. When the component being designed will be made to the designer's specifications, it is recommended that final dimensions be specified from a set of *preferred basic sizes*. Appendix A–2 lists such data for fractional inch dimensions, decimal inch dimensions, and metric dimensions.

American Standard screw threads. Threaded fasteners and machine elements having threaded connections are manufactured according to standard dimensions to ensure interchangeability of parts and to permit convenient manufacture with standard machines and tooling. Appendix A–3 gives the dimensions of American Standard Unified threads. Sizes smaller than $\frac{1}{4}$ in are given numbers from 0 to 12, while fractional-inch sizes are specified for $\frac{1}{4}$ in and larger sizes. Two series are listed: UNC is the designation for coarse threads and UNF designates fine threads. Standard designations are illustrated below.

6–32 UNC   (number size 6, 32 threads per inch, coarse thread)
12–28 UNF   (number size 12, 28 threads per inch, fine thread)
$\frac{1}{2}$–13 UNC   (fractional size $\frac{1}{2}$ in, 13 threads per inch, coarse thread)
$1\frac{1}{2}$–12 UNF   (fractional size $1\frac{1}{2}$ in, 12 threads per inch, fine thread)

Given in the tables are the basic major diameter (D), number of threads per inch (n), and the tensile stress area found from

$$A_t = 0.7854\left(D - \frac{0.9743}{n}\right)^2 \qquad (1\text{–}9)$$

When a threaded member is subjected to direct tension, the tensile stress area is used to compute the average tensile stress. It corresponds to the smallest area that would be produced by a transverse cut across the threaded rod. For convenience, some standards use the root area or gross area and adjust the value of the allowable stress.

Metric screw threads. Appendix A–3 gives similar dimensions for metric threads. Standard metric thread designations are of the form

$$M10 \times 1.5$$

where M stands for metric, the following number is the basic major diameter in mm, and the last number is the pitch between adjacent threads in mm. Thus the designation above would denote a metric thread with a basic major diameter of 10.0 mm and a pitch of 1.5 mm. Note that pitch = $1/n$.

Standard wood beams. Appendix A–4 gives the dimensions and section properties for many standard sizes of wood beams. Note that the nominal size is simply the "name" of the beam and it relates to the approximate rough size prior to finishing. Actual finished dimensions are significantly smaller than the nominal sizes. For example, a common "2 × 4" board is actually 1.5 in wide and 3.5 in high. Also note the sketch of the orientation of the beams for the standard designation of the X- and Y-axes. When used as a beam in bending, the long dimension should be vertical for maximum strength and stiffness.

Steel structural shapes.   Steel manufacturers provide a large array of standard structural shapes that are efficient in the use of material and that are convenient for specification and installation into building structures or machine frames. Included, as shown in Table 1–6, are standard angles (L-shapes), channels (C-shapes), wide-flange beams (W-shapes), American Standard beams (S-shapes), structural tubing, and pipe. Note that the W-shapes and S-shapes are often referred to in general conversation as "I-beams" because the shape of the cross section looks like the capital letter I.

Appendix tables A–5 through A–9 give geometric properties of selected structural shapes that cover a fairly wide range of sizes. Note that many more sizes are available as presented in Reference 2. The appendix tables give data for the area of the cross section (A), the weight per foot of length, the location of the centroid of the cross section, the moment of inertia (I), the section modulus (S), and the radius of gyration (r). Some of these properties may be new to you at this time and they will be defined as needed later in the book. The values of I and S are important in the analysis and design of beams. For column analysis, I and r are needed.

Steel angles (L-shapes).   Appendix A–5 shows sketches of the typical shapes of steel angles having equal or unequal leg lengths. Called L-shapes because of the appearance of the cross section, angles are often used as tension members of trusses and towers, framing members for machine structures, lintels over windows and doors in construction, stiffeners for large plates used in housings and beams, brackets, and ledge-type supports for equipment. Some refer to these shapes as "angle iron." The standard designation takes the form shown below, using one example size:

$$L4 \times 3 \times 1/2$$

where L refers to the L-shape, 4 is the length of the longer leg, 3 is the length of the shorter leg, and 1/2 is the thickness of the legs. Dimensions are in inches.

American Standard channels (C-shapes).   See Appendix A–6 for the appearance of channels and their geometric properties. Channels are used in applications similar to those described for angles. The flat web and the two flanges provide a generally stiffer shape than angles that are more resistant to twisting under load.

The sketch at the top of the table shows that channels have tapered flanges and webs with constant thickness. The slope of the flange taper is approximately 2 in in 12 in and this makes it difficult to attach other members to the flanges. Special tapered washers are available to facilitate fastening. Note the designation of the X- and Y-axes in the sketch, defined with the web of the channel vertical which gives it the characteristic "C" shape. This is most important when using channels as beams or columns. The X-axis is located on the horizontal axis of symmetry while the dimension *x*, given in

TABLE 1–6 Designations for steel and aluminum shapes

Name of shape	Shape	Symbol	Example designation
Angle		L	L4 × 3 × ½ Appendix A–5
Channel		C	C15 × 50 Appendix A–6
Wide - flange beam		W	W14 × 43 Appendix A–7
American Standard beam		S	S10 × 35 Appendix A–8
Structural tubing - square			4 × 4 × ¼ Appendix A–9
Structural tubing - rectangular			6 × 4 × ¼ Appendix A–9
Pipe			4-inch standard weight 4-inch Schedule 40 Appendix A–12
Aluminum Association channel		C	C4 × 1.738 Appendix A–10
Aluminum Association I-beam		I	I8 × 6.181 Appendix A–11

the table, locates the Y-axis relative to the back of the web. The centroid is at the intersection of the X- and Y-axes.

The form of the standard designation for channels is

$$C15 \times 50$$

where C indicates that it is a standard C-shape

      15 is the nominal (and actual) depth in inches with the web vertical

      50 is the weight per unit length in lb/ft

Wide-flange shapes (W-shapes).   Refer to Appendix A–7. This is the most common shape used for beams, as will be discussed in Chapters 7, 8, and 12. W-shapes have relatively thin webs and somewhat thicker, flat flanges with constant thickness. Most of the area of the cross section is in the flanges, farthest away from the horizontal centroidal axis (X-axis), thus making the moment of inertia very high for a given amount of material. Note that the properties of moment of inertia and section modulus are very much higher with respect to the X-axis than they are for the Y-axis. Therefore, W-shapes are typically used in the orientation shown in the sketch in Appendix A–7. Also, these shapes are best when used in pure bending without twisting because they are quite flexible in torsion.

The standard designation for W-shapes carries much information. Consider the example,

$$W14 \times 43$$

where W indicates that it is a W-shape

      14 is the nominal depth in inches

      43 is the weight per unit length in lb/ft

The term *depth* is the standard designation for the vertical height of the cross section when placed in the orientation shown in Appendix A–7. Note from the data in the table that the actual depth is often different from the nominal depth. For the W14 × 43, the actual depth is 13.66 in.

American Standard beams (S-shapes).   Appendix A–8 shows the properties for S-shapes. Much of the discussion given for W-shapes applies to S-shapes as well. Note that, again, the weight per foot of length is included in the designation such as the S10 × 35 that weighs 35 lb/ft. For most, but not all, of the S-shapes, the actual depth is the same as the nominal depth. The flanges of the S-shapes are tapered at a slope of approximately 2 in in 12 in, similar to the flanges of the C-shapes. The X- and Y-axes are defined as shown with the web vertical.

Often wide-flange shapes (W-shapes) are preferred over S-shapes because of their relatively wide flanges, the constant thickness of the flanges, and generally higher section properties for a given weight and depth.

Structural tubing (square and rectangular).   See Appendix A–9 for the appearance and properties for steel structural tubing. These shapes are usually formed from flat sheet and welded along the length. The section properties account for the corner radii. Note the sketches showing the X- and Y-axes. The standard designation takes the form,

$$6 \times 4 \times 1/4$$

where 6 is the depth of the longer side in inches

4 is the width of the shorter side in inches

1/4 is the wall thickness in inches

Square and rectangular tubing are very useful in machine structures because they provide good section properties for members loaded as beams in bending and for torsional loading (twisting) because of the closed cross section. The flat sides often facilitate fastening of members together or the attachment of equipment to the structural members. Some frames are welded into an integral unit that functions as a stiff spaceframe. Square tubing makes an efficient section for columns.

Pipe. Hollow circular sections, commonly called *pipe,* are very efficient for use as beams, torsion members, and columns. The placement of the material uniformly away from the center of the pipe enhances the moment of inertia for a given amount of material and gives the pipe uniform properties with respect to all axes through the center of the cross section. The closed cross-sectional shape gives it high strength and stiffness in torsion as well as in bending.

Appendix A–12 gives the properties for American National Standard schedule 40 welded and seamless wrought steel pipe. This is the type of pipe often used to transport water and other fluids, but it performs well also in structural applications. Note that the actual inside and outside diameters are somewhat different from the nominal size, except for the very large sizes. Construction pipe is often called *Standard Weight Pipe* and it has the same dimensions as the schedule 40 pipe for sizes from 1/2-in to 10 in. Other "schedules" and "weights" of pipe are available with larger and smaller wall thicknesses.

Other hollow circular sections are commonly available that are referred to as *tubing*. These are available in carbon steel, stainless steel, aluminum, copper, brass, titanium, and other materials. See References 2, 3, and 4 for a variety of types and sizes of pipe and tubing.

Aluminum Association standard channels and I-beams. Appendixes A–10 and A–11 give the dimensions and section properties of channels and I-beams developed by the Aluminum Association (Reference 1). These are extruded shapes having uniform thicknesses of the webs and flanges with generous radii where they meet. The proportions of these sections are somewhat different from those of the rolled steel sections described earlier. The extruded form offers advantages in the efficient use of material and in the joining of members. This book will use the following forms for the designation of aluminum sections:

$$C4 \times 1.738 \quad \text{or} \quad I8 \times 6.181$$

where C or I indicates the basic section shape

4 or 8 indicates the depth of the shape when in the orientation shown

1.738 or 6.181 indicates the weight per unit length in lb/ft

REFERENCES

1. Aluminum Association, *Aluminum Standards and Data,* Washington, DC, 1993.

2. American Institute of Steel Construction, *Manual of Steel Construction,* 9th ed., Chicago, 1989.

3. Avallone, Eugene A. and Theodore Baumeister III, eds., *Marks' Standard Handbook for Mechanical Engineers,* 9th ed., McGraw-Hill, New York, 1987.

4. Mott, R. L., *Applied Fluid Mechanics,* 4th ed., Merrill, an imprint of Macmillan Publishing Co., New York, 1994.

5. Oberg, E., F. D. Jones, and H. L. Horton, *Machinery's Handbook,* 24th ed., Industrial Press, New York, 1992.

6. Shigley, J. E., and Mischke, C. R., *Mechanical Engineering Design,* 5th ed., McGraw-Hill Book Company, New York, 1988.

7. Young, W. C., *Roark's Formulas for Stress and Strain,* 6th ed., McGraw-Hill Book Company, New York, 1989.

PROBLEMS

Definitions

1–1. Define *mass* and state the units for mass in both the U.S. Customary unit system and the SI metric unit system.

1–2. Define *weight* and state its units in both systems.

1–3. Define *stress* and state its units in both systems.

1–4. Define *direct normal stress.*

1–5. Explain the difference between compressive stress and tensile stress.

1–6. Define *direct shear stress.*

1–7. Explain the difference between single shear and double shear.

1–8. Draw a stress element subjected to direct tensile stress.

1–9. Draw a stress element subjected to direct compressive stress.

1–10. Draw a stress element subjected to direct shear stress.

1–11. Define *normal strain* and state its units in both systems.

1–12. Define *shearing strain* and state its units in both systems.

1–13. Define *Poisson's ratio* and state its units in both systems.

1–14. Define *modulus of elasticity in tension* and state its units in both systems.

1–15. Define *modulus of elasticity in shear* and state its units in both systems.

Mass-Weight Conversions

1–16. A truck carries 1800 kg of gravel. What is the weight of the gravel in newtons?

1–17. A four-wheeled truck having a total mass of 4000 kg is sitting on a bridge. If 60% of the weight is on the rear wheels and 40% is on the front wheels, compute the force exerted on the bridge at each wheel.

1–18. A total of 6800 kg of a bulk fertilizer is stored in a flat-bottomed bin having side dimensions 5.0 m $\times$ 3.5 m. Compute the loading on the floor in newtons per square meter, or pascals.

1–19. A mass of 25 kg is suspended by a spring that has a spring scale of 4500 N/m. How much will the spring be stretched?

1–20. Measure the length, width, and thickness of this book in millimeters.

1–21. Determine your own weight in newtons and your mass in kilograms.

1–22. Express the weight found in Problem 1–16 in pounds.

1–23. Express the forces found in Problem 1–17 in pounds.

1–24. Express the loading in Problem 1–18 in pounds per square foot.

1–25. For the data in Problem 1–19, compute the weight of the mass in pounds, the spring scale in pounds per inch, and the stretch of the spring in inches.

1–26. A cast iron base for a machine weighs 2750 lb. Compute its mass in slugs.

1–27. A roll of steel hanging on a scale causes a reading of 12 800 lb. Compute its mass in slugs.

1–28. Determine your own weight in pounds and your mass in slugs.

Unit Conversions

1–29. A pressure vessel contains a gas at 1200 psi. Express the pressure in pascals.

1–30. A structural steel has an allowable stress of 21 600 psi. Express this in pascals.

1–31. The stress at which a material will break under a direct tensile load is called the *ultimate strength*. The range of ultimate strengths for aluminum alloys ranges from about 14 000 to 76 000 psi. Express this range in pascals.

1–32. An electric motor shaft rotates at 1750 rpm. Express the rotational speed in radians per second.

1–33. Express an area of 14.1 in^2 in the units of square millimeters.

1–34. An allowable deformation of a certain beam is 0.080 in. Express the deformation in millimeters.

1–35. A base for a building column measures 18.0 in by 18.0 in on a side and 12.0 in high. Compute the cross-sectional area in both square inches and square millimeters. Compute the volume in cubic inches, cubic feet, cubic millimeters, and cubic meters.

1–36. Compute the area in square inches of a rod having a diameter of 0.505 in. Then convert the result to square millimeters.

Direct Tensile and Compressive Stresses

1–37.M Compute the stress in a round bar subjected to a direct tensile force of 3200 N if the diameter of the bar is 10 mm.

1–38.M Compute the stress in a rectangular bar having cross-sectional dimensions of 10 mm by 30 mm if a direct tensile force of 20 kN is applied.

1–39.E A link in a mechanism for an automated packaging machine is subjected to a tensile force of 860 lb. If the link is square, 0.40 in on a side, compute the stress in the link.

1–40.E A circular rod, with a diameter of $\frac{3}{8}$ in, supports a heater assembly weighing 1850 lb. Compute the stress in the rod.

1–41.M A shelf is being designed to hold crates having a total mass of 1840 kg. Two support rods like those shown in Figure 1–21 hold the shelf. Each rod has a diameter of 12.0 mm. Assume that the center of gravity of the crates is at the middle of the shelf. Compute the stress in the middle portion of the rods.

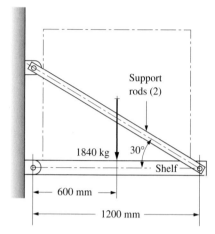

FIGURE 1–21 Shelf support for Problem 1–41.

1–42.E A concrete column base is circular, with a diameter of 8 in, and carries a direct compressive load of 70 000 lb. Compute the compressive stress in the concrete.

1–43.E Three short, square, wood blocks, $3\frac{1}{2}$ in on a side, support a machine weighing 29 500 lb. Compute the compressive stress in the blocks.

1–44.M A short link in a mechanism carries an axial compressive load of 3500 N. If it has a square cross section, 8.0 mm on a side, compute the stress in the link.

1–45.M A machine having a mass of 4200 kg is supported by three solid steel rods arranged as shown in Figure 1–22. Each rod has a diameter of 20 mm. Compute the stress in each rod.

1–46.M A centrifuge is used to separate liquids according to their densities using centrifugal force. Figure 1–23 illustrates one arm of a centrifuge having a bucket at its end to hold the liquid. In operation, the bucket and the liquid have a mass of 0.40 kg. The centrifugal force has the magnitude in newtons of

$$F = 0.010\,97 \cdot m \cdot R \cdot n^2$$

where m = rotating mass of bucket and liquid (kilograms)

R = radius to center of mass (meters)
n = rotational speed (revolutions per minute = 3000 rpm

Compute the stress in the round bar. Consider only the force due to the container.

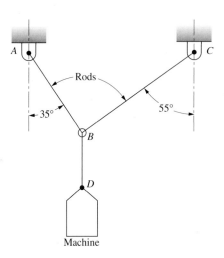

FIGURE 1–22 Support rods for Problem 1–45.

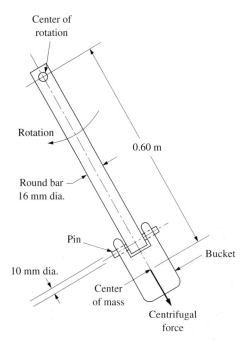

FIGURE 1–23 Centrifuge for Problem 1–46.

1–47.M A square bar carries a series of loads as shown in Figure 1–24. Compute the stress in each segment of the bar. All loads act along the central axis of the bar.

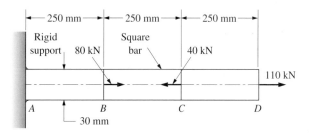

FIGURE 1–24 Bar carrying axial loads for Problem 1–47.

1–48.M Repeat Problem 1–47 for the circular bar in Figure 1–25.

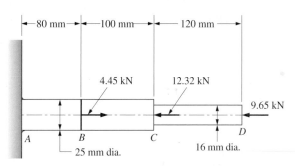

FIGURE 1–25 Bar carrying axial loads for Problem 1–48.

1–49.E Repeat Problem 1–47 for the pipe in Figure 1–26. The pipe is $1\frac{1}{2}$ in schedule 40 steel pipe.

1–50.E Compute the stress in member BD shown in Figure 1–27 if the applied force F is 2800 lb.

For Problems 51 and 52 using the trusses shown, compute the forces in all members and the stresses in the midsection, away from any joint. Refer to the Appendix for the cross-sectional area of the members indicated in the figures. Consider all joints to be pinned.

1–51.M Use Figure 1–28.

1–52.E Use Figure 1–29.

1–53.M Find the tensile stress in member AB shown in Figure 1–30.

1–54.E Figure 1–31 shows the shape of a test specimen used to measure the tensile properties of metals (as described in Chapter 2). An axial tensile

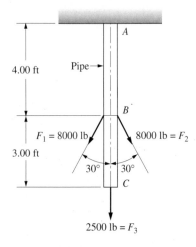

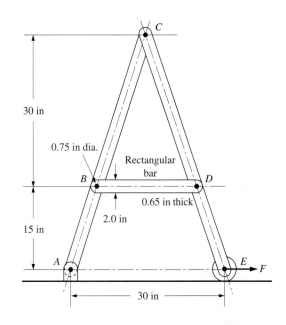

FIGURE 1–26   Pipe for Problem 1–49.

FIGURE 1–27   Frame for Problem 1–50.

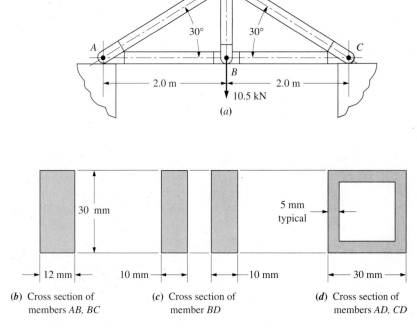

FIGURE 1–28   Truss for Problem 1–51.

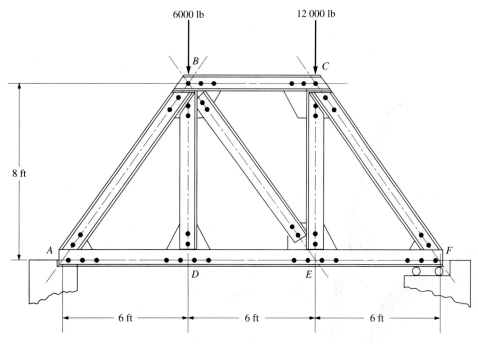

6000 lb   12 000 lb

8 ft

A   B   C   F

D   E

6 ft   6 ft   6 ft

Member specifications:

AD, DE, EF   L2 × 2 × ⅛ – doubled ——— ⅃L

BD, CE, BE   L2 × 2 × ⅛ – single ——— ⅃

AB, BC, CF   C3 × 4.1 – doubled ——— ⅃L

FIGURE 1–29   Truss for Problem 1–52.

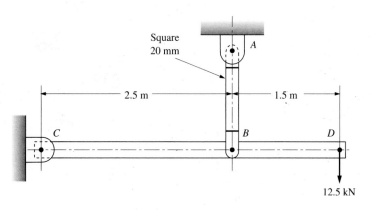

Square
20 mm   A

2.5 m   1.5 m

C   B   D

12.5 kN

FIGURE 1–30   Support for Problem 1–53.

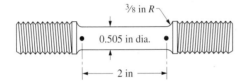

FIGURE 1–31 Tensile test specimen for Problem 1–54.

force is applied through the threaded ends and the test section is the reduced-diameter part near the middle. Compute the stress in the middle portion when the load is 12 600 lb.

1–55.E A short compression member has the cross section shown in Figure 1–32. Compute the stress in the member if a compressive force of 52 000 lb is applied in line with its centroidal axis.

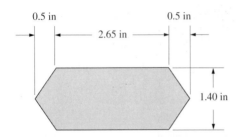

FIGURE 1–32 Short compression member for Problem 1–55.

1–56.M A short compression member has the cross section shown in Figure 1–33. Compute the stress in the member if a compressive force of 640 kN is applied in line with its centroidal axis.

Direct Shearing Stresses

1–57.M A pin connection like that shown in Figure 1–11 is subjected to a force of 16.5 kN. Determine the shear stress in the 12.0-mm-diameter pin.

1–58.M In a pair of pliers, the hinge pin is subjected to direct shear, as indicated in Figure 1–34. If the pin has a diameter of 3.0 mm and the force exerted at the handle, F_h, is 55 N, compute the stress in the pin.

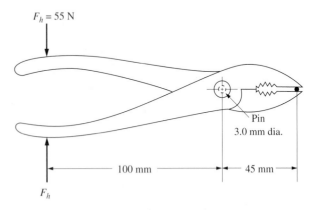

FIGURE 1–34 Pliers for Problem 1–58.

1–59.M For the centrifuge shown in Figure 1–23 and the data from Problem 1–46, compute the shear stress in the pin between the bar and the bucket.

1–60.E A notch is made in a piece of wood, as shown in Figure 1–35, in order to support an external load F of 1800 lb. Compute the shear stress in the wood.

1–61.M Figure 1–36 shows the shape of a slug to be punched from a sheet of aluminum 5.0 mm thick.

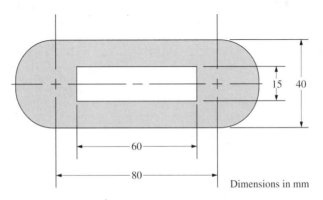

FIGURE 1–33 Short compression member for Problem 1–56.

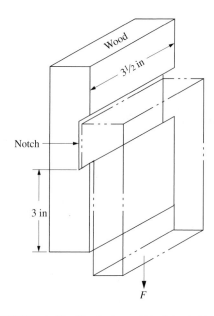

FIGURE 1–35 Notched wood block loaded in shear for Problem 1–60.

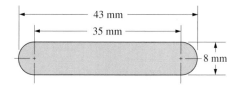

FIGURE 1–36 Shape of a slug for Problem 1–61.

Compute the shear stress in the aluminum if a punching force of 38.6 kN is applied.

1–62.E Figure 1–37 shows the shape of a slug to be punched from a sheet of steel 0.194 in thick. Compute the shear stress in the steel if a punching force of 45 000 lb is applied.

1–63.M The key in Figure 1–14 has the dimensions $b =$ 10 mm, $h = 8$ mm, and $L = 22$ mm. Determine the shear stress in the key when 95 N·m of torque is transferred from the 35-mm-diameter shaft to the hub.

1–64.E A key is used to connect a hub of a gear to a shaft, as shown in Figure 1–14. It has a rectangular cross section with $b = \frac{1}{2}$ in and $h = \frac{3}{8}$ in. The length is 2.25 in. Compute the shear stress in the key when it transmits 8000 lb·in of torque from the 2.0-in-diameter shaft to the hub.

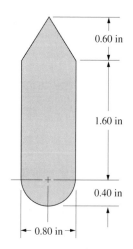

FIGURE 1–37 Shape of a slug for Problem 1–62.

1–65.E A set of two tubes is connected in the manner shown in Figure 1–38. Under a compressive load of 20 000 lb, the load is transferred from the upper tube through the pin to the connector, then

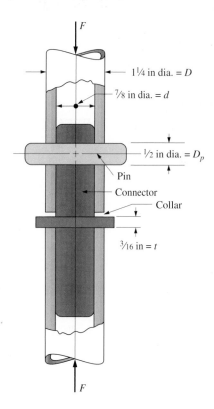

FIGURE 1–38 Connector for Problem 1–65.

through the collar to the lower tube. Compute the shear stress in the pin and in the collar.

1–66.E A small, hydraulic crane like that shown in Figure 1–39 carries an 800-lb load. Determine the shear stress that occurs in the pin at *B* which is in double shear. The pin diameter is $\frac{3}{8}$ in.

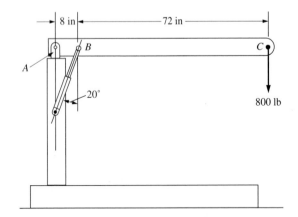

FIGURE 1–39 Hydraulic crane for Problem 1–66.

1–67.M A ratchet device on a jack stand for a truck has a tooth configuration as shown in Figure 1–40. For

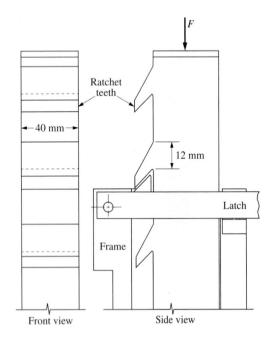

FIGURE 1–40 Ratchet for a jack stand for Problem 1–67.

a load of 88 kN compute the shear stress at the base of the tooth.

1–68.M Figure 1–41 shows an assembly in which the upper block is brazed to the lower block. Compute the shear stress in the brazing material if the force is 88.2 kN.

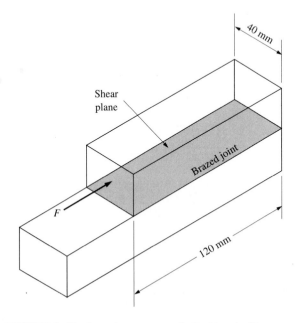

FIGURE 1–41 Brazed components for Problem 1–68.

1–69.M Figure 1–42 shows a bolt subjected to a tensile load. One failure mode would be if the circular shank of the bolt pulled out from the head, a shearing action. Compute the shear stress in the head for this mode of failure if a force of 22.3 kN is applied.

1–70.M Figure 1–43 shows a riveted lap joint connecting two steel plates. Compute the shear stress in the rivets due to a force of 10.2 kN applied to the plates.

1–71.M Figure 1–44 shows a riveted butt joint with cover plates connecting two steel plates. Compute the shear stress in the rivets due to a force of 10.2 kN applied to the plates.

Bearing Stress

1–72.E Compute the bearing stresses at the mating surfaces *A*, *B*, *C*, and *D*, in Figure 1–45.

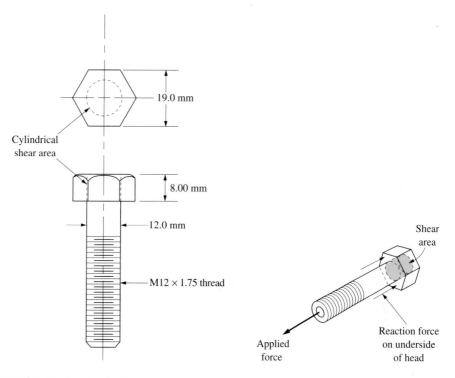

FIGURE 1-42 Bolt for Problem 1-69.

19.0 mm

Cylindrical
shear area

8.00 mm

12.0 mm

M12 × 1.75 thread

Shear
area

Applied
force

Reaction force
on underside
of head

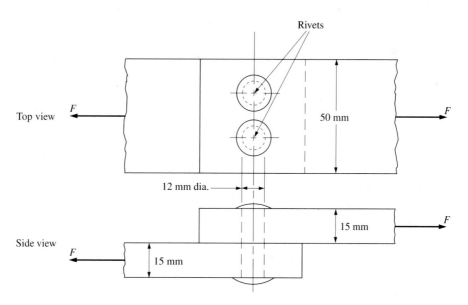

FIGURE 1-43 Riveted lap joint for Problem 1-70.

Rivets

Top view

F   F

50 mm

12 mm dia.

Side view

F   F

15 mm

15 mm

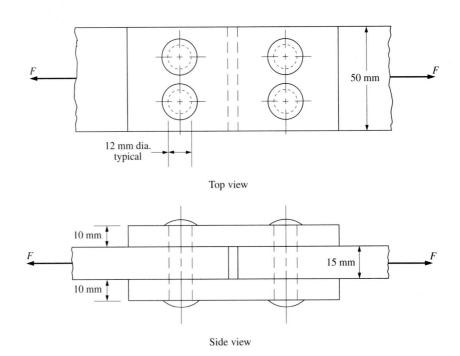

Top view

Side view

FIGURE 1–44 Riveted butt joint for Problem 1–71.

1–73.E A 2-in schedule 40 steel pipe is used as a leg for a machine. The load carried by the leg is 2350 lb.
  (a) Compute the bearing stress on the floor if the pipe is left open at its end.
  (b) Compute the bearing stress on the floor if a flat plate is welded to the bottom of the pipe having a diameter equal to the outside diameter of the pipe.

1–74.E A bolt and washer are used to fasten a wooden board to a concrete foundation as shown in Figure 1–46. A tensile force of 385 lb is created in the bolt as it is tightened. Compute the bearing stress (a) between the bolt head and the steel washer, and (b) between the washer and the wood.

1–75.E For the data of Problem 1–64, compute the bearing stress on the side of the key.

1–76.E For the data of Problem 1–65, compute the bearing stress on the tube at the interfaces with the pin and the collar.

1–77.M For the data of Problem 1–70, compute the bearing stress on the rivets.

1–78.M For the data of Problem 1–71, compute the bearing stress on the rivets.

1–79.M The heel of a woman's shoe has the shape shown in Figure 1–47. If the force on the heel is 535 N, compute the bearing stress on the floor.

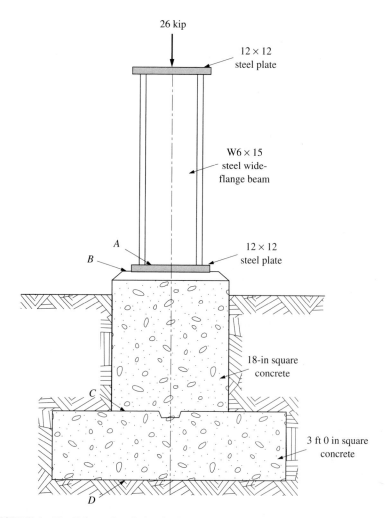

FIGURE 1–45 Column foundation for Problem 1–72.

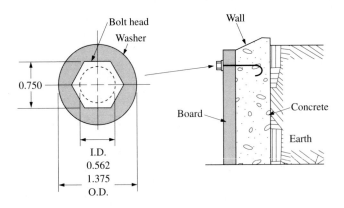

Washer dimensions (in)

FIGURE 1–46 Bolt and washer for Problem 1–74.

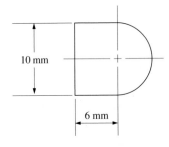

FIGURE 1–47 Shoe heel for Problem 1–79.

2

Design Properties of Materials

2–1 OBJECTIVES OF THIS CHAPTER

The study of strength of materials requires a knowledge of how external forces and moments affect the stresses and deformations developed in the material of a load-carrying member. In order to put this knowledge to practical use, however, a designer needs to know how such stresses and deformations can be withstood safely by the material. Thus, material properties as they relate to design must be understood along with the analysis required to determine the magnitude of stresses and deformations.

In this chapter we present information concerning the materials most frequently used to make components for structures and mechanical devices, emphasizing the design properties of the materials rather than their metallurgical structure or chemical composition. Although it is true that a thorough knowledge of the structure of materials is an aid to a designer, it is most important to know how the materials behave when carrying loads. This is the behavior on which we concentrate in this chapter.

First, we discuss metals, the most widely used materials in engineering design. The important properties of metals are described, along with the special characteristics of several different metals.

Nonmetals presented include wood, concrete, plastics, and composites. The manner in which the behavior of these materials differs from that of metals is discussed, along with some of their special properties.

After completing this chapter, you should be able to:

1. List typical uses for engineering materials.
2. Define *ultimate tensile strength*.
3. Define *yield point*.
4. Define *yield strength*.
5. Define *elastic limit*.
6. Define *proportional limit*.
7. Define *modulus of elasticity* and describe its relationship to the stiffness of materials.
8. Define *Hooke's law*.
9. Describe ductile and brittle behavior of materials.
10. Define *percent elongation* and describe its relationship to the ductility of materials.
11. Describe the Unified Numbering System (UNS) for metals and alloys.
12. Describe the four-digit designation system for steels.
13. Describe the important properties of carbon steels, alloy steels, stainless steels, and structural steels.
14. Describe the four-digit designation system for wrought and cast aluminum alloys.
15. Describe the aluminum temper designations.
16. Describe the design properties of copper, brass, bronze, zinc, magnesium, and titanium.
17. Describe the design properties of cast irons, including gray iron, ductile iron, austempered ductile iron, white iron, and malleable iron.
18. Describe the design properties of wood, concrete, plastics, and composites.

2–2 METALS IN MECHANICAL DESIGN

Metals are most widely used for load-carrying members in buildings, bridges, machines, and a wide variety of consumer products. Beams and columns in commercial buildings are made of structural steel or aluminum. In automobiles, a large number of steels are used, including carbon steel sheet for body panels, free-cutting alloys for machined parts, and high strength alloys for gears and heavily loaded parts. Cast iron is used in engine blocks, brake drums, and cylinder heads. Tools, springs, and other parts requiring high hardness and wear resistance are made from steel alloys containing a large amount of carbon. Stainless steels are used in transportation equipment, chemical plant products, and kitchen equipment where resistance to corrosion is required.

Aluminum sees many of the same applications as steel. Aluminum is used in many architectural products and frames for mobile equipment. Its corrosion resistance allows its use in chemical storage tanks, cooking utensils, marine equipment, and products such as highway signposts. Automotive pistons, trim, and die-cast housings for pumps and alternators are made of aluminum. Aircraft structures, engine parts, and sheet-metal skins use aluminum because of its high strength-to-weight ratio.

Copper and its alloys such as brass and bronze are used in electric conductors, heat exchangers, springs, bushings, marine hardware, and switch parts. Magnesium is often cast into truck parts, wheels, and appliance parts. Zinc sees similar service and may also be forged into machinery components and industrial hardware. Titanium has a high strength-to-weight ratio and good corrosion resistance, and thus is used in aircraft parts, pressure vessels, and chemical equipment.

Material selection requires consideration of many factors. Generally, strength, stiffness, ductility, weight, corrosion resistance, machinability, workability, weldability, appearance, cost, and availability must all be evaluated. Relative to the study of strength of materials, the first three of these factors are most important: strength, stiffness, and ductility.

Strength. Reference data listing the mechanical properties of metals will almost always include the *ultimate tensile strength* and *yield strength* of the metal. Comparison of the actual stresses in a part with the ultimate or yield strength of the material from which the part is made is the usual method of evaluating the suitability of the material to carry the applied loads safely. More is said about the details of stress analysis in Chapter 3 and subsequent chapters.

The ultimate tensile strength and yield strength are determined by testing a sample of the material in a tensile-testing machine such as that shown in Figure 2–1. A round bar or flat strip is placed in the upper and lower jaws. Figure 2–2 shows a

FIGURE 2–1 Universal testing machine for obtaining stress–strain data for materials. (Source: Tinius Olsen Testing Machine Co., Inc., Willow Grove, PA)

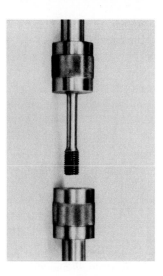

FIGURE 2–2 Tensile test specimen mounted in a holder. (Source: Tinius Olsen Testing Machine Co., Inc., Willow Grove, PA)

photograph of a typical tensile test specimen. A pulling force is applied slowly and steadily to the sample, stretching it until it breaks. During the test, a graph is made which shows the relationship between the stress in the sample and the strain or unit deformation. A typical stress–strain diagram for a low-carbon steel is shown in Figure 2–3. It can be seen that during the first phase of loading, the plot of stress versus strain is a straight line, indicating that stress is directly proportional to strain. After point *A* on the diagram, the curve is no longer a straight line. This point is called the *proportional*

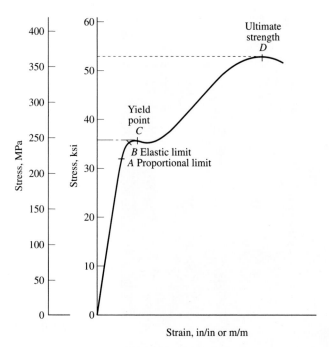

FIGURE 2–3 Typical stress–strain curve for steel.

Chapter 2 ▪ Design Properties of Materials

limit. As the load on the sample is continually increased, a point called the *elastic limit* is reached, marked *B* in Figure 2–3. At stresses below this point, the material will return to its original size and shape if the load is removed. At higher stresses, the material is permanently deformed. The *yield point* is the stress at which a noticeable elongation of the sample occurs with no apparent increase in load. The yield point is at *C* in Figure 2–3, about 36 000 psi (248 MPa). Applying still higher loads after the yield point has been reached causes the curve to rise again. After reaching a peak, the curve drops somewhat until finally the sample breaks, terminating the plot. The highest apparent stress taken from the stress–strain diagram is called the *ultimate strength*. In Figure 2–3 the ultimate strength would be about 53 000 psi (365 MPa).

The fact that the stress–strain curves in Figures 2–3 and 2–4 drop off after reaching a peak tends to indicate that the stress level decreases. Actually, it does not; the *true stress* continues to rise until ultimate failure of the material. The reason for the apparent decrease in stress is that the plot taken from a typical tensile test machine is actually *load versus elongation* rather than *stress versus strain*. The vertical axis is converted to stress by dividing the load (force) on the specimen by the *original* cross-sectional area of the specimen. When the specimen nears its breaking load, there is a reduction in diameter and consequently a reduction in the cross-sectional area. The reduced area required a lower force to continue stretching the specimen, even though the actual stress in the material is increasing. This results in the dropping curve shown in Figures 2–3 and 2–4. Because it is very difficult to monitor the decreasing diameter, and because experiments have shown that there is little difference between the true maximum stress and that found from the peak of the *apparent stress* versus strain curve, the peak is accepted as the ultimate tensile strength of the material.

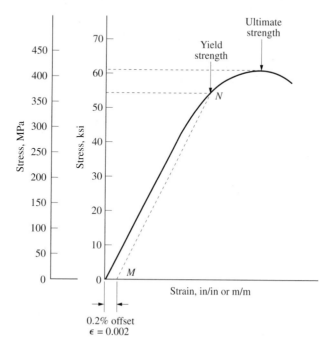

FIGURE 2–4 Typical stress–strain curve for aluminum.

A summary of the definitions of key strength properties of metals follows:

The proportional limit is the value of stress on the stress–strain curve at which the curve first deviates from a straight line.

The elastic limit is the value of stress on the stress–strain curve at which the material has deformed plastically; that is, it will no longer return to its original size and shape after removing the load.

The yield point is the value of stress on the stress–strain curve at which there is a significant increase in strain with little or no increase in stress.

The ultimate strength is the highest value of stress on the stress–strain curve.

Many metals do not exhibit a well-defined yield point like that in Figure 2–3. Some examples are high-strength alloy steels, aluminum, and titanium. However, these materials do in fact yield in the sense of deforming a sizable amount before fracture actually occurs. For these materials, a typical stress–strain diagram would look as shown in Figure 2–4. The curve is smooth with no pronounced yield point. For such materials, the yield strength is defined by a line like *M-N* drawn parallel to the straight-line portion of the test curve. Point *M* is usually determined by finding that point on the strain axis representing a strain of 0.002 in/in. This point is also called the point of 0.2% offset. The point *N*, where the offset line intersects the curve, defines the yield strength of the material, about 55 000 psi in Figure 2–4. The ultimate strength is at the peak of the curve, as was described before. *Yield strength* is used in place of yield point for these materials.

In summary, for materials that do not exhibit a pronounced yield point, the definition of *yield strength* is,

The yield strength is the value of stress on the stress–strain curve at which a straight line drawn from a strain value of 0.002 in/in (or m/m) and parallel to the straight portion of the stress–strain curve, intersects the curve.

In most wrought metals, the behavior of the materials in compression is similar to that in tension and so separate compression tests are not usually performed. However, for cast materials and nonhomogeneous materials such as wood and concrete, there are large differences between the tensile and compressive properties, and compressive testing should be done.

Stiffness. It is frequently necessary to determine how much a part will deform under load to ensure that excessive deformation does not destroy the usefulness of the part. This can occur at stresses well below the yield strength of the material, especially in very long members or in high-precision devices. Stiffness of a material is a function of its *modulus of elasticity,* sometimes called *Young's modulus.*

The modulus of elasticity, E, is a measure of the stiffness of a material determined by the slope of the straight-line portion of the stress–strain curve. It is the ratio of the change of stress to the corresponding change in strain.

This can be stated mathematically as

Modulus of Elasticity

$$E = \frac{\text{stress}}{\text{strain}} = \frac{\sigma}{\epsilon} \qquad (2-1)$$

Therefore, a material having a steeper slope on its stress–strain curve will be stiffer and will deform less under load than a material having a less steep slope. Figure 2–5 illustrates this concept by showing the straight-line portions of the stress–strain curves for steel, titanium, aluminum, and magnesium. It can be seen that if two otherwise identical parts were made of steel and aluminum, respectively, the aluminum part would deform about three times as much when subjected to the same load.

The design of typical load-carrying members in machines and structures is such that the stress is below the proportional limit; that is, in the straight-line portion of the stress–strain curve. Here we define *Hooke's law:*

> **When the level of stress in a material under load is below the proportional limit and there is a straight-line relationship between stress and strain, it is said that Hooke's law applies.**

Many of the formulas used for stress analysis are based on the assumption that Hooke's law applies. This concept is also useful for experimental stress analysis techniques in which strain is measured at a point. The corresponding stress at the point can be computed from a variation of Equation (2–1),

$$\sigma = E\epsilon \qquad (2-2)$$

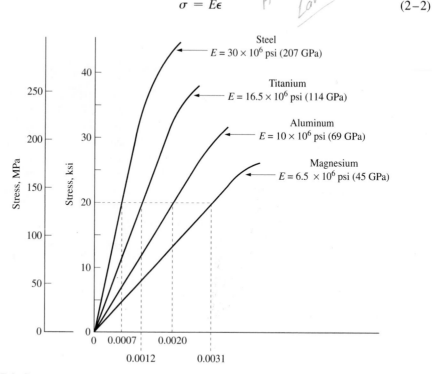

FIGURE 2–5 Modulus of elasticity for different metals.

Ductility. When metals break, their fracture can be classified as either ductile or brittle. A ductile material will stretch and yield prior to actual fracture, causing a noticeable decrease in the cross-sectional area at the fractured section. Conversely, a brittle material will fracture suddenly with little or no change in the area at the fractured section. Ductile materials are preferred for parts that carry repeated loads or are subjected to impact loading because they are usually more resistant to fatigue failure and because they are better at absorbing the impact energy.

Ductility in metals is usually measured during the tensile test by noting how much the material has elongated permanently after fracture. At the start of the test, a set of gage marks is scribed on the test sample as shown in Figure 2–6. Most tests use

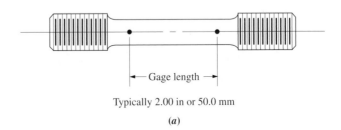

|←—— Gage length ——→|

Typically 2.00 in or 50.0 mm

(*a*)

(*b*)

FIGURE 2–6 Gage length for tensile test specimen. (a) Gage length marked on a test specimen. (b) Test specimen in a fixture used to mark gage length (Source: Tinius Olsen Testing Machine Co., Inc., Willow Grove, PA)

2.000 in or 50.0 mm for the gage length as shown in the figure. Very ductile structural steels sometimes use 8.000 in or 200.0 mm for the gage length. After the sample has been pulled to failure, the broken parts are fitted back together and the distance between the marks is again measured. From these data, the *percent elongation* is computed from

Percent Elongation

$$\text{percent elongation} = \frac{\text{final length} - \text{gage length}}{\text{gage length}} \times 100\% \qquad (2-3)$$

A metal is considered to be *ductile* if its percent elongation is greater than about 5.0%. A material with a percent elongation under 5.0% is considered to be *brittle* and does not exhibit the phenomenon of yielding. Failure of such materials is sudden, without noticeable deformation prior to ultimate fracture. In most structural and mechanical design applications, ductile behavior is desirable and the percent elongation of the material should be significantly greater than 5.0%. A high-percentage elongation indicates a highly ductile material.

In summary, the following definitions are used to describe ductility for metals:

The percent elongation is the ratio of the plastic elongation of a tensile specimen after ultimate failure within a set of gage marks to the original length between the gage marks. It is one measure of ductility.

A ductile material is one that can be stretched, formed, or drawn to a significant degree before fracture. A metal that exhibits a percent elongation greater than 5.0% is considered to be ductile.

A brittle material is one that fails suddenly under load with little or no plastic deformation. A metal that exhibits a percent elongation less than 5.0% is considered to be brittle.

Virtually all wrought forms of steel and aluminum alloys are ductile. But the higher-strength forms tend to have lower ductility and the designer is often forced to compromise strength and ductility in the specification of a material. Gray cast iron, many forms of cast aluminum, and some high-strength forms of wrought or cast steel are brittle.

Failure modes. For most designs, a machine element or structural member is considered to have failed when:

1. It actually breaks; that is, the stress exceeds the ultimate strength of the material.
2. The material deforms plastically; that is, it is stressed above its yield strength.
3. Excessive elastic deformation occurs that renders the member unsuitable for its intended use.

Deformation of the material before yielding occurs is dependent on the stiffness of the material, indicated by the modulus of elasticity. Methods for computing the total deformation of load-carrying members are presented in later chapters. There are no absolute standards for the level of deformation that would constitute failure. Rather, the designer must make a judgment based on the use of the structure or machine. Reference 11 presents some guidelines.

Classification of metals and alloys.   Various industry associations take responsibility for setting standards for the classification of metals and alloys. Each has its own numbering system, convenient to the particular metal covered by the standard. But this leads to confusion at times when there is overlap between two or more standards and when widely different schemes are used to denote the metals. Order has been brought to the classification of metals by the use of the Unified Numbering Systems (UNS) as defined in the Standard E 527-74 (Reapproved 1981), **Standard Practice for Numbering Metals and Alloys (UNS),** by the American Society for Testing and Materials (ASTM). Besides listing materials under the control of ASTM itself, the UNS coordinates designations of:

> The Aluminum Association (AA)
>
> American Iron and Steel Institute (AISI)
>
> Copper Development Association (CDA)
>
> Society of Automotive Engineers (SAE)

The primary series of numbers within UNS are listed in Table 2–1 along with the organization having responsibility for assigning numbers within each series.

TABLE 2–1   Unified numbering system (UNS)

Number series	Types of metals and alloys	Responsible organization
Nonferrous metals and alloys		
A00001–A99999	Aluminum and aluminum alloys	AA
C00001–C99999	Copper and copper alloys	CDA
E00001–E99999	Rare earth metals and alloys	ASTM
L00001–L99999	Low melting metals and alloys	ASTM
M00001–M99999	Miscellaneous nonferrous metals and alloys	ASTM
N00001–N99999	Nickel and nickel alloys	SAE
P00001–P99999	Precious metals and alloys	ASTM
R00001–R99999	Reactive and refractory metals and alloys	SAE
Z00001–Z99999	Zinc and zinc alloys	ASTM
Ferrous metals and alloys		
D00001–D99999	Steels, mechanical properties specified	SAE
F00001–F99999	Cast irons and cast steels	ASTM
G00001–G99999	Carbon and alloy steels (includes former SAE carbon and alloy steels)	AISI
H00001–H99999	H-steels; specified hardenability	AISI
J00001–J99999	Cast steels (except tool steels)	ASTM
K00001–K99999	Miscellaneous steels and ferrous alloys	ASTM
S00001–S99999	Heat and corrosion resistant (stainless) steels	ASTM
T00001–T99999	Tool steels	AISI

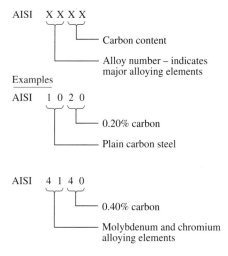

FIGURE 2–7 Steel designation system.

Many alloys within UNS retain the familiar numbers from the systems used for many years by the individual association. For example, the following section describes the four-digit designation system of the AISI for carbon and alloy steels. Figure 2–7 shows two examples: AISI 1020, a carbon steel, and AISI 4140, an alloy steel. These steels would carry the UNS designations, G10200 and G41400, respectively.

2–3 STEEL

The term *steel* refers to alloys of iron and carbon and, in many cases, other elements. Because of the large number of steels available, they will be classified in this section as carbon steels, alloy steels, stainless steels, and structural steels.

For *carbon steels* and *alloy steels,* a four-digit designation code is used to define each alloy. Figure 2–7 shows the significance of each digit. The four digits would be the same for steels classified by the American Iron and Steel Institute (AISI) and the Society of Automotive Engineers (SAE). Classification by the American Society for Testing and Materials (ASTM) will be discussed later.

Usually, the first two digits in a four-digit designation for steel will denote the major alloying elements, other than carbon, in the steel. The last two digits denote the average percent (or points) of carbon in the steel. For example, if the last two digits are 40, the steel would have about 0.4% carbon content. Carbon is given such a prominent place in the alloy designation because, in general, as carbon content increases, the strength and hardness of the steel also increases. Carbon content usually ranges from a low of 0.1% to about 1.0%. It should be noted that while strength increases with increasing carbon content, the steel also becomes more brittle.

Table 2–2 shows the major alloying elements, which correspond to the first two digits of the steel designation. Table 2–3 lists come common alloys along with the principal uses for each.

TABLE 2–2 Major alloying elements in steel alloys

Steel AISI No.	Alloying elements	Steel AISI No.	Alloying elements
10xx	Plain carbon	46xx	Molybdenum-nickel
11xx	Sulfur (free-cutting)	47xx	Molybdenum-nickel-chromium
13xx	Manganese	48xx	Molybdenum-nickel
14xx	Boron	5xxx	Chromium
2xxx	Nickel	6xxx	Chromium-vanadium
3xxx	Nickel-chromium	8xxx	Nickel-chromium-molybdenum
4xxx	Molybdenum	9xxx	Nickel-chromium-molybdenum (except 92xx)
41xx	Molybdenum-chromium	92xx	Silicon-manganese
43xx	Molybdenum-chromium-nickel		

TABLE 2–3 Common steel alloys and typical uses

Steel AISI No.	Typical uses
1020	Structural steel, bars, plate
1040	Machinery parts, shafts
1050	Machinery parts
1095	Tools, springs
1137	Shafts, screw machine parts (free-cutting alloy)
1141	Shafts, machined parts
4130	General-purpose, high-strength steel; shafts, gears, pins
4140	Same as 4130
4150	Same as 4130
5160	High-strength gears, bolts
8760	Tools, springs, chisels

Conditions for steels. The mechanical properties of carbon and alloy steels are very sensitive to the manner in which they are formed and to heat-treating processes. Appendix A–13 lists the ultimate strength, yield strength, and percent elongation for a variety of steels in a variety of conditions. Note that these are typical or example properties and may not be relied on for design. Material properties are dependent on many factors, including section size, temperature, actual composition, variables in processing, and fabrication techniques. It is the responsibility of the designer to investigate the possible range of properties for a material and to design load-carrying members to be safe regardless of the combination of factors present in a given situation.

Generally, the more severely a steel is worked, the stronger it will be. Some forms of steel, such as sheet, bar, and structural shapes, are produced by *hot rolling* while still at an elevated temperature. This produces a relatively soft, low-strength steel which has a very high ductility and is easy to form. Rolling the steel to final form while at or near room temperature is called *cold rolling* and produces a higher strength and somewhat lower ductility. Still higher strength can be achieved by *cold drawing,* drawing the material through dies while it is at or near room temperature. Thus, for these three popular methods of producing steel shapes, the cold-drawn (CD) form results in the highest strength, followed by the cold-rolled (CR) and hot-rolled (HR) forms. This can be seen in Appendix A–13 by comparing the strength of the same steel, say AISI 1040, in the hot-rolled and cold-drawn conditions.

Alloy steels are usually heat-treated to develop specified properties. Heat treatment involves raising the temperature of the steel to above about 1450 to 1650°F (depending on the alloy) and then cooling it rapidly by quenching in either water or oil. After quenching, the steel has a high strength and hardness, but it may also be brittle. For this reason a subsequent treatment called *tempering* (or *drawing*) is usually performed. The steel is reheated to a temperature in the range 400 to 1300°F and then cooled. The effect of tempering an alloy steel can be seen by referring to Figure 2–8. Thus the properties of a heat-treated steel can be controlled by specifying a tempering temperature. In Appendix A–13, the condition of heat-treated alloys is described in a manner like OQT 400. This means that the steel was heat-treated by quenching in oil and then tempered at 400°F. Similarly, WQT 1300 means water quenched and tempered at 1300°F.

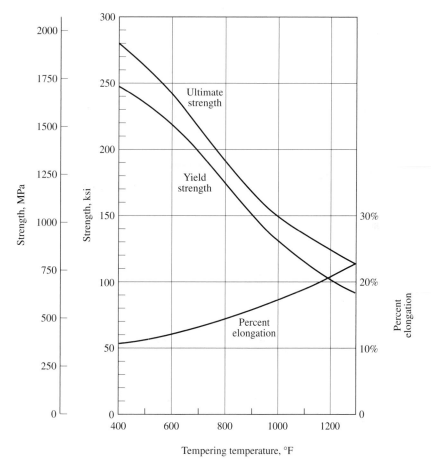

FIGURE 2–8 Effect of tempering temperature on the strength and ductility of an alloy steel.

The properties of heat-treated steels at tempering temperatures of 400 and 1300°F show the total range of properties that the heat-treated steel can have. However, typical practice would specify tempering temperatures no lower than 700°F because steels tend to be too brittle at the lower tempering temperatures. The properties of several alloys are listed in Appendix A–13 at tempering temperatures of 700°F, 900°F, 1100°F, and 1300°F to give you a feeling for the range of strengths available. These alloys are good choices for selecting materials in problems in later chapters. Strengths at intermediate temperatures can be found by interpolation.

Annealing and *normalizing* are thermal treatments designed to soften steel, give it more uniform properties, make it easier to form, or relieve stresses developed in the steel during such processes as welding, machining, or forming. Two of the types of annealing processes used are full anneal and stress-relief anneal. Figure 2–9 illustrates these thermal treatment processes, along with quenching and tempering.

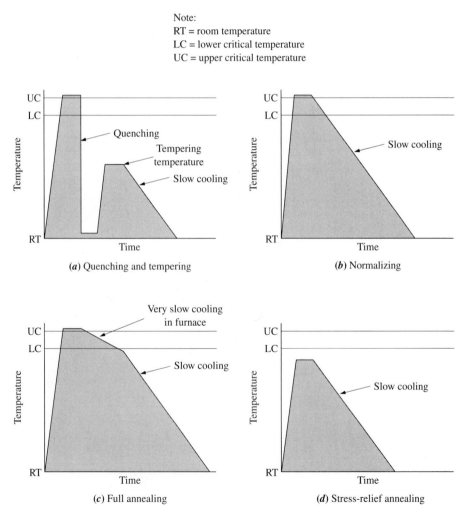

Note:
RT = room temperature
LC = lower critical temperature
UC = upper critical temperature

(a) Quenching and tempering

(b) Normalizing

(c) Full annealing

(d) Stress-relief annealing

FIGURE 2–9 Heat treatments for steel.

Normalizing of steel starts by heating it to approximately the same temperature (called the *upper critical temperature*) as would be required for through-hardening by quenching, as described before. But rather than quenching, the steel is cooled in still air to room temperature. This results in a uniform, fine-grained structure, improved ductility, better impact resistance, and improved machinability.

Full annealing involves heating to above the upper critical temperature followed by very slow cooling to the lower critical temperature and then cooling in still air to room temperature. This is one of the softest forms of the steel and makes it more workable for shearing, forming, and machining.

Stress-relief annealing consists of heating to below the lower critical temperature, holding to achieve uniform temperature throughout the part, and then cooling to room temperature. This relieves residual stresses and prevents subsequent distortion.

Stainless steels. *Stainless steels* get their name because of their corrosion resistance. The primary alloying element in stainless steels is chromium, being present at about 17% in most alloys. A minimum of 10.5% chromium is used, and it may range as high as 27%.

Although over 40 grades of stainless steel are available from steel producers, they are usually categorized into three series containing alloys with similar properties. The properties of some stainless steels are listed in Appendix A–14.

The 200 and 300 series steels have high strength and good corrosion resistance. They can be used at temperatures up to about 1200°F with good retention of properties. Because of their structure, these steels are essentially nonmagnetic. Good ductility and toughness and good weldability make them useful in chemical processing equipment, architectural products, and food-related products. They are not hardenable by heat treatment, but they can be strengthened by cold working. The range of cold working is typically given as *quarter hard, half hard, three-quarters hard,* and *full hard,* with strength increasing with higher hardness. But ductility decreases with increasing hardness. Appendix A–14 shows the properties of some stainless steel alloys at two conditions, annealed and full hard, indicating the extremes of strengths available. The annealed condition is sometimes called *soft.*

The AISI 400 series steels are used for automotive trim and for chemical processing equipment such as acid tanks. Certain alloys can be heat-treated so they can be used as knife blades, springs, ball bearings, and surgical instruments. These steels are magnetic.

Precipitation hardening steels, such as 17-4PH and PH13-8Mo, are hardened by holding at an elevated temperature, about 900 to 1100°F (480 to 600°C). Such steels are generally classed as *high-strength stainless steels* having yield strengths about 180 000 psi (1240 MPa) or higher.

Structural steels. *Structural steels* are produced in the forms of sheet, plate, bars, tubing, and structural shapes such as I-beams, wide-flange beams, channels, and angles. The American Society for Testing and Materials (ASTM) assigns a number designation to these steels which is the number of the standard that defines the required minimum properties. Appendix A–15 lists six frequently used grades of structural steels and their properties.

A very popular steel for structural applications is ASTM A36, a carbon steel used for many commercially available shapes, plates, and bars. It has a minimum yield point of 36 ksi (248 MPa), is weldable, and is used in bridges, buildings, and for general structural purposes.

ASTM A242 steel is called a high-strength, low-alloy steel. Produced as shapes, plates, and bars, it may be specified instead of A36 steel to allow the use of a smaller, lighter member. In sizes up to $\frac{3}{4}$ in thick, it has a minimum yield point of 50 ksi (345 MPa). From $\frac{3}{4}$ to $1\frac{1}{2}$ in thick, the miminum yield point of 46 ksi (317 MPa) is specified. Alloy A242 is for general structural purposes and is called a *weathering steel* since its corrosion resistance is about four times that of plain carbon steel. Cost, of course, must be considered before specifying this alloy.

Alloy A514 is a high-strength alloy, heat-treated by quenching and tempering to a yield point of 100 ksi (690 MPa) minimum. Produced as plates, it is used in welded bridges and similar structures.

Structural tubing is either round, square, or rectangular and is frequently made from ASTM A501 steel (hot formed) or ASTM A500 (cold formed) (see Appendixes A–9 and A–15).

Another general-purpose structural steel is ASTM A572, available as shapes, plates, and bars, and in grades 42 to 65. The grade number refers to the minimum yield point of the grade in ksi, and can be 42, 45, 50, 55, 60, and 65.

In summary, steels come in many forms and have a wide range of strengths and other properties. Selection of a suitable steel is indeed an art, supported by a knowledge of the significant features of each alloy.

2–4 CAST IRON

The attractive properties of cast iron include low cost, good wear resistance, good machinability, and its ability to be cast into complex shapes. Five varieties will be discussed here: gray iron, ductile iron, austempered ductile iron, white iron, and malleable iron.

Gray iron is used in automotive engine blocks, machinery bases, brake drums, and large gears. It is usually specified by giving a grade number corresponding to the minimum ultimate tensile strength. For example, grade 20 gray cast iron has a minimum ultimate strength of 20 000 psi (138 MPa); grade 60 has $s_u = 60\,000$ psi (414 MPa), and so on. The usual grades available are from 20 to 60. Gray iron is somewhat brittle, so that yield strength is not usually reported as a property. An outstanding feature of gray iron is that its compressive strength is very high, about three to five times as high as the tensile strength. This should be taken into account in design, especially when a part is subjected to bending stresses, as discussed in Chapter 8.

Because of variations in the rate of cooling after the molten cast iron is poured into a mold, the actual strength of a particular section of a casting is dependent on the thickness of the section. Figure 2–10 illustrates this for grade 40 gray iron. The range of in-place strength may range from as high as 52 000 psi (359 MPa) to as low as 27 000 psi (186 MPa).

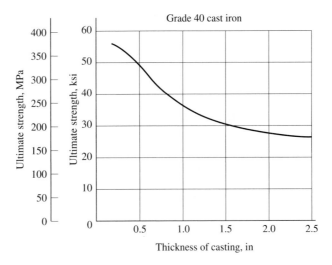

FIGURE 2–10 Strength versus thickness for grade 40 gray cast iron.

Ductile iron differs from gray iron in that it does exhibit yielding, has a greater percent elongation, and has generally higher tensile strength. Grades of ductile iron are designated by a three-number system such as 80-55-6. The first number indicates the minimum ultimate tensile strength in ksi; the second is the yield strength in ksi; and the third is the percent elongation. Thus grade 80-55-6 has an ultimate strength of 80 000 psi, a yield strength of 55 000 psi, and a percent elongation of 6%. Uses for ductile iron include crankshafts and heavily loaded gears.

The strength of ductile iron can be increased by nearly a factor of 2 by a process called *austempering*. First the castings are heated to between 1500 and 1700°F and held to achieve a uniform structure. Then they are quenched rapidly to a lower temperature, 450 to 750°F, and held again. After several hours of isothermal soaking, the castings are allowed to cool to room temperature.

Austempered ductile iron (ADI) has higher strength and better ductility than standard ductile irons, as can be seen in Appendix A–16. This allows parts to be smaller and lighter and makes ADI desirable for uses such as automotive gears, crankshafts, and structural members for construction and transportation equipment, replacing wrought or cast steels.

White iron is produced by rapidly chilling a casting of either gray iron or ductile iron during the solidification process. The chilling is typically applied to selected areas which become very hard and have a high wear resistance. The chilling does not allow the carbon in the iron to precipitate out during solidification, resulting in the white appearance. Areas away from the chilling medium solidify more slowly and acquire the normal properties of the base iron. One disadvantage of the chilling process is that the white iron is very brittle.

Malleable iron is used in automotive and truck parts, construction machinery, and electrical equipment. It does exhibit yielding, has tensile strengths comparable to ductile iron, and has ultimate compressive strengths which are somewhat higher than ductile iron. Generally, a five-digit number is used to designate malleable iron grades.

For example, grade 40010 has a yield strength of 40 000 psi (276 MPa) and a percent elongation of 10%.

Appendix A–16 lists the mechanical properties of several grades of gray iron, ductile iron, ADI, and malleable iron.

2–5 ALUMINUM

Alloys of aluminum are designed to achieve optimum properties for specific uses. Some are produced primarily as sheet, plate, bars, or wire. Standard structural shapes and special sections are often extruded. Several alloys are used for forging, while others are special casting alloys. Appendix A–17 lists the properties of selected aluminum alloys.

Aluminum in wrought form uses a four-digit designation to define the several alloys available. The first digit indicates the alloy group according to the principal alloying element. The second digit denotes a modification of the basic alloy. The last two digits identify a specific alloy within the group. A brief description of the seven major series of aluminum alloys follows.

1000 series, 99.0% aluminum or greater. Used in chemical and electrical fields. Excellent corrosion resistance, workability, and thermal and electrical conductivity. Low mechanical properties.

2000 series, copper alloying element. Heat-treatable with high mechanical properties. Lower corrosion resistance than most other alloys. Used in aircraft skins and structures.

3000 series, manganese alloying element. Non-heat-treatable, but moderate strength can be obtained by cold working. Good corrosion resistance and workability. Used in chemical equipment, cooking utensils, residential siding, and storage tanks.

4000 series, silicon alloying element. Non-heat-treatable with a low melting point. Used as welding wire and brazing alloy. Alloy 4032 used as pistons.

5000 series, magnesium alloying element. Non-heat-treatable, but moderate strength can be obtained by cold working. Good corrosion resistance and weldability. Used in marine service, pressure vessels, auto trim, builder's hardware, welded structures, TV towers, and drilling rigs.

6000 series, silicon and magnesium alloying elements. Heat-treatable to moderate strength. Good corrosion resistance, formability, and weldability. Used as heavy-duty structures, truck and railroad equipment, pipe, furniture, architectural extrusions, machined parts, and forgings. Alloy 6061 is one of the most versatile available.

7000 series, zinc alloying element. Heat-treatable to very high strength. Relatively poor corrosion resistance and weldability. Used mainly for aircraft structural members. Alloy 7075 is among the highest strength alloys available. It is produced in most rolled, drawn, and extruded forms and is also used in forgings.

Aluminum temper designations. Since the mechanical properties of virtually all aluminum alloys are very sensitive to cold working or heat treatment, suffixes are applied to the four-digit alloy designations to describe the temper. The most frequently used temper designations are described as follows:

O temper. Fully annealed to obtain the lowest strength. Annealing makes most alloys easier to form by bending or drawing. Parts formed in the annealed condition are frequently heat-treated later to improve properties.

H temper, strain-hardened. Used to improve the properties of non-heat-treatable alloys such as those in the 1000, 3000, and 5000 series. The H is always followed by a two- or three-digit number to designate a specific degree of strain hardening or special processing. The second digit following the H ranges from 0 to 8 and indicates a successively greater degree of strain hardening, resulting in higher strength. Appendix A–17 lists the properties of several aluminum alloys. Referring to alloy 3003 in that table shows that the yield strength is increased from 18 000 psi (124 MPa) to 27 000 psi (186 MPa) as the temper is changed from H12 to H18.

T temper, heat-treated. Used to improve strength and achieve a stable condition. The T is always followed by one or more digits indicating a particular heat treatment. For wrought products such as sheet, plate, extrusions, bars, and drawn tubes, the most frequently used designations are T4 and T6. The T6 treatment produces higher strength but generally reduces workability. In Appendix A–17 several heat-treatable alloys are listed in the O, T4, and T6 tempers to illustrate the change in properties.

Cast aluminum alloys are designated by a modified four-digit system of the form, XXX.X, in which the first digit indicates the main alloy group according to the major alloying elements. Table 2–4 shows the groups. The second two digits indicate the specific alloy within the group or indicate the aluminum purity. The last digit, after the decimal point, indicates the product form: 0 for castings, and 1 or 2 for ingots.

Aluminum is also sensitive to the manner in which it is produced, the size of the section, and temperature. Appendix A–17 lists typical properties and cannot be relied on for design. References 1 and 2 give extensive data on minimum strengths.

TABLE 2–4 Cast aluminum alloy groups

Group	Major alloying elements
1XX.X	99% or greater aluminum
2XX.X	Copper
3XX.X	Silicon, copper, magnesium
4XX.X	Silicon
5XX.X	Magnesium
6XX.X	(Unused series)
7XX.X	Zinc
8XX.X	Tin
9XX.X	Other elements

2-6 COPPER, BRASS, AND BRONZE

The name *copper* is properly used to denote virtually the pure metal having 99% or more copper. Its uses are primarily as electric conductors, switch parts, and motor parts which carry electric current. Copper and its alloys have good corrosion resistance, are readily fabricated, and have an attractive appearance. The main alloys of copper are beryllium copper, the brasses, and bronzes. Each has its special properties and applications.

Beryllium copper has very high strength and good electrical conductivity. Its uses include switch parts, fuse clips, electric connectors, bellows, Bourdon tubing for pressure gauges, and springs.

Brasses are alloys of copper and zinc. They have good corrosion resistance, workability, and a pleasing appearance, which leads to applications in automobile radiators, lamp bases, heat-exchanger tubes, marine hardware, ammunition cases, and home furnishings. Adding lead to brass improves its machinability, which makes it attractive to use for screw machine parts.

The major families of bronzes include phosphor bronze, aluminum bronze, and silicon bronze. Their high strength and corrosion resistance make them useful in marine applications, screws, blots, gears, pressure vessels, springs, bushings, and bearings.

The strength of copper and its alloys is dependent on the hardness that is achieved by cold working. Successively higher strengths would result from the tempers designated soft, quarter hard, half hard, three-quarters hard, hard, extra hard, spring, and extra spring tempers. The strengths of four copper alloys in the soft and hard tempers are listed in Appendix A–14.

2-7 ZINC, MAGNESIUM, AND TITANIUM

Zinc has moderate strength and toughness and excellent corrosion resistance. It is used in wrought forms such as rolled sheet and foil and drawn rod or wire. Dry-cell battery cans, builder's hardware, and plates for photoengraving are some of the major applications.

Many zinc parts are made by die casting because the melting point is less than 800°F (427°C), much lower than other die-casting metals. The as-cast finish is suitable for many applications, such as business machine parts, pump bodies, motor housings, and frames for light-duty machines. Where a decorative appearance is required, electroplating with nickel and chromium is easily done. Such familiar parts as radio grilles, lamp housings, and body moldings are made in this manner. Appendix A–14 lists the properties of one cast zinc alloy.

Magnesium is the lightest metal commonly used in load-carrying parts. Its density of only 0.066 lb/in^3 (1830 kg/m^3) is only about one-fourth that of steel and zinc, one-fifth that of copper, and two-thirds that of aluminum. It has moderate strength and lends itself well to applications where the final fabricated weight of the part or structure should be light. Ladders, hand trucks, conveyor parts, portable power tools, and lawn-mower housings use magnesium. In the automotive industry, body parts, blower wheels, pump bodies, and brackets are often made of magnesium. In aircraft, its lightness makes this metal attractive for floors, frames, fuselage skins, and wheels. The stiffness (modulus of elasticity) of magnesium is low, which is an advantage in parts where impact energy must be absorbed. Also, its lightness results in low-weight designs when compared with other metals on an equivalent rigidity basis. See Appendix A–14 for properties of one cast magnesium alloy.

Titanium has very high strength, and its density is only about half that of steel. Although aluminum has a lower density, titanium is superior to both aluminum and most steels on a strength-to-weight basis. It retains a high percentage of its strength at elevated temperatures and can be used up to about 1000°F (538°C). Most applications of titanium are in the aerospace industry in engine parts, fuselage parts and skins, ducts, spacecraft structures, and pressure vessels. Because of its corrosion resistance and high-temperature strength, the chemical industries use titanium in heat exchangers and as a lining for processing equipment. High cost is a major factor to be considered.

Appendix A–14 gives the properties of one titanium alloy containing aluminum and vanadium used in the aerospace, marine, and chemical process industries. This is a popular heat-treatable alloy where the term, *aged*, refers to a heating and quenching cycle followed by heating at a lower temperature.

2–8 NONMETALS IN ENGINEERING DESIGN

Wood and concrete are widely used in construction. Plastics and composites are found in virtually all fields of design, including consumer products, industrial equipment, automobiles, aircraft, and architectural products. To the designer, the properties of strength and stiffness are of primary importance with nonmetals, as they are with metals. Because of the structural differences in the nonmetals, their behavior is quite different from the metals.

Wood, concrete, composites, and many plastics have structures that are *anisotropic*. This means that the mechanical properties of the material are different, depending on the direction of the loading. Also, because of natural chemical changes, the properties vary with time and often with climatic conditions. The designer must be aware of these factors.

2–9 WOOD

Since wood is a natural material, its structure is dependent on the way it grows and not on manipulation by human beings, as is the case in metals. The long, slender, cylindrical shape of trees results in an internal structure composed of longitudinal cells. As the tree grows, successive rings are added outside the older wood. Thus the inner core, called heartwood, has different properties than the sapwood, near the outer surface.

The species of the wood also affects its properties, as different kinds of trees produce harder or softer, stronger or weaker wood. Even in the same species variability occurs because of different growing conditions, such as differences in soil and amount of sun and rain.

The cellular structure of the wood gives it the grain which is so evident when sawn into boards and timber. The strength of the wood is dependent on whether it is loaded perpendicular to or parallel to the grain. Also, going across the grain, the strength is different in a radial direction than in a tangential direction with respect to the original cylindrical tree stem from which it was cut.

Another important variable affecting the strength of wood is moisture content. Changes in relative humidity can vary the amount of water absorbed by the cells of the wood.

Most construction lumber is stress-graded by standard rules adopted by the U.S. Forest Products Laboratory. Appendix A–18 lists allowable stresses for several species and grades of lumber. These allowable stresses account for variability due to natural imperfections.

2–10 CONCRETE

The components of concrete are cement and an aggregate. The addition of water and the thorough mixing of the components tend to produce a uniform structure with cement coating all the aggregate particles. After curing, the mass is securely bonded together. Some of the variables involved in determining the final strength of the concrete are the type of cement used, the type and size of aggregate, and the amount of water added.

A higher quantity of cement in concrete yields a higher strength. Decreasing the quantity of water relative to the amount of cement increases the strength of the concrete. Of course, sufficient water must be added to cause the cement to coat the aggregate and to allow the concrete to be poured and worked before excessive curing takes place. The density of the concrete, affected by the aggregate, is also a factor. A mixture of sand, gravel, and broken stone is usually used for construction grade concrete.

Concrete is graded according to its compressive strength, which varies from about 2000 psi (14 MPa) to 7000 psi (48 MPa). The tensile strength of concrete is extremely low, and it is common practice to assume that it is zero. Of course, reinforcing concrete with steel bars allows its use in beams and wide slabs since the steel resists the tensile loads.

Concrete must be cured to develop its rated strength. It should be kept moist for at least 7 days, at which time it has about 75% of its rated compressive strength. Although its strength continues to increase for years, the strength at 28 days is often used to determine its rated strength.

The allowable working stresses in concrete are typically 25% of the rated 28-day strength. For example, concrete rated at 2000 psi (14 MPa) would have an allowable stress of 500 psi (3.4 MPa).

The specific weight of gravel-based concrete averages approximately 150 lb/ft^3. The modulus of elasticity is somewhat dependent on the specific weight and the rated strength. According to the American Concrete Institute, an estimate of the modulus can be computed from

$$E_c = 33\gamma^{3/2}\sqrt{s_c} \qquad (2-4)$$

where E_c = Modulus of elasticity in compression, psi
  γ = Specific weight, lb/ft^3
  s_c = Rated compressive strength of the concrete, psi

Using $\gamma = 150$ lb/ft^3, the range of expected values for modulus of elasticity, computed from Equation (2−4), is shown below.

Rated strength, s_c		Modulus of elasticity, E_c	
psi	MPa	psi	GPa
2000	13.8	2.7×10^6	18.6
3000	20.7	3.3×10^6	22.7
4000	27.6	3.8×10^6	26.2
5000	34.5	4.3×10^6	29.6
6000	41.4	4.7×10^6	32.4
7000	48.3	5.1×10^6	35.2

2−11 PLASTICS

Plastics are composed of long chain-like molecules called polymers. They are synthetic organic materials which can be formulated and processed in literally thousands of ways.

One classification that can be made is between *thermoplastic* materials and *thermosetting* materials. Thermoplastics can be softened repeatedly by heating with no change in properties or chemical composition. Conversely, after initial curing of thermosetting plastics, they cannot be resoftened. A chemical change occurs during curing with heat and pressure.

Some examples of thermoplastics include ABS, acetals, acrylics, cellulose acetate, TFE fluorocarbons, nylon, polyethylene, polypropylene, polystyrene, and vinyls. Thermosetting plastics include phenolics, epoxies, polyesters, silicones, urethanes, alkyds, allyls, and aminos.

A particular plastic is often selected for a combination of properties such as light weight, flexibility, color, strength, chemical resistance, low friction, or transparency. Since the available products are so numerous, only a brief table of properties of plastics is included as Appendix A−19. Table 2−5 lists the primary plastic materials used for six different types of applications. An extensive comparative study of the design properties of plastics can be found in References 4, 7, and 9.

2−12 COMPOSITES

Composites are materials having two or more constituents blended in a way that results in mechanical or adhesive bonding between the materials. To form a composite, a filler material is distributed in a matrix so that the filler reinforces the matrix. Typically, the filler is a strong, stiff material while the matrix has a relatively low density. When the two materials bond together, much of the load-carrying ability of the composite is produced by the filler material. The matrix serves to hold the filler in a favorable orientation relative to the manner of loading and to distribute the loads to the filler. The result is a somewhat optimized composite that has high strength and high stiffness with low weight.

TABLE 2–5 Applications of plastic materials

Applications	Desired properties	Suitable plastics
Housings, containers, ducts	High impact strength, stiffness, low cost, formability, environmental resistance, dimensional stability	ABS, polystyrene, polypropylene, polyethylene, cellulose acetate, acrylics
Low friction—bearings, slides	Low coefficient of friction; resistance to abrasion, heat, corrosion	TFE fluorocarbons, nylon, acetals
High-strength components, gears, cams, rollers	High tensile and impact strength, stability at high temperatures, machinable	Nylon, phenolics, TFE-filled acetals
Chemical and thermal equipment	Chemical and thermal resistance, good strength, low moisture absorption	Fluorocarbons, polypropylene, polyethylene, epoxies, polyesters, phenolics
Electrostructural parts	Electrical resistance, heat resistance, high impact strength, dimensional stability, stiffness	Allyls, alkyds, aminos, epoxies, phenolics, polyesters, silicones
Light-transmission components	Good light transmission in transparent and translucent colors, formability, shatter resistance	Acrylics, polystyrene, cellulose acetate, vinyls

A virtually unlimited variety of composite materials can be produced by combining different matrix materials with different fillers in different forms and in different orientations. Some typical materials are listed below.

Matrix materials. Among the more frequently used matrix materials are:

- Thermoplastic polymers: Polyethylene, nylon, polypropylene, polystyrene, polyamides
- Thermosetting polymers: Polyester, epoxy, phenolic polyimide
- Ceramics and glass
- Carbon and graphite
- Metals: Aluminum, magnesium, titanium

Forms of filler materials. Many forms of filler materials are used as listed here.

- Continuous fiber strand consisting of many individual filaments bound together
- Chopped strands in short lengths (0.75 to 50 mm or 0.03 to 2.00 in)
- Chopped strands randomly spread in the form of a mat
- Roving: A group of parallel strands
- Woven fabric made from roving or strands
- Metal filaments or wires
- Solid or hollow microspheres

- Metal, glass, or mica flakes
- Single crystal whiskers of materials such as graphite, silicon carbide, and copper

Types of filler materials. Fillers, also called fibers, come in many types based on both organic and inorganic materials. Some of the more popular fillers are listed below.

- Glass fibers in five different types:
  A-glass: Good chemical resistance because it contains alkalis such as sodium oxide
  C-glass: Special formulations for even higher chemical resistance than A-glass
  E-glass: Widely used glass with good electrical insulating ability and good strength
  S-glass: High strength, high temperature glass
  D-glass: Better electrical properties than E-glass
- Quartz fibers and high-silica glass: Good properties at high temperatures up to 2000°F (1095°C)
- Carbon fibers made from PAN-base carbon (PAN is polyacrylonitrile): Approximately 95% carbon with very high modulus of elasticity
- Graphite fibers: Greater than 99% carbon and even higher modulus of elasticity than carbon; the stiffest fibers typically used in composites
- Boron coated onto tungsten fibers: Good strength and higher modulus of elasticity than glass
- Silicon carbide coated onto tungsten fibers: Strength and stiffness similar to boron/tungsten but with higher temperature capability
- Aramid fibers: A member of the polyamide family of polymers; higher strength and stiffness with lower density as compared with glass; very flexible (Aramid fibers produced by the DuPont company carry the name *Kevlar*)

Advantages of composites. Designers typically seek to produce products that are safe, strong, stiff, lightweight, and highly tolerant of the environment in which the product will operate. Composites often excel in meeting these objectives when compared to alternative materials such as metals, wood, and unfilled plastics. Two parameters that are used to compare materials are *specific strength* and *specific modulus,* defined as,

> *Specific strength is the ratio of the tensile strength of a material to its specific weight.*

> *Specific modulus is the ratio of the modulus of elasticity of a material to its specific weight.*

Because the modulus of elasticity is a measure of the stiffness of a material, the specific modulus is sometimes called *specific stiffness.*

Although obviously not a length, both of these quantities have the *unit* of length, derived from the ratio of the units for strength or modulus of elasticity and the units for specific weight. In the U.S. Customary system, the units for tensile strength and modulus of elasticity are lb/in^2 while specific weight (weight per unit volume) is in lb/in^3. Thus, the unit for specific strength or specific modulus is inches. In the SI metric system, strength and modulus are expressed in N/m^2 (Pascals) while specific weight is in N/m^3. Then the unit for specific strength or specific modulus is meters.

Table 2–6 gives comparisons of the specific strength and specific stiffness of selected composite materials with certain steel, aluminum, and titanium alloys. Figure 2–11 shows a comparison of these materials using bar charts. Figure 2–12 is a plot of these data with specific strength on the vertical axis and specific modulus on the horizontal axis. When weight is critical, the ideal material would lie in the upper right part of this chart. Note that data in these charts and figures are for composites having the filler materials aligned in the most favorable direction to withstand the applied loads.

Advantages of composites can be summarized as follows:

1. Specific strengths for composite materials can range as high as five times those of high strength steel alloys.

2. Specific modulus values for composite materials can be as high as eight times those for either steel, aluminum, or titanium alloys.

3. Composite materials typically perform better than steel or aluminum in applications where cyclic loads are encountered leading to the potential for fatigue failure.

4. Where impact loads and vibrations are expected, composites can be specially formulated with materials that provide high toughness and a high level of damping.

5. Some composites have much higher wear resistance than metals.

TABLE 2–6 Comparison of specific strength and specific modulus for selected materials

Material	Tensile strength, s_u (ksi)	Specific weight, γ (lb/in^3)	Specific strength (in)	Specific modulus (in)
Steel ($E = 30 \times 10^6$ psi)				
AISI 1020 HR	55	0.283	0.194×10^6	1.06×10^8
AISI 5160 OQT 700	263	0.283	0.929×10^6	1.06×10^8
Aluminum ($E = 10.0 \times 10^6$ psi)				
6061-T6	45	0.098	0.459×10^6	1.02×10^8
7075-T6	83	0.101	0.822×10^6	0.99×10^8
Titanium ($E = 16.5 \times 10^6$ psi)				
Ti-6Al-4V Quenched and aged at 1000°F	160	0.160	1.00×10^6	1.03×10^8
Glass/epoxy composite ($E = 4.0 \times 10^6$ psi)				
34% fiber content	114	0.061	1.87×10^6	0.66×10^8
Aramid/epoxy composite ($E = 11.0 \times 10^6$ psi)				
60% fiber content	200	0.050	4.0×10^6	2.20×10^8
Boron/epoxy composite ($E = 30.0 \times 10^6$ psi)				
60% fiber content	270	0.075	3.60×10^6	4.00×10^8
Graphite/epoxy composite ($E = 19.7 \times 10^6$ psi)				
62% fiber content	278	0.057	4.86×10^6	3.45×10^8
Graphite/epoxy composite ($E = 48 \times 10^6$ psi)				
Ultrahigh modulus	160	0.058	2.76×10^6	8.28×10^8

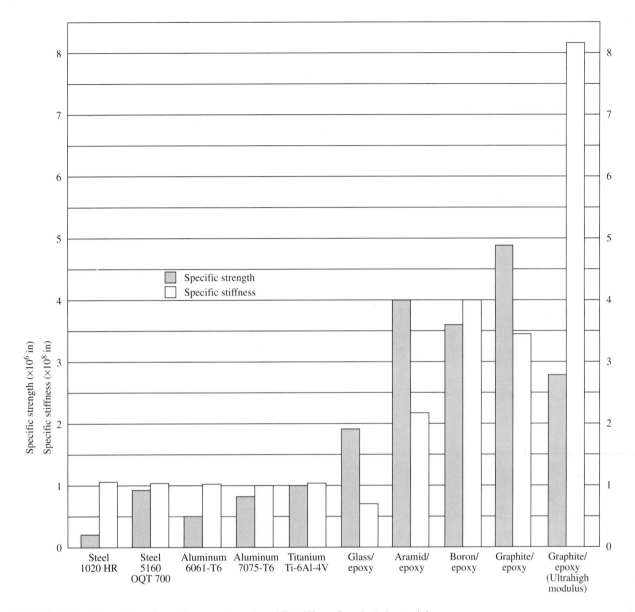

FIGURE 2–11 Comparison of specific strength and specific stiffness for selected materials.

6. Careful selection of the matrix and filler materials can provide superior corrosion resistance.

7. Dimensional changes due to changes in temperature are typically much less for composites than for metals. More is included on this topic in Chapter 4 in which the property, *coefficient of thermal expansion,* is defined.

8. Because composite materials have properties that are highly directional, designers can tailor the placement of reinforcing fibers in directions that

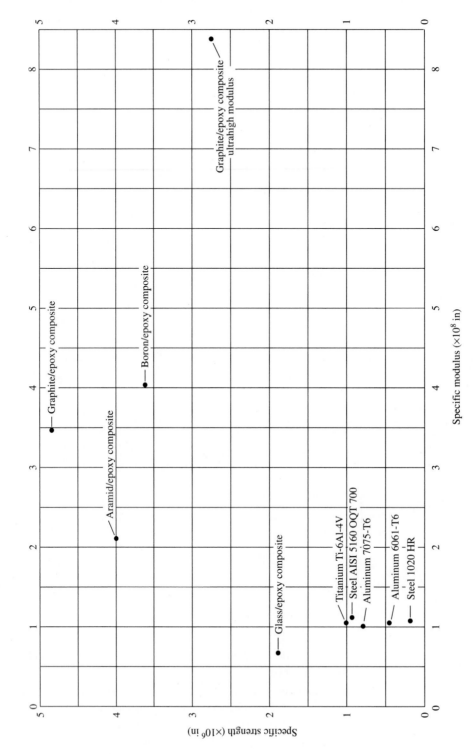

FIGURE 2–12 Specific strength versus specific modulus for selected metals and composites.

provide the required strength and stiffness under the specific loading conditions to be encountered.

9. Composite structures can often be made in complex shapes in one piece, thus reducing the number of parts in a product and the number of fastening operations required. The elimination of joints typically improves the reliability of such structures as well.

10. Composite structures are typically made in their final form directly or in a near-net shape, thus reducing the number of secondary operations required.

Limitations of composites. Designers must balance many properties of materials in their designs while simultaneously considering manufacturing operations, costs, safety, life, and service of the product. The following list gives some of the major concerns when using composites.

1. Material costs for composites are typically higher than for many alternative materials.

2. Fabrication techniques are quite different from those used to shape metals. New manufacturing equipment may be required along with additional training for production operators.

3. The performance of products made from some composite production techniques is subject to a wider range of variability than for most metal fabrication techniques.

4. The operating temperature limits for composites having a polymeric matrix are typically 500°F (260°C). [But ceramic or metal matrix composites can be used at higher temperatures such as those found in engines.]

5. The properties of composite materials are not isotropic. This means that properties vary dramatically with the direction of the applied loads. Designers must account for these variations to ensure safety and satisfactory operation under all expected types of loading.

6. At this time, there is a general lack of understanding of the behavior of composite materials and the details of predicting failure modes. While major advancements have been made in certain industries such as the aerospace and recreational equipment fields, there is a need for more general understanding about designing with composite materials.

7. The analysis of composite structures requires the detailed knowledge of more properties of the materials than would be required for metals.

8. Inspection and testing of composite structures is typically more complicated and less precise than for metal structures. Special non-destructive techniques may be required to ensure that there are no major voids in the final product that could seriously weaken the structure. Testing of the complete structure may be required rather than testing a sample of the material because of the interaction of different parts on each other and because of the directionality of the material properties.

9. Repair and maintenance of composite structures is a serious concern. Some of the initial production techniques require special environments of temperature and pressure that may be difficult to reproduce in the field when damage repair is required. Bonding of a repaired area to the parent structure may also be difficult.

Laminated composite construction.   Many structures made from composite materials are made from several layers of the basic material containing both the matrix and the reinforcing fibers. The manner in which the layers are oriented relative to one another affects the final properties of the completed structure.

As an illustration, consider that each layer is made from a set of parallel strands of the reinforcing filler material, such as E-glass fibers, embedded in the resin matrix, such as polyester. In this form, the material is sometimes called a *prepreg*, indicating that the filler has been pre-impregnated with the matrix prior to forming the structure and curing the assembly. To produce the maximum strength and stiffness in a particular direction, several layers or plies of the prepreg could be laid on top of one another with all of the fibers aligned in the direction of the expected tensile load. This is called a *unidirectional laminate.* After curing, the laminate would have a very high strength and stiffness when loaded in the direction of the strands, called the *longitudinal* direction. However, the resulting product would have a very low strength and stiffness in the direction perpendicular to the fiber direction, called the *transverse* direction. If any off-axis loads are encountered, the part may fail or deform significantly. Table 2–7 gives sample data for a unidirectional laminated carbon/epoxy composite.

To overcome the lack of off-axis strength and stiffness, laminated structures should be made with a variety of orientations of the layers. One popular arrangement is shown in Figure 2–13. Naming the longitudinal direction of the surface layer the *0° ply,* this structure is referred to as

$$0°, 90°, +45°, -45°, -45°, +45°, 90°, 0°$$

The symmetry and balance of this type of layering technique results in more nearly uniform properties in two directions. The term *quasi-isotropic* is sometimes used to describe such a structure. Note that the properties perpendicular to the faces of the layered structure (through the thickness) are still quite low because fibers do not extend in that direction. Also, the strength and stiffness in the primary directions are somewhat lower than if the plies were aligned in the same direction. Table 2–7 shows sample data

TABLE 2–7   Examples of the effect of laminate construction on strength and stiffness

	Tensile strength				Modulus of elasticity			
	Longitudinal		Transverse		Longitudinal		Transverse	
Laminate type	ksi	MPa	ksi	MPa	10^6 psi	GPa	10^6 psi	GPa
Unidirectional	200	1380	5	34	21	145	1.6	11
Quasi-isotropic	80	552	80	552	8	55	8	55

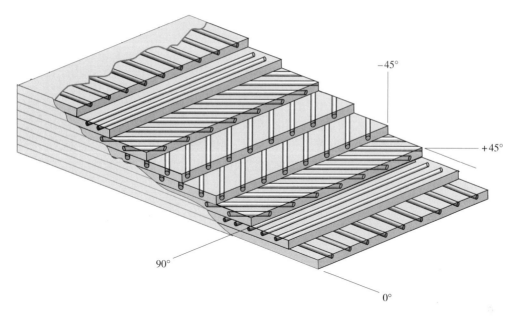

-45°

+45°

90°

0°

FIGURE 2–13 Multilayer laminated composite construction designed to produce quasi-isotropic properties.

for a quasi-isotropic laminate compared with one having unidirectional fibers in the same matrix.

Predicting composite properties. The following discussion summarizes some of the important variables needed to define the properties of a composite. The subscript c refers to the composite, m refers to the matrix, and f refers to the fibers. The strength and stiffness of a composite material depend on the elastic properties of the fiber and matrix components. But another parameter is the relative volume of the composite composed of fibers, V_f, and that composed of the matrix material, V_m. That is,

$$V_f = \text{Volume fraction of fiber in the composite}$$
$$V_m = \text{Volume fraction of matrix in the composite}$$

Note that for a unit volume, $V_f + V_m = 1$. Then, $V_m = 1 - V_f$.

We will use an ideal case to illustrate the way in which the strength and stiffness of a composite can be predicted. Consider a composite with unidirectional continuous fibers aligned in the direction of the applied load. The fibers are typically much stronger and stiffer than the matrix material. Furthermore, the matrix will be able to undergo a larger strain before fracture than the fibers can. Figure 2–14 shows these phenomena on a plot of stress versus strain for the fibers and the matrix. We will use the following notation for key parameters from Figure 2–14:

$$s_{uf} = \text{Ultimate strength of fiber}$$
$$\epsilon_{uf} = \text{Strain in the fiber corresponding to its ultimate strength}$$
$$\sigma'_m = \text{Stress in the matrix at the same strain as } \epsilon_{uf}$$

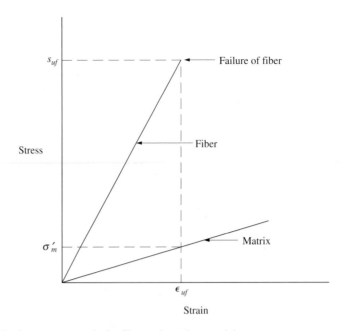

FIGURE 2–14 Stress versus strain for fiber and matrix materials.

The ultimate strength of the composite, s_{uc}, is at some intermediate value between s_{uf} and σ'_m, depending on the volume fraction of fiber and matrix in the composite, That is,

$$s_{uc} = s_{uf} V_f + \sigma'_m V_m \qquad (2\text{–}5)$$

At any lower level of stress, the relationship among the overall stress in the composite, the stress in the fibers, and the stress in the matrix follows a similar pattern.

$$\sigma_c = \sigma_f V_f + \sigma_m V_m \qquad (2\text{–}6)$$

Figure 2–15 illustrates this relationship on a stress–strain diagram.

   Both sides of Equation (2–6) can be divided by the strain at which these stresses occur. And, since for each material, $\sigma/\epsilon = E$, the modulus of elasticity for the composite can be shown as,

$$E_c = E_f V_f + E_m V_m \qquad (2\text{–}7)$$

The density of a composite can be computed in a similar fashion.

$$\rho_c = \rho_f V_f + \rho_m V_m \qquad (2\text{–}8)$$

Density is defined as *mass per unit volume*. A related property, specific weight, is defined as *weight per unit volume* and is denoted by the symbol γ (Greek letter gamma). The relationship between density and specific weight is simply $\gamma = \rho g$, where g is the

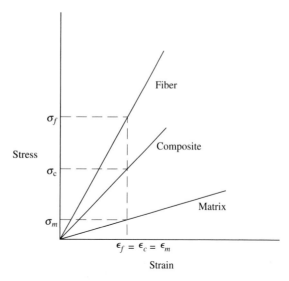

FIGURE 2–15 Relationship among stresses and strains for a composite and its fiber and matrix materials.

acceleration due to gravity. Multiplying each term in Eq. (2–8) by g gives the formula for the specific weight of a composite,

$$\gamma_c = \gamma_f V_f + \gamma_m V_m \qquad (2\text{–}9)$$

The form of Equations (2–6), (2–7), (2–8), and (2–9) is often called *the rule of mixtures.*

Table 2–8 lists example values for the properties of some matrix and filler materials. Remember that wide variations can occur in such properties, depending on the exact formulation and the condition of the materials.

TABLE 2–8 Example properties of matrix and filler materials

	Tensile strength		Tensile modulus		Specific weight	
	ksi	MPa	10^6 psi	GPa	lb/in^3	kN/m^3
Matrix materials:						
Polyester	10	69	0.40	2.76	0.047	12.7
Epoxy	18	124	0.56	3.86	0.047	12.7
Aluminum	45	310	10.0	69	0.100	27.1
Titanium	170	1170	16.5	114	0.160	43.4
Filler materials:						
S glass	600	4140	12.5	86.2	0.09	24.4
Carbon–PAN	470	3240	33.5	231	0.064	17.4
Carbon–PAN (high strength)	820	5650	40	276	0.065	17.7
Carbon (high-modulus)	325	2200	100	690	0.078	21.2
Aramid	500	3450	19.0	131	0.052	14.1

Compute the expected properties of ultimate tensile strength, modulus of elasticity, and specific weight of a composite made from unidirectional strands of carbon-PAN fibers in an epoxy matrix. The volume fraction of fibers is 30%. Use data from Table 2–8.

Solution **Objective** Compute the expected values of s_{uc}, E_c, and γ_c for the composite.

Given Matrix–epoxy: $s_{um} = 18$ ksi; $E_m = 0.56 \times 10^6$ psi; $\gamma_m = 0.047$ lb/in^3.
Fiber–carbon–PAN: $s_{uf} = 470$ ksi; $E_f = 33.5 \times 10^6$ psi; $\gamma_f = 0.064$ lb/in^3.
Volume fraction of fiber, $V_f = 0.30$. And, $V_m = 1.0 - 0.30 = 0.70$.

Analysis and Results

Ultimate tensile strength, s_{uc}, computed from Eq. (2–5)

$$s_{uc} = s_{uf} V_f + \sigma'_m V_m$$

To find σ'_m we first find the strain at which the fibers would fail at s_{uf}. Assume that the fibers are linearly elastic to failure. Then,

$$\epsilon_f = s_{uf}/E_f = (470 \times 10^3 \text{ psi})/(33.5 \times 10^6 \text{ psi}) = 0.014$$

At this same strain, the stress in the matrix is

$$\sigma'_m = E_m \epsilon = (0.56 \times 10^6 \text{ psi})(0.014) = 7840 \text{ psi}$$

Then, in Eq. (2–5),

$$s_{uc} = (470\,000 \text{ psi})(0.30) + (7840 \text{ psi})(0.70) = 146\,500 \text{ psi}$$

Modulus of elasticity computed from Eq. (2–7),

$$E_c = E_f V_f + E_m V_m = (33.5 \times 10^6)(0.30) + (0.56 \times 10^6)(0.70)$$
$$E_c = 10.4 \times 10^6 \text{ psi}$$

Specific weight computed from Eq. (2–9),

$$\gamma_c = \gamma_f V_f + \gamma_m V_m = (0.064)(0.30) + (0.047)(0.70) = 0.052 \text{ lb/in}^3$$

Summary of Results

$$s_{uc} = 146\,500 \text{ psi}$$
$$E_c = 10.4 \times 10^6 \text{ psi}$$
$$\gamma_c = 0.052 \text{ lb/in}^3$$

Comment Note that the resulting properties for the composite are intermediate between those for the fibers and the matrix.

REFERENCES

1. Aluminum Association, *Aluminum Standards and Data,* 11th ed., Washington, DC, 1993.

2. American Society for Testing and Materials, *Annual Book of Standards, 1994,* Philadelphia, PA, 1994.

3. ASM INTERNATIONAL, *Composites, Engineered Materials Handbook, Volume 1,* Metals Park, OH, 1987.

4. ASM INTERNATIONAL, *Engineering Plastics, Engineered Materials Handbook, Volume 2,* Metals Park, OH, 1988.

5. ASM INTERNATIONAL, *Metals Handbook, Volumes 1–17,* 9th ed., Metals Park, OH, 1990.

6. Avallone, Eugene A., and Theodore Baumeister III, eds., *Marks' Standard Handbook for Mechanical Engineers,* 9th ed., McGraw-Hill, New York, 1987.

7. Berins, Michael L., ed., *Plastics Engineering Handbook of the Society of the Plastics Industry, Inc.,* 5th ed., Van Nostrand Reinhold, New York, 1991.

8. *Design Guide for Advanced Composites Applications,* Advanstar Communications, Inc., Duluth, MN, 1993.

9. *Machine Design Magazine, 1994 Basics of Design Engineering Reference Volume, Materials Selection,* Penton Publishing, Inc., Cleveland, OH, 1994.

10. Mallick, P. K., *Fiber-Reinforced Composites: Materials, Manufacturing, and Design,* Marcel Dekker, New York, 1988.

11. Mott, Robert L., *Machine Elements in Mechanical Design,* 2nd ed., Merrill, an imprint of Macmillan Publishing Co., New York, 1992.

12. Strong, A. Brent, *Fundamentals of Composites Manufacturing: Materials, Methods, and Applications,* Society of Manufacturing Engineers, Dearborn, MI, 1989.

13. U.S. Department of Agriculture Forest Products Laboratory, *Handbook of Wood and Wood-Based Materials for Engineers, Architects, and Builders,* Hemisphere Publishing Corp., New York, 1989.

14. Weeton, John W., Dean M. Peters, and Karyn L. Thomas, eds., *Engineers' Guide to Composite Materials,* ASM INTERNATIONAL, Metals Park, OH, 1987.

PROBLEMS

2–1. Name four kinds of metals commonly used for load-carrying members.

2–2. Name 11 factors that should be considered when selecting a material for a product.

2–3. Define *ultimate tensile strength.*

2–4. Define *yield point.*

2–5. Define *yield strength.*

2–6. When is yield strength used in place of yield point?

2–7. Define *stiffness.*

2–8. What material property is a measure of its stiffness?

2–9. State Hooke's law.

2–10. What material property is a measure of its ductility?

2–11. How is a material classified as to whether it is ductile or brittle?

2–12. Name four types of steels.

2–13. What does the designation AISI 4130 for a steel mean?

2–14. What are the ultimate strength, yield strength, and percent elongation of AISI 1040 hot-rolled steel? Is it a ductile or a brittle material?

2–15. Which has a greater ductility: AISI 1040 hot-rolled steel or AISI 1020 hot-rolled steel?

2–16. What does the designation AISI 1141 OQT 700 mean?

2–17.E If the required yield strength of a steel is 150 ksi, could AISI 1141 be used? Why?

2–18.M What is the modulus of elasticity for AISI 1141 steel? For AISI 5160 steel?

2–19.E A rectangular bar of steel is 1.0 in by 4.0 in by 14.5 in. How much does it weigh in pounds?

2–20.M A circular bar is 50 mm in diameter and 250 mm long. How much does it weigh in newtons?

2–21.M If a force of 400 N is applied to a bar of titanium and an identical bar of magnesium, which would stretch more?

2–22. Name four types of structural steels and list the yield point for each.

2–23. What does the aluminum alloy designation 6061-T6 mean?

2–24.E List the ultimate strength, yield strength, modulus of elasticity, and density for 6061-O, 6061-T4, and 6061-T6 aluminum.

2–25. List five uses for bronze.

2–26. List three desirable characteristics of titanium as compared with aluminum or steel.

2–27. Name five varieties of cast iron.

2–28. Which type of cast iron is usually considered to be brittle?

2–29.E What are the ultimate strengths in tension and in compression for ASTM A48 grade 40 cast iron?

2–30. How does a ductile iron differ from gray iron?

2–31.E List the allowable stresses in bending, tension, compression, and shear for No. 2 grade Douglas fir.

2–32.E What is the normal range of compressive strengths for concrete?

2–33. Describe the difference between thermoplastic and thermosetting materials.

2–34. Name three suitable plastics for use as gears or cams in mechanical devices.

2–35. Describe the term *composite*.

2–36. Name five basic types of materials that are used as a matrix for composites.

2–37. Name five different thermoplastics used as a matrix for composites.

2–38. Name three different thermosetting plastics used as a matrix for composites.

2–39. Name three metals used as a matrix for composites.

2–40. Describe nine forms that filler materials take when used in composites.

2–41. Discuss the differences among *strands*, *roving*, and *fabric* as different forms of fillers for composites.

2–42. Name seven types of filler materials used for composites.

2–43. Name five different types of glass fillers used for composites and describe the primary features of each.

2–44. Which of the commonly used filler materials has the highest stiffness?

2–45. Which filler materials should be considered for high temperature applications?

2–46. What is a common brand name for aramid fibers?

2–47. Define *specific strength* of a composite.

2–48. Define *specific modulus* of a composite.

2–49. List ten advantages of composites when compared to metals.

2–50. List nine limitations of composites.

2–51. From the data for selected materials in Table 2–6, list the ten materials in order of specific strength from highest to lowest. For each, compute the ratio of its specific strength to that for AISI 1020 HR steel.

2–52. From the data for selected materials in Table 2–6, list the ten materials in order of specific modulus from highest to lowest. For each, compute the ratio of its specific modulus to that for AISI 1020 HR steel.

2–53. Describe a unidirectional laminate and its general strength and stiffness characteristics.

2–54. Describe a quasi-isotropic laminate and its general strength and stiffness characteristics.

2–55. Compare the generally expected specific strength and stiffness characteristics of a quasi-isotropic laminate with a unidirectional laminate.

2–56. Describe a laminated composite that carries the designation 0°, +45°, −45°, −45°, +45°, 0°.

2–57. Describe a laminated composite that carries the designation 0°, +30°, +45°, +45°, +30°, 0°.

2–58. Define the term *volume fraction of fiber* for a composite.

2–59. Define the term *volume fraction of matrix* for a composite.

2–60. If a composite has a volume fraction of fiber of 0.60, what is the volume fraction of matrix?

2–61. Write the equation for the expected ultimate strength of a composite in terms of the properties of its matrix and filler materials.

2–62. Write the equations for the *rule of mixtures* as applied to a unidirectional composite for the stress in the composite, its modulus of elasticity, its density, and its specific weight.

2–63.M Compute the expected properties of ultimate strength, modulus of elasticity, and specific weight of a composite made from unidirectional strands of high-strength carbon–PAN fibers in an epoxy matrix. The volume fraction of fibers is 50%. Compute the specific strength and specific stiffness. Use data from Table 2–8.

2–64.M Repeat Problem 2–63 with high modulus carbon fibers.

2–65.M Repeat Problem 2–63 with aramid fibers.

3

Design of Members under Direct Stresses

3–1 OBJECTIVES OF THIS CHAPTER

In Chapter 1 the concept of direct stress was presented together with examples of the calculation of direct tensile stress, direct compressive stress, direct shear stress, and bearing stress. The emphasis was on the understanding of the basic phenomena, units, terminology, and the magnitude of stresses encountered in typical structural and mechanical applications. Nothing was said about the acceptability of the stress levels which were computed or about the design of members to carry a given load.

In this chapter the primary emphasis is on *design* in which you, as the designer, must make decisions about whether or not a proposed design is satisfactory; what the shape and size of the cross section of a load-carrying member should be; and what material the member should be made from.

After completing this chapter, you should be able to:

1. Describe the conditions that must be met for satisfactory application of the direct stress formulas.
2. Define *design stress* and tell how to determine an acceptable value for it.
3. Define *design factor* and select appropriate values for it depending on the conditions present in a particular design.
4. Discuss the relationship among the terms *design stress, allowable stress,* and *working stress.*

5. Discuss the relationship among the terms *design factor, factor of safety,* and *margin of safety.*

6. Describe 11 factors that affect the specification of the design factor.

7. Describe various types of loads experienced by structures or machine members, including static load, repeated load, impact, and shock.

8. Design members subjected to direct tensile stress, direct compressive stress, direct shear stress, and bearing stress.

9. Determine when stress concentrations exist and specify suitable values for stress concentration factors.

10. Use stress concentration factors in design.

3–2 DESIGN OF MEMBERS UNDER DIRECT TENSION OR COMPRESSION

In Chapter 1 the direct stress formula was developed and stated as follows:

$$\sigma = \frac{F}{A} \tag{3–1}$$

where σ = direct normal stress: tension or compression
$\quad F$ = direct axial load
$\quad A$ = cross-sectional area of member subjected to F

For Equation (3–1) to be valid, the following conditions must be met:

1. The loaded member must be straight.

2. The loaded member must have a uniform cross section over the length under consideration.

3. The material from which the member is made must be homogeneous.

4. The load must be applied along the centroidal axis of the member so there is no tendency to bend it.

5. Compression members must be short so that there is no tendency to buckle. (See Chapter 14 for the special analysis required for long, slender members under compressive stress and for the method to decide when a member is to be considered long or short.)

It is important to recognize that the concept of stress refers to the internal resistance provided by a *unit area,* that is, an infinitely small area. Stress is considered to act at a point and may, in general, vary from point to point in a particular body. Equation (3–1) indicates that for a member subjected to direct axial tension or compression, the stress is uniform across the entire area if the five conditions are met. In many practical applications the minor variations that could occur in the local stress levels are accounted for by carefully selecting the allowable stress, as discussed later.

3–3 DESIGN NORMAL STRESSES

Failure occurs in a load-carrying member when it breaks or deforms excessively, rendering it unacceptable for the intended purpose. Therefore, it is essential that the level of applied stress never exceed the ultimate tensile strength or the yield strength of the material. Consideration of excessive deformation without yielding is discussed in later chapters.

> *Design stress is that level of stress which may be developed in a material while ensuring that the loaded member is safe.*

To compute design stress, two factors must be specified: the *design factor N* and the *property of the material on which the design will be based.* Usually, for metals, the design stress is based on either the yield strength s_y or the ultimate strength s_u of the material.

> *The* **design factor N** *is a number by which the reported strength of a material is divided to obtain the* **design stress** σ_d

The following equations can be used to compute the design stress for a certain value of *N*:

 **Design Stress**

$$\sigma_d = \frac{s_y}{N} \quad \text{based on yield strength} \tag{3–2}$$

or

$$\sigma_d = \frac{s_u}{N} \quad \text{based on ultimate strength} \tag{3–3}$$

The value of the design factor is normally determined by the designer, using judgment and experience. In some cases, codes, standards, or company policy may specify design factors or design stresses to be used. When the designer must determine the design factor, his or her judgment must be based on an understanding of how parts may fail and the factors that affect the design factor. Sections 3–4, 3–5, and 3–6 give additional information about the design factor and about the choice of methods for computing design stresses.

Other references may use the term *factor of safety* in place of *design factor.* Also, *allowable stress* or *working stress* may be used in place of *design stress.* The choice of terms for use in this book is made to emphasize the role of the designer in specifying the design stress.

Theoretically, a material could be subjected to a stress up to s_y before yield would occur. This condition corresponds to a value of the design factor of $N = 1$ in Equation (3–2). Similarly, with a design factor of $N = 1$ in Equation (3–3), the material would be on the brink of ultimate fracture. Thus $N = 1$ is the lowest value we can consider.

A different approach to evaluating the acceptability of a given design, used primarily in the aerospace industry, is the *margin of safety,* defined as follows:

Margin of Safety

$$\text{margin of safety} = \frac{\text{yield strength}}{\text{maximum stress}} - 1.0 \tag{3–4}$$

when the design is based on yielding of the material. When based on the ultimate strength, the margin of safety is

$$\text{margin of safety} = \frac{\text{ultimate strength}}{\text{maximum stress}} - 1.0 \qquad (3-5)$$

Then the lowest feasible margin of safety is 0.0.

In this book we use the concept of design stresses and design factors as opposed to the margin of safety.

3-4 DESIGN FACTOR

Many different aspects of the design problem are involved in the specification of the design factor. In some cases the precise conditions of service are not known. The designer must then make conservative estimates of the conditions, that is, estimates that would cause the resulting design to be on the safe side when all possible variations are considered. The final choice of a design factor depends on the following 11 conditions.

Codes and standards.   If the member being designed falls under the jurisdiction of an existing code or standard, obviously the design factor or design stress must be chosen to satisfy the code or standard. Examples of standard-setting bodies are:

American Institute of Steel Construction (AISC): buildings, bridges, and similar structures using steel

Aluminum Association (AA): buildings, bridges, and similar structures using aluminum

American Society of Mechanical Engineers (ASME): boilers, pressure vessels, and shafting

State building codes: buildings, bridges, and similar structures affecting the public safety

Department of Defense—Military Standards: aerospace vehicle structures and other military products

American National Standards Institute (ANSI): a wide variety of products

American Gear Manufacturers Association (AGMA): gears and gear systems

It is the designer's responsibility to determine which, if any, standards or codes apply to the member being designed and to ensure that the design meets those standards.

Material strength basis.   Most designs using metals are based on either yield strength or ultimate strength or both, as stated previously. This is because most theories of metal failure show a strong relationship between the stress at failure and these material properties. Also, these properties will almost always be reported for materials used in engineering design. The value of the design factor will be different, depending on which material strength is used as the basis for design, as will be shown later.

Type of material. A primary consideration with regard to the type of material is its ductility. The failure modes for brittle materials are quite different from those for ductile materials. Since brittle materials such as cast iron do not exhibit yielding, designs are always based on ultimate strength. Generally a metal is considered to be brittle if its percent elongation in a 2-in gage length is less than 5%. Except for highly hardened alloys, virtually all steels are ductile. Except for castings, aluminum is ductile. Other material factors that can affect the strength of a part are its uniformity and the confidence in the stated properties.

Manner of loading. Three main types of loading can be identified. A *static load* is one that is applied to a part slowly and gradually and that remains applied, or at least is applied and removed only infrequently during the design life of the part. *Repeated loads* are those that are applied and removed several thousand times during the design life of the part. Under repeated loading a part fails by the mechanism of fatigue at a stress level much lower than that which would cause failure under a static load. This calls for the use of a higher design factor for repeated loads than for static loads. Parts subject to *impact* or *shock* require the use of a large design factor for two reasons. First, a suddenly applied load causes stresses in the part that are several times higher than those that would be computed by standard formulas. Second, under impact loading the material in the part is usually required to absorb energy from the impacting body. The certainty with which the designer knows the magnitude of the expected loads also must be considered when specifying the design factor.

Possible misuse of the part. In most cases the designer has no control over actual conditions of use of the product he or she designs. Legally, it is the responsibility of the designer to consider any reasonably foreseeable use or *misuse* of the product and to ensure the safety of the product. The possibility of an accidental overload on any part of a product must be considered.

Complexity of stress analysis. As the manner of loading or the geometry of a structure or a part becomes more complex, the designer is less able to perform a precise analysis of the stress condition. Thus the confidence in the results of stress analysis computations has an effect on the choice of a design factor.

Environment. Materials behave differently in different environmental conditions. Consideration should be given to the effects of temperature, humidity, radiation, weather, sunlight, and corrosive atmospheres on the material during the design life of the part.

Size effect, sometimes called *mass effect*. Metals exhibit different strengths as the cross-sectional area of a part varies. Most material property data were obtained using standard specimens about 0.50 in (12.5 mm) in diameter. Parts with larger sections usually have lower strengths. Parts of smaller size, for example drawn wire, have significantly higher strengths. An example of the size effect is shown in Table 3–1.

Quality control. The more careful and comprehensive a quality control program is, the better a designer knows how the product will actually appear in service. With poor quality control, a larger design factor should be used.

TABLE 3–1 Size effect for AISI 4140 OQT 1100 steel

Specimen size		Tensile strength		Yield strength		Percent elongation % in 2 in
in	mm	ksi	MPa	ksi	MPa	
0.50	12.5	158	1089	149	1027	18
1.00	25.4	140	965	135	931	20
2.00	50.8	128	883	103	710	22
4.00	101.6	117	807	87	600	22

Hazard presented by a failure. The designer must consider the consequences of a failure to a particular part. Would a catastrophic collapse occur? Would people be placed in danger? Would other equipment be damaged? Such considerations may justify the use of a higher than normal design factor.

Cost. Compromises must usually be made in design in the interest of limiting cost to a reasonable value under market conditions. Of course, where danger to life or property exists, compromises should not be made that would seriously affect the ultimate safety of the product or structure.

3–5 DESIGN FACTOR GUIDELINES

Experience in design and knowledge about the conditions above must be applied to determine a design factor. Table 3–2 includes guidelines that will be used in this book for selecting design factors. These should be considered to be average values. Special conditions or uncertainty about conditions may justify the use of other values.

The design factor is used to determine design stress as shown in Equations (3–2) and (3–3).

If the stress in a part is already known and one wishes to choose a suitable material for a particular application, the computed stress is considered to be the design stress. The required yield or ultimate strength is then found from

$$s_y = N \cdot \sigma_d \quad (N \text{ based on yield strength})$$

or

$$s_u = N \cdot \sigma_d \quad (N \text{ based on ultimate strength})$$

TABLE 3–2 Design stress guidelines—direct normal stresses

Manner of loading	Ductile material	Brittle material
Static	$\sigma_d = s_y/2$	$\sigma_d = s_u/6$
Repeated	$\sigma_d = s_u/8$	$\sigma_d = s_u/10$
Impact or shock	$\sigma_d = s_u/12$	$\sigma_d = s_u/15$

3-6 METHODS OF COMPUTING DESIGN STRESS

As mentioned in Section 3–3, an important factor to be considered when computing the design stress is the manner in which a part may fail when subjected to loads. In this section we discuss failure modes relevant to parts subjected to tensile and compressive loads. Other kinds of loading are discussed later.

The failure modes and the consequent methods of computing design stresses can be classified according to the type of material and the manner of loading. Ductile materials, having more than 5% elongation, exhibit somewhat different modes of failure than do brittle materials. Static loads, repeated loads, and shock loads produce different modes of failure.

Ductile materials under static loads. Ductile materials will undergo large plastic deformations when the stress reaches the yield strength of the material. Under most conditions of use, this would render the part unfit for its intended use. Therefore, for ductile materials subjected to static loads, the design stress is usually based on yield strength. That is,

$$\sigma_d = \frac{s_y}{N}$$

As indicated in Table 3–2, a design factor of $N = 2$ would be a reasonable choice under average conditions.

Ductile materials under repeated loads. Under repeated loads, ductile materials fail by a mechanism called *fatigue*. The level of stress at which fatigue occurs is lower than the yield strength. By testing materials under repeated loads, the stress at which failure will occur can be measured. The terms *fatigue strength* or *endurance strength* are used to denote this stress level. However, fatigue-strength values are often not available. Also, factors such as surface finish, the exact pattern of loading, and the size of a part have a marked effect on the actual fatigue strength. To overcome these difficulties, it is often convenient to use a high value for the design factor when computing the design stress for a part subjected to repeated loads. It is also recommended that the ultimate strength be used as the basis for the design stress because tests show that there is a good correlation between fatigue strength and the ultimate strength. Therefore, for ductile materials subjected to repeated loads, the design stress can be computed from

$$\sigma_d = \frac{s_u}{N}$$

A design factor of $N = 8$ would be reasonable under average conditions. Also, stress concentrations, which are discussed in Section 3–9, must be accounted for since fatigue failures often originate at points of stress concentrations.

Where data are available for the endurance strength of the material, the design stress can be computed from

$$\sigma_d = \frac{s_n}{N}$$

where s_n is the symbol for endurance strength. See Reference 6.

Ductile materials under impact or shock loading.   The failure modes for parts subjected to impact or shock loading are quite complex. They depend on the ability of the material to absorb energy and on the flexibility of the part. Because of the general inability of designers to perform precise analysis of stresses under shock loading, large design factors are recommended. In this book we will use

$$\sigma_d = \frac{s_u}{N}$$

with $N = 12$ for ductile materials subjected to impact or shock loads.

Brittle materials.   Since brittle materials do not exhibit yielding, the design stress must be based on ultimate strength. That is,

$$\sigma_d = \frac{s_u}{N}$$

with $N = 6$ for static loads, $N = 10$ for repeated loads, and $N = 15$ for impact or shock loads.

Design stresses from selected codes.   Table 3–3 gives a summary of specifications for design stresses for structural steel as defined by the American Institute of Steel Construction (AISC) for structural steel and by the Aluminum Association for aluminum alloys. These data pertain to members loaded in tension under static loads such as those found in building-type structures. See References 1 and 2 for a more detailed discussion of these specifications.

TABLE 3–3   Design stress from selected codes — direct normal stresses — static loads on building-like structures

Structural Steel (AISC):
$\sigma_d = s_y/1.67 = 0.60\ s_y$ or $\sigma_d = s_u/2.00 = 0.50\ s_u$
   whichever is lower
Aluminum (Aluminum Association):
$\sigma_d = s_y/1.65 = 0.61\ s_y$ or $\sigma_d = s_u/1.95 = 0.51\ s_u$
   whichever is lower

Example Problem 3–1

A structural support for a machine will be subjected to a static tensile load of 16.0 kN. It is planned to fabricate the support from a square rod made from AISI 1020 hot-rolled steel. Specify suitable dimensions for the cross section of the rod.

Solution

Objective Specify the dimensions of the cross section of the rod.

Given $F = 16.0$ kN $= 16\,000$ N static load.
Material: AISI 1020 HR; $s_y = 207$ MPa; 25% elongation (ductile).
(Data from Appendix A–13)

Analysis Let $\sigma = \sigma_d = s_y/2$ (Table 3–2; ductile material, static load).
Stress analysis: $\sigma = F/A$; then required area $= A = F/\sigma_d$.
But $A = a^2$ ($a =$ dimension of side of square).
Minimum allowable dimension $a = \sqrt{A}$.

Results $\sigma_d = s_y/2 = 207$ MPa/2 $= 103.5$ MPa $= 103.5$ N/mm^2
Required area: $A = F/\sigma_d = (16\,000$ N$)/(103.5$ N/mm$^2) = 154.6$ mm^2.
Minimum dimension a: $a = \sqrt{A} = \sqrt{154.6 \text{ mm}^2} = 12.4$ mm.
Specify: $a = 14$ mm (Appendix A–2; preferred size).

Example Problem 3–2

A tensile member for a roof truss for a building is to carry a static axial tensile load of 19 800 lb. It has been proposed that a standard, equal-leg structural steel angle be used for this application using ASTM A36 structural steel. Use the AISC code. Specify a suitable angle from Appendix A–5.

Solution

Objective Specify a standard equal-leg steel angle.

Given $F = 19\,800$ lb static load.
Material: ASTM A36; $s_y = 36\,000$ psi; $s_u = 58\,000$ psi.
(Data from Appendix A–15)

Analysis Let $\sigma = \sigma_d = 0.60 \, s_y$ or $\sigma_d = 0.50 \, s_u$ (Table 3–3).
Stress analysis: $\sigma = F/A$; then required area $= A = F/\sigma_d$.

Results $\sigma_d = 0.60 \, s_y = 0.60 \,(36\,000$ psi$) = 21\,600$ psi
or $\sigma_d = 0.50 \, s_u = 0.50 \,(58\,000$ psi$) = 29\,000$ psi
Use lower value; $\sigma_d = 21\,600$ psi.
Required area: $A = F/\sigma_d = (19\,800$ lb$)/(21\,600$ lb/in$^2) = 0.917$ in^2.
This is the minimum allowable area.
Specify: L2 $\times$ 2 $\times$ 1/4 steel angle (Appendix A–5; lightest section).
$A = 0.938$ in^2; weight $= 3.19$ lb/ft.

Example Problem 3–3

A machine element in a packaging machine is subjected to a tensile load of 36.6 kN which will be repeated several thousand times over the life of the machine. The cross section of the element is 12 mm thick and 20 mm wide. Specify a suitable material from which to make the element.

Solution **Objective** Specify a material for a machine element.

Given $F = 36.6$ kN $= 36\,600$ N repeated load.
Cross section of machine element: rectangle; 12 mm $\times$ 20 mm.

Analysis Ductile material desirable for repeated loading.
Let $\sigma = \sigma_d = s_u/8$ (Table 3–2). Then required $s_u = 8\sigma$.
Stress analysis: $\sigma = F/A$.

Results Area $= A = (12$ mm$)(20$ mm$) = 240$ mm^2
$\sigma = F/A = (36\,600$ N$)/(240$ mm$^2) = 152.5$ N/mm$^2 = 152.5$ MPa
Required ultimate strength: $s_u = 8\sigma = 8$ (152.5 MPa) $= 1220$ MPa.
Specify: AISI 4140 OQT 900 steel (Appendix A–13).
$s_u = 1289$ MPa; 15% elongation; adequate strength, good ductility.

Comments Other materials could be selected. Required strength indicates that a heat-treated alloy steel is required. The one selected has the highest percent elongation of any listed in Appendix A–13. If the size of the element could be made somewhat larger, the required strength would be lower and a less costly steel may be found.

Example Problem 3–4 Figure 3–1 shows a design for the support for a heavy machine that will be loaded in axial compression. Gray cast iron, grade 20 has been selected for the support. Specify the allowable load on the support.

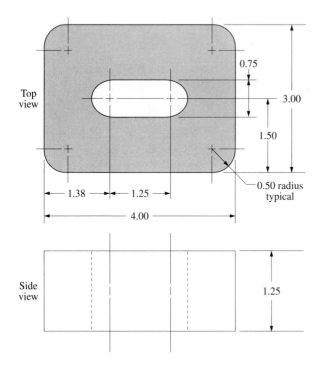

Top view

0.75

3.00

1.50

0.50 radius typical

1.38   1.25

4.00

Side view

1.25

Dimensions in inches

FIGURE 3–1 Machine support for Example Problem 3–4.

Objective   Specify the allowable axial compression load on the support.

Given   Materal: Gray cast iron, grade 20; $s_u = 80$ ksi in compression (Table A–16); material is brittle. Assume load will be static. Shape of support in Fig. 3–1. Compression member is *short* so no buckling occurs.

Analysis   Stress analysis: $\sigma = F/A$; area computed from Fig. 3–1.
Let $\sigma = \sigma_d = s_u/N$; use $N = 6$ (Table 3–2).
Then, allowable $F = \sigma_d/A$.

Results $\sigma_d = s_u/6 = 80\,000$ psi/6 $= 13\,300$ psi
The cross section of the support is the same as the top view. The net area can be calculated by taking the area of a 3.00 in by 4.00 in rectangle and subtracting the area of the slot and the four corner fillets.

$$\text{Rectangle: } A_R = (3.00 \text{ in})(4.00 \text{ in}) = 12.00 \text{ in}^2$$

$$\text{Slot: } \quad A_S = (0.75)(1.25) + \frac{\pi(0.75)^2}{4} = 1.38 \text{ in}^2$$

The area of each fillet can be computed by the difference between the area of a square with sides equal to the radius of the corner (0.50 in) and a quarter circle of the same radius. Then

$$\text{Fillet: } \quad A_F = r^2 - \frac{1}{4}(\pi r^2)$$

$$A_F = (0.50)^2 - \frac{1}{4}[\pi(0.50)^2] = 0.0537 \text{ in}^2$$

Then the total area is

$$A = A_R - A_S - 4A_F = 12.00 - 1.38 - 4(0.0537) = 10.41 \text{ in}^2$$

We now have the data needed to compute the allowable load.

$$P = A\sigma_d = (10.41 \text{ in}^2)(13\,300 \text{ lb/in}^2) = 138\,500 \text{ lb}$$

This completes the example problem.

**Example Problem
3–5**

Figure 3–2 shows a piece of manufacturing equipment called a C-frame press used to pressform sheet-metal products. The ram is driven down with a large force, closing the dies and forming the part. The pressforming action causes the open end of the press to tend to expand, an undesirable action if the deformation is excessive. As an aid in limiting expansion, tie rods are installed across the open part of the C-frame and tightened under a high tensile load. During operation, a peak tensile force of 40 000 lb is applied to the rods with moderate shock as the pressforming operation is completed. Specify an appropriate material for the rods and compute the required diameter.

Solution   **Objective**   Specify a material for the rods and the required diameter.

Given $F = 40\,000$ lb tension; moderate shock and repeated for each press cycle.

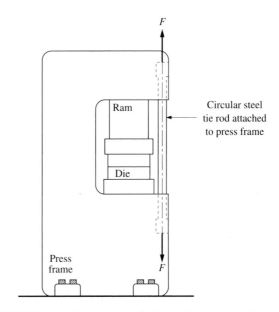

FIGURE 3–2 C-frame press for Example Problem 3–5.

Analysis Stress analysis: $\sigma = F/A$.
Material should have high strength and good ductility.
Let $\sigma = \sigma_d = s_u/N$; use $N = 12$ (Table 3–2).
Required area: $A = F/\sigma_d$ and $A = \pi D^2/4$.
Required diameter: $D = \sqrt{4A/\pi}$.

Results Several different steel alloys could meet the requirements. As a first trial, specify AISI 5160 OQT 1100; $s_u = 149$ ksi; 17% elongation.
Then $\sigma_d = s_u/N = 149\,000$ psi/12 = 12 400 psi.
Required area: $A = F/\sigma_d = (40\,000\ \text{lb})/(12\,400\ \text{lb/in}^2) = 3.23\ \text{in}^2$.
Required diameter: $D = \sqrt{4A/\pi} = \sqrt{4(3.23\ \text{in}^2)/\pi} = 2.03$ in.

Specify Convenient standard size: $D = 2.25$ in.
Material: AISI 5160 OQT 1100.

Comment Using a material with a slightly higher ultimate tensile strength would permit the use of a smaller diameter rod, say 2.00 in. Consider the same material (AISI 5160) with a lower tempering temperature, say OQT 1000. Using the data in Appendix Table A–13, interpolation between the values of s_u for OQT 900 and OQT 1100 would give $s_u = 172.5$ ksi and 14.5% elongation. This would still provide good ductility. Repeating the analysis of the required diameter gives:

$$\sigma_d = s_u/N = 172\,500\ \text{psi}/12 = 14\,400\ \text{psi}$$

Required area: $A = F/\sigma_d = (40\,000\ \text{lb})/(14\,400\ \text{lb/in}^2) = 2.78\ \text{in}^2$.
Required diameter: $D = \sqrt{4A/\pi} = \sqrt{4(2.78\ \text{in}^2)/\pi} = 1.88$ in.

Specify Convenient standard size: $D = 2.00$ in.
Material: AISI 5160 OQT 1000.

3–7 DESIGN SHEAR STRESS

When members are subjected to shear stresses, design must be based on the *design shear stress, τ_d.*

 **Design Shear Stress**

$$\tau_d = \frac{s_{ys}}{N} \quad \text{based on the yield strength in shear} \tag{3–6}$$

Yield strength in shear.

The yield strength in shear, s_{ys}, is the level of shear stress at which the material would exhibit the phenomenon of yield. That is, it would undergo a significant amount of shear deformation with little or no increase in applied shear-type loading.

Virtually all designs for members in shear would require that the actual shear stress be well below the value of s_{ys}, as indicated by Equation (3–6). The selection of design factor is taken from Table 3–4. Also consult Section 3–4 for other considerations in the selection of a design factor. Conditions that are more severe than normally encountered or where there is a significant amount of uncertainty about the magnitude of loads or material properties would justify higher design factors.

Of course, if the values of the yield strength in shear are available, they can be used in the design stress equations. But unfortunately, such values are frequently not reported and it is necessary to rely on estimates. For the yield strength in shear, a frequently used estimate is

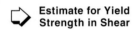 **Estimate for Yield Strength in Shear**

$$s_{ys} = \frac{s_y}{2} = 0.5s_y \tag{3–7}$$

This value is taken from the observation of a typical tensile test in which the shear stress is one-half of the direct tensile stress. This phenomenon, related to the *maximum shear stress theory of failure,* is somewhat conservative and will be discussed further in Chapter 10.

Ultimate strength in shear.

The ultimate strength in shear, s_{us}, is the level of shear stress at which the material would actually fracture.

There are some practical applications of shear stress where fracture of the shear-loaded member is *intended* and, therefore, an estimate of s_{us} is needed. Examples include the *shear pin* often used as an element in the drive train of machines having expensive

TABLE 3–4 Design stress guidelines for shear

Manner of loading	Design stress–Ductile materials $\tau_d = s_{ys}/N = 0.5\, s_y/N = s_y/2N$	
Static	Use $N = 2$	$\tau_d = s_y/4$
Repeated	Use $N = 4$	$\tau_d = s_y/8$
Impact	Use $N = 6$	$\tau_d = s_y/12$

Chapter 3 ▪ Design of Members under Direct Stress

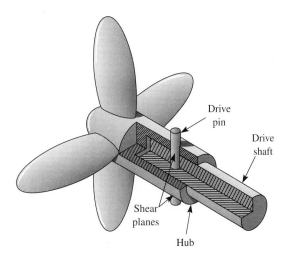

FIGURE 3–3 Propeller drive pin for Example Problem 3–6.

components. Figure 3–3 shows a propeller drive shaft for a boat in which the torque from the drive shaft is transmitted through the pin to the hub of the propeller. The pin must be designed to transmit a level of torque that is typically encountered in moving the boat through the water. However, if the propeller should encounter an obstruction such as a submerged log, it would be desirable to have the inexpensive pin fail rather than the costly propeller. See Example Problem 3–6.

   Another example where an estimate of the ultimate strength in shear is needed is the case of the punching operation as described in Chapter 1 and shown in Figure 1–5. Here, the punch is expected to completely cut (shear) the desired part from the larger sheet of material. Therefore, the sheared sides of the part must be stressed up to the ultimate strength in shear.

   When data for the ultimate strength in shear are known, they should be used. For example, Appendix A–16 gives some values for cast irons and Appendix A–17 gives data for aluminum alloys. But, for occasions when published data are not available, estimates can be computed from the relationships in Table 3–5, taken from Reference 4.

Brittle materials. Design shear stresses for brittle materials are based on the ultimate strength in shear because they do not exhibit yielding. A higher design factor should be used than for ductile materials because the materials are often less consistent in structure. However, published data on acceptable design factors are lacking. It is

TABLE 3–5 Estimates for the ultimate strength in shear

Formula	Material
$s_{us} = 0.65\ s_u$	Aluminum alloys
$s_{us} = 0.82\ s_u$	Steel
$s_{us} = 0.90\ s_u$	Malleable iron and copper alloys
$s_{us} = 1.30\ s_u$	Gray cast iron

recommended that testing be performed on actual prototypes of shear-loaded members made from brittle materials.

Example Problem 3–6 Figure 3–3 shows a boat propeller mounted on a shaft with a cylindrical drive pin inserted through the hub and the shaft. The torque required to drive the propeller is 1575 lb·in and the shaft diameter inside the hub is 3.00 in. Usually, the torque is steady and it is desired to design the pin to be safe for this condition. Specify a suitable material and the diameter for the pin.

Solution **Objective** Specify a material and the diameter for the pin.

Given Torque $= T = 1575$ lb·in (steady).
Shaft diameter $= D = 3.00$ in.

Analysis 1. The pin would be subjected to direct shear at the interface between the shaft and the inside of the hub, as shown in Figure 3–4. The applied

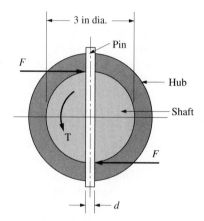

FIGURE 3–4 Cross section through propeller hub and shaft.

torque from the shaft results in two equal forces acting perpendicular to the axis of the pin on opposite sides of the shaft, forming a couple. That is,

$$T = FD$$

Then, $F = T/D$.

2. A material with a moderate-to-high strength is desirable so that the pin will not be overly large. Also it should have good ductility because of the likelihood of mild shock loading from time to time. Several materials could be chosen.

3. Stress analysis: $\tau = F/A$, where $A = \pi d^2/4$. This is one cross-sectional area for the pin.

4. Design shear stress: $\tau_d = s_y/4$ (Table 3–4).

5. Let $\tau = \tau_d$. Then the required $A = F/\tau_d$ and the required d is

$$d = \sqrt{4A/\pi}$$

6. Specify a convenient standard size for the pin.

Results   **1.** $F = T/D = (1575\ \text{lb·in})/(3.00\ \text{in}) = 525\ \text{lb}$

   2.   Specify AISI 1020 cold drawn as a trial ($s_y = 51\,000$ psi; 15% elongation). Note that the pin would have to be protected from corrosion.

   3, 4, 5. $\tau_d = s_y/4 = 51\,000\ \text{psi}/4 = 12\,750\ \text{psi}$
   $A = F/\tau_d = (525\ \text{lb})/(12\,750\ \text{lb/in}^2) = 0.0412\ \text{in}^2$
   $d = \sqrt{4A/\pi} = \sqrt{4(0.0412\ \text{in}^2)/\pi} = 0.229\ \text{in}$

   6.   From Appendix A–2, specify $d = 0.250$ in (1/4 in).
   Material: AISI 1020 cold drawn steel.

Comment   The size seems reasonable compared to the diameter of the shaft.

When the drive pin designed in Example Problem 3–6 is subjected to an overload such as that seen when striking a log, it is desirable for the pin to shear rather than to damage the propeller. The following example problem considers this situation.

Example Problem 3–7   Compute the torque required to shear the pin designed in Example Problem 3–6 and shown in Figures 3–3 and 3–4.

Solution   **Objective**   Compute the torque required to shear the pin.

   Given   Design as shown in Figures 3–3 and 3–4.
   $D = 3.00$ in; $d = 0.250$ in.
   Material: AISI 1020 cold drawn steel.

   Analysis   The analysis would be similar to the reverse of that for Example Problem 3–6, as summarized below.

   1. The pin material would fail when $\tau = s_{us} = 0.82\ s_u$ (Table 3–5).
   2. $\tau = F/A$. Then $F = \tau A = (0.82\ s_u)A$.
   Also, $A = \pi d^2/4$.
   3. $T = FD$

   Results   **1.** $s_{us} = 0.82\ s_u = 0.82\ (61\,000\ \text{psi}) = 50\,000\ \text{psi}$
   2. $A = \pi d^2/4 = A = \pi\ (0.250\ \text{in})^2/4 = 0.0491\ \text{in}^2$
   $F = \tau A = (50\,000\ \text{lb/in}^2)\ (0.0491\ \text{in}^2) = 2450\ \text{lb}$
   3. $T = FD = (2450\ \text{lb})\ (3.00\ \text{in}) = 7350\ \text{lb·in}$

Comment   Compared with the normally applied torque, this value is quite high. The ratio of normal torque to that required to shear the pin is

$$\text{Ratio} = 7350/1575 = 4.67$$

This indicates that the pin would not likely shear under anticipated conditions. However, it may be too high to protect the propeller. Testing of the propeller should be done.

3–8   DESIGN BEARING STRESS

Bearing stress is a localized phenomenon created when two load-carrying parts are placed in contact. The stress condition is actually a compressive stress, but because of the localized nature of the stress, different allowable stresses are used.

Steel.   According to the AISC, the allowable bearing stress in steel for flat surfaces or on the projected area of pins in reamed, drilled, or bored holes is

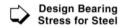 **Design Bearing Stress for Steel**

$$\sigma_{bd} = 0.90 s_y \tag{3–8}$$

When rollers or rockers are used to support a beam or other load-carrying member to allow for expansion of the member, the bearing stress is dependent on the diameter of the roller or rocker, d, and its length, L. The stress is inherently very high because the load is carried on only a small rectangular area. Theoretically, the contact between the flat surface and the roller is simply a line; but because of the elasticity of the materials, the actual area is rectangular. Instead of specifying an allowable bearing stress, the AISC standard allows the computation of the allowable bearing load, W_b, from

 **Allowable Bearing Load for Steel Roller**

$$W_b = \frac{s_y - 13}{20}(0.66\,dL) \tag{3–9}$$

where s_y is in ksi, d and L are in inches, and W_b is in kips.

Example Problem 3–8   A short beam, shown in Figure 3–5, is made from a rectangular steel bar, 1.25 in thick and 4.50 in high. At each end, the length resting on a steel plate is 2.00 in. If both the bar

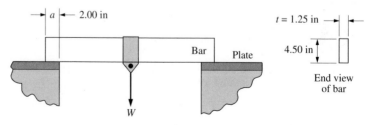

FIGURE 3–5   Beam for Example Problem 3–8.

and the plate are made from ASTM A36 structural steel, compute the maximum allowable load, W, which could be carried by the beam, based on only the bearing stress at the supports. The load is centered between the supports.

Solution

Objective Compute the allowable load, W, based on only bearing stress.

Given Loading in Figure 3−5.
Bearing area at each end: 2.00 in by 1.25 in (that is, $A_b = ta$).
Material: ASTM A36 structural steel ($s_y = 36\,000$ psi).

Analysis Bearing stress: $\sigma_b = R/A_b$
Where R is the reaction at the support and $R = W/2$.
Design stress Eq. (3−8):

$$\sigma_{bd} = 0.90\ s_y = 0.90\ (36\,000\text{ psi}) = 32\,400\text{ psi}$$

Then, $R = A_b\sigma_b$ and $W = 2R$.

Results Bearing area: $A_b = ta = (1.25\text{ in})(2.00\text{ in}) = 2.50\text{ in}^2$.
$R = A_b\sigma_b = (2.50\text{ in}^2)(32\,400\text{ lb/in}^2) = 81\,000\text{ lb}$.
$W = 2R = 2(81\,000\text{ lb}) = 162\,000\text{ lb} = 162\text{ kip}$

Comment This is a very large load, so it is unlikely that bearing is the mode of failure for this beam.

Example Problem 3−9 An alternative proposal for supporting the bar described in Example 3−8 is shown in Figure 3−6. To allow for expansion of the bar, the left end is supported on a 2.00 in-diameter roller made from AISI 1040 CD steel. Compute the allowable load, W, for this arrangement.

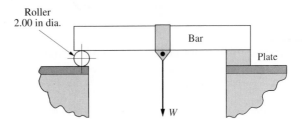

FIGURE 3−6 Beam for Example Problem 3−9.

Solution

Objective Compute the allowable load on the beam.

Given Loading in Figure 3−6.
Width of the beam ($t = 1.25$ in) rests on the roller ($d = 2.00$ in).
Beam material: ASTM A36 structural steel ($s_y = 36$ ksi).
Roller material: AISI 1040 CD ($s_y = 71$ ksi).

Analysis At the roller support, Equation (3−9) applies to determine the allowable load. Note that $R = W_b$ and $W = 2W_b$. Use $s_y = 36$ ksi, the weaker of the two materials in contact.

Results   The allowable bearing load is

$$W_b = \frac{36 - 13}{20}(0.66)(2.00)(1.25) = 1.90 \text{ kip}$$

This would be the allowable reaction at each support. The total load is

$$W = 2W_b = 2(1.90 \text{ kip}) = 3.80 \text{ kip}$$

Comments   Note that this is significantly lower than the allowable load for the surfaces found in Example Problem 3–8. Bearing at the roller support may, indeed, limit the load that could be safely supported.

Aluminum.   The Aluminum Association (1) bases the allowable bearing stresses on aluminum alloys for flat surfaces and pins on the *bearing yield strength*.

$$\sigma_{bd} = \frac{\sigma_{by}}{2.48} \tag{3–10}$$

The minimum values for bearing yield strength are listed in Reference 1. But many references, including the appendix tables in this book, do not include these data. An analysis of the data shows that for most aluminum alloys, the bearing yield strength is approximately 1.60 times larger than the tensile yield strength. Then Equation (3–10) can be restated as

**Design Bearing
Stress for
Aluminum**

$$\sigma_{bd} = \frac{1.60s_y}{2.48} = 0.65s_y \tag{3–11}$$

We will use this form for design bearing stress for aluminum in this book.

**Example Problem
3–10**

A rectangular bar is used as a hanger, as shown in Figure 3–7. Compute the allowable load on the basis of bearing stress at the pin connection if the bar and the clevis members are made from 6061-T6 aluminum. The pin is to be made from a stronger material.

Solution   **Objective**   Compute the allowable load on the hanger.

Given   Loading in Figure 3–7. Pin diameter $= d = 18$ mm.
Thickness of the hanger $= t_1 = 25$ mm; width $= w = 50$ mm.
Thickness of each part of clevis $= t_2 = 12$ mm.
Hanger and clevis material: aluminum 6061-T6 ($s_y = 145$ MPa).
Pin is stronger than hanger or clevis.

Analysis   For cylindrical pins in close-fitting holes, the bearing stress is based on the *projected* area in bearing, found from the diameter of the pin times the length over which the load is distributed.

$$\sigma_b = \frac{F}{A_b} = \frac{F}{dL}$$

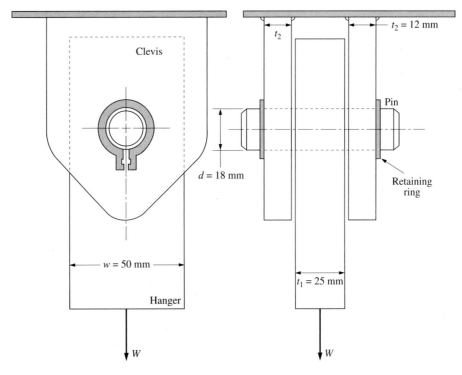

FIGURE 3–7 Hanger for Example Problem 3–10.

Let $\sigma_b = \sigma_{bd} = 0.65\ s_y$ for aluminum 6061-T6.
Bearing area for hanger: $A_{b1} = t_1\ d = (25\ mm)\,(18\ mm) = 450\ mm^2$.
This area carries the full applied load, W.
For each side of clevis: $A_{b2} = t_2\ d = (12\ mm)\,(18\ mm) = 216\ mm^2$.
This area carries 1/2 of the applied load, $W/2$.
Because A_{b2} is less than 1/2 of A_{b1}, bearing on the clevis governs.

Results $\sigma_{bd} = 0.65\ s_y = 0.65\ (145\ MPa) = 94.3\ MPa = 94.3\ N/mm^2$
$\sigma_b = \sigma_{bd} = (W/2)/A_{b2}$
Then, $W = 2(A_{b2})\,(\sigma_{bd}) = 2(216\ mm^2)\,(94.3\ N/mm^2) = 40\,740\ N$.

Comments   This is a very large force and other failure modes for the hanger would have to be analyzed. Failure could occur by shear of the pin or tensile failure of the hanger bar or the clevis members.

Masonry.   The lower part of a support system is often made from concrete, brick, or stone. Loads transferred to such supports usually require consideration of bearing stresses because the strengths of these materials are relatively low compared with metals. It should be noted that the actual strengths of the materials should be used whenever possible because of the wide variation of properties. Also, some building codes list allowable bearing stresses for certain kinds of masonry.

TABLE 3–6 Allowable bearing stresses on masonry

| | Allowable bearing stress | |
Material	psi	MPa
Sandstone and limestone	400	2.76
Brick in cement mortar	250	1.72
Concrete: on full area of support	$0.35\sigma_c$	
(σ_c = specified strength of concrete)		
σ_c = 1500 psi	525	3.62
σ_c = 2000 psi	700	4.83
σ_c = 2500 psi	875	6.03
σ_c = 3000 psi	1050	7.24
Concrete: on less than full area of support		
$\sigma_{bd} = 0.35\ \sigma_c\sqrt{A_2/A_1}$		
A_1 = bearing area		
A_2 = full area of support		
But maximum $\sigma_{bd} = 0.7\sigma_c$		

TABLE 3–7 Safe bearing capacity of soils

| | Safe bearing capacity | |
Nature of soil	psi	kPa
Solid hard rock	350	2400
Shale or medium rock	140	960
Soft rock	70	480
Hard clay or compact gravel	55	380
Soft clay or loose sand	15	100

In the absence of specific data, the AISC (2) recommends the allowable bearing stresses shown in Table 3–6.

Soils. The masonry or concrete supports are often placed on soils to transfer the loads directly to the earth. *Marks' Standard Handbook for Mechanical Engineers* (3) lists the values for the safe bearing capacity of soils, as shown in Table 3–7. Variations should be expected and test data should be obtained where possible.

Example Problem 3–11

Figure 1–45 shows a column resting on a foundation and carrying a load of 26 000 lb. Determine if the bearing stresses are acceptable for the concrete and the soil. The concrete has a specified strength of 2000 psi and the soil is compact gravel.

Solution

Objective Are bearing stresses on the concrete and the soil safe?

Given Foundation shown in Figure 1–45 in Chapter 1. Load = F = 26 000 lb.
For concrete: σ_c = 2000 psi.
For soil (compact gravel): σ_{bd} = 55 psi (Table 3–7).

Analysis and Results

For concrete: The load is transferred from the column to the concrete through the 12 in square steel plate ($A_1 = 144$ in²). But the concrete pier is 18 in square ($A_2 = 324$ in²). Therefore, the bearing load is on less than the full area of the concrete. Then, from Table 3–6,

$$\sigma_{bd} = 0.35\sigma_c \sqrt{\frac{A_2}{A_1}} = 0.35 \,(2000 \text{ psi}) \sqrt{\frac{324}{144}} = 1050 \text{ psi}$$

The bearing stress exerted on the concrete by the steel plate at the base of the column is

$$\sigma_b = \frac{F}{A_b} = \frac{26\,000 \text{ lb}}{(12 \text{ in})^2} = 180 \text{ psi}$$

Thus the bearing stress is acceptable.

For the soil (gravel) at the base of the foundation:

$$\sigma_b = \frac{F}{A_b} = \frac{26\,000 \text{ lb}}{(36 \text{ in})^2} = 20.1 \text{ psi}$$

This is acceptable because the allowable bearing stress for compact gravel is 55 psi.

3–9 STRESS CONCENTRATION FACTORS

In defining the method for computing stress due to a direct tensile or compressive load on a member, it was emphasized that the member must have a uniform cross section in order for the equation $\sigma = F/A$ to be valid. The reason for this restriction is that where a change in the geometry of a loaded member occurs, the actual stress developed is higher than would be predicted by the standard equation. This phenomenon is called *stress concentration* because detailed studies reveal that localized high stresses appear to concentrate around sections where geometry changes occur.

Figure 3–8 illustrates the case of stress concentration for the example of an axially loaded round bar in tension that has two diameters with a step between them. Note that there is a small fillet at the base of the step. Its importance will be discussed later. Below the sketch of the stepped bar is a graph of stress versus position on the bar. At section 1, where the bar diameter is D and well away from the step, the stress can be computed from

$$\sigma_1 = F/A_1 = F/(\pi D^2/4)$$

At section 2, where the bar diameter has the smaller value of d, the stress is

$$\sigma_2 = F/A_2 = F/(\pi d^2/4)$$

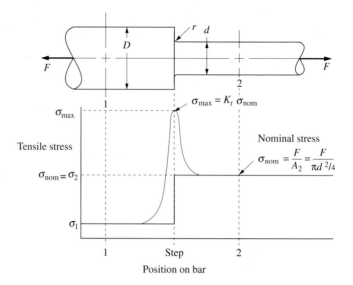

FIGURE 3–8 Stress distribution near a change in geometry.

Then the plot of stress versus position might be expected to appear as the straight lines with an abrupt step at the location of the change in diameter. But tests would show that the actual stress distribution would look more like the curved line: blending with the two straight lines well away from the step, but rising much higher near the step itself.

To account for the higher-than-predicted stress at the step, we modify the direct stress formula to include a *stress concentration factor, K_t,* to produce the form shown below.

$$\sigma_{max} = K_t \sigma_{nom} \qquad (3–12)$$

where, in this case, the nominal stress is based on the smaller section 2. That is,

$$\sigma_{nom} = \sigma_2 = F/A_2 = F/(\pi d^2/4)$$

Then the value of K_t represents the factor by which the actual stress is higher than the nominal stress computed by the standard formula.

Stress concentrations are most damaging for dynamic loading such as repeated loads, impact, or shock. Indeed, fatigue failures most often occur near locations of stress concentrations with small local cracks that grow with time until the remaining section is unable to withstand the load. Under static loading, the high stress near the discontinuity may cause local yielding that would result in a redistribution of the stress to an average value less than the yield strength, and thus the part would still be safe.

Values of stress concentration factors. The magnitude of the stress concentration factor, K_t, depends on the geometry of the member in the vicinity of the discontinuity. Most data have been obtained experimentally by carefully measuring the maximum stress, σ_{max}, using experimental stress analysis techniques such as strain gaging or photoelasticity. Computerized approaches using finite element analysis could also be used. Then, the value of K_t is computed from

Stress Concentration Factor

$$K_t = \sigma_{\max}/\sigma_{\text{nom}} \qquad (3\text{–}13)$$

where σ_{nom} is the stress that would be computed at the section of interest without considering the stress concentration. In the case now being discussed, direct tensile stress, $\sigma_{\text{nom}} = F/A$.

Appendix A–21 contains several charts that can be used to determine the value of K_t for a variety of geometries.

A–21–1:   Axially loaded round bar in tension with a circular groove

A–21–2:   Axially loaded round bar in tension with a step and a fillet

A–21–3:   Axially loaded flat plate in tension with a step and a fillet

A–21–4:   Flat plate with a central hole

A–21–5:   Round bar with a transverse hole

The chart in Appendix A–21–1 shows the typical pattern for presenting values for stress concentration factors. The vertical axis gives the value of K_t itself. The pertinent geometry factors are the diameter of the full round section, D, the diameter at the base of the groove, d_g, and the radius of the circular groove, r. From these data, *two* parameters are computed. The horizontal axis is the ratio of r/d_g. The family of curves in the chart is for different values of the ratio of D/d_g. The normal use of such a chart when the complete geometry is known is to enter the chart at the value of r/d_g, then project vertically to the curve of D/d_g, then horizontally to the vertical axis to read K_t. Interpolation between curves in the chart is often necessary. Note that the nominal stress for the grooved round bar is based on the stress at the *bottom of the groove,* the smallest area in the vicinity. While this is typical, it is important for you to know the basis of the nominal stress for any given stress concentration chart. The grooves with a circular bottom are often used to distribute oil or other lubricants to a shaft.

The chart in Appendix A–21–2 for the stepped round bar has three geometry factors: the larger diameter, D, the smaller diameter, d, and the radius of the fillet at the step where the change of diameter takes place. Note that the value of K_t rises rapidly for small values of the fillet radius. As a designer, you should provide the largest practical radius for such a fillet to maintain a relatively low maximum stress at the step.

An important use for the chart in Appendix A–21–2 is the analysis of stress concentration factors for round bars having retaining ring grooves, as shown in Figure 3–9. The typical geometry of the groove, specified by the ring manufacturer, is shown in Figure 3–10. The bottom of the groove is flat and the fillet at each end is quite small to provide a sizeable vertical surface against which to seat the ring. The result is that the ring groove acts like two steps close together. Then Appendix A–21–2 can be used to determine the value of K_t. Sometimes the geometry of the ring groove will result in K_t values that are off the top of the chart. In such cases, an estimate of $K_t = 3.0$ is reasonable but additional data should be sought.

The chart in Appendix A–21–4 contains three curves, all related to a flat plate with a central hole. Curve A is for the case where the plate is subjected to direct tensile

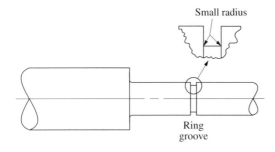

FIGURE 3–9 Stepped shaft with a ring groove.

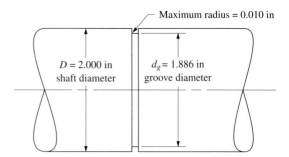

FIGURE 3–10 Sample geometry for a retaining ring groove in a round bar.

stress across its entire cross section in the vicinity of the hole. Curve B is for the case where a close-fitting pin is inserted in the hole and the tensile load is applied through the pin. The resulting stress concentration factors are somewhat larger because of the more concentrated load. Curve C is for the case of the plate in bending and this will be discussed in Chapter 8. In each case, however, note that the nominal stress is based on the *net section* through the plate at the location of the hole. For tensile loading, the *net area* is used for σ_{nom}. That is,

$$\sigma_{nom} = F/A_{net} = F/(w - d)t$$

where w = width of the plate
      t = thickness
      d = diameter of the hole

The chart in Appendix A–21–5 for the round bar with a transverse hole contains a variety of data for different loading patterns: tension, bending, and torsion. For now we are concerned with only Curve A for the case of axial tension. Torsion is discussed in Chapter 5 and bending, in Chapter 8. Also note that the stress concentration factor is based on the *gross section,* not on the net section at the hole. This means that K_t includes the effects of both the removal of material and the discontinuity, resulting in values that are quite high. However, it makes the use of the chart much easier for you, the designer, because you do not have to compute the area of net section.

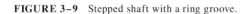

Example Problem 3–12   The stepped bar shown in Figure 3–8 is subjected to an axial tensile force of 12 500 lb. Compute the maximum tensile stress in the bar for the following dimensions:

$$D = 1.50 \text{ in}; \quad d = 0.75 \text{ in}; \quad r = 0.060 \text{ in}$$

Solution   **Objective**   Compute the maximum tensile stress.

            Given $F = 12\,500$ lb; $D = 1.50$ in; $d = 0.75$ in; $r = 0.060$ in

Analysis   Because of the change in diameter, use Equation (3–12).
Use the chart in Appendix A–21–2 to find the value of K_t using r/d and D/d as parameters.

Results $\sigma_{max} = K_t\sigma_{nom}$

$\sigma_{nom} = \sigma_2 = F/A_2 = F/(\pi d^2/4) = (12\,500\text{ lb})/[\pi(0.75\text{ in})^2/4]$

$\sigma_{nom} = 28\,294\text{ lb/in}^2$

$r/d = 0.06/0.75 = 0.080$ and $D/d = 1.50/0.75 = 2.00$

Read $K_t = 2.12$ from Appendix A–21–2.
Then $\sigma_{max} = K_t\sigma_{nom} = 2.12\,(28\,294\text{ psi}) = 59\,983\text{ psi}$.

Comment   The actual maximum stress of approximately 60 000 psi is more than double the value that would be predicted by the standard formula.

REFERENCES

1. Aluminum Association, *Specifications for Aluminum Structures,* Washington, DC, 1986.

2. American Institute of Steel Construction, *Manual of Steel Construction,* 9th ed., Chicago, IL, 1989.

3. Avallone, E. A., and T. Baumeister III, eds., *Marks' Standard Handbook for Mechanical Engineers,* 9th ed., 1987.

4. Deutschman, A. D., W. J. Michels, and C. E. Wilson, *Machine Design Theory and Practice,* Macmillan, New York, 1975.

5. Juvinall, R. C., and K. M. Marshek, *Fundamentals of Machine Component Design,* 2nd ed., Wiley, New York, 1983.

6. Mott, R. L., *Machine Elements in Mechanical Design,* 2nd ed., Merrill, An Imprint of Macmillan, New York, 1992.

7. Peterson, R. E., *Stress Concentration Factors,* Wiley, New York, 1974.

8. Shigley, J. E., and C. R. Mischke, *Mechanical Engineering Design,* 5th ed., McGraw-Hill, New York, 1988.

9. Spotts, M. F., *Design of Machine Elements,* 6th ed., Prentice-Hall, Englewood Cliffs, NJ, 1985.

10. Young, W. C., *Roark's Formulas for Stress and Strain,* 6th ed., McGraw-Hill, New York, 1989.

PROBLEMS

3–1.M Specify a suitable aluminum alloy for a round bar having a diameter of 10 mm subjected to a static direct tensile force of 8.50 kN.

3–2.M A rectangular bar having cross-sectional dimensions of 10 mm by 30 mm is subjected to a direct tensile force of 20.0 kN. If the force is to be repeated many times, specify a suitable steel material.

3–3.E A link in a mechanism for an automated packaging machine is subjected to a direct tensile force of 1720 lb, repeated many times. The link is square, 0.40 in on a side. Specify a suitable steel for the link.

3–4.E A circular steel rod having a diameter of $\frac{3}{8}$ in supports a heater assembly and carries a static tensile load of 1850 lb. Specify a suitable structural steel for the rod.

3–5.E A tension member in a wood roof truss is to carry a static tensile force of 5200 lb. It has been proposed to use a standard 2 × 4 made from southern pine, No. 2 grade. Would this be acceptable?

3–6.E For the data in Problem 3–5, suggest an alternative design that would be safe for the given loading. A different size member or a different material may be specified.

3–7.E A guy wire for an antenna tower is to be aluminum, having an allowable stress of 12 000 psi. If the expected maximum load on the wire is 6400 lb, determine the required diameter of the wire.

3–8.M A hopper having a mass of 1150 kg is designed to hold a load of bulk salt having a mass of 6350 kg. The hopper is to be suspended by four rectangular straps, each carrying one-fourth of the load. Steel plate with a thickness of 8.0 mm is to be used to make the straps. What should be the width in order to limit the stress to 70 MPa?

3–9.M A shelf is being designed to hold crates having a total mass of 1840 kg. Two support rods like that shown in Figure 3–11 will hold the shelf. Assume that the center of gravity of the crates is at the middle of the shelf. Specify the required diameter of the circular rods to limit the stress to 110 MPa.

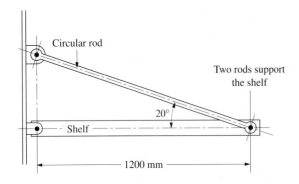

FIGURE 3–11 Shelf support rods for Problem 3–9.

3–10.E A concrete column base is circular, with a diameter of 8.0 in, and carries a static direct compressive load of 70 000 lb. Specify the required rated strength of the concrete according to the recommendations in Section 2–10.

3–11.E Three short wood blocks made from standard 4 × 4 posts support a machine weighing 29 500 lb and share the load equally. Specify a suitable type of wood for the blocks.

3–12.M A circular pier to support a column is to be made of concrete having a rated strength of 3000 psi (20.7 MPa). Specify a suitable diameter for the pier if it is to carry a direct compressive load of 1.50 MN.

3–13.E An aluminum ring has an outside diameter of 12.0 mm and an inside diameter of 10 mm. If the ring is short and is made of 2014-T6, compute the force required to produce ultimate compressive failure in the ring. Assume that s_u is the same in both tension and compression.

3–14.M A wooden cube 40 mm on a side is made from No. 2 hemlock. Compute the allowable compressive force that could be applied to the cube either parallel or perpendicular to the grain.

3–15.E A round bar of ASTM A242 structural steel is to be used as a tension member to stiffen a frame. If a maximum static load of 4000 lb is expected, specify a suitable diameter of the rod.

3–16.E A portion of a casting made of ASTM A48, grade 20 gray cast iron has the shape shown in Figure 1–32 and is subjected to compressive force in line with the centroidal axis of the section. If the member is short and carries a load of 52 000 lb, compute the stress in the section and the design factor.

3–17.M A part for a truck suspension system is to carry a compressive load of 135 kN with the possibility of shock loading. Malleable iron ASTM A220 grade 45008 is to be used. The cross section is to be rectangular with the long dimension twice the short dimension. Specify suitable dimensions for the part.

3–18.E A rectangular plastic link in an office printer is to be made from glass-filled acetal copolymer (see Appendix A–19). It is to carry a tensile force of 110 lb. Space limitations permit a maximum thickness for the link of 0.20 in. Specify a suitable width of the link if a design factor of 8 based on the tensile strength of the plastic is expected.

3–19.M Figure 1–33 shows the cross section of a short compression member that is to carry a static load of 640 kN. Specify a suitable material for the member.

3–20.M Figure 1–25 shows a bar carrying several static loads. If the bar is made from ASTM A36 structural steel, is it safe?

3–21.M In Figure 1–30, specify a suitable aluminum alloy for the member AB if the load is to be repeated many times. Consider only the square part near the middle of the member.

3–22.M Figure 3–12 shows the design of the lower end of the member *AB* in Figure 1–30. Use the load shown in Figure 1–30 and assume that it is static. Member 1 is made from aluminum alloy 6061-T4; member 2 is made from aluminum alloy 2014-T4; pin 3 is made from aluminum alloy 2014-T6. Perform the follwing analyses:

(a) Evaluate member 1 for safety in tension in the area of the pin holes.

(b) Evaluate member 1 for safety in bearing at the pin.

(c) Evaluate member 2 for safety in bearing at the pin.

(d) Evaluate the pin 3 for safety in bearing.

(e) Evaluate the pin 3 for safety in shear.

3–23.E A standard steel angle, L2 $\times$ 2 $\times \frac{1}{4}$, serves as a tension member in a truss carrying a static load. If the angle is made from ASTM A36 structural steel, compute the allowable tensile load based on the AISC specifications.

3–24.M A machine weighs 90 kN and rests on four legs. Two concrete supports serve as the foundation. The load is symmetrical, as shown in Figure 3–13. Evaluate the proposed design with regard to the safety of the steel plate, the concrete foundation, and the soil for bearing.

3–25.M A column for a building is to carry 160 kN. If the column is to be supported on a square concrete foundation that sits on soft rock, determine the required dimensions of the foundation block.

3–26.E A base for moving heavy machinery is designed as shown in Figure 3–14. How much load could

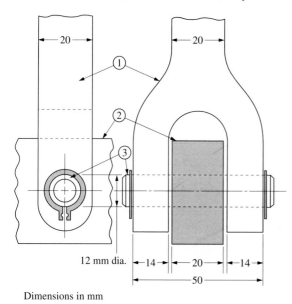

Dimensions in mm

FIGURE 3–12 Clevis for beam support for Problem 3–22.

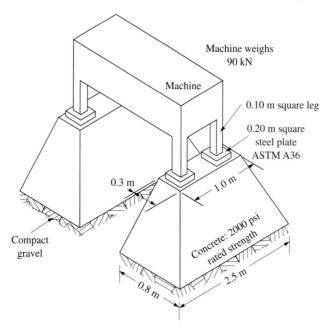

FIGURE 3–13 Machine supports for Problem 3–24.

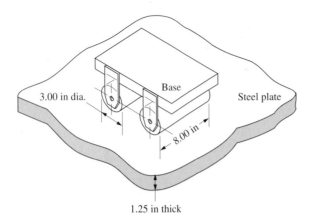

FIGURE 3–14 Roller base for moving machinery for Problems 3–26 and 3–27.

the base support based on the bearing capacity of the steel plate if it is 1.25 in thick and made from ASTM A36 structural steel?

3–27.E Repeat Problem 3–26 but use ASTM A242 high-strength low-alloy steel plate.

3–28.E Figure 3–15 shows an alternative design for the machinery moving base described in Problem 3–26. Compute the allowable load for this design if it rests on (a) ASTM A36 steel or (b) ASTM A242 steel.

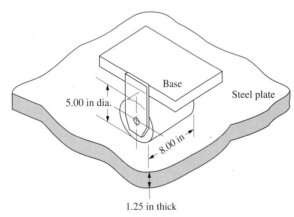

FIGURE 3–15 Roller base for moving machinery for Problem 3–28.

3–29.E A heavy table for industrial use has four legs made from $2 \times 2 \times \frac{1}{4}$ square steel tubing. Compute the bearing stress exerted by each leg on the floor if a total load of 10 000 lb is placed on the

table such that the load is balanced among all four legs. Then suggest a redesign for the legs if it is desired to keep the bearing stress below 400 psi.

3–30.C One end of a beam is supported on a rocker having a radius of 200 mm and a width of 150 mm. If the rocker and the plate on which it rests are made of ASTM A36 structural steel, specify the maximum allowable reaction at this end of the beam.

3–31.E A gear transmits 5000 lb·in of torque to a circular shaft having a diameter of 2.25 in. A square key 0.50 in on a side connects the shaft to the hub of the gear, as shown in Figure 1–14. The key is made from AISI 1020 cold-drawn steel. Determine the required length of the key, L, to be safe for shear and bearing. Use a design factor of 2.0 based on yield for shear and the AISC allowable bearing stress.

3–32.E A support for a beam is made as shown in Figure 3–16. Determine the required thickness of the projecting ledge a if the maximum shear stress is to be 6000 psi. The load on the support is 21 000 lb.

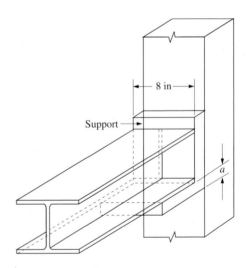

FIGURE 3–16 Beam support for Problem 3–32.

3–33.E A section of pipe is supported by a saddle-like structure which, in turn, is supported on two steel pins, as illustrated in Figure 3–17. If the load on the saddle is 42 000 lb, determine the required diameter and length of the pins. Use AISI 1040 cold-drawn steel. Consider both shear and bearing.

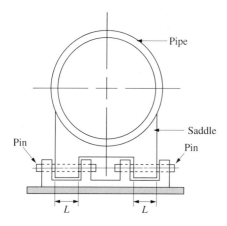

FIGURE 3–17   Pipe saddle for Problem 3–33.

3–34.M The lower control arm on an automotive suspension system is connected to the frame by a round steel pin 16 mm in diameter. Two sides of the arm transfer loads from the frame to the arm, as sketched in Figure 3–18. How much shear force could the pin withstand if it is made of AISI 1040 cold-drawn steel and a design factor of 6 based on the yield strength in shear is desired?

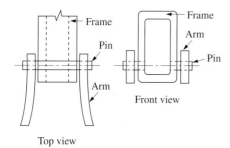

FIGURE 3–18   Pin for automotive suspension system for Problem 3–34.

3–35.M A centrifuge is used to separate liquids according to their densities using centrifugal force. Figure 1–23 illustrates one arm of a centrifuge having a bucket at its end to hold the liquid. In operation, the bucket and the liquid have a mass of 0.40 kg. The centrifugal force has the magnitude in newtons of

$$F = 0.010\,97 \cdot m \cdot R \cdot n^2$$

where m = rotating mass of bucket and liquid (kg)
   R = radius to center of mass (meters)
   n = rotational speed (rpm)

The centrifugal force places the pin holding the bucket in direct shear. Compute the stress in the pin due to a rotational speed of 3000 rpm. Then, specify a suitable steel for the pin, considering the load to be repeated.

3–36.M A circular punch is used to punch a 20.0-mm-diameter hole in a sheet of AISI 1020 hot-rolled steel having a thickness of 8.0 mm. Compute the force required to punch out the slug.

3–37.M Repeat Problem 3–36, but the material is aluminum 6061-T4.

3–38.M Repeat Problem 3–36, but the material is C14500 copper, hard.

3–39.M Repeat Problem 3–36, but the material is AISI 430 full hard stainless steel.

3–40.M Determine the force required to punch a slug the shape shown in Figure 1–36 from a sheet of AISI 1020 hot-rolled steel having a thickness of 5.0 mm.

3–41.E Determine the force required to punch a slug the shape shown in Figure 1–37 from a sheet of aluminum 3003-H18 having a thickness of 0.194 in.

3–42.E A notch is made in a piece of wood, as shown in Figure 1–35, to support an external load of 1800 lb. Compute the shear stress in the wood. Is the notch safe? (See Appendix A–18.)

3–43.E Compute the force required to shear a straight edge of a sheet of AISI 1040 cold-drawn steel having a thickness of 0.105 in. The length of the edge is 7.50 in.

3–44.E Repeat Problem 3–43 but the material is AISI 5160 OQT 700 steel.

3–45.E Repeat Problem 3–43 but the material is AISI 301 full hard stainless steel.

3–46.E Repeat Problem 3–43 but the material is C36000 brass, hard.

3–47.E Repeat Problem 3–43, but the material is aluminum 5154-H32.

3–48.M For the lever shown in Figure 3–19, called a *bell-crank*, compute the required diameter of the pin A if the load is repeated and the pin is made from C17000 copper, hard. The loads are repeated many times.

3–49.M For the structure shown in Figure 3–20, determine the required diameter of each pin if it is made from AISI 1020 cold-drawn steel. Each pin is in double shear and the load is static.

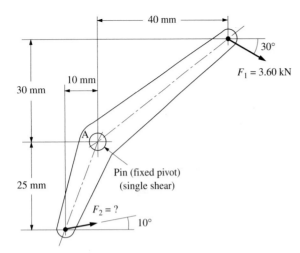

FIGURE 3–19 Bellcrank for Problem 3–48.

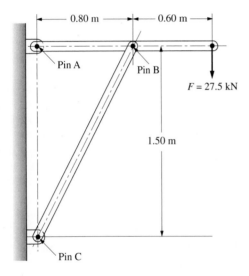

FIGURE 3–20 Pin-connected structure for Problem 3–49.

3–50.M For the structure shown in Figure 1–28, determine the required diameter of each pin if it is made from ASTM A572 high-strength low-alloy columbium-vanadium structural steel, grade 50. Each pin is in double shear and the load is static.

3–51.E A pry bar as shown in Figure 3–21 is used to provide a large mechanical advantage for lifting heavy machines. An operator can exert a force of 280 lb on the handle. Compute the lifting force and the shear stress on the wheel axle.

3–52.E Figure 3–22 illustrates a type of engineering chain used for conveying applications. All components are made from AISI 1040 cold-drawn steel. Evaluate the allowable tensile force (repeated) on the chain with respect to:
(a) Shear of the pin
(b) Bearing of the pin on the side plates
(c) Tension in the side plates.

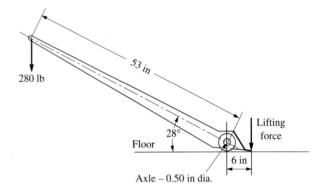

FIGURE 3–21 Pry bar for Problem 3–51.

3–53.E Figure 3–23 shows an anvil for an impact hammer held in a fixture by a circular pin. If the force is to be 500 lb, specify a suitable diameter for the steel pin if it is to be made from AISI 1040 WQT 900.

3–54.E Figure 3–24 shows a steel flange forged integrally with the shaft, which is to be loaded in torsion. Eight bolts serve to couple the flange to a mating flange. Assume that each bolt carries an equal load. Compute the maximum permissible torque on the coupling if the shear stress in the bolts is not to exceed 6000 psi.

3–55.M Figure 3–25 shows a circular shaft subjected to a repeated axial tensile load of 25 kN. The shaft is made from AISI 4140 OQT 1100 steel. Determine the design factor at the hole and at the fillet.

3–56.M A valve stem in an automotive engine is subjected to an axial tensile load of 900 N due to the valve spring, as shown in Figure 3–26. Compute the maximum stress in the stem at the place where the spring force acts against the shoulder.

3–57.M A round shaft has two grooves in which rings are placed to retain a gear in position, as shown in Figure 3–27. If the shaft is subjected to an axial

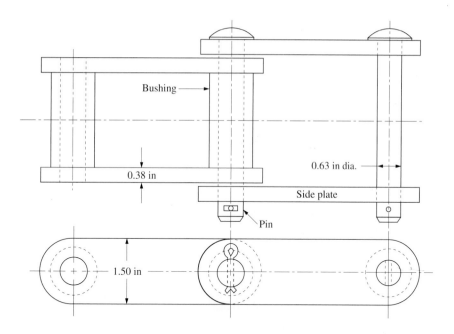

FIGURE 3–22 Conveyor chain for Problem 3–52.

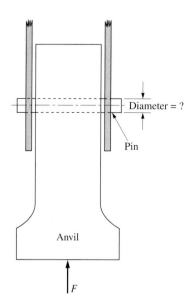

FIGURE 3–23 Impact hammer for Problem 3–53.

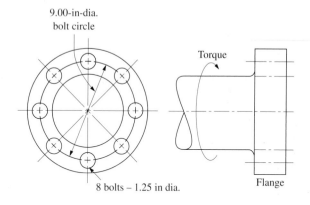

FIGURE 3–24 Flange coupling for Problem 3–54.

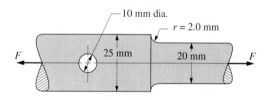

FIGURE 3–25 Shaft for Problem 3–55.

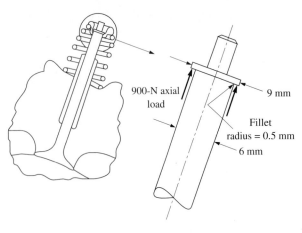

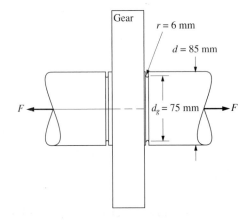

FIGURE 3–26   Valve stem for Problem 3–56.

FIGURE 3–27   Shaft for Problem 3–57.

tensile force of 36 kN, compute the maximum tensile stress in the shaft.

3–58.E   Figure 3–28 shows the proposed design for a tensile rod. The larger diameter is known, $D = 1.00$ in, along with the hole diameter, $a = 0.50$ in. It has also been decided that the stress concentration factor at the fillet is to be 1.7. The smaller diameter, d, and the fillet radius, r, are to be specified such that the stress at the fillet is the same as that at the hole.

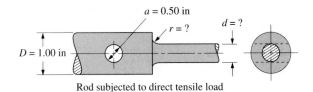

Rod subjected to direct tensile load

FIGURE 3–28   Tensile rod for Problem 3–58.

Deformation and Thermal Stress

4–1 OBJECTIVES OF THIS CHAPTER

The study of strength of materials involves both the determination of stresses in load-carrying members and the deflection or deformation of those members. In general terms, this requires the study of both *stress* and *strain,* as defined in Chapter 1. The material presented in Chapters 1, 2, and 3 enabled you to compute the magnitude of stresses created in members subjected to direct axial forces, either tensile or compressive. This chapter extends your knowledge of such members to include deformation.

Two kinds of deformation are presented in this chapter, *elastic deformation* due to the application of external loads, and *thermal deformation* due to changes in temperature. When a material is heated it tends to expand, and when it is cooled it tends to contract. If these thermal deformations are permitted to occur without restraint, no stresses are created. But if the member is held in such a way as to restrain movement, stresses are developed. Such stresses are called *thermal stresses.*

The principles of elastic deformation can also be used to solve some fairly complex problems in which members made of more than one material are subjected to loads. Such members are often *statically indeterminate;* that is, the internal forces and stresses cannot be found by simple equations of statics. We will show the combined use of statics, stress analysis, and elastic deformation analysis to solve such problems.

After completing this chapter, you should be able to:

1. Compute the amount of elastic deformation for a member carrying an axial tensile or compressive load.

2. Design axially loaded members to limit their deformation to a set value.

3. Define *coefficient of thermal expansion,* and select the appropriate value for use in computing thermal deformation.

4. Compute the amount of thermal deformation for a member subjected to changes in temperature when the deformation is unrestrained.

5. Compute the thermal stress resulting from the restraint of a member subjected to changes in temperature.

6. Compute the stress in components of a composite structure having members made of more than one material subjected to axial loads.

4–2 ELASTIC DEFORMATION IN TENSION AND COMPRESSION MEMBERS

Deformation refers to some change in the dimensions of a load-carrying member. Being able to compute the magnitude of deformation is important in the design of precision mechanisms, machine tools, building structures, and machine structures.

An example of where deformation is important is shown in Figure 3–2, in which circular steel tie rods are attached to a C-frame punch press. The tie rods are subjected to tension when in operation. Since they contribute to the rigidity of the press, the amount that they deform under load is something the designer needs to be able to determine.

To develop the relationship from which deformation can be computed for members subjected to axial tension or compression, some concepts from Chapter 1 must be reviewed. *Strain* is defined as the ratio of the total deformation to the original length of a member. Using the symbols ϵ for strain, δ for total deformation, and L for length, the formula for strain becomes

$$\epsilon = \frac{\delta}{L} \qquad (4\text{–}1)$$

The stiffness of a material is a function of its modulus of elasticity E, defined as

$$E = \frac{\text{stress}}{\text{strain}} = \frac{\sigma}{\epsilon} \qquad (4\text{–}2)$$

Solving for strain gives

$$\epsilon = \frac{\sigma}{E} \qquad (4\text{–}3)$$

Now Equations (4–1) and (4–3) can be equated:

$$\frac{\delta}{L} = \frac{\sigma}{E} \qquad (4\text{–}4)$$

Solving for deformation gives

$$\delta = \frac{\sigma L}{E} \qquad (4\text{--}5)$$

Since this formula applies to members that are subjected to either direct tensile or compressive forces, the direct-stress formula can be used to compute the stress σ. That is, $\sigma = F/A$, where F is the applied load and A is the cross-sectional area of the member. Substituting this into Equation (4–5) gives

▷ **Axial Deformation**

$$\delta = \frac{\sigma L}{E} = \frac{FL}{AE} \qquad (4\text{--}6)$$

Equation (4–6) can be used to compute the total deformation of any load-carrying member, provided that it meets the conditions defined for direct tensile and compressive stress. That is, the member must be straight and have a constant cross section; the material must be homogeneous; the load must be directly axial; and the stress must be below the proportional limit of the material. Recall that the value of the proportional limit is close to the yield strength, s_y.

Example Problem 4–1   The tie rods in the press in Figure 3–2 are made of the steel alloy AISI 5160 OQT 1000. Each rod has a diameter of 2.00 in and an initial length of 68.5 in. An axial tensile load of 40 000 lb is exerted on each rod during operation of the press. Compute the deformation of the rods.

Solution   **Objective**   Compute the deformation of the tie rods.

Given   Rods are steel, AISI 5160 OQT 1000; diameter $= D = 2.00$ in.
Length $= L = 68.5$ in. Axial force $= F = 40\,000$ lb.

Analysis   Equation (4–6) will be used. But the stress in the rods must be checked to ensure that it is below the proportional limit.

Results   *Axial tensile stress:* $\sigma = F/A$.
Area $= A = \dfrac{\pi D^2}{4} = \dfrac{\pi (2.0 \text{ in})^2}{4} = 3.14 \text{ in}^2$.
Then, $\sigma = \dfrac{40\,000 \text{ lb}}{3.14 \text{ in}^2} = 12\,700$ psi.
Appendix A–13 lists the yield strength of the steel to be 132 ksi.
Therefore, the stress is well below the proportional limit.
Axial deformation: Use Equation (4–6). All data are known except the modulus of elasticity, E. From the footnotes of Appendix A–13 we find $E = 30 \times 10^6$ psi. Then,

$$\delta = \frac{FL}{AE} = \frac{(40\,000 \text{ lb})(68.5 \text{ in})}{(3.14 \text{ in}^2)(30 \times 10^6 \text{ lb/in}^2)} = 0.029 \text{ in}$$

Comment   The acceptability of this amount of axial deformation would have to be determined by an analysis of the entire press system.

A large pendulum is composed of a 10.0-kg ball suspended by an aluminum wire having a diameter of 1.00 mm and a length of 6.30 m. The aluminum is the alloy 7075-T6. Compute the elongation of the wire due to the weight of the 10-kg ball.

Solution   **Objective**   Compute the elongation of the wire.

Given   Wire is aluminum alloy 7075-T6; diameter $= D = 1.00$ mm.
Length $= L = 6.30$ m; ball has a mass of 10.0 kg.

Analysis   The force on the wire is equal to the weight of the ball which must be computed from $w = mg$. Then the stress in the wire must be checked to ensure that it is below the proportional limit. Finally, because the stress will then be known, Equation (4–5) will be used to compute the elongation of the wire.

Results   ***Force on the wire:*** $F = w = m{\cdot}g = 10.0$ kg$\cdot$9.81 m/s^2 = 98.1 N.
Axial tensile stress: $\sigma = F/A$

$$A = \frac{\pi D^2}{4} = \frac{\pi(1.00 \text{ mm})^2}{4} = 0.785 \text{ mm}^2$$

$$\sigma = \frac{F}{A} = \frac{98.1 \text{ N}}{0.785 \text{ mm}^2} = 125 \text{ N/mm}^2 = 125 \text{ MPa}$$

Appendix A–17 lists the yield strength of 7075-T6 aluminum alloy to be 503 MPa. The stress is well below the proportional limit.
Elongation: For use in Equation (4–5), all data are known except the modulus of elasticity, E. The footnote of Appendix A–17 lists the value of $E = 72$ GPa $= 72 \times 10^9$ Pa. Then,

$$\delta = \frac{\sigma L}{E} = \frac{(125 \text{ MPa}) (6.30 \text{ m})}{72 \text{ GPa}} = \frac{(125 \times 10^6 \text{ Pa}) (6.30 \text{ m})}{72 \times 10^9 \text{ Pa}}$$

$$\delta = 10.9 \times 10^{-3} \text{ m} = 10.9 \text{ mm}$$

A tension link in a machine must have a length of 610 mm and will be subjected to a repeated axial load of 3000 N. It has been proposed that the link be made of steel and that it have a square cross section. Determine the required dimensions of the link if the elongation under load must not exceed 0.05 mm.

Solution   **Objective**   Determine the required dimensions of the square cross section of the link to limit the elongation, δ, to 0.05 mm or less.

Given   Axial loading on the link $= F = 3000$ N; length $= L = 610$ mm.
Link will be steel; then $E = 207$ GPa $= 207 \times 10^9$ N/m^2. (Appendix A–13)

Analysis   In Equation (4–6) for the axial deformation, let $\delta = 0.05$ mm. Then all other data are known except for the cross-sectional area, A. We can solve for A. Let each side of the square cross section be d. Then $A = d^2$ and we can compute the minimum acceptable value for d from $d = \sqrt{A}$. After specifying a convenient size for d, we must check to ensure that the stress is safe and below the proportional limit.

Results   Solving for A from Equation (4-6) and substituting values gives,

$$A = \frac{PL}{E\delta} = \frac{(3000 \text{ N}) (610 \text{ mm})}{(207 \times 10^9 \text{ N/m}^2) (0.05 \text{ mm})} = 176.8 \times 10^{-6} \text{ m}^2$$

Converting to mm^2

$$A = 176.8 \times 10^{-6} \text{ m}^2 \times \frac{(10^3 \text{ mm})^2}{\text{m}^2} = 176.8 \text{ mm}^2$$

and
$$d = \sqrt{A} = \sqrt{176.8 \text{ mm}^2} = 13.3 \text{ mm}$$

Appendix A-2 lists the next larger preferred size to be 14.0 mm.
The actual cross-sectional area is $A = d^2 = (14.0 \text{ mm})^2 = 196 \text{ mm}^2$.
Stress: $\sigma = F/A = 3000 \text{ N}/196 \text{ mm}^2 = 15.3 \text{ N/mm}^2 = 15.3 \text{ MPa}$.
For a repeated load, Table 3-2 recommends that the design stress be
$\sigma_d = s_u/8$. Letting $\sigma_d = \sigma$, the required value for the ultimate strength is

$$s_u = 8(\sigma) = 8(15.3 \text{ MPa}) = 123 \text{ MPa}$$

Comments   Referring to Appendix A-13, we can see that virtually any steel has an ultimate strength far greater than 123 MPa. Unless there were additional design requirements, we should specify the least expensive steel. Then we will specify:

$d = 14.0$ mm

AISI 1020 hot rolled steel should be low cost. $s_u = 448$ MPa

Note that this design was limited by the allowable elongation and that the resulting stress is relatively low.

Example Problem 4-4   Figure 1-26 shows a steel pipe being used as a bracket to support equipment through cables attached as shown. The forces are $F_1 = 8000$ lb and $F_2 = 2500$ lb. Select the smallest standard schedule 40 steel pipe that will limit the stress to no more than 18 000 psi. Then for the pipe selected, determine the total downward deflection of point C at the bottom of the pipe as the loads are applied.

Solution   **Objective**   Specify a suitable standard schedule 40 steel pipe size and determine the elongation of the pipe.

Given   Loading in Figure 1-26; $F_1 = 8000$ lb (two forces); $F_2 = 2500$ lb.
Pipe length from A to B: $L_{A-B} = 4.00 \text{ ft}(12 \text{ in/ft}) = 48.0$ in.
Pipe length from B to C: $L_{B-C} = 3.00 \text{ ft}(12 \text{ in/ft}) = 36.0$ in.
Maximum allowable stress = 18 000 psi; $E = 30 \times 10^6$ psi (steel).

Analysis   The maximum axial tensile load on the pipe is the sum of F_2 plus the vertical components of each of the F_1 forces. This occurs throughout the part of the pipe from A to B. The size of pipe, and the resulting cross-sectional area, must result in a stress in that section of 18 000 psi or less. From B to C, the axial tensile load is $F_{B-C} = 2500$ lb. Because the load is different in the two sections, the computation of the elongation of the pipe must be done in two separate calculations. That is,

$$\delta_C = \delta_{total} = \delta_{A-B} + \delta_{B-C}$$

Results   *Axial loads:*

$$F_{A-B} = F_2 + 2F_1 \cos 30°$$
$$= 2500 \text{ lb} + 2(8000 \text{ lb}) \cos 30°$$
$$= 16\,400 \text{ lb}$$
$$F_{B-C} = F_2 = 2500 \text{ lb}$$

Stress analysis and calculation of required cross-sectional area: Letting $\sigma = 18\,000$ psi, the required cross-sectional area of the metal in the pipe is

$$A = \frac{F_{A-B}}{\sigma} = \frac{16\,400 \text{ lb}}{18\,000 \text{ lb/in}^2} = 0.911 \text{ in}^2$$

From Appendix A–12, listing the properties of steel pipe, the standard size with the next larger cross-sectional area is the 2-in schedule 40 pipe with $A = 1.075 \text{ in}^2$.

Deflection:

$$\delta_C = \delta_{total} = \delta_{A-B} + \delta_{B-C}$$
$$\delta_{A-B} = \left(\frac{F_{AB} L}{AE}\right)_{A-B} = \frac{(16\,400 \text{ lb})(48 \text{ in})}{(1.075 \text{ in}^2)(30 \times 10^6 \text{ lb/in}^2)} = 0.024 \text{ in}$$
$$\delta_{B-C} = \left(\frac{F_{BC} L}{AE}\right)_{B-C} = \frac{(2500 \text{ lb})(36 \text{ in})}{(1.075 \text{ in}^2)(30 \times 10^6 \text{ lb/in}^2)} = 0.003 \text{ in}$$

Then

$$\delta_C = \delta_{A-B} + \delta_{B-C} = 0.027 \text{ in}$$

Comments   In summary, when the bracket shown in Figure 1–26 is made from a standard 2-in schedule 40 steel pipe, point C moves downward 0.027 in under the influence of the applied loads.

4–3   DEFORMATION DUE TO TEMPERATURE CHANGES

A machine or a structure could undergo deformation or be subjected to stress by changes in temperature in addition to the application of loads. Bridge members and other structural components see temperatures as low as −30°F (−34°C) to as high as 110°F (43°C) in some areas. Vehicles and machinery operating outside experience similar temperature variations. Frequently, a machine part will start at room temperature and then become quite hot as the machine operates. Examples are parts of engines, furnaces, metal-cutting machines, rolling mills, plastics molding and extrusion equipment, food-processing equipment, air compressors, hydraulic and pneumatic devices, and high-speed automation equipment.

As a metal part is heated, it tends to expand. If the expansion is unrestrained, the dimensions of the part will grow but no stress will be developed in the metal. However, in some cases the part is unrestrained, preventing the change in dimensions. Under such circumstances, stresses will occur.

Different materials change dimensions at different rates when subjected to temperature changes. Most materials expand with increasing temperature, but a few contract and some virtually stay the same size. The *coefficient of thermal expansion* governs the thermal deformation and thermal stress experienced by a material.

> **The coefficient of thermal expansion, α, is the property of a material that indicates the amount of unit change in dimension with a unit change in temperature.**

The lowercase Greek letter alpha, α, is used as the symbol for the coefficient of thermal expansion.

The units for α are derived from its definition. Stated slightly differently, α is the measure of the change in length of a material per unit length for a 1.0-degree change in temperature. The units, then, for α in the U.S. Customary system would be

$$\text{in/(in} \cdot {}^\circ \text{F)} \quad \text{or} \quad 1/{}^\circ \text{F} \quad \text{or} \quad {}^\circ \text{F}^{-1}$$

In SI units, α would be

$$\text{m/(m} \cdot {}^\circ \text{C)} \quad \text{or} \quad \text{mm/(mm} \cdot {}^\circ \text{C)} \quad \text{or} \quad 1/{}^\circ \text{C} \quad \text{or} \quad {}^\circ \text{C}^{-1}$$

For use in computations, the last form of each unit type is most convenient. However, the first form will help you remember the physical meaning of the term.

It follows from the definition of the coefficient of thermal expansion that the change in length δ of a member can be computed from the equation

Thermal Expansion

$$\delta = \alpha \cdot L \cdot \Delta t \tag{4-7}$$

where L = original length of the member
Δt = change in temperature

Table 4–1 gives representative values for the coefficient of thermal expansion for several metals, plate glass, pine wood, and concrete. The actual value for any material varies somewhat with temperature. Those reported in Table 4–1 are approximately average values over the range of temperatures from 32°F (0°C) to 212°F (100°C).

Table 4–2 gives values for α for six selected plastic materials. Note that the actual values depend strongly on temperature and on the inclusion of any filler material in the plastic resin. For each plastic listed, the approximate values for α are given for the unfilled resin and for one filled with 30% glass.

TABLE 4–1 Coefficients of thermal expansion, α, for some metals, plate glass, wood, and concrete

Material	α	
	$°F^{-1}$	$°C^{-1}$
Steel, AISI		
1020	6.5×10^{-6}	11.7×10^{-6}
1040	6.3×10^{-6}	11.3×10^{-6}
4140	6.2×10^{-6}	11.2×10^{-6}
Structural steel	6.5×10^{-6}	11.7×10^{-6}
Gray cast iron	6.0×10^{-6}	10.8×10^{-6}
Stainless steel		
AISI 301	9.4×10^{-6}	16.9×10^{-6}
AISI 430	5.8×10^{-6}	10.4×10^{-6}
AISI 501	6.2×10^{-6}	11.2×10^{-6}
Aluminum alloys		
2014	12.8×10^{-6}	23.0×10^{-6}
6061	13.0×10^{-6}	23.4×10^{-6}
7075	12.9×10^{-6}	23.2×10^{-6}
Brass, C36000	11.4×10^{-6}	20.5×10^{-6}
Bronze, C22000	10.2×10^{-6}	18.4×10^{-6}
Copper, C14500	9.9×10^{-6}	17.8×10^{-6}
Magnesium, ASTM AZ63A-T6	14.0×10^{-6}	25.2×10^{-6}
Titanium, Ti-6A1-4V	5.3×10^{-6}	9.5×10^{-6}
Plate glass	5.0×10^{-6}	9.0×10^{-6}
Wood (pine)	3.0×10^{-6}	5.4×10^{-6}
Concrete	6.0×10^{-6}	10.8×10^{-6}

TABLE 4–2 Coefficients of thermal expansion, α, for selected plastics

Material	α	
	$°F^{-1}$	$°C^{-1}$
ABS–resin	53×10^{-6}	95.4×10^{-6}
ABS/glass	16×10^{-6}	28.8×10^{-6}
Acetal–resin	45×10^{-6}	81.0×10^{-6}
Acetal/glass	22×10^{-6}	39.6×10^{-6}
Nylon 6/6–resin	45×10^{-6}	81.0×10^{-6}
Nylon 6/6/glass	13×10^{-6}	23.4×10^{-6}
Polycarbonate–resin	37×10^{-6}	66.6×10^{-6}
Polycarbonate/glass	13×10^{-6}	23.4×10^{-6}
Polyester–resin	53×10^{-6}	95.4×10^{-6}
Polyester/glass	12×10^{-6}	21.6×10^{-6}
Polystyrene–resin	36×10^{-6}	64.8×10^{-6}
Polystyrene/glass	19×10^{-6}	34.2×10^{-6}

Composites were described in Chapter 2 as materials that combine a matrix with reinforcing fibers made from a variety of materials such as glass, aramid polymer, carbon, or graphite. The matrix materials can be polymers such as polyester or epoxy, ceramics, or some metals such as aluminum. The value for α for the fibers is typically

TABLE 4–3 Coefficients of thermal expansion, α, for selected composites

| | α | | | |
| | Longitudinal | | Transverse | |
Material	°F^{-1}	°C^{-1}	°F^{-1}	°C^{-1}
E-glass/epoxy–unidirectional	3.5×10^{-6}	6.30×10^{-6}	11×10^{-6}	19.8×10^{-6}
Aramid/epoxy–unidirectional	-1.1×10^{-6}	-1.98×10^{-6}	38×10^{-6}	68.4×10^{-6}
Carbon/epoxy–unidirectional	0.05×10^{-6}	0.09×10^{-6}	9×10^{-6}	16.2×10^{-6}
Carbon/epoxy–quasi-isotropic	1.6×10^{-6}	2.88×10^{-6}	1.6×10^{-6}	2.88×10^{-6}

much smaller than for the matrix. In addition, there are numerous ways in which the fibers can be placed in the matrix. The coefficient of thermal expansion for composites, therefore, is very difficult to generalize. Table 4–3 gives representative values for a few forms of composites. Refer to Chapter 2 for the description of the terms *unidirectional* and *quasi-isotropic*. Particularly with unidirectional placement of the fibers in the matrix, there is a dramatic difference in the value of the coefficient of thermal expansion as a function of the orientation of the material. In the longitudinal direction, aligned with the fibers, the low value of α for the fibers tends to produce a low overall value. But in the transverse direction, the fibers are not very effective, and the overall value of α is much higher. Notice, too, that for the particular unidirectional aramid/epoxy composite listed, the value of α is actually negative, meaning that this composite gets *smaller* with increasing temperature.

Example Problem 4–5

A rod made from AISI 1040 steel is used as a link in a steering mechanism of a large truck. If its nominal length is 56 in, compute its change in length as the temperature changes from −30°F to 110°F.

Solution

Objective Compute the change in length for the link.

Given Link is made from AISI 1040 steel; length = L = 56 in.
Original temperature = t_1 = −30°F.
Final temperature = t_2 = 110°F.

Analysis Use Equation (4–7). From Table 4–1, $\alpha = 6.3 \times 10^{-6}$°F^{-1}.

$$\Delta t = t_2 - t_1 = 110°F - (-30°F) = 140°F$$

Results $\delta = \alpha \cdot L \cdot \Delta t = (6.3 \times 10^{-6}$°F$^{-1})(56 \text{ in})(140°F) = 0.049 \text{ in}$

Comment The significance of this amount of deformation would have to be evaluated within the overall design of the steering mechanism for the truck.

Example Problem 4–6

A pushrod in the valve mechanism of an automotive engine has a nominal length of 203 mm. If the rod is made of AISI 4140 steel, compute the elongation due to a temperature change from −20°C to 140°C.

Solution

Objective Compute the change in length for the pushrod.

Given   Link is made from AISI 4140 steel; length $= L = 203$ mm.
Original temperature $= t_1 = -20°C$.
Final temperature $= t_2 = 140°C$.

Analysis   Use Equation (4–7). From Table 4–1, $\alpha = 11.2 \times 10^{-6}°C^{-1}$.

$$\Delta t = t_2 - t_1 = 140°C - (-20°C) = 160°C$$

Results $\delta = \alpha \cdot L \cdot \Delta t = (11.2 \times 10^{-6}°C^{-1})(203 \text{ mm})(160°C) = 0.364 \text{ mm}$

Comment   It would be important to account for this expansion in the design of the valve mechanism.

Example Problem 4–7   An aluminum frame of 6061 alloy for a window is 4.350 m long and holds a piece of plate glass 4.347 m long when the temperature is 35°C. At what temperature would the aluminum and glass be the same length?

Solution   **Objective**   Compute the temperature at which the aluminum frame and the glass would be the same length.

Given   Aluminum is 6061 alloy; from Table 4–1 we find that $\alpha_a = 23.4 \times 10^{-6}°C^{-1}$.
For plate glass, $\alpha_g = 9.0 \times 10^{-6}°C^{-1}$.
At $t_1 = 35°C$, $L_{a1} = 4.350$ m; $L_{g1} = 4.347$ m.

Analysis   The temperature would have to decrease in order for the aluminum and glass to reach the same length, since aluminum contracts at a greater rate than glass. As the temperature decreases, the change in temperature Δt would be the same for both the aluminum and the glass. After the temperature change, the length of the aluminum would be

$$L_{a2} = L_{a1} - \alpha_a \cdot L_{a1} \cdot \Delta t$$

where the subscript a refers to the aluminum, 1 refers to the initial condition, and 2 refers to the final condition. The length of the glass would be

$$L_{g2} = L_{g1} - \alpha_g \cdot L_{g1} \cdot \Delta t$$

But when the glass and the aluminum have the same length,

$$L_{a2} = L_{g2}$$

Then

$$L_{a1} - \alpha_a \cdot L_{a1} \cdot \Delta t = L_{g1} - \alpha_g \cdot L_{g1} \cdot \Delta t$$

Solving for Δt gives

$$\Delta t = \frac{L_{a1} - L_{g1}}{\alpha_a \cdot L_{a1} - \alpha_g \cdot L_{g1}}$$

Results

$$\Delta t = \frac{4.350 \text{ m} - 4.347 \text{ m}}{(23.4 \times 10^{-6}\,°C^{-1})(4.350 \text{ m}) - (9.0 \times 10^{-6}\,°C^{-1})(4.347 \text{ m})}$$

$$= \frac{0.003}{(0.000102) - (0.000039)}\,°C = 48°C$$

Then, $t_2 = t_1 - \Delta t = 35°C - 48°C = -13°C$

Comments Since this is well within the possible ambient temperature for a building, a dangerous condition could be created by this window. The window frame and glass would contract without stress until a temperature of $-13°C$ was reached. If the temperature continued to decrease, the frame would contract faster than the glass and would generate stress in the glass. Of course, if the stress is great enough, the glass would fracture, possibly causing injury. The window should be reworked so there is a larger difference in size between the glass and the aluminum frame.

4–4 THERMAL STRESS

In the preceding section, parts that were subjected to changes in temperature were unrestrained, so that they could grow or contract freely. If the parts were held in such a way that deformation was resisted, stresses would be developed.

Consider a steel structural member in a furnace that is heated while the members to which it is attached are kept at a lower temperature. Assuming the ideal case, the supports would be considered rigid and immovable. Thus all expansion of the steel member would be prevented.

If the steel part were allowed to expand, it would elongate by an amount $\delta = \alpha \cdot L \cdot \Delta t$. But since it is restrained, this represents the apparent total strain in the steel. Then the unit strain would be

$$\epsilon = \frac{\delta}{L} = \frac{\alpha \cdot L \cdot \Delta t}{L} = \alpha(\Delta t) \qquad (4–8)$$

The resulting stress in the part can be found from

$$\sigma = E\epsilon$$

or

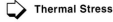 **Thermal Stress**

$$\sigma = E\alpha(\Delta t) \qquad (4–9)$$

Example Problem 4–8 A steel structural member in a furnace is made from AISI 1020 steel and undergoes an increase in temperature of 95°F while being held rigid at its ends. Compute the resulting stress in the steel.

Solution **Objective** Compute the thermal stress in the steel.

Given Steel is AISI 1020; from Table 4–1, $\alpha = 6.5 \times 10^{-6}\,°F^{-1}$.
$E = 30 \times 10^6$ psi; $\Delta t = 95°F$.

Analysis   Use Equation (4–9); $\sigma = E\alpha(\Delta t)$.

Results $\sigma = (30 \times 10^6 \text{ psi})(6.5 \times 10^{-6}\,°\text{F}^{-1})(95°\text{F}) = 18\,500 \text{ psi}$

Comment   Appendix A–13 shows the yield strength of annealed AISI 1020 steel, its weakest form, to be 43 000 psi. Therefore, the structural member would be safe.

Example Problem 4–9   An aluminum rod of alloy 2014-T6 in a machine is held at its ends while being cooled from 95°C. At what temperature would the tensile stress in the rod equal half of the yield strength of the aluminum if it is originally at zero stress?

Solution   **Objective**   Compute the temperature when $\sigma = s_y/2$.

Given   Aluminum is alloy 2014-T6; from Table 4–1, $\alpha = 23.0 \times 10^{-6}\,°\text{C}^{-1}$. From Appendix A–17, $s_y = 414$ MPa; $E = 73$ GPa; $t_1 = 95°$C.

Analysis   Use Equation (4–9) and solve for Δt.

$$\sigma = E\alpha(\Delta t)$$

$$\Delta t = \frac{\sigma}{E\alpha}$$

Let the stress be,

$$\sigma = \frac{s_y}{2} = \frac{414 \text{ MPa}}{2} = 207 \text{ MPa}$$

Results $\Delta t = \dfrac{\sigma}{E\alpha} = \dfrac{207 \text{ MPa}}{(73 \text{ GPa})(23.0 \times 10^{-6}\,°\text{C}^{-1})}$

$$= \frac{207 \times 10^6 \text{ Pa}}{(73 \times 10^9 \text{ Pa})(23.0 \times 10^{-6}\,°\text{C}^{-1})} = 123°\text{C}$$

Since the rod had zero stress when its temperature was 95°C, the temperature at which the stress would be 207 MPa would be

$$t = 95°\text{C} - 123°\text{C} = -28°\text{C}$$

Comments   The stress level of one-half of the yield strength is a reasonable design stress for a statically loaded member. Thus, if the temperature were to be lower than $-28°$C, the safety of the rod would be jeopardized.

4–5   MEMBERS MADE OF MORE THAN ONE MATERIAL

When two or more materials in a load-carrying member share the load, a special analysis is required to determine what portion of the load each material takes. Consideration of the elastic properties of the materials is required.

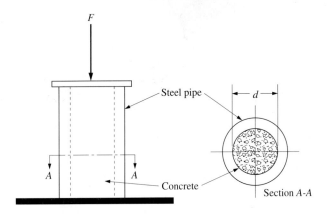

FIGURE 4–1 Steel and concrete post.

Figure 4–1 shows a steel pipe filled with concrete and used to support part of a large structure. The load is distributed evenly across the top of the support. It is desired to determine the stress in both the steel and the concrete.

Two concepts must be understood in deriving the solution to this problem.

1. The total load F is shared by the steel and the concrete such that $F = F_s + F_c$.

2. Under the compressive load F, the composite support deforms and the two materials deform in equal amounts. That is, $\delta_s = \delta_c$.

Now since the steel and the concrete were originally the same length,

$$\frac{\delta_s}{L} = \frac{\delta_c}{L}$$

But

$$\frac{\delta_s}{L} = \epsilon_s \quad \text{and} \quad \frac{\delta_c}{L} = \epsilon_c$$

Also,

$$\epsilon_s = \frac{\sigma_s}{E_s} \quad \text{and} \quad \epsilon_c = \frac{\sigma_c}{E_c}$$

Then

$$\frac{\sigma_s}{E_s} = \frac{\sigma_c}{E_c}$$

Solving for σ_s yields

$$\sigma_s = \frac{\sigma_c E_s}{E_c} \qquad (4-10)$$

This equation gives the relationship between the two stresses.

Now consider the loads,

$$F_s + F_c = F$$

Since both materials are subject to axial stress,

$$F_s = \sigma_s A_s \quad \text{and} \quad F_c = \sigma_c A_c$$

where A_s and A_c are the areas of the steel and concrete, respectively. Then

$$\sigma_s A_s + \sigma_c A_c = F \qquad (4\text{--}11)$$

Substituting Equation (4–10) into Equation (4–11) gives

$$\frac{A_s \sigma_c E_s}{E_c} + \sigma_c A_c = F$$

Now, solving for σ_c gives

$$\sigma_c = \frac{F E_c}{A_s E_s + A_c E_c} \qquad (4\text{--}12)$$

Equations (4–10) and (4–12) can now be used to compute the stresses in the steel and the concrete.

Example Problem 4–10   For the support shown in Figure 4–1, the pipe is a standard 6-in schedule 40 steel pipe completely filled with concrete. If the load P is 155 000 lb, compute the stress in the concrete and the steel. For steel use $E = 30 \times 10^6$ psi. For concrete use $E = 3.3 \times 10^6$ psi for a rated strength of $s_c = 3000$ psi (see Section 2–10).

Solution   **Objective**   Compute the stress in the concrete and the steel.

Given   Load $= F = 155\,000$ lb; $E_s = 30 \times 10^6$ psi; $E_c = 3.3 \times 10^6$ psi.
From Appendix A–12, for a 6-in schedule 40 pipe:
$A_s = 5.581$ in²; inside diameter $= d = 6.065$ in.

Analysis   Use Equation (4–12) to compute the stress in the concrete, σ_c. Then use Equation (4–10) to compute σ_s. All data are known except A_c. But,

$$A_c = \frac{\pi d^2}{4} = \frac{\pi (6.065 \text{ in})^2}{4} = 28.89 \text{ in}^2$$

Results   Then in Equation (4–12),

$$\sigma_c = \frac{(155\,000 \text{ lb})(3.3 \times 10^6 \text{ psi})}{(5.581 \text{ in}^2)(30 \times 10^6 \text{ psi}) + (28.89 \text{ in}^2)(3.3 \times 10^6 \text{ psi})} = 1946 \text{ psi}$$

Using Equation (4–10) gives

$$\sigma_s = \frac{\sigma_c E_s}{E_c} = \frac{(1946 \text{ psi})(30 \times 10^6 \text{ psi})}{3.3 \times 10^6 \text{ psi}} = 17\,696 \text{ psi}$$

Comments These stresses are fairly high. If it was desired to have at least a design factor of 2.0 based on the yield strength of the steel and 4.0 on the rated strength of the concrete, the required strengths would be,

Steel: $s_y = 2(17\,696 \text{ psi}) = 35\,392 \text{ psi}$

Concrete: Rated $\sigma_c = 4(1946 \text{ psi}) = 7784 \text{ psi}$

The steel could be similar to AISI 1020 annealed or any stronger condition. A rated strength of 3000 psi for the concrete would not be satisfactory.

Summary. The analysis presented for Example Problem 4–10 can be generalized for any situation in which loads are shared by two or more members of two different materials provided all members undergo equal strains. Replacing the subscripts s and c in the preceding analysis by the more general subscripts 1 and 2, we can state Equations (4–12) and (4–10) in the forms,

$$\sigma_2 = \frac{F E_2}{A_1 E_1 + A_2 E_2} \qquad (4\text{–}13)$$

$$\sigma_1 = \frac{\sigma_2 E_1}{E_2} \qquad (4\text{–}14)$$

PROBLEMS

Elastic Deformation

4–1.E A post of No. 2 grade hemlock is 6.0 ft long and has a square cross section with side dimensions of 3.50 in. How much would it be shortened when it is loaded in compression up to its allowable load parallel to the grain?

4–2.M Determine the elongation of a strip of plastic 0.75 mm thick by 12 mm wide by 375 mm long if it is subjected to a load of 90 N and is made of (a) glass-reinforced ABS or (b) phenolic (see Appendix A–19).

4–3.E A hollow aluminum cylinder made of 2014-T4 has an outside diameter of 2.50 in and a wall thickness of 0.085 in. Its length is 14.5 in. What axial compressive force would cause the cylinder to shorten by 0.005 in? What is the resulting stress in the aluminum?

4–4.E A metal bar is found in a stock bin and it appears to be made from either aluminum or magnesium. It has a square cross section with side dimensions of 0.25 in. Discuss two ways that you could determine what the material is.

4–5.M A tensile member is being designed for a car. It must withstand a repeated load of 3500 N and not elongate more than 0.12 mm in its 630-mm length. Use a design factor of 8 based on ultimate strength, and compute the required diameter of a round rod to satisfy these requirements using (a) AISI 1020 hot-rolled steel, (b) AISI 4140 OQT 700 steel, and (c) aluminum alloy 6061-T6. Compare the mass of the three options.

4–6.M A steel bolt has a diameter of 12.0 mm in the unthreaded portion. Determine the elongation in a length of 220 mm if a force of 17.0 kN is applied.

4–7.M In an aircraft structure, a rod is designed to be 1.25 m long and have a square cross section 8.0 mm on a side. Determine the amount of elongation that would occur if it is made of (a) titanium Ti-6Al-4V and (b) AISI 501 OQT 1000 stainless steel. The load is 5000 N.

4–8.E A tension member in a welded steel truss is 13.0 ft long and subjected to a force of 35 000 lb. Choose an equal-leg angle made of ASTM A36 steel that will limit the stress to 21 600 psi. Then compute the elongation in the angle due the force. Use $E = 29.0 \times 10^6$ psi for structural steel.

4–9.E A link in a mechanism is a rectangular steel bar that is subjected alternately to a tensile load of 450 lb and a compressive load of 50 lb. Its dimensions are: length = 8.40 in, width = 0.25 in, thickness = 0.125 in. Compute the elongation and compression of the link.

4–10.E A steel bar is attached at the top and is subjected to three axial forces, as shown in Figure 4–2. It has a circular cross section with an area of 0.50 in². Determine the deflection of the free end.

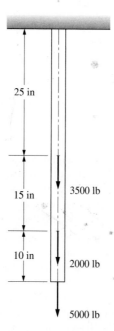

FIGURE 4–2 Bar in axial tension for Problem 4–10.

4–11.E A link in an automated packaging machine is a hollow tube made from 6061-T6 aluminum. Its dimensions are: outside diameter = 1.250 in, in-

side diameter = 1.126 in, length = 36.0 in. Compute the force required to produce a deflection of the bar of 0.050 in. Would the stress produced by the force just found be safe if the load is applied repeatedly?

4–12.E A tension member in a truss is subjected to a static load of 2500 lb. Its dimensions are: length = 8.75 ft, outside diameter = 0.750 in, inside diameter = 0.563 in. First specify a suitable aluminum alloy that would be safe. Then compute the elongation of the member.

4–13.M A hollow 6061-T4 aluminum tube, 40 mm long, is used as a spacer in a machine and is subjected to an axial compressive force of 18.2 kN. The tube has an outside diameter of 56.0 mm and an inside diameter of 48.0 mm. Compute the deflection of the tube and the resulting compressive stress.

4–14.E A guy wire is made from AISI 1020 CD steel and has a length of 135 ft. Its diameter is 0.375 in. Compute the stress in the wire and its deflection when subjected to a tensile force of 1600 lb.

4–15.M Compute the total elongation of the bar shown in Figure 1–24 if it is made from titanium Ti-6Al-4V.

4–16.E During a test of a metal bar it was found that an axial tensile force of 10 000 lb resulted in an elongation of 0.023 in. The bar had these original dimensions: length = 10.000 in, diameter = 0.750 in. Compute the modulus of elasticity of the metal. What kind of metal was it probably made from?

4–17.M The bar shown in Figure 1–25 carries three loads. Compute the deflection of point D relative to point A. The bar is made from polycarbonate plastic.

4–18.E A column is made from a cylindrical concrete base supporting a standard hollow square, 4 × 4 × 1/2 steel tube, with a length of 8.60 ft. The base is 3.0 ft long and has a diameter of 8.00 in. First, specify the concrete from Section 2–10 with a rated strength suitable to carry a compressive load of 64 000 lb. Then, assuming that no buckling occurs, compute the total amount that the column would be shortened.

4–19.E A 10.5 ft length of 14-gauge copper electrical wire (C14500, hard) is rigidly attached to a beam at its top. The wire diameter is 0.064 in. How much would it stretch if a person weighing 120 lb hung from the bottom? How much would it stretch if the person weighs 200 lb?

4–20.E A measuring tape used by carpenters is 25.00 ft long and is made from flat strip steel with these dimensions: width = 0.750 in, thickness = 0.006 in. Compute the elongation of the tape and the stress in the steel if a tensile force of 25.0 lb is applied.

4–21.E A wooden post is made from a standard 4 × 4 (Appendix A–4). If it is made from No. 2 grade southern pine, compute the axial compressive load it could carry before reaching its allowable stress in compression parallel to the grain. Then, if the post is 10.75 ft long, compute the amount that it would be shortened under that load.

4–22.M Ductile iron, ASTM A536, grade 60-40-18, is formed into a hollow square shape, 200 mm outside dimension and 150 mm inside dimension. Compute the load that would produce an axial compressive stress in the iron of 200 MPa. Then, for that load, compute the amount that the member would be shortened from its original length of 1.80 m.

4–23.M A brass wire (C36000, hard) has a diameter of 3.00 mm and is initially 3.600 m long. At this condition, the lower end, with a plate for applying a load, is 6.0 mm from the floor. How many kilograms of lead would have to be added to the plate to just cause it to touch the floor? What would be the stress in the wire at that time?

4–24.M Compute the elongation of the square bar AB in Figure 1–30 if it is 1.25 m long and made from aluminum 6061-T6.

Thermal Deformation

4–25.E A concrete slab in a highway is 80 ft long. Determine the change in length of the slab if the temperature changes from −30°F to +110°F.

4–26.M A steel rail for a railroad siding is 12.0 m long and made from AISI 1040 HR steel. Determine the change in length of the rail if the temperature changes from −34°C to +43°C.

4–27.M Determine the stress that would result in the rail described in Problem 4–26 if it were completely restrained from expanding.

4–28.M The pushrods that actuate the valves on a six-cylinder engine are AISI 1040 steel and are 625 mm long and 8.0 mm in diameter. Calculate the change in length of the rods if their tempera-

ture varies from −40°C to +116°C and the expansion is unrestrained.

4–29.M If the pushrods described in Problem 4–28 were installed with zero clearance with other parts of the valve mechanism at 25°C, compute the following:
(a) The clearance between parts at −40°C.
(b) The stress in the rod due to a temperature rise to 116°C.
Assume that mating parts are rigid.

4–30.E A bridge deck is made as one continuous concrete slab to 140 ft long at 30°F. Determine the required width of expansion joints at the ends of the bridge if no stress is to be developed when the temperature varies from +30°F to +110°F.

4–31.E When the bridge deck of Problem 4–30 was installed, the width of the expansion joint at each end was only 0.25 in. What stress would be developed if the supports are rigid? For the concrete use s_c = 4000 psi and find E from Section 2–10.

4–32.E For the bridge deck in Problem 4–30, assume that the deck is to be just in contact with its support at the temperature of 110°F. If the deck is to be installed when the temperature is 60°F, what should the gap be between the deck and its supports?

4–33.M A ring of AISI 301 stainless steel is to be placed on a shaft having a temperature of 20°C and a diameter of 55.200 mm. The inside diameter of the ring is 55.100 mm. To what temperature must the ring be heated to make it 55.300 mm in diameter and thus allow it to be slipped onto the shaft?

4–34.M When the ring of Problem 4–33 is placed on the shaft and then cooled back to 20°C, what tensile stress will be developed in the ring?

4–35.M A heat exchanger is made by arranging several brass (C36000) tubes inside a stainless steel (AISI 430) shell. Initially, when the temperature is 10°C, the tubes are 4.20 m long and the shell is 4.50 m long. Determine how much each will elongate when heated to 85°C.

4–36.E In Alaska, a 40-ft section of AISI 1020 steel pipe may see a variation in temperature from −50°F when it is at ambient temperature to +140°F when it is carrying heated oil. Compute the change in the length of the pipe under these conditions.

4–37.M A square bar of magnesium is 30 mm on a side and 250.0 mm long at 20°C. It is placed between two rigid supports set 250.1 mm apart. The bar is then heated to 70°C while the supports do not move. Compute the resulting stress in the bar.

4–38.M A square rod, 8.0 mm on a side, is made from AISI 1040 cold-drawn steel and has a length of 175 mm. It is placed snugly between two unmoving supports with no stress in the rod. Then the temperature is increased by 90°C. What is the final stress in the rod?

4–39.E A square bar is made from 6061-T4 aluminum alloy. At 75°F its length is 10.500 in. It is placed between rigid supports with a distance between them of 10.505 in. If the supports do not move, describe what would happen to the bar when its temperature is raised to 400°F.

4–40.E A straight carpenter's level is supported on two bars; one made of a polyester resin and the other made of titanium Ti-6Al-4V. The distance between the bars is 24.00 in. At a temperature of 65°F, the level is perfectly level and each bar has a length of 30.00 in. What would be the angle of tilt of the level when the temperature is increased to 212°F?

4–41.E When manufactured, a steel (AISI 1040) measuring tape was exactly 25.000 ft long at a temperature of 68°F. Compute the error that would result if the tape is used at −15°F.

4–42.M Figure 4–3 shows two bars of different materials separated by 0.50 mm when the temperature is 20°C. At what temperature would they touch?

4–43.M A stainless steel (AISI 301) wire is stretched between rigid supports so that a stress of 40 MPa is induced in the wire at a temperature of 20°C. What would be the stress at a temperature of −15°C?

4–44.M For the conditions described in Problem 4–43, at what temperature would the stress in the wire be zero?

Members Made from Two Materials

4–45.M A short post is made by welding steel plates into a square, as shown in Figure 4–4, and then filling the area inside with concrete. Compute the stress in the steel and in the concrete if $b = 150$ mm, $t = 10$ mm, and the post carries an axial load of 900 kN. See Section 2–10 for concrete properties. Use $s_c = 6000$ psi.

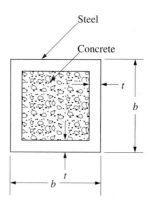

Steel

Concrete

FIGURE 4–4 Post for Problems 4–45 and 4–46.

4–46.E A short post is made by filling a standard 6 × 6 × 1/2 steel tube with concrete, as shown in Figure 4–4. The steel has an allowable stress of 21 600 psi. The concrete has a rated strength of 6000 psi but, in this application, the stress is to be limited to 1500 psi. See Section 2–10 for the modulus of elasticity for the concrete. Compute the allowable load on the post.

4–47.E A short post is being designed to support an axial compressive load of 500 000 lb. It is to be made by welding $\frac{1}{2}$-in thick plates of A36 steel into a square and filling the area inside with concrete, as shown in Figure 4–4. It is required to deter-

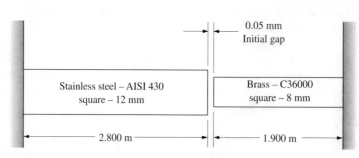

0.05 mm Initial gap

Stainless steel – AISI 430 square – 12 mm

Brass – C36000 square – 8 mm

2.800 m

1.900 m

FIGURE 4–3 Problem 4–42.

mine the dimension of the side of the post b in order to limit the stress in the steel to no more than 21 600 psi and in the concrete to no more than 1500 psi. See Section 2–10 for concrete properties. Use s_c = 6000 psi.

4–48.M Two disks are connected by four rods, as shown in Figure 4–5. All rods are 6.0 mm in diameter and have the same length. Two rods are steel (E = 207 GPa), and two are aluminum (E = 69 GPa). Compute the stress in each rod when an axial force of 11.3 kN is applied to the disks.

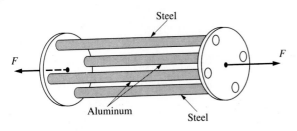

FIGURE 4–5 Problem 4–48.

4–49.M An array of three wires is used to suspend a casting having a mass of 2265 kg in such a way that the wires are symmetrically loaded (see Figure 4–6). The outer two wires are AISI 430 stainless steel,

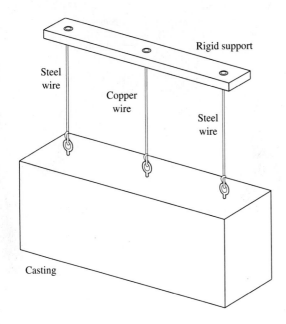

FIGURE 4–6 Problem 4–49.

full hard. The middle wire is hard copper, C17000. All three wires have the same diameter and length. Determine the required diameter of the wires if none is to be stressed beyond one-half of its yield strength.

4–50.M Figure 4–7 shows a load being applied to an inner cylindrical member that is initially 0.12 mm longer than a second concentric hollow pipe. What would be the stress in both members if a total load of 350 kN is applied?

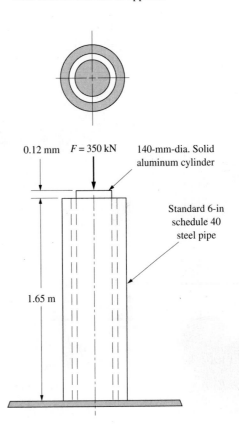

FIGURE 4–7 Aluminum bar in a steel pipe under an axial compression load for Problem 4–50.

4–51.M Figure 4–8 shows an aluminum cylinder being capped by two end plates that are held in position with four steel tie rods. A clamping force is created by tightening the nuts on the ends of the tie rods. Compute the stress in the cylinder and the tie rods if the nuts are turned one full turn from the hand-tight condition.

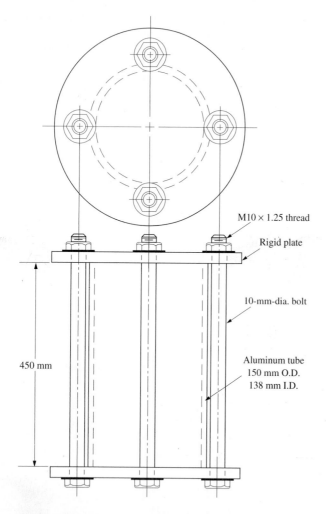

FIGURE 4–8 Tie rods on a cylinder for Problem 4–51.

4–52.E A column for a building is made by encasing a W6 × 15 wide-flange shape in concrete, as shown in Figure 4–9. The concrete helps to protect the steel from the heat of a fire and also shares in

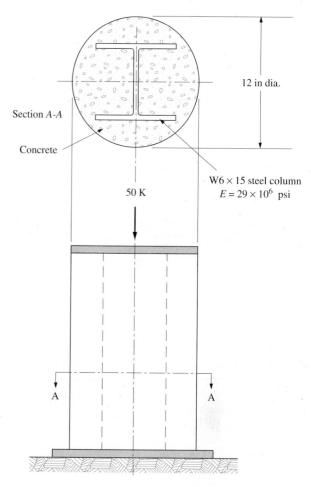

FIGURE 4–9 Steel column encased in concrete for Problem 4–52.

carrying the load. What stress would be produced in the steel and the concrete by a total load of 50 kip? See Section 2–10 for concrete properties. Use s_c = 2000 psi.

5

Torsional Shear Stress and Torsional Deflection

5–1 OBJECTIVES OF THIS CHAPTER

Torsion refers to the loading of a member that tends to twist it. Such a load is called a *torque, twisting moment,* or *couple.* When a torque is applied to a member, such as a round shaft, *shearing stress* is developed within the shaft and a *torsional deflection* is created, resulting in an angle of twist of one end of the shaft relative to the other.

After completing this chapter, you should be able to:

1. Define *torque* and compute the magnitude of torque exerted on a member subjected to torsional loading.
2. Define the relationship among the three critical variables involved in power transmission: power, torque, and rotational speed.
3. Manipulate the units for power, torque, and rotational speed in both the SI metric system and the U.S. Customary system.
4. Compute the maximum shear stress in a member subjected to torsional loading.
5. Define the *polar moment of inertia* and compute its value for solid and hollow round shafts.
6. Compute the shear stress at any point within a member loaded in torsion.
7. Specify a suitable design shear stress for a member loaded in torsion.

8. Define the *polar section modulus* and compute its value for solid and hollow round shafts.

9. Determine the required diameter of a shaft to carry a given torque safely.

10. Compare the design of solid and hollow shafts on the basis of the mass of the shafts required to carry a certain torque while limiting the torsional shear stress to a certain design value.

11. Apply stress concentration factors to members in torsion.

12. Compute the angle of a twist of a member loaded in torsion.

13. Define the *shear modulus of elasticity*.

14. Discuss the method of computing torsional shear stress and torsional deflection for members with noncircular cross sections.

15. Describe the general shapes of members having relatively high torsional stiffness.

5–2 TORQUE, POWER, AND ROTATIONAL SPEED

A necessary task in approaching the calculation of torsional shear stress and deflection is the understanding of the concept of *torque* and the relationship among the three critical variables involved in power transmission: *torque, power,* and *rotational speed.*

Figure 5–1 shows a socket wrench with an extension shaft being used to tighten a bolt. The *torque,* applied to both the bolt and the extension shaft, is the product of the

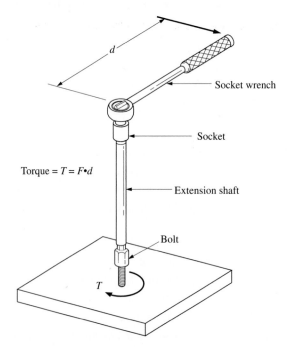

FIGURE 5–1 Wrench applying a torque to a bolt.

applied force and the distance from the line of action of the force to the axis of the bolt. That is,

Torque

$$\text{torque} = T = F \times d \tag{5-1}$$

Thus torque is expressed in the units of *force times distance,* which is N·m in the SI metric system and lb·in or lb·ft in the U.S. Customary system.

Example Problem 5-1

For the wrench in Figure 5–1, compute the magnitude of the torque applied to the bolt if a force of 50 N is exerted at a point 250 mm out from the axis of the socket.

Solution

Using Equation (5–1) yields

$$T = F \times d = (50 \text{ N})(250 \text{ mm}) \times \frac{1 \text{ m}}{1000 \text{ mm}} = 12.5 \text{ N·m}$$

Figure 5–2 shows a drive system for a boat. Power developed by the engine flows through the transmission and the drive shaft to the propeller, where it drives the boat forward. The crankshaft inside the engine, the various power transmission shafts in the transmission, and the drive shaft are all subjected to torsion. The magnitude of the torque in a power transmission shaft is dependent on the amount of power it carries and on the speed of rotation, according to the following relation:

$$\text{power} = \text{torque} \times \text{rotational speed}$$

Power

$$P = T \times n \tag{5-2}$$

This is a very useful relationship because if any two values, P, n, or T, are known, the third can be computed.

Careful attention must be paid to units when working with torque, power, and rotational speed. Appropriate units in the SI metric system and the U.S. Customary system are reviewed next.

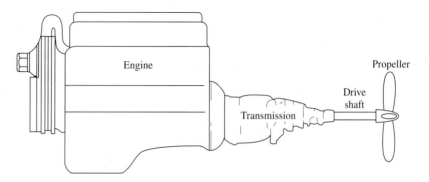

FIGURE 5–2   Drive system for a boat.

SI metric system.

Power is defined as the rate of transferring energy.

In the SI metric system, the *joule* is the standard unit for energy and it is equivalent to the N·m, the standard unit for torque. That is,

$$1.0 \text{ J} = 1.0 \text{ N·m}$$

Then power is defined as

 **SI Units for Power**

$$\text{power} = \frac{\text{energy}}{\text{time}} = \frac{\text{joule}}{\text{second}} = \frac{\text{J}}{\text{s}} = \frac{\text{N·m}}{\text{s}} = \text{watt} = \text{W} \qquad (5\text{–}3)$$

Note that 1.0 J/s is defined to be 1.0 watt (1.0 W). The watt is a rather small unit of power, so the kilowatt (1.0 kW = 1000 W) is often used.

The standard unit for rotational speed in the SI metric system is *radians per second,* rad/s. Frequently, however, rotational speed is expressed in revolutions per minute, rpm. The conversion required is illustrated below, converting 1750 rpm to rad/s.

$$n = \frac{1750 \text{ rev}}{\text{min}} \times \frac{2\pi \text{ rad}}{\text{rev}} \times \frac{1 \text{ min}}{60 \text{ s}} = 183 \text{ rad/s}$$

When using *n* in rad/s in Equation (5–2), the radian is considered to be *no unit at all,* as illustrated in the following example problem.

Example Problem 5–2 The drive shaft for the boat shown in Figure 5–2 transmits 95 kW of power while rotating at 525 rpm. Compute the torque in the shaft.

Solution **Objective** Compute the torque in the shaft.

Given $P = 95 \text{ kW} = 95\,000 \text{ W} = 95\,000 \text{ N·m/s}$; $n = 525 \text{ rpm}$

Analysis Equation (5–2) will be solved for *T* and used to compute torque.

$$P = Tn; \quad \text{then, } T = P/n$$

But *n* must be in rad/s, found as follows:

$$n = \frac{525 \text{ rev}}{\text{min}} \times \frac{2\pi \text{ rad}}{\text{rev}} \times \frac{1 \text{ min}}{60 \text{ s}} = 55.0 \text{ rad/s}$$

Results The torque is

$$T = \frac{P}{n} = \frac{95\,000 \text{ N·m}}{\text{s}} \times \frac{1}{55.0 \text{ rad/s}} = 1727 \text{ N·m}$$

Comment Note that the radian unit is ignored in such calculations.

U.S. Customary units. Typical units for torque, power, and rotational speed in the U.S. Customary unit system are

$$T = \text{torque (lb·in)}$$

$$n = \text{rotational speed (rpm)}$$

$$P = \text{power (horsepower, hp)}$$

Note that 1.0 hp = 6600 lb·in/s. Then the unit conversions required to ensure consistent units are

$$\text{power} = T(\text{lb·in}) \times n\left(\frac{\text{rev}}{\text{min}}\right) \times \frac{1 \text{ min}}{60 \text{ s}} \times \frac{2\pi \text{ rad}}{\text{rev}} \times \frac{1 \text{ hp}}{6600 \text{ lb·in/s}}$$

or

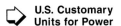

**U.S. Customary
Units for Power**

$$\text{power} = \frac{Tn}{63\,000} \qquad (5\text{–}4)$$

**Example Problem
5–3**

Compute the power, in the unit of horsepower, being transmitted by a shaft if it is developing a torque of 15 000 lb·in and rotating at 525 rpm.

Solution

Objective Compute the power transmitted by the shaft.

Given $T = 15\,000$ lb·in; $n = 525$ rpm

Analysis Equation (5–4) will be used directly because T and n are in the proper units of lb·in and rpm. Power will be in horsepower.

Results The power is

$$P = \frac{Tn}{63\,000} = \frac{(15\,000)(525)}{63\,000} = 125 \text{ hp}$$

5–3 TORSIONAL SHEAR STRESS IN MEMBERS WITH CIRCULAR CROSS SECTIONS

When a member is subjected to an externally applied torque, an internal resisting torque must be developed in the material from which the member is made. The internal resisting torque is the result of stresses developed in the material.

Figure 5–3(a) shows a circular bar subjected to a torque, T. Section N would be rotated relative to section M as shown. If an element on the surface of the bar were isolated, it would be subjected to shearing forces on the sides parallel to cross sections M and N, as shown. These shearing forces result in shear stresses on the element. For the stress element to be in equilibrium, equal shearing stresses must exist on the top and bottom faces of the element.

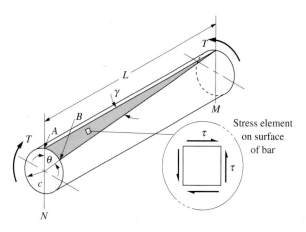

FIGURE 5–3 Torsional shear stress in a circular bar.

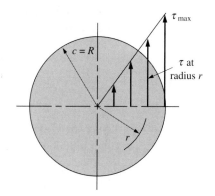

FIGURE 5–4 Distribution of shear stress on a cross section of the bar.

The shear stress element shown in Figure 5–3 is fundamentally the same as that shown in Figure 1–15 in the discussion of direct shear stress. While the manner in which the stresses are created differs, the nature of torsional shear stress is the same as direct shear stress when an infinitesimal element is considered.

When the circular bar is subjected to the externally applied torque, the material in each cross section is deformed in a manner such that the fibers on the outside surface experience the maximum strain. At the central axis of the bar, no strain at all is produced. Between the center and the outside, there is a linear variation of strain with radial position r. Because stress is directly proportional to strain, we can say that the maximum stress occurs at the outside surface, that there is a linear variation of stress with radial position r, and that a zero stress level occurs at the center. Figure 5–4 illustrates these observations.

The derivation will be shown in the next section for the formula for the maximum shear stress on the outer surface of the bar. For now, we state the *torsional shear stress formula* as

Torsional Shear Stress Formula

$$\tau_{max} = \frac{Tc}{J} \tag{5–5}$$

where T = applied torque at the section of interest
  c = radius of the cross section
  J = polar moment of inertia of the circular cross section

The formula for J for a solid circular cross section is developed in Section 5–5. At this time, we show the results of the derivation as,

Polar Moment of Inertia for Circular Bar

$$J = \frac{\pi D^4}{32} \tag{5–6}$$

where D is the diameter of the shaft; that is, $D = 2R$.

Because of the linear variation of stress and strain with position in the bar as shown in Figure 5–4, the stress, τ, at any radial position, r, can be computed from

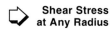

Shear Stress at Any Radius

$$\tau = \tau_{max} \frac{r}{c} \qquad (5-7)$$

Equations (5–5), (5–6), and (5–7) can be used to compute the shear stress at any point in a circular bar subjected to an externally applied torque. The following example problems illustrate the use of these equations.

Example Problem 5–4   For the socket wrench extension shown in Figure 5–1, compute the maximum torsional shear stress in the middle portion where the diameter is 9.5 mm. The applied torque is 10.0 N·m.

Solution   **Objective**   Compute the maximum torsional shear stress in the extension.

Given   Torque = T = 10.0 N·m; diameter = D = 9.5 mm

Analysis   Use Equation (5–6) to compute J and Equation (5–5) to compute the maximum shear stress. Also, $c = D/2 = 9.5$ mm/2 = 4.75 mm.

Results
$$J = \frac{\pi D^4}{32} = \frac{\pi (9.5 \text{ mm})^4}{32} = 800 \text{ mm}^4$$

$$\tau_{max} = \frac{Tc}{J} = \frac{(10 \text{ N·m})(4.75 \text{ mm})}{800 \text{ mm}^4} \times \frac{10^3 \text{ mm}}{\text{m}} = 59.4 \text{ N/mm}^2 = 59.4 \text{ MPa}$$

Comment   This level of stress would occur at all points on the surface of the circular part of the extension.

Example Problem 5–5   Calculate the maximum torsional shear stress that would be developed in a solid circular shaft, having a diameter of 1.25 in, if it is transmitting 125 hp while rotating at 525 rpm.

Solution   **Objective**   Compute the maximum torsional shear stress in the shaft.

Given   Power = P = 125 hp; rotational speed = n = 525 rpm
Shaft diameter = D = 1.25 in

Analysis   Solve Equation (5–4) for the torque, T. Use Equation (5–6) to compute J and Equation (5–5) to compute the maximum shear stress. Also, $c = D/2 = 1.25$ in/2 = 0.625 in.

Results   Equation (5–4),

$$\text{Power} = P = \frac{Tn}{63\,000}$$

Solving for the torque T gives

$$T = \frac{63\,000P}{n}$$

Recall that this equation will give the value of the torque directly in lb·in when P is in horsepower and n is in rpm. Then

$$T = \frac{63\,000(125)}{525} = 15\,000 \text{ lb·in}$$

$$J = \frac{\pi D^4}{32} = \frac{\pi(1.25 \text{ in})^4}{32} = 0.240 \text{ in}^4$$

Then

$$\tau_{max} = \frac{Tc}{J} = \frac{(15\,000 \text{ lb·in})(0.625 \text{ in})}{0.240 \text{ in}^4} = 39\,100 \text{ psi}$$

Comment   This level of stress would occur at all points on the surface of the shaft.

5–4   DEVELOPMENT OF THE TORSIONAL SHEAR STRESS FORMULA

The standard form of the torsional shear stress formula for a circular bar subjected to an externally applied torque was shown as Equation (5–5) and its use was illustrated in Example Problems 5–4 and 5–5. This section will show the development of that formula. Reference should be made to Figures 5–3 and 5–4 for the general nature of the torsional loading and a visualization of the effect the torque has on the behavior of the circular bar.

In this development, it is assumed that the material for the bar behaves in accordance with Hooke's law; that is, stress is directly proportional to strain. Also, the properties of the bar are homogeneous and isotropic; that is, the material reacts the same regardless of the direction of the applied loads. Also, it is assumed that the bar is of constant cross section in the vicinity of the section of interest.

Considering two cross sections M and N, at different places on the bar, section N would be rotated through an angle θ relative to section M. The fibers of the material would undergo a strain that would be maximum at the outside surface of the bar and vary linearly with radial position to zero at the center of the bar. Because, for elastic materials obeying Hooke's law, stress is proportional to strain, the maximum stress would also occur at the outside of the bar, as shown in Figure 5–4. The linear variation of stress, τ, with radial position in the cross section, r, is also shown. Then, by proportion using similar triangles,

$$\frac{\tau}{r} = \frac{\tau_{max}}{c} \tag{5–8}$$

Then the shear stress at any radius can be expressed as a function of the maximum shear stress at the outside of the shaft,

$$\tau = \tau_{max} \times \frac{r}{c} \tag{5-9}$$

It should be noted that the shear stress τ acts uniformly on a small ring-shaped area, dA, of the shaft, as illustrated in Figure 5–5. Now since force equals stress times area, the force on the area dA is

$$dF = \tau\, dA = \underbrace{\tau_{max}\frac{r}{c}}_{\text{stress}} \times \underbrace{dA}_{\text{area}}$$

The next step is to consider that the torque dT developed by this force is the product of dF and the radial distance to dA. Then

$$dT = dF \times r = \underbrace{\tau_{max}\frac{r}{c}\,dA}_{\text{force}} \times \underbrace{r}_{\text{radius}} = \tau_{max}\frac{r^2}{c}\,dA$$

This equation is the internal resisting torque developed on the small area dA. The total torque on the entire area would be the sum of all the individual torques on all areas of the cross section. The process of summing is accomplished by the mathematical technique of integration, illustrated as follows:

$$T = \int_A dT = \int_A \tau_{max}\frac{r^2}{c}\,dA$$

In the process of integration, constant terms such as τ_{max} and c can be brought outside the integral sign, leaving

$$T = \frac{\tau_{max}}{c}\int_A r^2\,dA \tag{5-10}$$

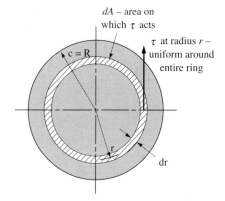

dA – area on which τ acts

τ at radius r – uniform around entire ring

$c = R$

r

dr

FIGURE 5–5 Shear stress τ at radius r acting on the area dA.

In mechanics, the term $\int r^2 \, dA$ is given the name *polar moment of inertia* and is identified by the symbol J. The derivation of J is shown in the next section. Equation (5–10) can then be written

$$T = \tau_{max} \frac{J}{c}$$

or

$$\tau_{max} = \frac{Tc}{J} \tag{5–11}$$

The method of evaluating J is developed in the next section.

Equation (5–11), which is identical to Equation (5–5), can be used to compute the maximum shear stress on a circular bar subjected to torsion. The maximum shear stress occurs anywhere on the outer surface of the bar.

5–5 POLAR MOMENT OF INERTIA FOR SOLID CIRCULAR BARS

Refer to Figure 5–5 showing a solid circular cross section. To evaluate J from

$$J = \int r^2 \, dA$$

it must be seen that dA is the area of a small ring located at a distance r from the center of the section and having a thickness dr.

For a small magnitude of dr, the area is that of a strip having a length equal to the circumference of the ring times the thickness.

$$dA = 2\pi r \times dr$$

thickness of the ring

circumference of a ring at the radius r

Then the polar moment of inertia for the entire cross section can be found by integrating from $r = 0$ at the center of the bar to $r = R$ at the outer surface.

$$J = \int_0^R r^2 \, dA = \int_0^R r^2 (2\pi r) \, dr = \int_0^R 2\pi r^3 \, dr = \frac{2\pi R^4}{4} = \frac{\pi R^4}{2}$$

It is usually more convenient to use diameter rather than radius. Then since $R = D/2$,

$$J = \frac{\pi (D/2)^4}{2} = \frac{\pi D^4}{32} \tag{5–12}$$

5-6 TORSIONAL SHEAR STRESS AND POLAR MOMENT OF INERTIA FOR A HOLLOW CIRCULAR BAR

It will be shown later that there are many advantages to using a hollow circular bar, as compared with a solid bar, to carry a torque. In this section, we discuss the method of computing the maximum shear stress and the polar moment of inertia for a hollow bar.

Figure 5–6 shows the basic geometry for a hollow bar. The variables are:

R_i = inside radius

D_i = inside diameter

R_o = outside radius = c

D_o = outside diameter

The logic and details of the development of the torsional shear stress formula as shown in Section 5–4 apply as well to a hollow bar as to the solid bar. The difference between them is in the evaluation of the polar moment of inertia, as will be shown later. Therefore, Equation (5–5) or (5–11) can be used to compute the maximum torsional shear stress in either the solid or the hollow bar.

Also, as illustrated in Figure 5–6, the maximum shear stress occurs at the outer surface of the bar and there is a linear variation of stress with radial position inside the bar. The minimum shear stress occurs at the inside surface. The shear stress at any radial position can be computed from Equation (5–7) or (5–9).

Polar moment of inertia for a hollow bar. The process of developing the formula for the polar moment of inertia for a hollow bar is similar to that used for the solid bar.

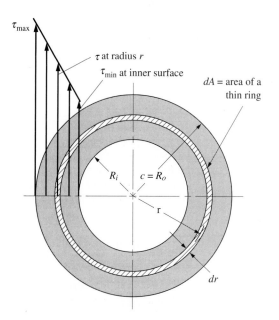

FIGURE 5–6 Notation for variables used to derive J for a hollow round bar.

Refer again to Figure 5–6 for the geometry. Starting with the basic definition of the polar moment of inertia,

$$J = \int r^2 \, dA$$

as before, $dA = 2\pi r \, dr$. But for the hollow bar, r varies only from R_i to R_o. Then

$$J = \int_{R_i}^{R_o} r^2 (2\pi r) \, dr = 2\pi \int_{R_i}^{R_o} r^3 \, dr = \frac{2\pi(R_o^4 - R_i^4)}{4}$$

$$J = \frac{\pi}{2}(R_o^4 - R_i^4)$$

Substituting $R_o = D_o/2$ and $R_i = D_i/2$ gives

**Polar Moment of
Inertia for a
Hollow Bar**

$$J = \frac{\pi}{32}(D_o^4 - D_i^4) \tag{5–13}$$

This is the equation for the polar moment of inertia for a hollow circular bar.

Summary of relationships for torsional shear stresses in hollow circular bars.

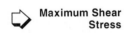
**Maximum Shear
Stress**

$$\tau_{\max} = \frac{Tc}{J} \tag{5–11}$$

$\tau_{\max}$ occurs at the outer surface of the bar, where c is the radius of the bar.

**Shear Stress at
Any Radial
Position r**

$$\tau = \tau_{\max}\frac{r}{c} = \frac{Tr}{J} \tag{5–9}$$

**Polar Moment
of Inertia for
Hollow Bars**

$$J = \frac{\pi}{32}(D_o^4 - D_i^4) \tag{5–13}$$

**Example Problem
5–6**

For the propeller drive shaft of Figure 5–2, compute the torsional shear stress when it is transmitting a torque of 1.76 kN·m. The shaft is a hollow tube having an outside diameter of 60 mm and an inside diameter of 40 mm. Find the stress at both the outer and inner surfaces.

Solution   **Objective**   Compute the torsional shear stress at the outer and inner surfaces of the hollow propeller drive shaft.

Given   Shaft shown in Figure 5–2. Torque $= T = 1.76$ kN·m $= 1.76 \times 10^3$ N·m.
Outside diameter $= D_o = 60$ mm; inside diameter $= D_i = 40$ mm.

Chapter 5 ▪ Torsional Shear Stress and Torsional Deflection

Analysis   The final calculation for the torsional shear stress at the outer surface will be made using Equation (5–11). Equation (5–9) will be used to compute the stress at the inner surface. The polar moment of inertia will be computed using Equation (5–13). And, $c = D_o/2 = 30$ mm.

Results   At the outer surface,

$$\tau_{max} = \frac{Tc}{J}$$

$$J = \frac{\pi}{32}(D_o^4 - D_i^4) = \frac{\pi}{32}(60^4 - 40^4) \text{ mm}^4 = 1.02 \times 10^6 \text{ mm}^4$$

$$\tau_{max} = \frac{Tc}{J} = \frac{(1.76 \times 10^3 \text{ N·m})(30 \text{ mm})}{1.02 \times 10^6 \text{ mm}^4} \times \frac{10^3 \text{ mm}}{m}$$

$$= 51.8 \text{ N/mm}^2 = 51.8 \text{ MPa}$$

At the inner surface, $r = D_i/2 = 40 \text{ mm}/2 = 20$ mm.

$$\tau = \tau_{max}\frac{r}{c} = 51.8 \text{ MPa} \times \frac{20 \text{ mm}}{30 \text{ mm}} = 34.5 \text{ MPa}$$

Comment   You should visualize these stress values plotted on the cross section shown in Figure 5–6.

5–7   DESIGN OF CIRCULAR MEMBERS UNDER TORSION

In a design problem, the loading on a member is known, and it is required to determine the geometry of the member to ensure that it will carry the loads safely. Material selection and the determination of design stresses are integral parts of the design process. *The techniques developed in this section are for circular members only, subjected only to torsion.* Of course, both solid and hollow circular members are covered. Torsion in noncircular members is covered in a later section of this chapter. The combination of torsion with bending and axial loads is presented in Chapters 10 and 11.

The basic torsional shear stress equation, Equation (5–11), was expressed as

$$\tau_{max} = \frac{Tc}{J} \tag{5–11}$$

In design, we can substitute a certain design stress τ_d for τ_{max}. As in the case of members subjected to direct shear stress and made of ductile materials, the design stress is related to the yield strength of the material in shear. That is,

$$\tau_d = \frac{s_{ys}}{N}$$

where N is the design factor chosen by the designer based on the manner of loading. Table 5–1 can be used as a guide to determine the value of N.

Manner of loading	Design factor	Design shear stress $\tau_d = s_y/2N$
Static torsion	2	$\tau_d = s_y/4$
Repeated torsion	4	$\tau_d = s_y/8$
Torsional impact or shock	6	$\tau_d = s_y/12$

Where the data for s_{ys} are not available, the value can be estimated as $s_y/2$. This will give a reasonable, and usually conservative, estimate for ductile metals, especially steel. Then

Design Shear Stress

$$\tau_d = \frac{s_{ys}}{N} = \frac{s_y}{2N} \qquad (5\text{–}14)$$

The torque T would be known in a design problem. Then, in Equation (5–11), only c and J are left to be determined. Notice that both c and J are properties of the geometry of the member that is being designed. For solid circular members (shafts), the geometry is completely defined by the diameter. It has been shown that

$$c = \frac{D}{2}$$

and

$$J = \frac{\pi D^4}{32}$$

It is now convenient to note that if the quotient J/c is formed, a simple expression involving D is obtained.

In the study of strength of materials, the term J/c is given the name *polar section modulus,* and the symbol Z_p is used to denote it.

Polar Section Modulus– Solid Shafts

$$Z_p = \frac{J}{c} = \frac{\pi D^4}{32} \times \frac{1}{D/2} = \frac{\pi D^3}{16} \qquad (5\text{–}15)$$

Substituting Z_p for J/c in Equation (5–11) gives

Maximum Shear Stress

$$\tau_{\max} = \frac{T}{Z_p} \qquad (5\text{–}16)$$

To use this equation in design, we can let $\tau_{\max} = \tau_d$ and then solve for Z_p.

Required Polar Section Modulus

$$Z_p = \frac{T}{\tau_d} \qquad (5\text{–}17)$$

Equation (5–17) will give the required value of the polar section modulus of a circular shaft to limit the torsional shear stress to τ_d when subjected to a torque T. Then Equation (5–15) can be used to find the required diameter of a solid circular shaft. Solving for D gives us

**Required
Diameter**

$$D = \sqrt[3]{\frac{16Z_p}{\pi}}$$ (5–18)

If a hollow shaft is to be designed,

$$Z_p = \frac{J}{c} = \frac{\pi}{32}(D_o^4 - D_i^4) \times \frac{1}{D_o/2}$$

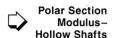
**Polar Section
Modulus–
Hollow Shafts**

$$Z_p = \frac{\pi}{16}\frac{D_o^4 - D_i^4}{D_o}$$ (5–19)

In this case, one of the diameters *or* the relationship between the two diameters would have to be specified in order to solve for the complete geometry of the hollow shaft.

PROGRAMMED EXAMPLE PROBLEM

**Example Problem
5–7**

The final drive to a conveyor that feeds coal to a railroad car is a shaft loaded in pure torsion and carrying 800 N·m of torque. A proposed design calls for the shaft to have a solid circular cross section. Complete the design by first specifying a suitable steel for the shaft and then specifying the diameter.

Solution **Objective**   1. Specify a suitable steel for the shaft.

2. Specify the shaft diameter.

Given   Applied torque $= T =$ 800 N·m
Shaft drives a coal conveyor.

Analysis   *The completion of this problem is presented in a programmed format. You should answer each question as it is posed before looking at the next panel beyond the line across the page. This process is intended to involve you in the decision-making activities required of a designer.*
First, as an aid in selecting a suitable material, what manner of loading will the shaft experience in service?

The drive for a coal conveyor is likely to experience very rough service as coal is dumped onto the conveyor. Therefore, the design should be able to accommodate impact and shock loading. Now, what properties should the steel for the shaft possess?

A highly ductile material should be used because such materials withstand shock loading much better than more brittle materials. The steel should have a moderately high strength so that the required diameter of the shaft is reasonable. It may be important to choose a steel with good machinability because the shaft is likely to require machining during its manufacture. What is a typically used measure of ductility for steels?

It was mentioned in Chapter 2 that the *percent elongation* for a steel is an indication of its ductility. To withstand impact and shock, a steel having a value somewhat higher than 10% elongation should be specified. Now specify a suitable steel.

There are many steels that can be used satisfactorily. Let's specify AISI 1141 OQT 1300. List pertinent data from Appendix A–13.

You should have found that $s_y = 469$ MPa and that the 28% elongation indicates a high ductility. Also note, as stated in Chapter 2, the 1100 series steels have good machinability because of a relatively high sulfur content in the alloy.

We will be using Equations (5–16), (5–17), and (5–18) to continue the design process with the ultimate goal of specifying a suitable diameter for the shaft. We know the applied torque is 800 N·m. The next task should be to determine an acceptable design shear stress. How should you do that?

Table 5–1 calls for $\tau_d = s_y/2N$ with $N = 6$; that is, $\tau_d = s_y/12$. Then, $\tau_d = s_y/12 = 469$ MPa/12 $= 39.1$ MPa $= 39.1$ N/mm^2. What should be the next step?

We can use Equation (5–17) to compute the required value of the polar section modulus for the cross section of the shaft. Do that now.

You should have the required $Z_p = 20.5 \times 10^3$ mm^3, found from

$$Z_p = \frac{T}{\tau_d} = \frac{800 \text{ N·m}}{39.1 \text{ N/mm}^2} \times \frac{10^3 \text{ mm}}{\text{m}} = 20.5 \times 10^3 \text{ mm}^3$$

What is the next step?

We can compute the minimum acceptable diameter for the shaft using Equation (5–18). Do that now.

You should have $D_{min} = 47.1$ mm, found from

$$D = \sqrt[3]{\frac{16Z_p}{\pi}} = \sqrt[3]{\frac{16(20.5 \times 10^3) \text{ mm}^3}{\pi}} = 47.1 \text{ mm}$$

It would be appropriate to specify a convenient size for the shaft diameter that is slightly larger than this value. Use Appendix A–2 as a guide and specify a diameter.

Specifying $D = 50$ mm is preferred.

Summary of Results

The shaft will be made from AISI 1141 OQT 1300 steel with a diameter of 50 mm.

Comments The maximum shear stress at the outer surface of the 50-mm diameter shaft is actually less than the design stress because we specified a preferred diameter slightly greater than the minimum required diameter of 47.1 mm. Let's now compute the actual maximum stress in the shaft. First, we will compute the polar section modulus for the 50-mm diameter shaft.

$$Z_p = \frac{\pi D^3}{16} = \frac{\pi (50)^3 \text{ mm}^3}{16} = 24.5 \times 10^3 \text{ mm}^3$$

Then, the maximum shear stress is

$$\tau_{max} = \frac{T}{Z_p} = \frac{800 \text{ N·m}}{24.5 \times 10^3 \text{ mm}^3} \times \frac{10^3 \text{ mm}}{1 \text{ m}} = 32.6 \text{ N/mm}^2 = 32.6 \text{ MPa}$$

We will now demonstrate, by example, that hollow shafts are more efficient than solid shafts. Here the term, efficiency, is used as a measure of the mass of material in a shaft required to carry a given torque with a given shear stress level. The following example problem shows the design of a hollow shaft with a slightly larger outside diameter that has the same maximum shear stress as the 50-mm diameter solid shaft just designed. Then, the mass of the hollow shaft is compared with that of the solid shaft.

Example Problem 5–8 An alternative design for the shaft described in Example Problem 5–7 would be to use a hollow tube for the shaft. Assume that a tube having an outside diameter of 60 mm is available in the same material as specified for the solid shaft (AISI 1141 OQT 1300). Compute what maximum inside diameter the tube can have that would result in the same stress in the steel as the 50-mm solid shaft.

Solution

Objective Compute the maximum allowable inside diameter for the hollow shaft.

Given From Example Problem 5–7, maximum shear stress = τ_{max} = 32.6 MPa. D_o = 60 mm. Applied torque = T = 800 N·m.

Analysis Because torsional shear stress is inversely proportional to the polar section modulus, it is necessary that the hollow tube have the same value for Z_p as does the 50-mm diameter solid shaft. That is, Z_p = 24.5 × 10³ mm³. Now, what is the formula for Z_p for a hollow shaft?

$$Z_p = \frac{\pi}{16} \frac{D_o^4 - D_i^4}{D_o}$$

The outside diameter, D_o, is known to be 60 mm. We can then solve for the required inside diameter, D_i. Do that now.

You should have

$$D_i = \left(D_o^4 - \frac{16 Z_p D_o}{\pi} \right)^{1/4}$$

Now compute the maximum allowable inside diameter.

$$D_i = \left[(60)^4 - \frac{(16)(24.5 \times 10^3)(60)}{\pi} \right]^{1/4} \text{ mm} = 48.4 \text{ mm}$$

Summary of Results
Final design of the hollow shaft:
$D_o = 60$ mm; $D_i = 48.4$ mm.
Material: AISI 1141 OQT 1300 steel.
Maximum shear stress at outer surface $= \tau_{max} = 32.6$ MPa.

Comment Figure 5–7 shows a comparison of the hollow shaft with the solid shaft having the same maximum shear stress. The designs are drawn full size. It appears to the eye that the hollow shaft uses less material. The following section will demonstrate that this is true.

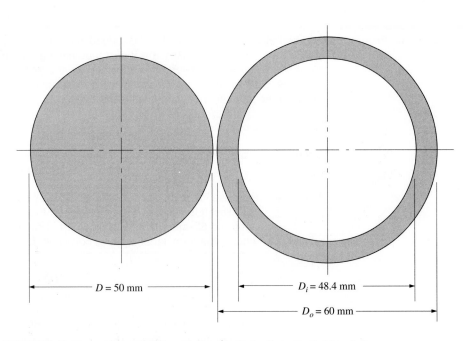

$D = 50$ mm

$D_i = 48.4$ mm

$D_o = 60$ mm

FIGURE 5–7 Comparison of solid and hollow shafts for Example Problem 5–8.

5–8 COMPARISON OF SOLID AND HOLLOW CIRCULAR MEMBERS

In many design situations, economy of material usage is a major criterion of performance for a product. In aerospace applications, every reduction in the mass of the aircraft or space vehicle allows increased payload. Automobiles achieve higher fuel economy when they are lighter. Also, since raw materials are purchased on a price per unit mass basis, a lighter part generally costs less.

Providing economy of material usage for load-carrying members requires that all the material in the member be stressed to a level approaching the safe design stress. Then every portion is carrying its share of the load.

Example Problems 5–7 and 5–8 can be used to illustrate this point. Recall that both designs shown in Figure 5–7 result in the same maximum torsional shear stress in the steel shaft. The hollow shaft is slightly larger in outside diameter, but it is the *volume* of metal that determines the mass of the shaft. Consider a length of shaft 1.0 m long. For the solid shaft, the volume is the cross-sectional area times the length.

$$V_s = AL = \frac{\pi D^2}{4} L$$

$$= \frac{\pi (50 \text{ mm})^2}{4} \times 1.0 \text{ m} \times \frac{1 \text{ m}^2}{(10^3 \text{ mm})^2} = 1.96 \times 10^{-3} \text{ m}^3$$

The mass is the volume times the density, ρ. Appendix A–13 gives the density of steel to be 7680 kg/m^3. Then the mass of the solid shaft is $V_s \times \rho$.

$$M_s = 1.96 \times 10^{-3} \text{ m}^3 \times 7680 \text{ kg/m}^3 = 15.1 \text{ kg}$$

Now for the hollow shaft the volume is

$$V_H = AL = \frac{\pi}{4}(D_o^2 - D_i^2)(L)$$

$$= \frac{\pi}{4}(60^2 - 48.4^2) \text{ mm}^2 \times 1.0 \text{ m} \times \frac{1 \text{ m}^2}{(10^3 \text{ mm})^2}$$

$$= 0.988 \times 10^{-3} \text{ m}^3$$

The mass of the hollow shaft is $V_H \times \rho$.

$$M_H = 0.988 \times 10^{-3} \text{ m}^3 \times 7680 \text{ kg/m}^3 = 7.58 \text{ kg}$$

Thus it can be seen that the hollow shaft has almost exactly *one-half the mass* of the solid shaft, even though both are subjected to the same stress level for a given applied torque. Why?

The reason for the hollow shaft being lighter is that a greater portion of its material is being stressed to a higher level than in the solid shaft. Figure 5–4 shows a sketch of the stress distribution in the solid shaft. The maximum stress, 32.6 MPa, occurs at the outer surface. The stress then varies linearly with the radius for other points within the shaft to *zero* at the center. From this it can be seen that the material near the middle of the shaft is not being used efficiently.

Contrast this with the sketch of the hollow shaft in Figure 5–6. Again the stress at the outer surface is the maximum, 32.6 MPa. The stress at the inner surface of the hollow shaft can be found from Equation (5–6).

$$\tau = \tau_{max} \frac{r}{c}$$

At the inner surface, $r = R_i = D_i/2 = 48.4$ mm$/2 = 24.2$ mm. Also, $c = R_o = D_o/2 = 60$ mm$/2 = 30$ mm. Then

$$\tau = 32.6 \text{ MPa} \frac{24.2}{30} = 26.3 \text{ MPa}$$

The stress at points between the inner and outer surfaces varies linearly with the radius to each point. Thus it can be seen that all of the material in the hollow shaft shown in Figure 5–6 is being stressed to a fairly high but safe level. This illustrates why the hollow section requires less material.

Of course, the specific data used in the illustration above cannot be generalized to all problems. However, it can be said that for torsional loading of circular members, a hollow section can be designed that is lighter than a solid section while subjecting the material to the same maximum torsional shear stress.

5–9 STRESS CONCENTRATIONS IN TORSIONALLY LOADED MEMBERS

Torsionally loaded members, particularly power transmission shafts, are often made with changes in the geometry at various positions. Figure 5–8 shows an example. This

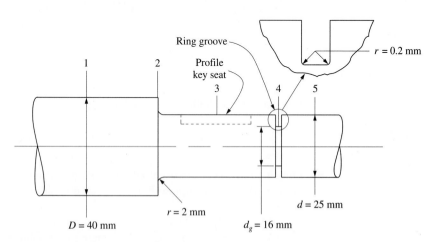

FIGURE 5–8 Shaft with stress concentrations.

is a part of a shaft where a power transmission element, such as a gear, would be mounted. The bore in the hub of the gear would have a diameter that would allow it to slide over the right part of the shaft where the shaft diameter is $d = 25$ mm. A square or rectangular key would be placed in the keyseat and there would be a corresponding keyway in the hub of the gear so it could pass over the key. The gear would then be moved onto the shaft from the right until it stopped against the shoulder at Section 2, created by the increase in the shaft diameter to $D = 40$ mm. To keep the gear in position, a retaining ring is inserted into the ring groove at Section 4.

Changes in the cross section of a member loaded in torsion cause the local stress near the changes to be higher than would be predicted by using the torsional shear stress formula. The actual level of stress in such cases is determined experimentally. Then a *stress concentration factor* is determined which allows the maximum stress in similar designs to be computed from the relationship

$$\tau_{\max} = K_t \tau_{\text{nom}} = K_t(T/Z_p) \tag{5-20}$$

The term τ_{nom} is the nominal stress due to torsion which would be developed in the parts if the stress concentration were not present. Thus the standard torsional shear stress formulas [Equations (5-5) and (5-16)] can be used to compute the nominal stress. The value of K_t is a factor by which the actual maximum stress is greater than the nominal stress.

Referring again to Figure 5-8, you should note that there would be several levels of stress at different places along the length of the bar, even if the applied torque is the same throughout. The differing diameters and the presence of stress concentrations cause the varying stress levels. The stress at Section 1, where $D = 40$ mm, would be relatively low because there is a large diameter and a correspondingly large polar section modulus. At Section 2, the diameter of the shaft reduces to $d = 25$ mm and the step produces a stress concentration that tends to raise the local stress level. Then the keyseat at Section 3 sets up a different stress concentration. At Section 4, two major factors occur that both tend to increase the local stress. Cutting the ring groove reduces the diameter to $d_g = 16$ mm and also produces two closely spaced steps with relatively small fillet radii at the bottom of the groove. At Section 5, well away from the ring groove, the stress would be equal to the nominal stress in the 25-mm diameter shaft. Example Problem 5-10 illustrates all of these situations by performing actual calculations of the stresses at all five sections of the shaft.

First, let's talk more about the nature of stress concentration factors. The following list of Appendix charts gives data for several typical cases.

Appendix A-21-5: Round Bar with Transverse Hole in Torsion

Appendix A-21-6: Grooved Round Bar in Torsion

Appendix A-21-7: Stepped Round Bar in Torsion

Appendix A-21-11: Shafts with Keyseats (see also Figure 5-9)

Round bar with transverse hole.   One purpose for drilling a hole in a shaft is to insert a pin through the hole and through the corresponding hole in the hub of a machine element such as a gear, pulley, or chain sprocket. The pin serves to locate the

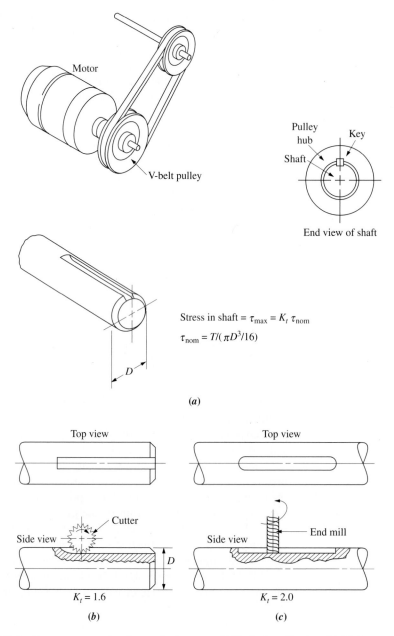

Stress in shaft = $\tau_{\max} = K_t \tau_{\text{nom}}$

$\tau_{\text{nom}} = T/(\pi D^3/16)$

(a)

$K_t = 1.6$

(b)

$K_t = 2.0$

(c)

FIGURE 5–9 Stress concentration factors for keyseats. (a) Typical application. (b) Sled runner type keyseat made with a circular milling cutter. (c) Profile type keyseat made with an end mill.

machine element axially on the shaft while also transmitting torque from the shaft to the element or from the element to the shaft. The hole in the shaft is an abrupt change in geometry and it causes a stress concentration. Appendix A–21–5 is a chart for this case from which K_t can be determined. Curve C is for the case of torsionally loaded shafts. Note that the formula for the nominal stress in the shaft is based on the full, gross, circular cross section of the shaft.

Grooved round bar.   Round-bottomed grooves are cut into round bars for the purpose of installing seals or for distributing lubricating oil around a shaft. The stress concentration factor is dependent on the ratio of the shaft diameter to the diameter of the groove and on the ratio of the groove radius to the groove diameter. The groove is cut with a tool having a rounded nose to produce the round-bottomed groove. The radius should be as large as possible to minimize the stress concentration factor. Note that the nominal stress is based on the diameter *at the base of the groove*. See Appendix A–21–6.

Stepped round bar.   Shafts are often made with two or more diameters, resulting in a stepped shaft like that shown in Appendix A–21–7. The face of the step provides a convenient means of locating one side of an element mounted on the shaft, such as a bearing, gear, pulley, or chain sprocket. Care should be exercised in defining the radius at the bottom of the step, called the *fillet radius*. Sharp corners are to be avoided, as they cause extremely high stress concentration factors. The radius should be as large as possible while being compatible with the elements mounted on the shaft.

Retaining rings seated in grooves cut into the shaft are often used to locate machine elements, as shown in Figure 5–8. The grooves are typically flat bottomed with small radii at the sides. Some designers treat such grooves as two steps on the shaft close together and use the stepped shaft chart (Appendix A–21–7) to determine the stress concentration factor. Because of the small radius at the base of the groove, the relative radius is often quite small, resulting in high values of K_t off the chart. In such cases a value of $K_t = 3.0$ is sometimes used.

Shafts with keyseats.   Power transmission elements typically transmit torque to and from shafts through keys fitted into keyseats cut into the shaft, as shown in Figure 5–9. The V-belt pulley mounted on the end of the motor shaft shown is an example. Two types of keyseats are in frequent use: the *sled-runner* and the *profile* keyseats.

A circular milling cutter having a thickness equal to the width of the keyseat is used to cut the sled-runner keyseat, typically on the end of a shaft, as shown in Figure 5–9(b). As the cutter ends its cut, it leaves a gentle radius, as shown in the side view, resulting in $K_t = 1.6$ as a design value.

A profile keyseat is cut with an end mill having a diameter equal to the width of the keyseat. Usually used at a location away from the ends of the shaft, it leaves a square corner at the ends of the keyseat when viewed from the side, as shown in Figure 5–9(c). This is more severe than the sled-runner and a value of $K_t = 2.0$ is used. Note that the stress concentration factors account for both the removal of material from the shaft and the change in geometry.

The use of stress concentration factors is illustrated in the following example problem.

Figure 5–10 shows a portion of a shaft in which a circular groove has been machined. For an applied torque of 4500 lb·in, compute the torsional shear stress at Section 1 in the full-diameter part of the shaft and at Section 2 where the groove is located.

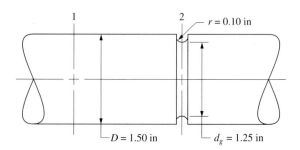

FIGURE 5–10   Shaft with a circular groove for Example Problem 5–9.

Solution   **Objective**   Compute the stress at Sections 1 and 2.

Given   Applied torque = T = 4500 lb·in. Shaft geometry in Figure 5–10.

Analysis   Assuming that Section 1 is well away from the groove, there is no significant stress concentration. Then the standard torsional shear stress formula [Equation (5–16)] can be used. At Section 2 where the groove is located, Equation (5–20) must be used. The value of the stress concentration factor can be determined from Appendix A–21–6.

Results   At Section 1: $\tau_{max} = T/Z_p$.

$$Z_p = \pi D^3/16 = \pi(1.50 \text{ in})^3/16 = 0.663 \text{ in}^3$$

$$\tau_{max} = \frac{4500 \text{ lb·in}}{0.663 \text{ in}^3} = 6790 \text{ psi}$$

At Section 2: $\tau_{max} = K_t T/Z_p$

$$Z_p = \pi d_g^3/16 = \pi(1.25 \text{ in})^3/16 = 0.383 \text{ in}^3$$

To evaluate K_t, two ratios must be computed, as called for in Appendix A–21–6.

$$D/d_g = (1.50 \text{ in})/(1.25 \text{ in}) = 1.20$$
$$r/d_g = (0.10 \text{ in})/(1.25 \text{ in}) = 0.08$$

Then, reading from the chart in Appendix A–21–6, $K_t = 1.55$.

Chapter 5 ■ Torsional Shear Stress and Torsional Deflection

We can now compute the maximum shear stress.

$$\tau_{max} = \frac{K_t T}{Z_p} = \frac{(1.55)(4500 \text{ lb·in})}{0.383 \text{ in}^3} = 18\,200 \text{ psi}$$

Comments Note that the stress at the groove is substantially higher than that in the full-diameter part of the shaft. Also, the use of the stress concentration factor is essential to predict the actual maximum stress level at the groove.

Example Problem 5–10 Figure 5–8 shows a portion of a shaft where a gear is to be mounted. The gear will be centered over the keyset at Section 3. It will rest against the shoulder at Section 2 and be held in position with a retaining ring placed in the groove at Section 4. A repeated torque of 20 N·m is applied throughout the shaft. Compute the maximum shear stress in the shaft at Sections 1, 2, 3, 4, and 5. Then specify a suitable steel material for the shaft.

Solution **Objective** 1. Compute the stresses at Sections 1, 2, 3, 4, and 5.

2. Specify a suitable steel for the shaft.

Given Shaft geometry shown in Figure 5–8. $T = 20$ N·m repeated.

Analysis The stress in each section will be analyzed separately in a panel set off by a horizontal line across the page. You should perform the indicated calculations before looking at the results shown. In each case, the analysis requires the application of Equation (5–20).

$$\tau_{max} = K_t T / Z_p$$

The torque will always be taken to be 20 N·m. You must evaluate the stress concentration factor and the appropriate polar section modulus for each section. Note that $K_t = 1.0$ where there is no change in the geometry. Now compute the stress at Section 1.

Section 1. There is no change in geometry, so $K_t = 1.0$. The shaft diameter is $D = 40$ mm. Then,

$$Z_p = \frac{\pi D^3}{16} = \frac{\pi (40 \text{ mm})^3}{16} = 12\,570 \text{ mm}^3$$

$$\tau_1 = \frac{20 \text{ N·m}}{12\,570 \text{ mm}^3} \times \frac{10^3 \text{ mm}}{m} = 1.59 \frac{N}{mm^2} = 1.59 \text{ MPa}$$

Now compute the stress at Section 2.

Section 2. The stepped shaft and the shoulder fillet produce a stress concentration here that must be evaluated using Appendix A–21–7. The polar section modulus must be based on the smaller diameter; $d = 25$ mm. The results are

$$\tau_{nom} = \frac{T}{\pi d^3 / 16} = \frac{20 \text{ N·m}}{[\pi (25)^3 / 16] \text{ mm}^3} \times \frac{10^3 \text{ mm}}{m}$$
$$= 6.52 \text{ N/mm}^2 = 6.52 \text{ MPa}$$

The value of K_t depends on the ratios D/d and r/d.

$$\frac{D}{d} = \frac{40 \text{ mm}}{25 \text{ mm}} = 1.60$$

$$\frac{r}{d} = \frac{2 \text{ mm}}{25 \text{ mm}} = 0.08$$

Then from Appendix A–21–7, $K_t = 1.45$. Then

$$\tau_2 = (1.45)(6.52 \text{ MPa}) = 9.45 \text{ MPa}$$

Now compute the stress at Section 3.

Section 3. The profile-type keyseat presents a stress concentration factor of 2.0. The nominal stress is the same as that computed at the shoulder fillet. Then

$$\tau_3 = K_t \tau_{\text{nom}} = (2.0)(6.52 \text{ MPa}) = 13.04 \text{ MPa}$$

Now compute the stress at Section 4.

Section 4. Section 4 is the location of the ring groove. Here the nominal stress is computed on the basis of the root diameter of the groove.

$$\tau_{\text{nom}} = \frac{T}{\pi d_g^3/16} = \frac{20 \text{ N·m}}{[\pi(16)^3/16] \text{ mm}^3} \times \frac{10^3 \text{ mm}}{m} = 24.9 \frac{N}{mm^2}$$
$$= 24.9 \text{ MPa}$$

The value of K_t depends on d/d_g and r/d_g.

$$\frac{d}{d_g} = \frac{25 \text{ mm}}{16 \text{ mm}} = 1.56$$

$$\frac{r}{d_g} = \frac{0.2 \text{ mm}}{16 \text{ mm}} = 0.013$$

Referring to Appendix A–21–7, the stress concentration factor is off the chart. This is the type of case for which $K_t = 3.0$ is reasonable.

$$\tau_4 = K_t \tau_{\text{nom}} = (3.0)(24.9 \text{ MPa}) = 74.7 \text{ MPa}$$

Now compute the stress at Section 5.

Section 5. Section 5 is in the smaller portion of the shaft, where no stress concentration occurs. Then $K_t = 1.0$ and,

$$\tau = \frac{T}{Z_p} = \frac{T}{\pi d^3/16}$$

Notice that this is identical to the nominal stress computed for Sections 2 and 3. Then at Section 5,

$$\tau_5 = 6.52 \text{ MPa}$$

Summary of Results

A wide range of stress levels exists in the vicinity of the place on the shaft where the gear is to be mounted.

$\tau_1 = 1.59$ MPa $D = 40$ mm. $K_t = 1.0$.

$\tau_2 = 9.45$ MPa $d = 25$ mm. $K_t = 1.45$. Step.

$\tau_3 = 13.04$ MPa $d = 25$ mm. $K_t = 2.0$. Keyseat.

$\tau_4 = 74.7$ MPa $d_g = 16$ mm. $K_t = 3.0$. Ring groove.

$\tau_5 = 6.52$ MPa $d = 25$ mm. $K_t = 1.0$.

The specification of a suitable material must be based on the stress at Section 4 at the ring groove. Let the design stress, τ_d, be equal to that stress level and determine the required yield strength of the material.

You should have a required yield strength of $s_y = 598$ MPa. For the repeated torque, $N = 4$ is recommended in Table 5–1, resulting in

$$\tau_d = s_y/2N = s_y/8$$

Then, solving for s_y gives,

$$s_y = 8(\tau_d) = 8(74.7 \text{ MPa}) = 598 \text{ MPa}$$

Now specify a suitable material.

From Appendix A–13, two suitable steels for this requirement are AISI 1040 WQT 900 and AISI 4140 OQT 1300. Both have adequate strength and a high ductility as measured by the percent elongation. Certainly, other alloys and heat treatments could be used.

Comment Review the results of this example problem. It illustrates the importance of considering the details of the design of a shaft at any local area where stress concentrations may occur.

5–10 TWISTING—ELASTIC TORSIONAL DEFORMATION

Stiffness in addition to strength is an important design consideration for torsionally loaded members. The measure of torsional stiffness is the angle of twist of one part of a shaft relative to another part when a certain torque is applied.

In mechanical power transmission applications, excessive twisting of a shaft may cause vibration problems which would result in noise and improper synchronization of moving parts. One guideline for torsional stiffness is related to the desired degree of precision, as listed in Table 5–2 (see Refs. 1 and 3).

In structural design, load-carrying members are sometimes loaded in torsion as well as tension or bending. The rigidity of the structure then depends on the torsional stiffness of the components. Any load applied off from the axis of a member and transverse to the axis will produce torsion. This section will discuss twisting of circular

TABLE 5-2 Recommended torsional stiffness: angle of twist per unit length

Application	Torsional deflection	
	deg/in	rad/m
General machine part	1×10^{-3} to 1×10^{-2}	6.9×10^{-4} to 6.9×10^{-3}
Moderate precision	2×10^{-5} to 4×10^{-4}	1.4×10^{-5} to 2.7×10^{-4}
High precision	1×10^{-6} to 2×10^{-5}	6.9×10^{-7} to 1.4×10^{-5}

members, both solid and hollow. Noncircular sections will be covered in a later section. It is very important to note that the behavior of an open-section shape such as a channel or angle is much different from that of a closed section such as a pipe or rectangular tube. In general, the open sections have very low torsional stiffness.

To aid in the development of the relationship for computing the angle of twist of a circular member, consider the shaft shown in Figure 5–3. One end of the shaft is held fixed while a torque T is applied to the other end. Under these conditions the shaft will twist between the two ends through an angle θ.

The derivation of the angle-of-twist formula depends on some basic assumptions about the behavior of a circular member when subjected to torsion. As the torque is applied, an element along the outer surface of the member, which was initially straight, rotates through a small angle γ (gamma). Likewise, a radius of the member in a cross section rotates through a small angle θ. In Figure 5–3, the rotations γ and θ are both related to the arc length AB on the surface of the bar. From geometry, for small angles, the arc length is the product of the angle in radians and the radius from the center of the rotation. Therefore, the arc length AB can be expressed as either

$$AB = \gamma L$$

or

$$AB = \theta c$$

where c is the outside radius of the bar. These two expressions for the arc length AB can be equated to each other,

$$\gamma L = \theta c$$

Solving for γ gives

$$\gamma = \frac{\theta c}{L} \tag{5–21}$$

The angle γ is a measure of the maximum shearing strain in an element on the outer surface of the bar. It was discussed in Chapter 1 that the shearing strain, γ, is related to the shearing stress, τ, by the modulus of elasticity in shear, G. That was expressed as Equation (1–7),

$$G = \frac{\tau}{\gamma} \tag{1–7}$$

At the outer surface, then,

$$\tau = G\gamma$$

But the torsional shear stress formula [Equation (5–11)] states

$$\tau = \frac{Tc}{J}$$

Equating these two expressions for γ gives

$$G\gamma = \frac{Tc}{J}$$

Now, substituting from Equation (5–21) for γ, we obtain

$$\frac{G\theta c}{L} = \frac{Tc}{J}$$

We can now cancel c and solve for θ:

 **Angle of Twist**

$$\theta = \frac{TL}{JG} \tag{5–22}$$

The resulting angle of twist, θ, is in radians. When consistent units are used for all terms in the calculation, all units will cancel, leaving a dimensionless number. This should be interpreted as the angle, θ, in radians.

Equation (5–22) can be used to compute the angle of twist of one section of a circular bar, either solid or hollow, with respect to another section where L is the distance between them, provided that the torque, T, the polar moment of inertia, J, and the shear modulus of elasticity, G, are the same over the entire length, L. If any of these factors vary in a given problem, the bar can be subdivided into segments over which they are constant to compute angles of rotation for those segments. Then the computed angles can be combined algebraically to get the total angle of twist. This principle, called *superposition,* will be illustrated in example problems.

The shear modulus of elasticity, G, is a measure of the torsional stiffness of the material of the bar. Table 5–3 gives values for G for selected materials.

TABLE 5–3   Shear modulus of elasticity, G

| | Shear modulus, G | |
Material	GPa	psi
Plain carbon and alloy steels	80	11.5×10^6
Stainless steel type 304	69	10.0×10^6
Aluminum 6061-T6	26	3.75×10^6
Beryllium copper	48	7.0×10^6
Magnesium	17	2.4×10^6
Titanium alloy	43	6.2×10^6

Example Problem 5–11

Determine the angle of twist in degrees between two sections 250 mm apart in a steel rod having a diameter of 10 mm when a torque of 15 N·m is applied. Figure 5–3 shows a sketch of the arrangement.

Solution

Objective   Compute the angle of twist in degrees.

Given   Applied torque = T = 15 N·m. Circular bar: diameter = D = 10 mm.
Length = L = 250 mm.

Analysis   Equation (5–22) can be used. Compute $J = \pi D^4/32$.
G = 80 GPa = 80 × 10⁹ N/m² (Table 5–3).

Results $\theta = \dfrac{TL}{JG}$

The value of J is

$$J = \frac{\pi D^4}{32} = \frac{\pi(10\ \text{mm})^4}{32} = 982\ \text{mm}^4$$

Then

$$\theta = \frac{TL}{JG} = \frac{(15\ \text{N·m})\,(250\ \text{mm})}{(982\ \text{mm}^4)\,(80 \times 10^9\ \text{N/m}^2)} \times \frac{(10^3\ \text{mm})^3}{1\ \text{m}^3} = 0.048\ \text{rad}$$

Note that all units cancel. Expressing the angle in degrees,

$$\theta = 0.048\ \text{rad} \times \frac{180\ \text{deg}}{\pi\ \text{rad}} = 2.73\ \text{deg}$$

Example Problem 5–12

Determine the required diameter of a round shaft made of aluminum alloy 6061-T6 if it is to twist not more than 0.08 deg in 1.0 ft of length when a torque of 75 lb·in is applied.

Solution

Objective   Compute the required diameter, D, of the round shaft.

Given   Applied torque = T = 75 lb·in. Length = L = 1.0 ft = 12 in.
Maximum angle of twist = 0.08 deg. Aluminum 6061-T6.

Analysis   Equation (5–22) can be solved for J because J is the only term involving the unknown diameter, D. Then, solve for D from $J = \pi D^4/32$.
G = 3.75 × 10⁶ psi (Table 5–3).

Results $\theta = \dfrac{TL}{JG}$

$J = \dfrac{TL}{\theta G}$

The angle of twist must be expressed in radians.

$$\theta = 0.08\ \text{deg} \times \frac{\pi\ \text{rad}}{180\ \text{deg}} = 0.0014\ \text{rad}$$

Then

$$J = \frac{TL}{\theta G} = \frac{(75 \text{ lb·in})(12 \text{ in})}{(0.0014)(3.75 \times 10^6 \text{ lb/in}^2)} = 0.171 \text{ in}^4$$

Now since $J = \pi D^4/32$,

$$D = \left(\frac{32J}{\pi}\right)^{1/4} = \left[\frac{(32)(0.171 \text{ in}^4)}{\pi}\right]^{1/4} = 1.15 \text{ in}$$

Comment This is the minimum acceptable diameter. You should specify a convenient, preferred size, say 1.25 in. The resulting angle of twist will then be less than 0.08 deg over a 1.0 ft length.

Example Problem 5–13

Figure 5–11 shows a steel rod to which three disks are attached. The rod is fixed against rotation at its left end, but free to rotate in a bearing at its right end. Each disk is 300 mm in diameter. Downward forces act at the outer surfaces of the disks so that torques are applied to the rod. Determine the angle of twist of Section *A* relative to the fixed Section *E*.

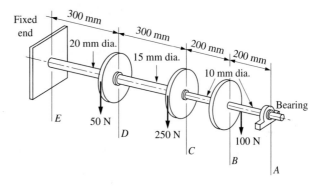

FIGURE 5–11   Rod for Example Problem 5–13.

Solution

Objective Compute the angle of twist of the rod at *A* relative to *E*.

Given Rod is steel. $G = 80$ GPa. (Table 5–3)
Geometry of rod and loading shown in Figure 5–11.
For each disk, diameter = $D = 300$ mm. Radius = $R = 150$ mm.

Analysis The design of the system shown in Figure 5–11 is a rod with a total length of 1000 mm or 1.0 m. But there are four segments of the rod having different lengths, diameters, or applied levels to torque. Therefore, Equation (5–22) must be applied to each segment separately to compute the angle of twist for each. Then, the total angle of twist in Section *A* relative to *E* will be the algebraic sum of the four angles.

    The completion of this Example Problem will be shown in programmed format. You should perform each indicated operation yourself before moving to the next panel.

The first operation is to compute the magnitude and direction of the applied torques at each disk, *B*, *C*, and *D*. Do that now. Recall the definition of torque from Equation (5–1).

For directions, we will assume a view point along the rod from the right end. The magnitude of the torque on each disk is the product of the force acting on the periphery of the disk times the radius of the disk. Then, considering clockwise to be positive,

Torque on disk *B*, clockwise:

$$T_B = (100 \text{ N})(150 \text{ mm}) = 15\,000 \text{ N·mm} = 15 \text{ N·m}$$

Torque on disk *C*, counterclockwise:

$$T_C = -(250 \text{ N})(150 \text{ mm}) = -37\,500 \text{ N·mm} = -37.5 \text{ N·m}$$

Torque on disk *D*, counterclockwise:

$$T_D = -(50 \text{ N})(150 \text{ mm}) = -7500 \text{ N·mm} = -7.5 \text{ N·m}$$

Now determine the level of torque in each segment of the rod. You should visualize a free-body diagram of any part of the rod between the ends of each segment by "cutting" the rod and computing the magnitude of the torque applied to the rod to the right of the cut section. The internal torque in the rod must be equal in magnitude and opposite in direction to the externally applied torque to maintain equilibrium.

It is suggested that you start at the right end at *A*. The bearing allows free rotation of the rod at that end. Then move to the left and compute the torque for segments *AB*, *BC*, *CD*, and *DE*. What is the level of torque in segment *AB*?

For the segment *AB*, up to, but not including *B*, the torque in the rod is zero because the bearing allows free rotation. Now, consider the torque applied by the disk at *B* and determine the torque in the segment *BC*.

Cutting the rod anywhere to the right of *C* in the segment *BC* would result in an externally applied torque of 15 N·m clockwise, due to the torque on disk *B*. Therefore, the torque throughout the segment *BC* is

$$T_{BC} = 15 \text{ N·m} \quad \text{(CW)}$$

We will consider this torque to be clockwise (CW) and positive because it tends to cause a clockwise rotation of the rod.

Now determine the torque in segment *CD*, called T_{CD}.

Cutting the rod anywhere between *C* and *D* would result in both T_C and T_D acting on the rod to the right of the cut section. But they act in opposite sense, one clockwise and one counterclockwise. Thus the net torque applied to the rod is the difference between them. That is,

$$T_{CD} = -T_C + T_B = -37.5 \text{ N·m} + 15 \text{ N·m} = -22.5 \text{ N·m} \quad \text{(CCW)}$$

Now continue this process for the final segment, *DE*.

Between *D* and *E* in the rod, the torque is the resultant of all the applied torques at *D*, *C*, and *B*.

$$T_{DE} = -T_D - T_C + T_B = -7.5 \text{ N·m} - 37.5 \text{ N·m} + 15 \text{ N·m} = -30 \text{ N·m}$$
$$\text{(CCW)}$$

The fixed support at *E* must be capable of providing a reaction torque of 30 N·m to maintain the rod in equilibrium.

In summary, the distribution of torque in the rod can be shown in graphical form as in Figure 5–12. Notice that the applied torques T_B, T_C, and T_D are the *changes* in torque that occur at *B*, *C*, and *D* but that they are not the magnitudes of the torque *in the rod* at those points.

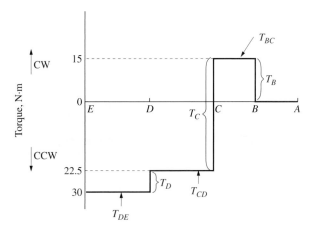

FIGURE 5–12 Torque distribution in rod for Example Problem 5–13.

Now compute the angle of twist in each segment by applying Equation (5–22) successively. Start with segment *AB*.

Segment AB

$$\theta_{AB} = T_{AB}\left(\frac{L}{JG}\right)_{AB}$$

Since $T_{AB} = 0$, $\theta_{AB} = 0$. There is no twisting of the rod between *A* and *B*.
Now continue with segment *BC*.

Segment BC

$$\theta_{BC} = T_{BC}\left(\frac{L}{JG}\right)_{BC}$$

We know $T_{BC} = 15$ N·m, $L = 200$ mm, and $G = 80$ GPa for steel. For the 10-mm-diameter rod,

$$J = \frac{\pi D^4}{32} = \frac{\pi (10 \text{ mm})^4}{32} = 982 \text{ mm}^4$$

Then

$$\theta_{BC} = \frac{(15 \text{ N·m})(200 \text{ mm})}{(982 \text{ mm}^4)(80 \times 10^9 \text{ N/m}^2)} \times \frac{(10^3)^3 \text{ mm}^3}{\text{m}^3} = 0.038 \text{ rad}$$

This means that Section *B* is rotated 0.038 rad clockwise relative to Section *C*, since θ_{BC} is the total angle of twist in the segment *BC*.
Now continue with segment *CD*.

Segment CD

$$\theta_{CD} = T_{CD}\left(\frac{L}{JG}\right)_{CD}$$

Here $T_{CD} = -22.5$ N·m, $L = 300$ mm, and the rod diameter is 15 mm. Then

$$J = \frac{\pi D^4}{32} = \frac{\pi (15 \text{ mm})^4}{32} = 4970 \text{ mm}^4$$

$$\theta_{CD} = \frac{-(22.5 \text{ N·m})(300 \text{ mm})}{(4970 \text{ mm}^4)(80 \times 10^9 \text{ N/m}^2)} \times \frac{(10^3)^3 \text{ mm}^3}{\text{m}^3} = -0.017 \text{ rad}$$

Section *C* is rotated 0.017 rad counterclockwise relative to Section *D*.
Now, finally, complete the analysis for segment *DE*.

Segment DE

$$\theta_{DE} = T_{DE}\left(\frac{L}{JG}\right)_{DE}$$

Here $T_{DE} = -30$ N·m, $L = 300$ mm, and $D = 20$ mm. Then

$$J = \frac{\pi D^4}{32} = \frac{\pi (20 \text{ mm})^4}{32} = 15\,700 \text{ mm}^4$$

$$\theta_{DE} = \frac{-(30 \text{ N·m})(300 \text{ mm})}{(15\,700 \text{ mm}^4)(80 \times 10^9 \text{ N/mm}^2)} \times \frac{(10^3)^3 \text{ mm}^3}{\text{m}^3} = -0.007 \text{ rad}$$

Section *D* is rotated 0.007 rad counterclockwise relative to *E*.
The final operation is to compute the total angle of twist from *E* to *A* by summing the angles of twist for all segments algebraically. Do that now.

Total angle of twist from *E* to *A*

$$\theta_{AE} = \theta_{AB} + \theta_{BC} - \theta_{CD} - \theta_{DE}$$
$$= 0 + 0.038 - 0.017 - 0.007 = 0.014 \text{ rad}$$

Summary and Comment It should help your visualization of what is happening in the rod throughout its length by plotting a graph of the angle of twist as a function of position. This is done in Figure 5–13 by setting the zero reference point as *E*. The straight line from *E* to *D* shows the linear change in angle with position to the value of −0.007 rad (counterclockwise). From there, the angle grows by an additional −0.017 rad between *D*

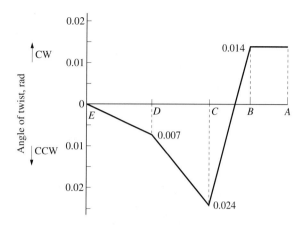

FIGURE 5–13 Angle of twist versus position on the rod for Example Problem 5–13.

and *C*. In segment *BC*, the relative rotation is clockwise with a magnitude of 0.038 rad, ending at the final value of 0.014 rad at *B*. And, because there is no torque in the segment *AB*, the angle of rotation remains at that value.

5–11 TORSION IN NONCIRCULAR SECTIONS

The behavior of noncircular sections when subjected to torsion is vastly different from that of circular sections, for which the discussions earlier in this chapter applied. There is a large variety of shapes that can be imagined, and the analysis of stiffness and strength is different for each. The development of the relationships involved will not be done here. Compilations of the pertinent formulas occur in References 1 to 5 listed at the end of this chapter, and a few are given in this section.

Some generalizations can be made. Solid sections having the same cross-sectional area are stiffer when their shape more closely approaches a circle (see Figure 5–14). Conversely, a member made up of long, thin sections that do not form a closed, tube-like shape are very weak and flexible in torsion. Examples of flexible sections are common structural shapes such as wide-flange beams, standard I-beams, channels, angles, and tees, as illustrated in Figure 5–15. Pipes, solid bars, and structural rectangular tubes have high rigidity, or stiffness (see Figure 5–16).

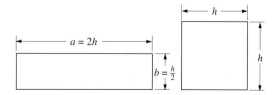

FIGURE 5–14 Comparison of stiffness for rectangle and square sections in torsion. Square is two times stiffer than rectangle even though both have the same area: $ab = h^2$.

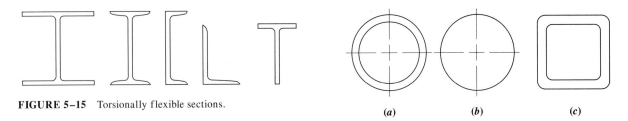

FIGURE 5–15 Torsionally flexible sections.

FIGURE 5–16 Torsionally stiff sections.

An interesting illustration of the lack of stiffness of open, thin sections is shown in Figure 5–17. The thin plate (a), the angle (b), and the channel (c) have the same thickness and cross-sectional area, and all have nearly the same torsional stiffness. Likewise, if the thin plate were formed into a circular shape (d), but with a slit remaining, its stiffness would remain low. However, closing the tube completely as in Fig. 5–16(a) by welding or by drawing a seamless tube would produce a relatively stiff member. Understanding these comparisons is an aid to selecting a reasonable shape for members loaded in torsion.

Figure 5–18 shows seven cases of noncircular cross sections that are commonly encountered in machine design and structural analysis. The computation of the maximum shear stress and the angle of twist can be made by slightly modifying the formulas used for circular cross sections as given here.

$$\tau_{max} = \frac{T}{Q} \tag{5–23}$$

$$\theta = \frac{TL}{GK} \tag{5–24}$$

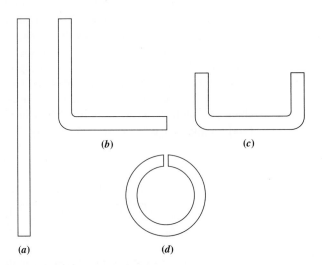

FIGURE 5–17 Sections having nearly equal (and low) torsional stiffness.

Chapter 5 ■ Torsional Shear Stress and Torsional Deflection

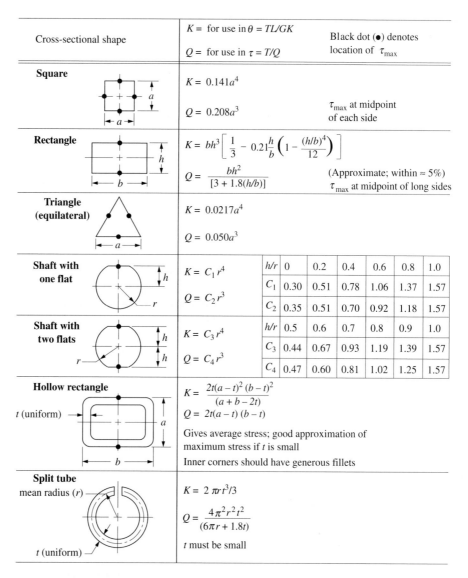

Cross-sectional shape	$K =$ for use in $\theta = TL/GK$ $Q =$ for use in $\tau = T/Q$	Black dot (●) denotes location of τ_{max}
Square	$K = 0.141a^4$ $Q = 0.208a^3$	τ_{max} at midpoint of each side
Rectangle	$K = bh^3\left[\dfrac{1}{3} - 0.21\dfrac{h}{b}\left(1 - \dfrac{(h/b)^4}{12}\right)\right]$ $Q = \dfrac{bh^2}{[3 + 1.8(h/b)]}$	(Approximate; within ≈ 5%) τ_{max} at midpoint of long sides
Triangle (equilateral)	$K = 0.0217a^4$ $Q = 0.050a^3$	

Shaft with one flat — $K = C_1 r^4$, $Q = C_2 r^3$

h/r	0	0.2	0.4	0.6	0.8	1.0
C_1	0.30	0.51	0.78	1.06	1.37	1.57
C_2	0.35	0.51	0.70	0.92	1.18	1.57

Shaft with two flats — $K = C_3 r^4$, $Q = C_4 r^3$

h/r	0.5	0.6	0.7	0.8	0.9	1.0
C_3	0.44	0.67	0.93	1.19	1.39	1.57
C_4	0.47	0.60	0.81	1.02	1.25	1.57

Hollow rectangle — t (uniform)

$$K = \frac{2t(a-t)^2(b-t)^2}{(a+b-2t)}$$
$$Q = 2t(a-t)(b-t)$$

Gives average stress; good approximation of maximum stress if t is small

Inner corners should have generous fillets

Split tube — mean radius (r), t (uniform)

$$K = 2\pi r t^3/3$$
$$Q = \frac{4\pi^2 r^2 t^2}{(6\pi r + 1.8t)}$$

t must be small

FIGURE 5–18 Methods for determining values for K and Q for several types of cross sections. (Source: *Machine Elements in Mechanical Design,* 2nd ed., Robert L. Mott, copyright © Macmillan Publishing Co., New York. Reprinted by permission of the publisher.)

The term Q is analogous to the polar section modulus Z_p used for circular bars. The torsional stiffness is indicated by K, analogous to the polar moment of inertia J.

Note in Figure 5–18 that the points of maximum shear stress for the noncircular cross sections are indicated by a prominent black dot.

An example is now given which illustrates the high degree of flexibility of a slit tube in comparison with a closed tube.

A tube is made by forming a flat steel sheet, 4.0 mm thick, into the circular shape having an outside diameter of 90 mm. The final step is to weld the seam along the length of the tube. Figures 5–17(a), 5–17(d), and 5–16(a) show the stages of the process. Perform the following calculations to compare the behavior of the closed, welded tube with that of the open tube.

(a) Compute the torque that would create a stress of 10 MPa in the closed, welded tube.

(b) Compute the angle of twist of a 1.0-m length of the closed tube for the torque found in part (a).

(c) Compute the stress in the open tube for the torque found in part (a).

(d) Compute the angle of twist of a 1.0-m length of the open tube for the torque found in part (a).

(e) Compare the stress and deflection of the open tube with those of the closed tube.

Solution   *The solution will be shown in a programmed format with each part of the solution, (a)–(e), given as a separate panel. Each part can be approached as a different problem with* **Objective**, **Given**, **Analysis**, *and* **Results** *sections.*
     Complete part (a) now.

Objective   Compute the torque on the closed tube that would create a torsional shear stress of 10 MPa.

Given   Tube is steel. D_o = 90 mm. Wall thickness = t = 4.0 mm.
     $D_i = D_o - 2t$ = 90 mm − 2(4.0 mm) = 82 mm.

Analysis   Use the maximum shear stress Equation (5–11) and solve for T.

Results
$$\tau_{max} = \frac{Tc}{J} \qquad (5\text{–}11)$$

Then

$$T = \frac{\tau_{max} J}{c}$$

We can compute J from Equation (5–13),

$$J = \frac{\pi}{32}(D_o^4 - D_i^4) \qquad (5\text{–}13)$$

Using D_o = 90 mm = 0.09 m and D_i = 82 mm = 0.082 m,

$$J = \frac{\pi}{32}(0.09^4 - 0.082^4) \text{ m}^4 = 2.00 \times 10^{-6} \text{ m}^4$$

Now, letting τ_{max} = 10 MPa = 10×10^6 N/m², we have

$$T = \frac{\tau_{max} J}{c} = \frac{(10 \times 10^6 \text{ N/m}^2)(2.00 \times 10^{-6} \text{ m}^4)}{0.045 \text{ m}} = 444 \text{ N·m}$$

That is, a torque of 444 N·m applied to the closed welded tube would produce a maximum torsional shear stress of 10 MPa in the tube. Note that this is a very low stress level for steel.

Now complete part (b) of the problem.

Objective   For the closed tube used in (a), compute the angle of twist.

Given $J = 2.00 \times 10^{-6}$ m⁴. Length $= L = 1.0$ m. Torque $= T = 444$ N·m. $G = 80$ GPa.

Analysis   Use Equation (5–22) to compute the angle of twist, θ.

Results $$\theta = \frac{TL}{GJ} = \frac{(444 \text{ N·m})(1.0 \text{ m})}{(80 \times 10^9 \text{ N/m}^2)(2.00 \times 10^{-6} \text{ m}^4)} = 0.00278 \text{ rad}$$

Converting θ to degrees gives

$$\theta = 0.00278 \text{ rad} \frac{180 \text{ deg}}{\pi \text{ rad}} = 0.159 \text{ deg}$$

Again, note that this is a very small angle of twist.

Now consider a tube shaped as in parts (a) and (b) except that it is not closed, as shown in Figure 5–17(d). Compute the torsional shear stress in the open tube due to the torque of 444 N·m.

Objective   Compute the shear stress in the open tube.

Given   Torque $= T = 444$ N·m. $D_o = 90$ mm. Wall thickness $= t = 4.0$ mm.

Analysis   Equation (5–23) can be used to compute the maximum shear stress for the open tube before it is welded, treating it as a noncircular cross section. The formula for Q is given in Figure 5–18.

$$Q = \frac{4\pi^2 r^2 t^2}{6\pi r + 1.8t}$$

The mean radius is

$$r = \frac{D_o}{2} - \frac{t}{2} = \frac{90 \text{ mm}}{2} - \frac{4 \text{ mm}}{2} = 43 \text{ mm}$$

Then

$$Q = \frac{4\pi^2(43)^2(4)^2}{6\pi(43) + 1.8(4)} \text{ mm}^3 = 1428 \text{ mm}^3$$

The stress in the open tube is, then,

$$\tau_{max} = \frac{T}{Q} = \frac{444 \text{ N·m}}{1428 \text{ mm}^3} \frac{1000 \text{ mm}}{\text{m}} = 311 \text{ MPa}$$

Now complete part (d) by computing the angle of twist of the open tube.

Objective Compute the angle of twist of the open tube for $T = 444$ N·m.

Given Length $= L = 1.0$ m. Torque $= T = 444$ N·m. $G = 80$ GPa.
$D_o = 90$ mm. Wall thickness $= t = 4.0$ mm.

Analysis The angle of twist for the open tube can be computed using Equation (5–24). Figure 5–18 gives us the formula for the torsional rigidity constant, K. Using $r = 43$ mm $= 0.043$ m and $t = 4$ mm $= 0.004$ m yields

$$K = \frac{2\pi r t^3}{3} = \frac{2\pi(0.043)(0.004)^3}{3} = 5.764 \times 10^{-9} \text{ m}^4$$

Then the angle of twist is

$$\theta = \frac{TL}{GK} = \frac{(444 \text{ N·m})(1.0 \text{ m})}{(80 \times 10^9 \text{ N/m}^2)(5.764 \times 10^{-9} \text{ m}^4)} = 0.963 \text{ rad}$$

Converting θ to degrees, we obtain

$$\theta = 0.963 \text{ rad} \frac{180 \text{ deg}}{\pi \text{ rad}} = 55.2 \text{ deg}$$

Now complete the final part (e), comparing the open and closed tubes.

Objective Compare the stress and deflection of the open tube with those of the closed tube.

Given From parts (a), (b), (c), and (d):

Stress in the closed tube $= \tau_c = 10$ MPa.
Angle of twist of the closed tube $= \theta_c = 0.159$ deg.
Stress in the open tube $= \tau_o = 311$ MPa.
Angle of twist of the open tube $= \theta_o = 55.2$ deg.

Analysis Ratio of stresses:

$$\frac{\tau_o}{\tau_c} = \frac{311 \text{ MPa}}{10 \text{ MPa}} = 31.1 \text{ times greater}$$

Ratio of angle of twist:

$$\frac{\theta_o}{\theta_c} = \frac{55.2 \text{ deg}}{0.159 \text{ deg}} = 347 \text{ times greater}$$

Comments These comparisons dramatically show the advantage of using closed sections to carry torsion.

In addition, the actual stress level computed for the open tube is probably greater than the allowable stress for many steels. Recall that the design shear stress is

$$\tau_d = \frac{0.5 s_y}{N}$$

Then the required yield strength of the material to be safe at a stress level of 311 MPa and a design factor of 2.0 is

$$s_y = \frac{N\tau_d}{0.5} = \frac{2(311\ \text{MPa})}{0.5} = 1244\ \text{MPa}$$

Only a few highly heat-treated steels listed in Appendix A−13 have this level of yield strength.

REFERENCES

1. Blodgett, O. W., *Design of Weldments,* James F. Lincoln Arc Welding Foundation, Cleveland, OH, 1963.

2. Boresi, A. P., O. M. Sidebottom, F. B. Seely, and J. O. Smith, *Advanced Mechanics of Materials,* 3rd ed., New York, 1978.

3. Mott, R. L., *Machine Elements in Mechanical Design,* 2nd ed., Macmillan Publishing Co., New York, 1992.

4. Popov, E. P., *Engineering Mechanics of Materials,* Prentice-Hall, Englewood Cliffs, NJ, 1990.

5. Young, W. C., *Roark's Formulas for Stress and Strain,* 6th ed., McGraw-Hill, New York, 1989.

PROBLEMS

5–1.M Compute the torsional shear stress that would be produced in a solid circular shaft having a diameter of 20 mm when subjected to a torque of 280 N·m.

5–2.M A hollow shaft has an outside diameter of 35 mm and an inside diameter of 25 mm. Compute the torsional shear stress in the shaft when it is subjected to a torque of 560 N·m.

5–3.E Compute the torsional shear stress in a shaft having a diameter of 1.25 in when carrying a torque of 1550 lb·in.

5–4.E A steel tube is used as a shaft carrying 5500 lb·in of torque. The outside diameter is 1.75 in, and the wall thickness is $\frac{1}{8}$ in. Compute the torsional shear stress at the outside and the inside surfaces of the tube.

5–5.M A movie projector drive mechanism is driven by a 0.08-kW motor whose shaft rotates at 180 rad/s. Compute the torsional shear stress in its 3.0-mm-diameter shaft.

5–6.M The impeller of a fluid agitator rotates at 42 rad/s and requires 35 kW of power. Compute the torsional shear stress in the shaft that drives the im-

peller if it is hollow and has an outside diameter of 40 mm and an inside diameter of 25 mm.

5–7.E A drive shaft for a milling machine transmits 15.0 hp at a speed of 240 rpm. Compute the torsional shear stress in the shaft if it is solid with a diameter of 1.44 in. Would the shaft be safe if the torque is applied with shock and it is made from AISI 4140 OQT 1300 steel?

5–8.E Repeat Problem 5–7 if the shaft contains a profile keyseat.

5–9.E Figure 5–19 shows the end of the vertical shaft for a rotary lawnmower. Compute the maximum torsional shear stress in the shaft if it is transmitting 7.5 hp to the blade when rotating 2200 rpm. Specify a suitable steel for the shaft.

5–10.E Figure 5–20 shows a stepped shaft in torsion. The larger section also has a hole drilled through.
   (a) Compute the maximum shear stress at the step for an applied torque of 7500 lb·in.
   (b) Determine the largest hole that could be drilled in the shaft and still maintain the stress near the hole at or below that at the step.

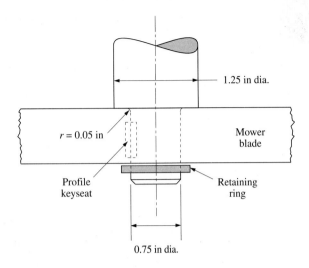

FIGURE 5–19 Shaft for Problem 5–9.

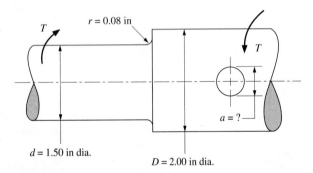

FIGURE 5–20 Shaft for Problem 5–10.

5–11.M Compute the torsional shear stress and the angle of twist in degrees in an aluminum tube, 600 mm long, having an inside diameter of 60 mm and an outside diameter of 80 mm when subjected to a steady torque of 4500 N·m. Then specify a suitable aluminum alloy for the tube.

5–12.M Two designs for a shaft are being considered. Both have an outside diameter of 50 mm and are 600 mm long. One is solid but the other is hollow with an inside diameter of 40 mm. Both are made from steel. Compare the torsional shear stress, angle of twist, and the mass of the two designs if they are subjected to a torque of 850 N·m.

5–13.M Determine the required inside and outside diameters for a hollow shaft to carry a torque of 1200 N·m with a maximum torsional shear stress of 45 MPa. Make the ratio of the outside diameter to the inside diameter approximately 1.25.

5–14.E A gear drive shaft for a milling machine transmits 7.5 hp at a speed of 240 rpm. Compute the torsional shear stress in the 0.860-in-diameter solid shaft.

5–15.E The input shaft for the gear drive described in Problem 5–14 also transmits 7.5 hp, but rotates at 1140 rpm. Determine the required diameter of the input shaft to give it the same stress as the output shaft.

5–16.E Determine the stress that would result in a $1\frac{1}{2}$-in schedule 40 steel pipe if a plumber applies a force of 80 lb at the end of a wrench handle 18 in long.

5–17.E A rotating sign makes 1 rev every 5 s. In a high wind, a torque of 30 lb·ft is required to maintain rotation. Compute the power required to drive the sign. Also compute the stress in the final drive shaft if it has a diameter of 0.60 in. Specify a suitable steel for the shaft to provide a design factor of 4 based on yield strength in shear.

5–18.M A short, cylindrical bar is welded to a rigid plate at one end, and then a torque is applied at the other. If the bar has a diameter of 15 mm and is made of AISI 1020 cold-drawn steel, compute the torque that must be applied to it to subject it to a stress equal to its yield strength in shear. Use $s_{ys} = s_y/2$.

5–19.E A propeller drive shaft on a ship is to transmit 2500 hp at 75 rpm. It is to be made of AISI 1040 WQT 1300 steel. Use a design factor of 6 based on the yield strength in shear. The shaft is to be hollow, with the inside diameter equal to 0.80 times the outside diameter. Determine the required diameter of the shaft.

5–20.E If the propeller shaft of Problem 5–19 was to be solid instead of hollow, determine the required diameter. Then compute the ratio of the weight of the solid shaft to that of the hollow shaft.

5–21.M A power screwdriver uses a shaft with a diameter of 5.0 mm. What torque can be applied to the screwdriver if the limiting stress due to torsion is 80 MPa?

5–22.M An extension for a socket wrench similar to that shown in Figure 5–1 has a diameter of 6.0 mm and a length of 250 mm. Compute the stress and angle of twist in the extension when a torque of 5.5 N·m is applied. The extension is steel.

5–23.M Compute the angle of twist in a steel shaft 15 mm in diameter and 250 mm long when a torque of 240 N·m is applied.

5–24.M Compute the angle of twist in an aluminum tube that has an outside diameter of 80 mm and an inside diameter of 60 mm when subjected to a torque of 2250 N·m. The tube is 1200 mm long.

5–25.E A steel rod with a length of 8.0 ft and a diameter of 0.625 in is used as a long wrench to unscrew a plug at the bottom of a pool of water. If it requires 40 lb·ft of torque to loosen the plug, compute the angle of twist of the rod.

5–26.E For the rod described in Problem 5–25, what must the diameter be if only 2.0 deg of twist is desired when 40 lb·ft of torque is applied?

5–27.M Compute the angle of twist of the free end relative to the fixed end of the steel bar shown in Figure 5–21.

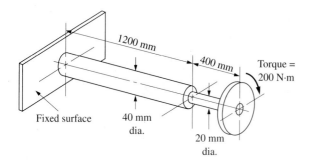

FIGURE 5–21 Bar for Problem 5–27.

5–28.M A meter for measuring torque uses the angle of twist of a shaft to indicate torque. The shaft is to be 150 mm long and made of 6061-T6 aluminum alloy. Determine the required diameter of the shaft if it is desired to have a twist of 10.0 deg when a torque of 5.0 N·m is applied to the meter. For the shaft thus designed, compute the torsional shear stress and then compute the resulting design factor for the shaft. Is it satisfactory? If not, what would you do?

5–29.M A beryllium copper wire having a diameter of 1.50 mm and a length of 40 mm is used as a small torsion bar in an instrument. Determine what angle of twist would result in the wire when it is stressed to 250 MPa.

5–30.M A fuel line in an aircraft is made of a titanium alloy. The tubular line has an outside diameter of 18 mm and an inside diameter of 16 mm. Compute the stress in the tube if a length of 1.65 m must be twisted through an angle of 40 deg during installation. Determine the design factor

based on the yield strength in shear if the tube is Ti-6Al-4V, aged.

5–31.M For the shaft shown in Figure 5–22 compute the angle of twist of pulleys B and C relative to A. The steel shaft has a diameter of 35 mm throughout its length. The torques are $T_1 = 1500$ N·m, $T_2 = 1000$ N·m, $T_3 = 500$ N·m. The lengths are $L_1 = 500$ mm, $L_2 = 800$ mm.

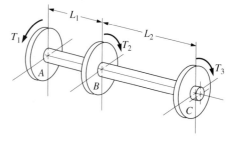

FIGURE 5–22 Shaft for Problem 5–31.

5–32.M A torsion bar in a light truck suspension is to be 820 mm long and made of steel. It is subjected to a torque of 1360 N·m and must be limited to 2.2 deg of twist. Determine the required diameter of the solid round bar. Then compute the stress in the bar.

5–33.M A steel drive shaft for an automobile is a hollow tube 1525 mm long. Its outside diameter is 75 mm, and its inside diameter is 55 mm. If the shaft transmits 120 kW of power at a speed of 225 rad/s, compute the torsional shear stress in the shaft and the angle of twist of one end relative to the other.

5–34.M A rear axle of an automobile is a solid steel shaft having the configuration shown in Figure 5–23.

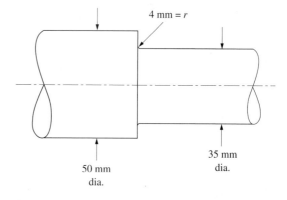

FIGURE 5–23 Axle for Problem 5–34.

Considering the stress concentration due to the shoulder, compute the torsional shear stress in the axle when it rotates at 70.0 rad/s, transmitting 60 kW of power.

5–35.M An output shaft from an automotive transmission has the configuration shown in Figure 5–24. If the shaft is transmitting 105 kW at 220 rad/s, compute the maximum torsional shear stress in the shaft. Account for the stress concentration at the place where the speedometer gear is located.

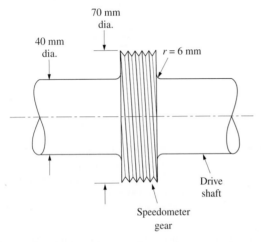

FIGURE 5–24 Shaft for Problem 5–35.

The indicated figures for Problems 5–36 through 5–39 show portions of shafts from power transmission equipment. Compute the maximum repeated torque that could safely be applied to each shaft if it is to be made from AISI 1141 OQT 1100 steel.

5–36.M Use Figure 5–25.

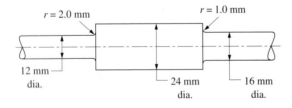

FIGURE 5–25 Shaft for Problem 5–36.

5–37.E Use Figure 5–26.
5–38.M Use Figure 5–27.
5–39.E Use Figure 5–28.

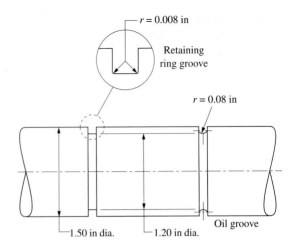

FIGURE 5–26 Shaft for Problem 5–37.

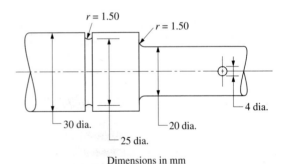

Dimensions in mm

FIGURE 5–27 Shaft for Problem 5–38.

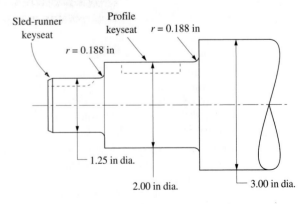

FIGURE 5–28 Shaft for Problem 5–39.

Chapter 5 ▪ Torsional Shear Stress and Torsional Deflection

Noncircular Sections

5–40.M Compute the torque that would produce a torsional shear stress of 50 MPa in a square steel rod 20 mm on a side.

5–41.M For the rod described in Problem 5–40 compute the angle of twist that would be produced by the torque found in the problem over a length of 1.80 m.

5–42.E Compute the torque that would produce a torsional shear stress of 7500 psi in a square aluminum rod, 1.25 in on a side.

5–43.E For the rod described in Problem 5–42, compute the angle twist that would be produced by the torque found in the problem over a length of 48 in.

5–44.E Compute the torque that would produce a torsional shear stress of 7500 psi in a rectangular aluminum bar 1.25 in thick by 3.00 in wide.

5–45.E For the bar described in Problem 5–44, compute the angle of twist that would be produced by the torque found in the problem over a length of 48 in.

5–46.M An extruded aluminum bar is in the form of an equilateral triangle 30 mm on a side. What torque is required to cause an angle of twist in the bar of 0.80 deg over a length of 2.60 m?

5–47.M For the triangular bar described in Problem 5–46, what stress would be developed in the bar when carrying the torque found in the problem?

5–48.E As shown in Figure 5–29, a portion of a steel shaft has a flat machined on one side. Compute

the torsional shear stress in both the circular section and the one with the flat when a torque of 850 lb·in is applied.

5–49.E For the steel shaft shown in Figure 5–29, compute the angle of twist of one end relative to the other if a torque of 850 lb·in is applied uniformly along the length.

5–50.E Repeat Problem 5–48 with all the data the same except that two flats are machined on the shaft, resulting in a total measurement across the flats of 1.25 in.

5–51.E Repeat Problem 5–49 for a shaft having two flats, resulting in a total measurement across the flats of 1.25 in.

5–52.M A square stud 200 mm long and 8 mm on a side is made from titanium, Ti-6Al-4V, aged. What angle of twist will result when a wrench is applying a pure torque that causes the stress to equal the yield strength of the material in shear?

5–53.E A standard square structural steel tube has cross-sectional dimensions of $4 \times 4 \times \frac{1}{4}$ in and is 8.00 ft long. Compute the torque required to twist the tube 3.00 deg.

5–54.E For the tube in Problem 5–53, compute the maximum stress in the tube when it is twisted 3.00 deg. Would this be safe if the tube is made from ASTM A501 structural steel and the load was static?

5–55.E Repeat Problem 5–53 for a rectangular tube, $6 \times 4 \times \frac{1}{4}$.

5–56.E Repeat Problem 5–54 for a rectangular tube, $6 \times 4 \times \frac{1}{4}$.

5–57.E A standard 6-in schedule 40 steel pipe has approximately the same cross-sectional area as a square tube, $6 \times 6 \times \frac{1}{4}$, and thus both would weigh about the same for a given length. If the same torque were applied to both, compare the resulting torsional shear stress and angle of twist for the two shapes.

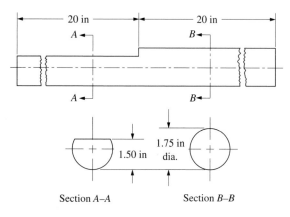

FIGURE 5–29 Problems 5–48 and 5–49.

COMPUTER PROGRAMMING ASSIGNMENTS

1. Given the need to design a solid circular shaft for a given torque, a given material yield strength, and a given design factor, compute the required diameter for the shaft.

Enhancements to Assignment 1

(a) For a given power transmitted and speed of rotation, compute the applied torque.

(b) Include a table of materials from which the designer can select one. Then automatically look up the yield strength.

(c) Include the design factor table, Table 5–1. Then prompt the designer to specify the manner of loading only and determine the appropriate design factor from the table within the program.

2. Repeat assignment 1, except design a hollow circular shaft. Three possible solution procedures exist:

(a) For a given outside diameter, compute the required inside diameter.

(b) For a given inside diameter, compute the required outside diameter.

(c) For a given ratio of D_i/D_o, find both D_i and D_o.

Enhancements to Assignment 2

(a) Compute the mass of the resulting design for a given length and material density.

(b) If the computer has graphics capability, draw the resulting cross section and dimension it.

3. Enter the stress concentration factor curves into a program, allowing the automatic computation of K_t for given factors such as fillet radius, diameter ratio, hole diameter, and so on. Any of the cases shown in Appendix A–21–5, A–21–6, or A–21–7 could be used. This program could be run by itself or made an enhancement of other stress analysis programs.

4. Compute the angle of twist from Equation (5–21) for given T, L, G, and J.

Enhancements to Assignment 4

(a) Compute J for given dimensions for the shaft, either solid or hollow.

(b) Include a table of values for G, from Table 5–3, in the program.

5. Compute the required diameter of a solid circular shaft to limit the angle of twist to a specified amount.

6. Compute the angle of twist for one end of a multisection shaft relative to the other, similar to Example Problem 5–13. Allow the lengths, diameters, materials, and torques to be different in each section.

7. Write a program to compute the values for the effective section modulus, Q, and the torsional stiffness constant, K, from Figure 5–18 for any or all cases.

Enhancement to Assignment 7

Determine the equations for C_1, C_2, C_3, and C_4 in terms of the ratio, h/r, for shafts with flats. Use a curve-fitting routine.

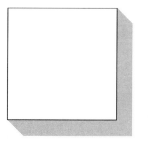

6

Shearing Forces and Bending Moments in Beams

6–1 OBJECTIVES OF THIS CHAPTER

Much of the discussion in the next six chapters deals with beams.

> *A beam is a member that carries loads transversely, that is, perpendicular to its long axis.*

Such loads cause *shearing stresses* to be developed in the beam and give the beam its characteristic *bent* shape, resulting in *bending stresses* as well.

To compute shearing stresses and bending stresses, it is necessary to be able to determine the magnitude of internal *shearing forces* and *bending moments* developed in beams due to a variety of loads. That is the primary objective of this chapter, after completion of which you should be able to:

1. Define the term *beam* and recognize when a load-carrying member is a beam.
2. Describe several kinds of beam loading patterns: *concentrated loads, uniformly distributed loads, linearly varying distributed loads,* and *concentrated moments.*
3. Describe several kinds of beams according to the manner of support: *simple beam, overhanging beam, cantilever,* and *composite beam* having more than one part.
4. Draw free-body diagrams for beams and parts of beams showing all external forces and reactions.

181

5. Compute the magnitude of reaction forces and moments and determine their directions.

6. Define *shearing force* and determine the magnitude of shearing force anywhere within a beam.

7. Draw free-body diagrams of *parts* of beams and show the internal shearing forces.

8. Draw complete shearing force diagrams for beams carrying a variety of loading patterns and with a variety of support conditions.

9. Define *bending moment* and determine the magnitude of bending moment anywhere within a beam.

10. Draw free-body diagrams of *parts* of beams and show the internal bending moments.

11. Draw complete bending moment diagrams for beams carrying a variety of loading patterns and with a variety of support conditions.

12. Use the *laws of beam diagrams* to relate the load, shear, and bending moment diagrams to each other and to draw the diagrams.

13. Draw free-body diagrams of parts of composite beams and structures and draw the shearing force and bending moment diagrams for each part.

14. Properly consider *concentrated moments* in the analysis of beams.

6–2 BEAM LOADING, SUPPORTS, AND TYPES OF BEAMS

Recall the definition of a beam.

A beam is a member that carries loads transversely, that is, perpendicular to its long axis.

When analyzing a beam to determine reactions, internal shearing forces, and internal bending moments, it is helpful to classify the manner of loading, the type of supports, and the type of beam.

Beams are subjected to a variety of loading patterns, including,

Normal concentrated loads

Inclined concentrated loads

Uniformly distributed loads

Varying distributed loads

Concentrated moments

Support types include,

Simple, roller-type support

Pinned support

Fixed support

Beam types include,

Simply supported beams; or simple beams

Overhanging beams

Cantilever beams; or cantilevers

Compound beams

Continuous beams

Understanding all of these terms will help you communicate features of beam designs and perform the required analyses. Brief descriptions of each will now be given with illustrations to help you visualize them.

Loading Patterns

In this section we will show that the nature of the loading pattern determines the variation of the shearing force and bending moment along the length of the beam. Here we define the five most frequently encountered loading patterns and give examples of each. More complex loading patterns can often be analyzed by considering them to be combinations of two or more of these basic types.

Normal concentrated loads

A normal concentrated load is one that acts perpendicular (normal) to the major axis of the beam at only a point or over a very small length of the beam.

Figure 6–1(a) shows the typical manner of representing a beam carrying normal concentrated loads. Each load is shown as a vector acting on the beam, perpendicular to its long

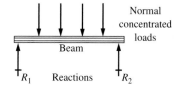

(*a*) Schematic representation of beam, loads, and reactions

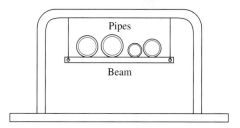

(*b*) Pictorial representation of beam and loads

FIGURE 6–1   Simple beam with normal concentrated loads.

axis. Part (b) is a sketch of a situation that would produce concentrated loads. The weight of the pipes and their contents determines the magnitudes of the loads. While we often visualize loads acting downward due to gravity, actual loads can act in any direction. Particularly in mechanical machinery, forces produced by linkages, actuators, springs, clamps, and other devices can act in any direction. Figure 6–2 shows a simple example.

Normal concentrated loads tend to cause pure bending of the beam. Most of the problems in this chapter will feature this type of loading. The analysis of the bending stresses produced is presented in Chapter 8.

Inclined concentrated loads

An inclined concentrated load is one that acts effectively at a point but whose line of action is at some angle to the main axis of the beam.

Figure 6–3 shows an example. The inclined load exerted by the spring causes a combination of bending and axial stresses to be developed in the beam. Chapter 11 presents the analysis techniques for such a loading.

Uniformly distributed loads

Loads of constant magnitude acting perpendicular to the axis of a beam over a significant part of the length of the beam are called uniformly distributed loads.

An example of this type of load would be the weight of wet snow of uniform depth on a roof carried by a flat, horizontal roof beam. Also, the materials that make up the roof

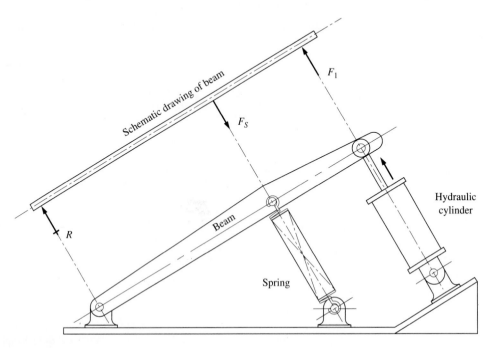

FIGURE 6–2 Machine lever as a simple beam carrying normal concentrated loads.

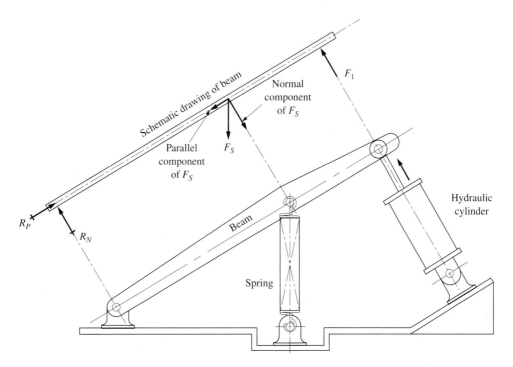

FIGURE 6–3 Machine lever as a simple beam carrying normal concentrated loads.

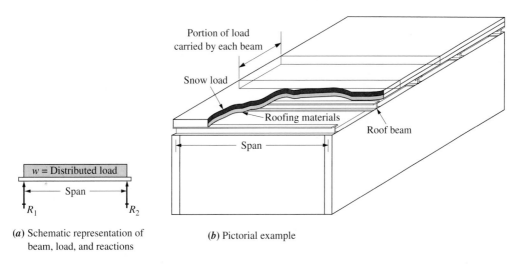

(*a*) Schematic representation of beam, load, and reactions

(*b*) Pictorial example

FIGURE 6–4 Simple beam with uniformly distributed load.

structure itself are often uniformly distributed. Figure 6–4 illustrates such a loading pattern and shows the manner we use in this book to represent uniformly distributed loads in problems. The gray shaded rectangular area defines the extent of the load along the length of the beam. The magnitude of the load is indicated by a "rate" of loading, w, in units of force per unit length. Typical units would be lb/in, kN/m, or K/ft. Recall that 1 K = 1 kip = 1000 lb. For example, if the loading on the beam shown in

Figure 6–4 was $w = 150$ lb/in, you should visualize that each 1.0 in of length of the beam carries 150 lb of load.

Varying distributed loads

Loads of varying magnitude acting perpendicular to the axis of a beam over a significant part of the length of the beam are called varying distributed loads.

Figures 6–5 and 6–6 show examples of structures carrying varying distributed loads. When the load varies linearly, we quantify such loads by giving the value of w at each end of the sloped line representing the load. For more complex, nonlinear variations, other schemes of giving the magnitude of the load would have to be devised.

Concentrated moments. A moment is an action that tends to cause rotation of an object. Moments can be produced by a pair of parallel forces acting in opposite

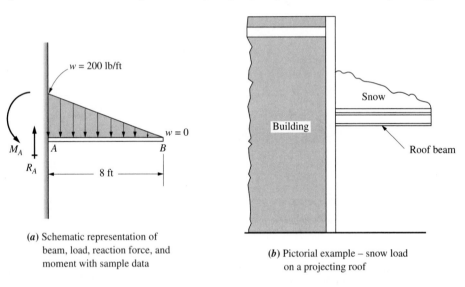

(*a*) Schematic representation of beam, load, reaction force, and moment with sample data

(*b*) Pictorial example – snow load on a projecting roof

FIGURE 6–5 Example of linearly varying distributed load on a cantilever.

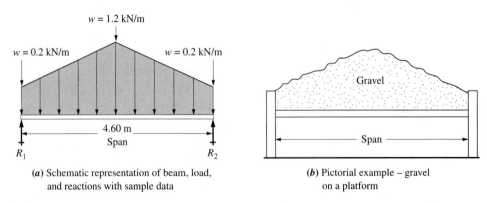

(*a*) Schematic representation of beam, load, and reactions with sample data

(*b*) Pictorial example – gravel on a platform

FIGURE 6–6 Example of linearly varying distributed load on a simple beam.

directions; this is called a *couple*. The action on a crank or a lever also produces a moment.

When a moment acts on a beam at a point in a manner that tends to cause it to undergo pure rotation, it is called a concentrated moment.

Figure 6–7 shows an example. The forces acting at the ends of the vertical arms form a couple and tend to twist the beam into the shape shown. The fact that the two forces making up the couple are equal and opposite results in no net horizontal force being applied to the beam.

Concentrated moments can also be applied to a beam by any force acting parallel to its axis with a line of action some distance from the axis. This is illustrated in Figure 6–8. The difference here is that there is an unbalanced horizontal force also applied to the beam.

Support Types

All beams must be supported in a stable manner to hold them in equilibrium. All externally applied loads and moments must be reacted by one or more supports. Different types of supports offer different types of reaction capability.

Simple support or roller

A simple support is one that can resist only forces acting perpendicular to the beam.

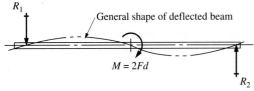

(*a*) Schematic representation of horizontal part of compound beam carrying a concentrated moment

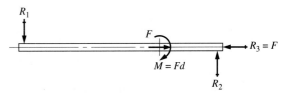

(*a*) Schematic representation of horizontal part of compound beam showing concentrated moment and horizontal reaction

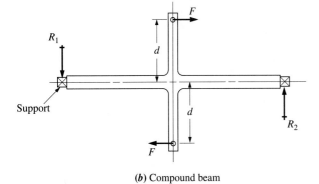

(*b*) Compound beam

FIGURE 6–7 Concentrated moment on a compound beam.

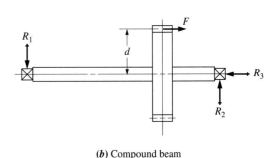

(*b*) Compound beam

FIGURE 6–8 Concentrated moment on a compound beam.

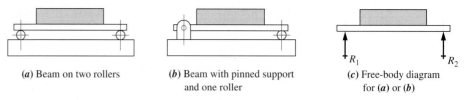

(a) Beam on two rollers

(b) Beam with pinned support and one roller

(c) Free-body diagram for (a) or (b)

FIGURE 6–9 Examples of simple supports.

One of the best illustrations of simple supports is the pair of theoretically frictionless rollers shown at the ends of the beam in Figure 6–9(a). They provide upward support against the downward action of the load on the beam. As the beam tends to bend under the influence of the applied loading and the reactions, the bending deformation would not be resisted by the rollers. But if there were any horizontal components of the load, the rollers would roll and the beam would be unrestrained. Therefore, using two rollers alone is not adequate.

Pinned support. An example of a pinned support is a hinge that can resist forces in two directions but which allows rotation about the axis of the pin in the hinge. Figure 6–9(b) shows the same beam as in Figure 6–9(a) with the roller at the left end replaced by a pinned support. This system would provide adequate support while allowing the beam to bend freely. Any horizontal force would be resisted in the pinned joint.

Fixed support

A fixed support is one that is held solidly such that it resists forces in any direction and also prohibits rotation of the beam at the support.

One manner of creating a fixed support is to produce a close-fitting, socket-like hole in a rigid structure into which the end of the beam is inserted. The fixed support resists moments as well as forces because it prevents rotation. Figure 6–10 shows two examples of the use of fixed supports.

Beam Types

The type of beam is indicated by the types of supports and their placement.

Simple beam. A simple beam is one that carries only loads acting perpendicular to its axis and that is supported only at its ends by simple supports acting perpendicular to the axis. Figure 6–1 is an example of a simple beam. When all loads act downward, the beam would deflect in the classic bent, concave upward shape. This is referred to as positive bending.

Overhanging beam. An overhanging beam is one in which the loaded beam extends outside the supports. Figure 6–11 is an example. The loads on the overhangs tend to bend them downward, producing negative bending.

Cantilever. A cantilever has only one end supported, as shown in Figure 6–12 depicting a crane boom attached rigidly to a stiff vertical column. It is essential that the support be fixed because it must provide vertical support for the externally applied loads

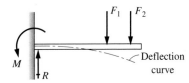

(a) Schematic representation of fixed support for a cantilever

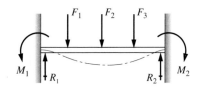

(c) Schematic representation of beam with two fixed supports

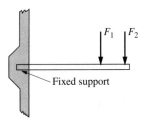

(b) Pictorial representation

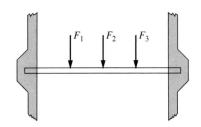

(d) Pictorial representation

FIGURE 6–10 Beams with fixed supports.

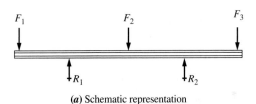

(a) Schematic representation

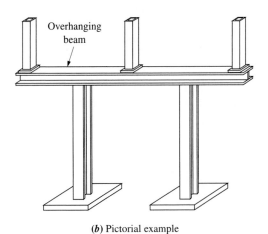

(b) Pictorial example

FIGURE 6–11 Overhanging beam.

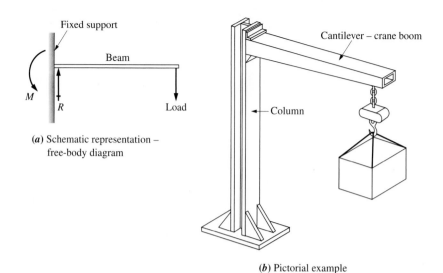

(a) Schematic representation –
free-body diagram

(b) Pictorial example

FIGURE 6–12 Cantilever beam.

along with a moment reaction to resist the moment produced by the loads. Figures 6–5 and 6–10(a) are other examples of cantilevers.

Compound beam.   While the beams depicted so far were single, straight members, we will use the term *compound beam* to refer to one having two or more parts extending in different directions. Figures 6–7 and 6–8 are examples. Such beams are typically analyzed in parts to determine the internal shearing forces and bending moments throughout. Often, the place where one part joins another is a critical point of interest.

Continuous beam.   Except for the beam shown in Figure 6–10, parts (c) and (d), beams discussed before had one or two supports and only two unknown reactions. The principles of statics allow us to compute all reaction forces and moments from the classical equations of equilibrium because there are two unknowns and two independent equations available from which to solve for those unknowns. Such beams are called *statically determinate*. In contrast, *continuous beams* have additional supports, requiring different approaches to analyze the reaction forces and moments. These are called *statically indeterminate* beams. Figure 6–13 shows an example of a continuous beam on three supports. Figure 6–14 shows a beam with one fixed end and one simply sup-

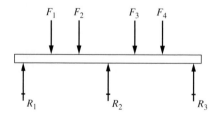

FIGURE 6–13 Continuous beam on three supports.

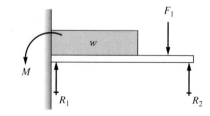

FIGURE 6–14 Supported cantilever.

ported end. Notice that there are two unknown forces and one unknown moment. Figures 6-10(c) and (d) show a beam with two fixed ends which is also statically indeterminate because there are two reaction forces and two moments that must be determined. Chapter 13 presents the analysis techniques for statically indeterminate beams.

6-3 BEAM SUPPORTS AND REACTIONS AT SUPPORTS

The first step in analyzing a beam to determine its safety under a given loading arrangement is to show completely the loads and support reactions on a free-body diagram. It is very important to be able to construct free-body diagrams from the physical picture or description of the loaded beam. This was done in each case for Figures 6-1 through 6-14.

After constructing the free-body diagram, it is necessary to compute the magnitude of all support reactions. It is assumed that the methods used to find reactions were studied previously. Therefore, only a few examples are shown here as a review and as an illustration of the techniques used throughout this book.

The following general procedure is recommended for solving for reactions on simple or overhanging beams.

Guidelines for solving for reactions

1. Draw the free-body diagram.
2. Use the equilibrium equation $\Sigma M = 0$ by summing moments about the point of application of one support reaction. The resulting equation can then be solved for the other reaction.
3. Use $\Sigma M = 0$ by summing moments about the point of application of the second reaction to find the first reaction.
4. Use $\Sigma F = 0$ to check the accuracy of your calculations.

Example Problem 6-1

Figure 6-15 shows the free-body diagram for the beam carrying pipes, which was originally shown in Figure 6-1. Compute the reaction forces in the support rods.

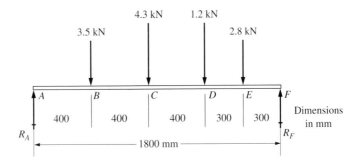

FIGURE 6-15 Beam loading.

Solution **Objective** Compute the reaction forces at the ends of the beam.

Given Pictorial of the beam shown in Figure 6–1. Free-body diagram showing the loading is Figure 6–15. Loads are applied at points labeled *B*, *C*, *D*, and *E*. Reactions act at points *A* and *F* and are called R_A and R_F.

Analysis The *Guidelines for solving for reactions* will be used to compute the reactions. Figure 6–15 is the free-body diagram, so we will start with Step 2.

Results To find the reaction R_F, sum moments about points *A*.

$$\Sigma M_A = 0 = 3.5(400) + 4.3(800) + 1.2(1200) + 2.8(1500) - R_F(1800)$$

Note that all forces are in kilonewtons and distances in millimeters. Now solve for R_F.

$$R_F = \frac{3.5(400) + 4.3(800) + 1.2(1200) + 2.8(1500)}{1800} = 5.82 \text{ kN}$$

Now, to find R_A, sum moments about point *F*.

$$\Sigma M_F = 0 = 2.8(300) + 1.2(600) + 4.3(1000) + 3.5(1400) - R_A(1800)$$

$$R_A = \frac{2.8(300) + 1.2(600) + 4.3(1000) + 3.5(1400)}{1800} = 5.98 \text{ kN}$$

Now use $\Sigma F = 0$ for the vertical direction as a check.

Downward forces: $(3.5 + 4.3 + 1.2 + 2.8)$ kN $= 11.8$ kN
Upward reactions: $(5.82 + 5.98)$ kN $= 11.8$ kN (check)

Comment Remember to show the reaction forces R_A and R_F at their proper points on the beam.

Example Problem 6–2 Compute the reactions on the beam shown in Figure 6–16(a).

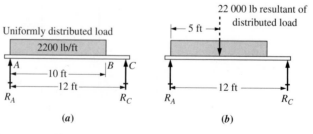

    (*a*)   (*b*)

FIGURE 6–16 Beam loading.

Solution **Objective** Compute the reaction forces at the ends of the beam.

Given Pictorial of the beam shown in Figure 6–16(a). Distributed load of 2200 lb/ft is applied over 10 ft at the left end of the beam. Reactions act at points A and C and are called R_A and R_C.

Analysis The *Guidelines for solving for reactions* will be used to compute the reactions. Figure 6–16(b) is an equivalent free-body diagram with the resultant of the distributed load shown acting at the centroid of the load.

Results $\sum M_A = 0 = 22\,000\text{ lb (5 ft)} - R_C(12\text{ ft})$

$$R_C = \frac{22\,000\text{ lb (5 ft)}}{12\text{ ft}} = 9167\text{ lb}$$

$\sum M_C = 0 = 22\,000\text{ lb (7 ft)} - R_A(12\text{ ft})$

$$R_A = \frac{22\,000\text{ lb (7 ft)}}{12\text{ ft}} = 12\,833\text{ lb}$$

Finally, as a check, in the vertical direction,

$$\sum F = 0$$

Downward forces: 22 000 lb

Upward forces: $R_A + R_C = 12\,833 + 9167 = 22\,000\text{ lb}$ (check)

Comment Note that the resultant is used only to find reactions. Later, when we find shearing forces and bending moments, the distributed load itself must be used.

Example Problem 6–3 Compute the reactions for the overhanging beam in Figure 6–17.

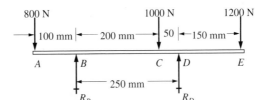

FIGURE 6–17 Beam loading.

Solution **Objective** Compute the reaction forces at points B and D.

Given Loading on the beam shown in Figure 6–17. Reactions are R_A and R_C.

Analysis The *Guidelines for solving for reactions* will be used to compute the reactions.

Results First, summing moments about point B,

$$\sum M_B = 0 = 1000(200) - R_D(250) + 1200(400) - 800(100)$$

Notice that forces that tend to produce clockwise moments about B are considered positive in this calculation. Now solving for R_D gives

$$R_D = \frac{1000(200) + 1200(400) - 800(100)}{250} = 2400 \text{ N}$$

Summing moments about D will allow computation of R_B.

$$\Sigma M_D = 0 = 1000(50) - R_B(250) + 800(350) - 1200(150)$$

$$R_B = \frac{1000(50) + 800(350) - 1200(150)}{250} = 600 \text{ N}$$

Check with $\Sigma F = 0$ in the vertical direction:

Downward forces: $(800 + 1000 + 1200) \text{ N} = 3000 \text{ N}$

Upward forces: $R_B + R_D = (600 + 2400) \text{ N} = 3000 \text{ N}$ (check)

Example Problem 6–4   Compute the reactions for the cantilever beam in Figure 6–18.

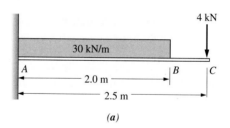

(a)

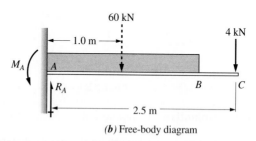

(b) Free-body diagram

FIGURE 6–18   Beam loading.

Solution   **Objective**   Compute the reactions at point A at the wall.

Given   Loading on the beam shown in Figure 6–18(a).

Analysis   The *Guidelines for solving for reactions* will be used.

Chapter 6 ▪ Shearing Forces and Bending Moments in Beams

In the case of cantilever beams, the reactions at the wall are composed of an upward force R_A which must balance all downward forces on the beam and a reaction moment M_A which must balance the tendency for the applied loads to rotate the beam. These are shown in Figure 6–18(b). Also shown is the resultant, 60 kN, of the distributed load.

Results   Then, by summing forces in the vertical direction, we obtain

$$R_A = 60 \text{ kN} + 4 \text{ kN} = 64 \text{ kN}$$

Summing moments about point A yields

$$M_A = 60 \text{ kN } (1.0 \text{ m}) + 4 \text{ kN } (2.5 \text{ m}) = 70 \text{ kN·m}$$

6–4 SHEARING FORCES

It will be shown later that the two kinds of stresses developed in a beam are shearing stresses and bending stresses. In order to compute these stresses, it will be necessary to know the magnitude of shearing forces and bending moments at all points in the beam. Therefore, although you may not yet understand the ultimate use of these factors, it is necessary to learn how to determine the variation of shearing forces and bending moments in beams for many types of loading and support combinations.

We define shearing forces as follows:

Shearing forces are internal forces developed in the material of a beam to balance externally applied forces in order to ensure equilibrium in all parts of the beam.

The presence of shearing forces can be visualized by considering any segment of the beam as a free body with all external loads applied. Figure 6–19 shows an example. The beam as a whole is in equilibrium under the action of the 1000-N load and the two 500-N reaction forces at the supports. And, any segment of the beam must also be in equilibrium.

A segment is made by making a cut at a section of interest and looking at the part of the beam to one side of the cut. Normally, we will consider the segment of interest to be to the left of the cut as shown in Figure 6–19(b) for a segment having a length of 0.5 m. Then, for the segment to be in equilibrium, there must, in general, be an *internal* force acting perpendicular to the axis of the beam at the cut section. In this case, the internal force must be 500 N acting downward. This is the shearing force and we will use the symbol V for it. That is, $V = 500$ N. We can generalize this process for finding shearing forces by stating the following rule:

> The magnitude of the shearing force in any part of a beam is equal to the algebraic sum of all external forces acting to the left of the section of interest.

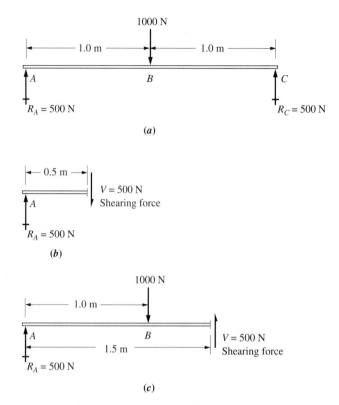

FIGURE 6–19 Use of free-body diagrams to find shearing forces in beams.

Note that, while the free-body diagram in Figure 6–19(b) is in equilibrium with respect to vertical forces, it is *not* yet in equilibrium with regard to rotation. The reaction force R_A and the shearing force V form a couple that tends to rotate the segment clockwise. We will show in the next section that there must also be an internal moment, called a *bending moment,* to maintain equilibrium.

Continuing the discussion of shearing forces, note that for any segment of the beam in Figure 6–19 taken from the left reaction at A up to the point of application of the 1000-N load at B, the free-body diagram would look like that in part (b) of the figure. Thus, the shearing force at any point in the beam from A to B would be 1000 N.

Now consider a segment of the beam 1.5 m long, as shown in Figure 6–19(c). For that part to be in equilibrium, an internal shearing force of 500 N upward must exist in the beam. This would be the same if the beam were cut anywhere between B and C.

Shearing force diagrams.   It is useful to plot the values of shearing force versus position on the beam in the manner shown in Figure 6–20. Such a plot is called a *shearing force diagram* and the following is a discussion of the method for creating it. General rules are also stated for constructing the diagram for any beam carrying only normal concentrated loads.

- The shearing force diagram is a graph with the vertical axis depicting the value of the shearing force at any section of the beam. This axis should be labeled as

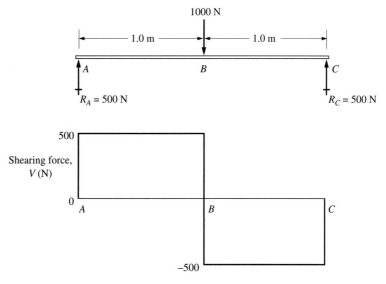

FIGURE 6–20 Shearing force diagram.

shown in Figure 6–20, giving the name of the quantity being plotted, *shearing force,* its symbol, *V,* and the units, in this case newtons (N). The horizontal axis gives the position on the beam and it is customary to draw it parallel to the drawing of the beam so the correspondence between the actual beam loading and the shearing forces can be seen.

- If any part of the beam extends to the left of the reaction at *A,* the shearing force would be zero because there would be no external force. The same can be stated about points to the right of point *C* at the right end of the beam. Therefore, a general rule is:

> Shearing force diagrams start and end at zero at the ends of the beam.

- Then, at *A* where the left reaction acts, the internal shearing force abruptly changes to 500 N acting downward to balance the upwardly directed reaction force. We will adopt the following sign convention for shearing forces:

> Internal shearing forces acting downward are considered to be positive.
> Internal shearing forces acting upward are considered to be negative.

Then the shearing force diagram abruptly rises from zero to 500 N at *A.* This can be stated mathematically as

$$V_A = 0 + 500 \text{ N} = 500 \text{ N}$$

A general rule is:

An upwardly directed concentrated load or reaction causes an abrupt equal increase in the value of the shearing force.

- As demonstrated in Figure 6–19(b), the shearing force remains at the value of 500 N anywhere between A and B. This is because there are no additional external loads applied. We can state this observation in the form,

$$V_{A-B} = 500 \text{ N}$$

The subscript, $A-B$, indicates that the value is for the entire part of the beam from A to B. The general rule is:

On any segment of the beam where no loads are applied, the value of the shearing force remains constant, resulting in a straight horizontal line on the shearing force diagram.

- At point B, where the 1000 N load is applied, it was shown in Figure 6–19(c) that the internal shearing force abruptly changed from 500 N downward (positive) to 500 N upward (negative). The total change in shearing force is 1000 N. That is,

$$V_B = 500 \text{ N} - 1000 \text{ N} = -500 \text{ N}$$

The general rule is:

A concentrated load on a beam causes an abrupt change in the shearing force in the beam by an amount equal to the magnitude of the load and in the direction of the load.

- Between B and C, there are no applied loads so the shearing force diagram plots as a straight horizontal line at -500 N. That is,

$$V_{B-C} = -500 \text{ N}$$

- At C the upwardly directed reaction force of 500 N causes an abrupt change in the value of the shearing force of the same amount, bringing the plot back to zero. That is,

$$V_C = -500 \text{ N} + 500 \text{ N} = 0$$

This is consistent with the first rule stated earlier.

The construction of the shearing force diagram is now complete. While the production of the free-body diagrams of segments of the beam was useful in developing the concept of shearing force, it is not generally necessary to do that. The rules just given can be summarized as a set of general guidelines for drawing shearing force diagrams.

Guidelines for drawing shearing force diagrams for beams carrying normal concentrated loads

1. Draw vertical and horizontal axes for the diagram in relation to the drawing of the beam loading diagram in a manner similar to that shown in Figure 6–20.
2. Label the vertical axis as *Shearing force, V,* and give the units for force.
3. Project lines from each applied load or reaction force on the beam loading diagram down to the shearing force diagram. Label points of interest for reference. We will label points where loads or reaction forces act with letters, starting from the left end of the beam.
4. Construct the shearing force graph by starting at the left end of the beam and proceeding to the right, applying the following rules.
5. *Shearing force diagrams start and end at zero at the ends of the beam.*
6. *An upwardly directed concentrated load or reaction causes an abrupt equal increase in the value of the shearing force.*
7. *On any segment of the beam where no loads are applied, the value of the shearing force remains constant, resulting in a straight horizontal line on the shearing force diagram.*
8. *A concentrated load on a beam causes an abrupt change in the shearing force in the beam by an amount equal to the magnitude of the load and in the direction of the load.*
9. Show the value of the shearing force at key points on the diagram, generally at points where forces or reactions act.

Looking at the complete shearing force diagram of Figure 6–20, it can be seen that the greatest value of shear is 500 N. Notice that even though there is an applied load of 1000 N, the maximum shearing force in the beam is only 500 N.

Another beam carrying concentrated loads will now be considered in an example problem. The general approach used earlier will be applicable to any beam carrying concentrated loads.

Example Problem 6–5   Draw the complete shearing force diagram for the beam shown in Figure 6–21.

Solution   **Objective**   Draw the complete shearing force diagram.

Given   The beam loading, including the values of the reaction forces, is shown in Figure 6–21. It is a simply supported beam carrying all normal concentrated loads. This is the same beam for which the reaction forces were computed in Example Problem 6–1.

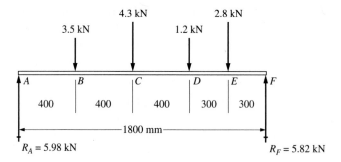

FIGURE 6–21   Beam loading.

Analysis   *The guidelines for drawing shearing force diagrams for beams carrying normal concentrated loads* will be used.

Results   The completed diagram is shown in Figure 6–22. The process used to determine each part of the diagram is described below.

Point A: The reaction R_A is encountered immediately. Then,

$$V_A = R_A = 5.98 \text{ kN}$$

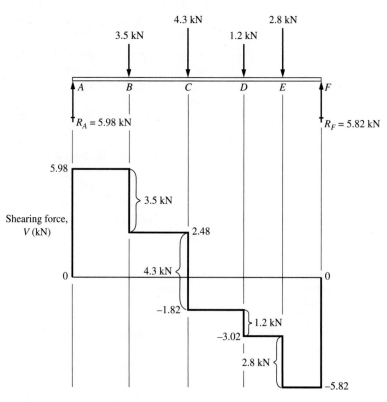

FIGURE 6–22   Shearing force diagram.

Between A and B: Because there are no loads applied, the shearing force remains constant. That is,

$$V_{A-B} = 5.98 \text{ kN}$$

Point B: The applied load of 3.5 kN causes an abrupt decrease in V.

$$V_B = 5.98 \text{ kN} - 3.5 \text{ kN} = 2.48 \text{ kN}$$

Between B and C: The shearing force remains constant.

$$V_{B-C} = 2.48 \text{ kN}$$

Point C: The applied load of 4.3 kN causes an abrupt decrease in V.

$$V_C = 2.48 \text{ kN} - 4.3 \text{ kN} = -1.82 \text{ kN}$$

Between C and D: The shearing force remains constant.

$$V_{C-D} = -1.82 \text{ kN}$$

Point D: The applied load of 1.2 kN causes an abrupt decrease in V.

$$V_D = -1.82 \text{ kN} - 1.2 \text{ kN} = -3.02 \text{ kN}$$

Between D and E: The shearing force remains constant.

$$V_{D-E} = -3.02 \text{ kN}$$

Point E: The applied load of 2.8 kN causes an abrupt decrease in V.

$$V_E = -3.02 \text{ kN} - 2.8 \text{ kN} = -5.82 \text{ kN}$$

Between E and F: The shearing force remains constant.

$$V_{E-F} = -5.82 \text{ kN}$$

Point F: The reaction force of 5.82 kN causes an abrupt increase in V.

$$V_F = -5.82 \text{ kN} + 5.82 \text{ kN} = 0$$

Comment   Note that values for the shearing forces at key points are shown right on the diagram at those points.

Shearing force diagrams for distributed loads.   The variation of shearing force with position on the beam for distributed loads is different from that for concentrated loads. The free-body diagram method is useful to help visualize these variations.

Consider the beam in Figure 6–23, carrying a uniformly distributed load of 1500 N/m over part of its length. It is desired to determine the magnitude of the shearing force at several points in the beam in order to draw a shearing force diagram. Let's

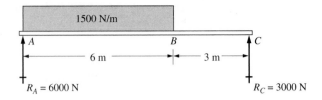

FIGURE 6–23 Beam loading.

choose points at intervals of 2 m across the beam, including the two ends of the beam. At the left end of the beam, just to the right of point A, the shearing force in the beam must be 6000 N in order to balance the reaction force R_A. Now, if a segment of the beam 2 m long was considered to be cut from the rest of the beam, the free-body diagram shown in Figure 6–24(a) would be obtained. There must be a shearing force in the beam to balance the external forces in order for the segment to be in equilibrium. The reaction R_A of 6000 N acts upward, while the total distributed load of 3000 N acts downward. The shearing force V must be 3000 N downward.

Analyzing a segment 4 m long in a similar manner would produce the free-body diagram shown in Figure 6–24(b). In this case $V = 0$, since the external loads themselves are balanced.

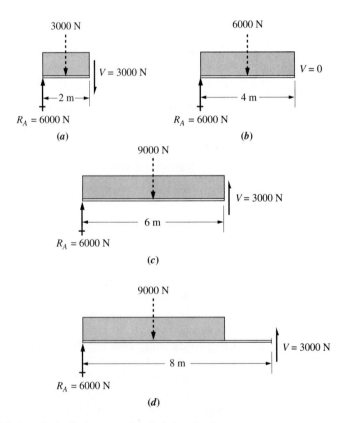

FIGURE 6–24 Free-body diagrams used to find shearing forces.

For a segment 6 m long, Figure 6–24(c) would be the free-body diagram. Now a shearing force V of 3000 N upward must exist. At 8 m, Figure 6–24(d) shows that $V = 3000$ N upward again. This would be true throughout the last 3 m of the length of the beam, since there are no external loads applied in this part.

In summary, the shearing forces calculated were:

$$\begin{aligned}
\text{At point } A:\quad & V = 6000 \text{ N downward} \\
\text{At 2 m:}\quad & V = 3000 \text{ N downward} \\
\text{At 4 m:}\quad & V = 0 \\
\text{At 6 m:}\quad & V = 3000 \text{ N upward} \\
\text{Between } B \text{ and } C:\quad & V = 3000 \text{ N upward}
\end{aligned}$$

By convention, downward shearing forces are considered positive, and upward shearing forces are negative. If these values are plotted on a graph of shearing force versus position on the beam, the shearing force diagram shown in Figure 6–25 would be produced. Notice that for the portion of the beam carrying the uniformly distributed load, the shearing force curve is a straight line. This is typical for such loads. The following general rules can be derived from this example. For the part of a beam carrying a uniformly distributed load:

---

1. Over that part of a beam carrying a uniformly distributed load, the shearing force diagram is a straight line.
2. The *change in shearing force* between any two points is equal to the area under the load diagram between those points.
3. The slope of the straight-line shearing force curve is equal to the rate of loading on the beam, that is, the load per unit length.

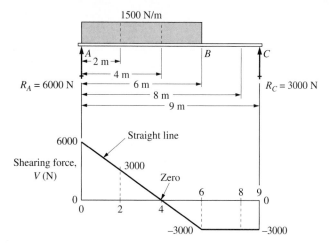

FIGURE 6–25 Shearing force diagram.

In the present example, rule 2 is illustrated by noting that between points A and B the shearing force decreases by 9000 N, from positive 6000 N to negative 3000 N. This is the area under the load curve as computed from

$$(1500 \text{ N/m}) (6 \text{ m}) = 9000 \text{ N}$$

Rule 3 states that the shearing force decreases 1500 N over each meter of length of the beam.

The general principles developed for both concentrated loads and distributed loads must be applied in the solution of some problems. The following steps should be applied:

Guidelines for drawing shearing force diagrams

1. Solve for the reaction forces at the supports.
2. Make a sketch of the beam. It helps to make the sketch approximately to scale.
3. Draw lines vertically down from key points on the loaded beam to the area below, where the shearing force diagram will be drawn.
4. Draw the horizontal axis of the shearing force diagram equal to the length of the beam. Label the vertical shear axis, showing the units for the shearing forces to be plotted.
5. Starting at the left end of the beam, plot the variation in shearing force across the entire beam. Remember that:
   a. The shearing force changes abruptly at each point where a concentrated load is applied. The change in shearing force is equal to the load.
   b. The shearing force curve is a straight, horizontal line between points where no loads are applied.
   c. The shearing force curve is a straight, inclined line between points where uniformly distributed loads are applied. The slope of the line is equal to the rate of loading.
   d. The change in shearing force between points equals the area under the load curve between the points.
6. Show the value of the shearing force at each point where major changes occur, such as concentrated loads and at the beginning and end of distributed loads.

6–5 BENDING MOMENTS

Bending moments, in addition to shearing forces, are developed in beams as a result of the application of loads acting perpendicular to the beam. It is these bending moments that cause the beam to assume its characteristic curved, or "bent," shape. Pushing on the middle of a thin stick, such as a ruler supported at its ends, illustrates this.

The determination of the magnitude of bending moments in a beam is another application of the principle of static equilibrium. In the preceding section, we analyzed the forces in the vertical direction to determine the shearing forces in the beam that must be developed to maintain all parts of the beam in equilibrium. To do this, it was helpful to consider parts of the beam as free bodies in order to visualize what happens inside the beam. A similar approach helps illustrate bending moments.

Figure 6–26 shows a simply supported beam carrying a concentrated load at its center. The entire beam is in equilibrium, and so is any part of it. Look at the free-body diagrams shown in parts (b), (c), (d), and (e) of Figure 6–26. By summing moments about the point where the beam is cut, the magnitude of the bending moment inside required to keep the segment in equilibrium can be found. The first 0.5-m segment is shown in Figure 6–26(b). Summing moments about point B gives

$$M_B = 500 \text{ N } (0.5 \text{ m}) = 250 \text{ N·m}$$

The shearing force, shown earlier in Figure 6–19, is also shown.

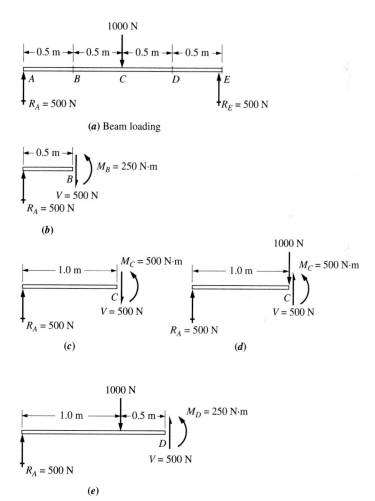

FIGURE 6–26 Free-body diagrams used to find bending moments.

In Figure 6–26(c), a segment 1.0 m long, which includes the left half of the beam but *not* the 1000-N load, is drawn as a free body. Summing moments about C gives

$$M_C = 500 \text{ N } (1.0 \text{ m}) = 500 \text{ N·m}$$

If the 1000-N load had been considered as shown in Figure 6–26(d), the result would be the same, since the load acts right at point C and, therefore, has no moment about that point.

Figure 6–26(e) shows 1.5 m of the beam isolated as a free body. Summing moments about point D gives

$$M_D = 500 \text{ N } (1.5 \text{ m}) - 1000 \text{ N } (0.5 \text{ m}) = 250 \text{ N·m}$$

If we consider the entire beam as a free body and sum moments about point E at the right end of the beam, we will get

$$M_E = 500 \text{ N } (2.0 \text{ m}) - 1000 \text{ N } (1.0 \text{ m}) = 0$$

A similar result would be obtained for point A at the left end. In fact, a general rule is:

> The bending moments at the ends of a simply supported beam are zero.

In summary, for the beam in Figure 6–26, the bending moments are:

Point A:	0
Point B:	250 N·m
Point C:	500 N·m
Point D:	250 N·m
Point E:	0

Figure 6–27 shows these values plotted on a bending moment diagram below the shear diagram developed earlier for the same beam. Notice that between A and C the bending moment values fall on a straight line. Similarly, between C and E, the points fall on a straight line. This is typical for segments of beams carrying only concentrated loads. A general rule, then, is,

> The bending moment curve will be a straight line over those segments where the shearing force curve has a constant value.

Figure 6–27 also illustrates another general rule:

> The change in moment between two points on a beam is equal to the area under the shearing force curve between the same two points.

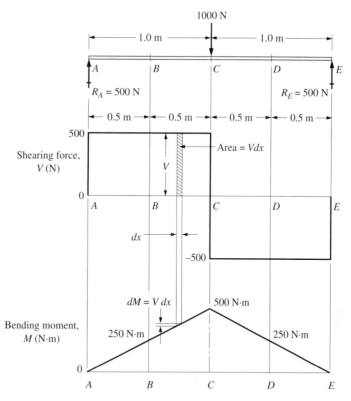

FIGURE 6–27 Shearing force and bending moment diagrams.

This rule, called the *area rule,* can be applied over a segment of any length in a beam to determine the change in bending moment.

Development of the Area Rule for Bending Moment Diagrams.
In general, the area rule can be stated

$$dM = Vdx \qquad (6-1)$$

where dM is the change in moment due to a shearing force V acting over a small length segment dx. Figure 6–27 illustrates Equation (6–1).

Over a larger length segment, the process of integration can be used to determine the total change in moment over the segment. Between two points, A and B,

$$\int_{M_A}^{M_B} dM = \int_{x_A}^{x_B} Vdx \qquad (6-2)$$

If the shearing force V is constant over the segment, as it is in segment A-B in Figure 6–27, Equation (6–2) becomes

$$\int_{M_A}^{M_B} dM = V\int_{x_A}^{x_B} dx \qquad (6-3)$$

Completing the integration gives

$$M_B - M_A = V(x_B - x_A) \qquad (6\text{--}4)$$

This result is consistent with the rule stated above. Note that $M_B - M_A$ is the *change* in moment between points A and B. The right side of Equation (6–4) is the area under the shearing force curve between A and B.

Illustration of the Area Rule. Once the principle behind the area rule is understood, it is not necessary to perform the process of integration to solve problems where the areas can be computed by simple geometry. Using the data from Figure 6–27, for example, between A and B,

$$M_B - M_A = V(x_B - x_A) = (500 \text{ N})(0.5 \text{ m} - 0) = 250 \text{ N·m}$$

That is, the bending moment increased by 250 N·m over the span A-B. But at A the moment $M_A = 0$. Then

$$M_B = M_A + 250 \text{ N·m} = 0 + 250 \text{ N·m} = 250 \text{ N·m}$$

Similarly, between B and C,

$$M_C - M_B = V(x_C - x_B) = (500 \text{ N})(1.0 \text{ m} - 0.5 \text{ m}) = 250 \text{ N·m}$$

Then

$$M_C = M_B + 250 \text{ N·m}$$

But $M_B = 250$ N·m. Then

$$M_C = 250 \text{ N·m} + 250 \text{ N·m} = 500 \text{ N·m}$$

Between C and D, note that $V = -500$ N. Then

$$M_D - M_C = V(x_D - x_C) = (-500 \text{ N})(1.5 \text{ m} - 1.0 \text{ m}) = -250 \text{ N·m}$$
$$M_D = M_C - 250 \text{ N·m} = 500 \text{ N·m} - 250 \text{ N·m} = 250 \text{ N·m}$$

Betwee D and E,

$$M_E - M_D = V(x_E - x_D) = (-500 \text{ N})(2.0 \text{ m} - 1.5 \text{ m}) = -250 \text{ N·m}$$
$$M_E = M_D - 250 \text{ N·m} = 250 \text{ N·m} - 250 \text{ N·m} = 0$$

These results are identical to those found by the free-body diagram method. We will use the area rule for generating the bending moment diagram from the known shearing force diagram for the remaining problems in this section and whenever area calculations can simply be made.

Draw the complete bending moment diagram for the beam shown in Figure 6–21. Show the values for the moment at points *A, B, C, D, E,* and *F.*

Solution

Objective Draw the complete bending moment diagram.

Given Beam loading in Figure 6–21.

Analysis The *area rule* and the other two rules defined above will be used. This requires that the shearing force curve be produced, and Example Problem 6–5 has already done that with the result shown in Figure 6–22. Now, Figure 6–28 includes the shearing force diagram along with the display of the results of the bending moment diagram, the details of which are described next.

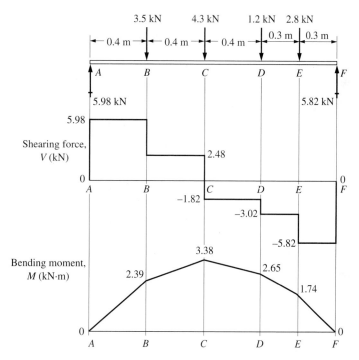

FIGURE 6–28 Shearing force and bending moment diagrams.

Results The bending moment diagram is drawn directly below the shearing force diagram so the relationship between points on the beam loading diagram can be related to both. It is helpful to extend vertical lines from the points of interest on the beam (points *A* to *F* in this example) down to the horizontal axis of the bending moment diagram.

It is most convenient to start drawing the bending moment diagram at the left end of the beam and work toward the right, considering each segment separately. For beams like this one having all concentrated loads, the segments should be chosen as those for which the shearing force is constant. In this case, the segments are *AB, BC, CD, DE,* and *EF.*

At point A: We use the rule that the bending moment is zero at the ends of a simply supported beam. That is, $M_A = 0$.

Point B: For this and each subsequent point, we apply the area rule. The general pattern is

$$M_B = M_A + [\text{Area}]_{AB}$$

where $[\text{Area}]_{AB}$ = area under the shearing force curve between A and B. Using data from the shearing force curve,

$$[\text{Area}]_{AB} = V_{AB} \times \text{width of the segment } AB$$

But, $V_{AB} = 5.98$ kN over the segment AB that is 0.40 m long. Then,

$$[\text{Area}]_{AB} = 5.98 \text{ kN}(0.40 \text{ m}) = 2.39 \text{ kN·m}$$

Finally, then,

$$M_B = M_A + [\text{Area}]_{AB} = 0 + 2.39 \text{ kN·m} = 2.39 \text{ kN·m}$$

This value is plotted at point B on the bending moment diagram. Then a straight line is drawn from M_A to M_B because the shearing force is constant over that segment. Bending moment values at C, D, E, and F are found in a similar manner.

Point C: $M_C = M_B + [\text{Area}]_{BC}$

$\qquad [\text{Area}]_{BC} = 2.48 \text{ kN}(0.40 \text{ m}) = 0.99 \text{ kN·m}$

$\qquad M_C = 2.39 \text{ kN·m} + 0.99 \text{ kN·m} = 3.38 \text{ kN·m}$

Point D: $M_D = M_C + [\text{Area}]_{CD}$

$\qquad [\text{Area}]_{CD} = -1.82 \text{ kN}(0.40 \text{ m}) = -0.73 \text{ kN·m}$

$\qquad M_D = 3.38 \text{ kN·m} - 0.73 \text{ kN·m} = 2.65 \text{ kN·m}$

Note that $[\text{Area}]_{CD}$ is negative because it is below the axis.

Point E: $M_E = M_D + [\text{Area}]_{DE}$

$\qquad [\text{Area}]_{DE} = -3.02 \text{ kN}(0.30 \text{ m}) = -0.91 \text{ kN·m}$

$\qquad M_E = 2.65 \text{ kN·m} - 0.91 \text{ kN·m} = 1.74 \text{ kN·m}$

Point F: $M_F = M_E + [\text{Area}]_{EF}$

$\qquad [\text{Area}]_{EF} = -5.82 \text{ kN}(0.30 \text{ m}) = -1.74 \text{ kN·m}$

$\qquad M_F = 1.74 \text{ kN·m} - 1.74 \text{ kN·m} = 0 \text{ kN·m}$

Summary and comments The bending moment values are shown on the diagram at their respective points so users of the diagram can see the relative values. The fact that $M_F = 0$ is a check on the calculations because the rule for simply supported beams states that the bending moment at F must be zero. The objective of drawing the bending moment diagram is often to locate the point where the maximum bending moment exists. Here we see that $M_C = 3.38$ kN·m is the maximum value.

Chapter 6 ▪ Shearing Forces and Bending Moments in Beams

Maximum bending moment rule. Example Problem 6–6 illustrated a useful rule that can be stated as,

> The maximum bending moment will occur at a point where the shearing force curve crosses the horizontal axis.

The area rule leads to this rule. To illustrate, in Example Problem 6–6, the areas under the shearing force curve in the first two segments are positive (above the axis) and, therefore, the bending moment is *increasing* up to point C. But the areas to the right of point C are negative (below the axis) and the bending moment *decreases*. Therefore, the maximum bending moment must occur at point C. In cases where the shearing force curve crosses the axis more than once, all points where the curve crosses the axis must be investigated to determine which is the largest.

Bending moment diagrams for distributed loads. The preceding examples showed the computation of bending moments and the plotting of bending moment diagrams for beams carrying only concentrated loads. Now we will consider distributed loads. The free-body diagram approach will be used again as an aid to visualizing the variation in bending moment as a function of position on the beam.

The beam shown in Figure 6–29 will be used to illustrate the typical results for distributed loads. This is the same beam for which the shearing force was determined, as shown in Figures 6–23 to 6–25. The free-body diagrams for segments of the beam taken in increments of 2 m will be used to compute the bending moments (refer to Figure 6–30).

For a segment at the left end of the beam, 2 m long, we can determine the bending moment in the beam by summing moments about that point due to all external loads to the left of the section, as shown in Figure 6–23(a). Notice that the resultant of the distributed load is shown to be acting at the middle of the 2-m segment. Then because the segment is in equilibrium,

$$M_2 = 6000 \text{ N (2 m)} - 3000 \text{ N (1 m)} = 9000 \text{ N·m}$$

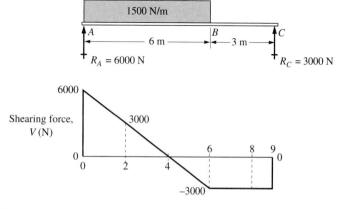

FIGURE 6–29 Beam loading and shearing force diagram.

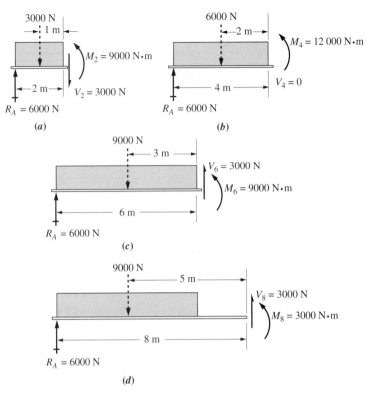

FIGURE 6–30 Free-body diagrams used to find bending moments.

The symbol M_2 is used to indicate the bending moment at the point 2 m out on the beam.

Using a similar method at points 4 m, 6 m, and 8 m out on the beam, as shown in Figure 6–30(b), (c), and (d), we would get

$$M_4 = 6000 \text{ N } (4 \text{ m}) - 6000 \text{ N } (2 \text{ m}) = 12\,000 \text{ N·m}$$
$$M_6 = 6000 \text{ N } (6 \text{ m}) - 9000 \text{ N } (3 \text{ m}) = 9000 \text{ N·m}$$
$$M_8 = 6000 \text{ N } (8 \text{ m}) - 9000 \text{ N } (5 \text{ m}) = 3000 \text{ N·m}$$

Remember that at each end of the beam the bending moment is zero. Now we have several points that can be plotted on a bending moment diagram below the shearing force diagram, as shown in Figure 6–31. First look at that section of the beam where the distributed load is applied, the first 6 m. By joining the points plotted for bending moment with a smooth curve, the typical shape of a bending moment curve for a distributed load is obtained. For the last 3 m, where no loads are applied, the curve is a straight line, as was the case in earlier examples.

Some important observations can be made from Figure 6–31, which can be generalized into rules for drawing bending moment diagrams.

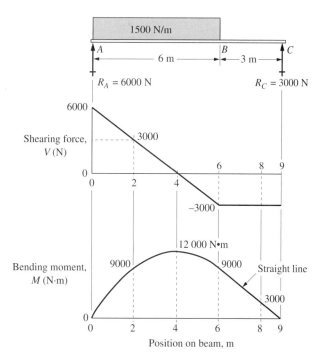

FIGURE 6–31 Complete beam loading, shearing force, and bending moment diagrams.

<table>
<tr><td>**Rules for drawing bending moment diagrams**</td><td>

1. At the ends of a simply supported beam, the bending moment is zero.

2. The *change in bending moment* between two points on a beam is equal to the area under the shearing force curve between those two points. Thus, when the area under the shearing force curve is positive (above the axis), the bending moment is increasing, and vice versa.

3. The maximum bending moment occurs at a point where the shearing force curve crosses its zero axis.

4. On a section of the beam where distributed loads act, the bending moment diagram will be curved.

5. On a section of the beam where no loads are applied, the bending moment diagram will be a straight line.

6. The *slope* of the bending moment curve at any point is equal to the magnitude of the shearing force at that point.

</td></tr>
</table>

Consider these rules applied to the beam in Figure 6–31. *Rule 1* is obviously satisfied since the moment at each end is zero. *Rule 2* can be used to check the points plotted in the moment diagram at the 2-m intervals. For the first 2 m, the area under the shearing force curve is composed of a rectangle and a triangle. Then the area is

$$A_{0-2} = 3000 \text{ N} (2 \text{ m}) + \frac{1}{2}(3000 \text{ N})(2 \text{ m}) = 9000 \text{ N·m}$$

This is the change in moment from point 0 to point 2 on the beam. For the segment from 2 to 4, the area under the shearing force curve is a triangle. Then

$$A_{2-4} = \frac{1}{2}(3000 \text{ N})(2 \text{ m}) = 3000 \text{ N·m}$$

Since this is the change in moment from point 2 to point 4,

$$M_4 = M_2 + A_{2-4} = 9000 \text{ N·m} + 3000 \text{ N·m} = 12\,000 \text{ N·m}$$

Similarly for the remaining segments,

$$A_{4-6} = \frac{1}{2}(-3000 \text{ N})(2 \text{ m}) = -3000 \text{ N·m}$$
$$M_6 = M_4 + A_{4-6} = 12\,000 \text{ N·m} - 3000 \text{ N·m} = 9000 \text{ N·m}$$
$$A_{6-8} = (-3000 \text{ N})(2 \text{ m}) = -6000 \text{ N·m}$$
$$M_8 = M_6 + A_{6-8} = 9000 \text{ N·m} - 6000 \text{ N·m} = 3000 \text{ N·m}$$
$$A_{8-9} = (-3000 \text{ N})(1 \text{ m}) = (3000 \text{ N·m})$$
$$M_9 = M_8 + A_{8-9} = 3000 \text{ N·m} - 3000 \text{ N·m} = 0$$

Here the fact that $M_9 = 0$ is a check on the process because *rule 1* must be satisfied.

Rule 3 is illustrated at point 4. Where the maximum bending moment occurs, the shearing force curve crosses the zero axis.

Rule 6 will probably take some practice to get familiar with, but it is extremely helpful in the process of sketching moment diagrams. Usually sketching is sufficient. The use of the six rules stated above will allow you to quickly sketch the shape of the diagram and compute key values.

In applying *rule 6*, remember the basic concepts about the slope of a curve or line, as illustrated in Figure 6–32. Seven different segments are shown, with curves for both the shearing force diagram and the moment diagram, including those most often encountered in constructing these diagrams. Thus, when drawing a portion of a diagram in which a particular shape is observed in the shearing force curve, the corresponding shape for the moment curve should be as illustrated in Figure 6–33. Applying this approach to the moment diagram in Figure 6–31, note that the curve from point 0 to point 4 is like that of type 5 in Figure 6–32. Between points 4 and 6, the curve is like type 6. Between points 6 and 9, the straight line with a negative slope, type 3, is used.

6–6 SHEARING FORCES AND BENDING MOMENTS FOR CANTILEVER BEAMS

The manner of support of a cantilever beam causes the analysis of its shearing forces and bending moments to be somewhat different from that for simply supported beams. The most notable difference is that, at the place where the beam is supported, it is fixed and can therefore resist moments. Thus, at the fixed end of the beam, the bending

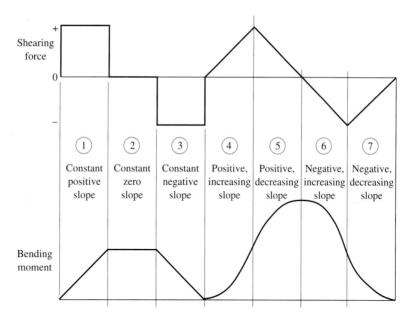

FIGURE 6-32 General shapes for moment curves relating to corresponding shearing force curves.

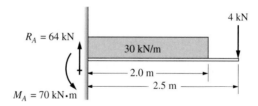

FIGURE 6-33 Beam loading and reactions.

moment is not zero, as it was for simply supported beams. In fact, the bending moment at the fixed end of the beam is usually the *maximum*.

Consider the cantilever beam shown in Figure 6–33. Earlier, in Example Problem 6–4, it was shown that the support reactions at point A are a vertical force $R_A = 64$ kN and a moment $M_A = 70$ kN·m. These are equal to the values of the shearing force and bending moment at the left end of the beam. According to convention, the upward reaction force R_A is positive, and the counterclockwise moment M_A is negative, giving the starting values for the shearing force diagram and bending moment diagram shown in Figure 6–34. The rules developed earlier about shearing force and bending moment diagrams can then be used to complete the diagrams.

The shearing force decreases in a straight-line manner from 64 kN to 4 kN in the interval A to B. Note that the change in shearing force is equal to the amount of the distributed load, 60 kN. The shearing force remains constant from B to C, where no loads are applied. The 4-kN load at C returns the curve to zero.

The bending moment diagram starts at -70 kN·m because of the reaction moment M_A. Between points A and B, the curve has a positive but decreasing slope (type 5

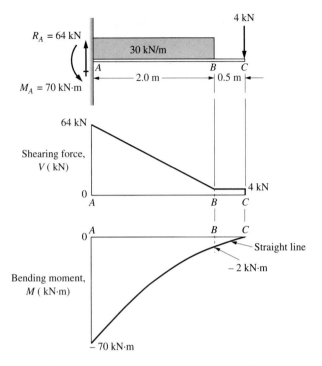

FIGURE 6–34 Complete load, shearing force, and bending moment diagrams.

in Figure 6–32). The change in moment between A and B is equal to the area under the shearing force curve curve between A and B. The area is

$$A_{A-B} = 4 \text{ kN (2 m)} + \frac{1}{2}(60 \text{ kN})(2 \text{ m}) = 68 \text{ kN·m}$$

Then the moment at B is

$$M_B = M_A + A_{A-B} = -70 \text{ kN·m} + 68 \text{ kN·m} = -2 \text{ kN·m}$$

Finally, from B to C,

$$M_C = M_B + A_{B-C} = -2 \text{ kN·m} + 4 \text{ kN (0.5 m)} = 0$$

Since point C is a *free* end of the beam, the moment there must be zero.

6–7 BEAMS WITH LINEARLY VARYING DISTRIBUTED LOADS

Figures 6–5 and 6–6 in Section 6–2 show two examples of beams carrying linearly varying distributed loads. We demonstrate here the method of drawing the general shape of the shearing force and bending moment diagrams for such beams, and how to determine the magnitude of the maximum shearing force and maximum bending moment. For many practical problems these are the primary objectives. Later, in Section 6–9, a mathematical approach is shown that yields a more complete definition of the shape of the shearing force and bending moment diagrams.

Refer now to Figure 6–35 which shows the load diagram for the cantilever from Figure 6–5. The rate of loading varies linearly from $w = -200$ lb/ft (downward) at the support point A to $w =$ zero at the right end B. This straight-line curve is called a *first degree curve* because the loading varies directly with position on the beam, x. For such a loading, the reaction at A, called R_A, is the resultant of the total distributed load, found by computing the area under the triangular-shaped load curve. That is,

$$R_A = \frac{1}{2}(-200 \text{ lb/ft}) (8 \text{ ft}) = -800 \text{ lb}$$

The bending moment at the support, called M_A, must equal the moment of all of the applied load to the right of A. This can be found by considering the resultant to act at the centroid of the distributed load. For the triangular-shaped load curve, the centroid is $1/3$ of the length of the beam from point A. Calling this distance, x, we can say,

$$x = L/3 = (8 \text{ ft})/3 = 2.667 \text{ ft}$$

Then the moment at A is the product of the resultant times x. That is,

$$M_A = R_A x = (800 \text{ lb}) (2.667 \text{ ft}) = 2133 \text{ lb·ft}$$

These values, $R_A = 800$ lb and $M_A = 2133$ lb·ft, are the maximum values for shearing force and bending moment, respectively. In most cases, that is the objective of the analysis. If so, the analysis can be concluded.

But, if the shapes of the shearing force and bending moment diagrams are desired, they can be sketched using the principles developed earlier in this chapter. Figure 6–36 shows the results. The shearing force diagram starts at A with the value of 800 lb, equal to the reaction, R_A. The value of the shearing force then decreases for points to the right of A as additional loads are applied. Note that the shearing force curve is not a

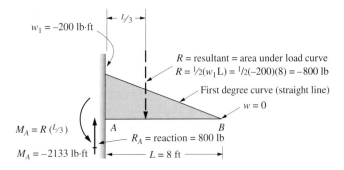

FIGURE 6–35 Load diagram, reaction, and moment for cantilever carrying a linearly varying distributed load.

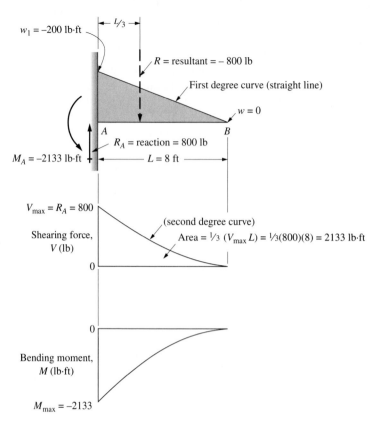

FIGURE 6–36 Load, shearing force, and bending moment diagrams for beam loading in Figure 6–35.

straight line because the rate of loading decreases as we proceed from *A* toward *B*. At *B* the rate of loading is zero, resulting in the zero value of shearing force at *B*. The slope of the shearing force curve at any point is equal to the rate of loading at the corresponding point on the load diagram. Thus, the shearing force curve starts with a relatively large negative slope at *A* and then a progressively smaller negative slope as we approach *B*. This is generally called a *second degree curve* because the value varies with the *square* of the distance *x*.

The bending moment diagram can be sketched by first noting that $M_A = -2133$ lb·ft. The curve has a relatively large positive slope at *A* because of the large positive value for shearing force at that point. Then the slope progressively decreases, increasing distance to zero at point *B*. The fact that the value of bending moment equals zero at *B* can be demonstrated, also, by computing the area under the shearing force curve from *A* to *B*. Appendix A–1 includes the formulas for computing the area under a second degree curve of the type shown in the shearing force diagram. That is,

$$\text{Area} = (1/3)\,(800 \text{ lb})\,(8 \text{ ft}) = 2133 \text{ lb·ft}$$

This is the *change* in bending moment from *A* to *B*, bringing the bending moment curve to zero at *B*.

6–8 FREE-BODY DIAGRAMS OF PARTS OF STRUCTURES

Examples considered thus far have been for generally straight beams with all transverse loads, that is, loads acting perpendicular to the main axis of the beam. Many machine elements and structures are more complex, having parts that extend away from the main beam-like part.

For example, consider the simple post with an extended arm shown in Figure 6–37 composed of vertical and horizontal parts. The vertical post is rigidly secured at its base. At the end of the extended horizontal arm, a downward load is applied. An example of such a loading is a support system for a sign over a highway. Another would be a support post for a basketball basket in which the downward force could be a player hanging from the rim after a slam dunk. A mechanical design application is a bracket supporting machine parts during processing.

Under such conditions, it is convenient to analyze the structure or machine element by considering each part separately and creating a free-body diagram for each part. At the actual joints between parts, one part exerts forces and moments on the other. Using this approach, you will be able to design each part on the basis of how it is loaded, using the basic principles of beam analysis in this chapter and those that follow.

Post with an extended arm. The objective of the analysis is to draw the complete shearing force and bending moment diagrams for the horizontal and vertical parts of the post/arm structure shown in Figure 6–37. The first step is to "break" the arm from the post at the right-angle corner.

Figure 6–38 shows the horizontal arm as a free body with the applied load, *F*, acting at its right end. The result appears similar to the cantilever analyzed earlier in this

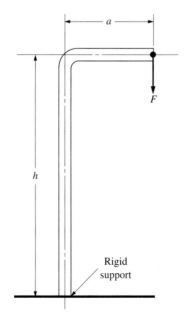

FIGURE 6–37 Post with an extended arm.

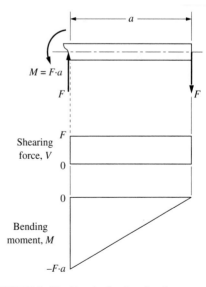

FIGURE 6–38 Free-body, shearing force and bending moment diagrams for the horizontal arm.

chapter. We know that the arm is in equilibrium as a part of the total structure and, therefore, it must be in equilibrium when considered by itself. Then, at the left end where it joins the vertical post, there must be a force equal to F acting vertically upward to maintain the sum of the vertical forces equal to zero. But the two vertical forces form a couple that tends to rotate the arm in a clockwise direction. To maintain rotational equilibrium, there must be a counterclockwise moment internal to the arm at its left end with a magnitude of $M = F \cdot a$, where a is the length of the arm. Having completed the free-body diagram, the shearing force and bending moment diagrams can be drawn as shown in Figure 6–38. The shearing force is equal to F throughout the length of the arm. The maximum bending moment occurs at the left end of the arm where $M = F \cdot a$.

The free-body diagram for the vertical post is shown in Figure 6–39. At the top of the post, a downward force and a clockwise moment are shown, exerted on the vertical post by the horizontal arm. Notice the action-reaction pair that exists at joints between parts. Equal but oppositely directed loads act on the two parts. Completing the free-body diagram for the post requires an upward force and a counterclockwise moment at its lower end, provided by the attachment means at its base. Finally, Figure 6–39 shows the shearing force and bending moment diagrams for the post, drawn vertically to relate the values to positions on the post. No shearing force exists because there are no transverse forces acting on the post. Where no shearing force exists, no change in bending moment occurs and there is a uniform bending moment throughout the post.

Beam with an L-shaped bracket. Figure 6–40 shows an L-shaped bracket extending below the main beam carrying an inclined force. The main beam is sup-

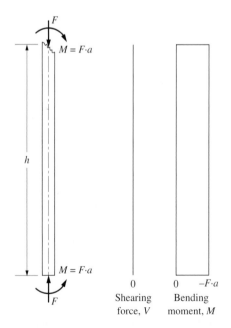

FIGURE 6–39 Free-body shearing force and bending moment diagrams for the vertical post.

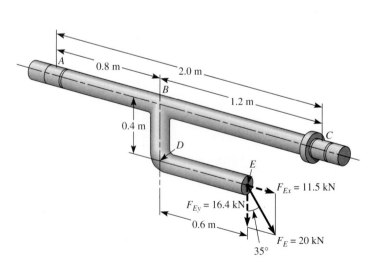

FIGURE 6–40 Beam with an L-shaped bracket.

ported by simple supports at A and C. Support C is designed to react to any unbalanced horizontal force. The objective is to draw the complete shearing force and bending moment diagrams for the main beam and the free-body diagrams for all parts of the bracket.

Three free-body diagrams are convenient to use here: one for the horizontal part of the bracket, one for the vertical part of the bracket, and one for the main beam itself. But first it is helpful to resolve the applied force into its vertical and horizontal components, as indicated by the dashed vectors at the end of the bracket.

Figure 6–41 shows the three free-body diagrams. Starting with the part DE shown in (a), the applied forces at E must be balanced by the oppositely directed forces at D for equilibrium in the vertical and horizontal directions. But rotational equilibrium must be produced by an internal moment at D. Summing moments with respect to point D shows that

$$M_D = F_{Ey} \cdot d = (16.4 \text{ kN})(0.6 \text{ m}) = 9.84 \text{ kN·m}$$

In Figure 6–41(b) the forces and moments at D have the same values but opposite directions from those at D in part (a) of the figure. Vertical and horizontal equilibrium conditions show the forces at B to be equal to those at D. The moment at B can be found

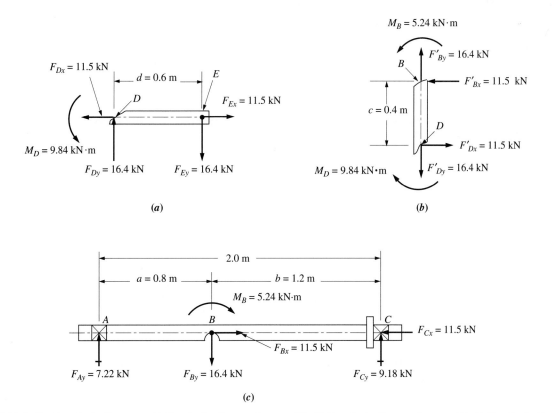

(a)

(b)

(c)

FIGURE 6–41 Free-body diagrams. (a) Free-body diagram for part DE. (b) Free-body diagram for part BD. (c) Free-body diagram for part ABC, the main beam.

by summing moments about B as follows:

$$\left(\sum M\right)_B = 0 = M_D - F_{Dx} \cdot c - M_B$$

Then

$$M_B = M_D - F_{Dx} \cdot c = 9.84 \text{ kN} \cdot \text{m} - (11.5 \text{ kN})(0.4 \text{ m}) = 5.24 \text{ kN} \cdot \text{m}$$

Now the main beam ABC can be analyzed. The forces and moment are shown applied at B with the values taken from point B on part BD. We now must solve for the reactions at A and C. First summing moments about point C yields

$$\left(\sum M\right)_C = 0 = F_{By} \cdot b - F_{Ay} \cdot (a + b) - M_B$$

Notice that the moment M_B applied at B must be included. Solving for F_{Ay} gives

$$F_{Ay} = \frac{(F_{By} \cdot b) - M_B}{a + b} = \frac{(16.4 \text{ kN})(1.2 \text{ m}) - 5.24 \text{ kN} \cdot \text{m}}{2.0 \text{ m}} = 7.22 \text{ kN}$$

Similarly, summing moments about point A gives

$$\left(\sum M\right)_A = 0 = F_{By} \cdot a - F_{Cy} \cdot (a + b) + M_B$$

Notice that the moment M_B applied at B is positive because it acts in the same sense as the moment due to F_{By}. Solving for F_{Cy} gives

$$F_{Cy} = \frac{(F_{By} \cdot a) + M_B}{a + b} = \frac{(16.4 \text{ kN})(0.8 \text{ m}) + 5.24 \text{ kN} \cdot \text{m}}{2.0 \text{ m}} = 9.18 \text{ kN}$$

A check on the calculation for these forces can be made by summing forces in the vertical direction and noting that the sum equals zero.

The completion of the free-body diagram for the main beam requires the inclusion of the horizontal reaction at C equal to the horizontal force at B.

Figure 6–42 shows the shearing force and bending moment diagrams for the main beam ABC. The shearing force diagram is drawn in the conventional manner with changes in shearing force occurring at each point of load application. The difference from previous work is in the moment diagram. The following steps were used:

1. The moment at A equals zero because A is a simple support.

2. The increase in moment from A to B equals the area under the shearing force curve between A and B, 5.78 kN·m.

3. At point B the moment M_B is considered to be a *concentrated moment* resulting in an abrupt change in the value of the bending moment by the amount of the

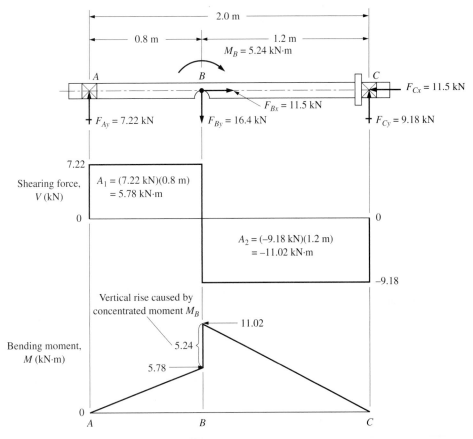

FIGURE 6–42 Shearing force and bending moment diagrams for main beam of Figure 6–40.

applied moment, 5.24 kN·m, thus resulting in the peak value of 11.02 kN·m. The convention used here is:

> **a.** When a concentrated moment is *clockwise*, the moment diagram *rises*.
> **b.** When a concentrated moment is *counterclockwise*, the moment diagram *drops*.

4. Between B and C, the moment decreases to zero because of the negative shearing force and the corresponding negative area under the shearing force curve.

This example is concluded.

6–9 MATHEMATICAL ANALYSIS OF BEAM DIAGRAMS

For most practical problems, the preparation of the load, shearing force, and bending moment diagrams using the techniques shown earlier in this chapter are adequate and convenient. A wide variety of beam types and loadings can be analyzed with sufficient detail to permit the logical design of the beams to ensure safety and to limit deflections to acceptable values. The methods of accomplishing these objectives are presented in Chapters 8–12.

But there are some types of loading and some types of design techniques that can benefit from the representation of the load, shearing force, and bending moment diagrams by mathematical equations. This section presents the methods of creating such equations.

The following is a set of guidelines for writing sets of equations that completely define the load, shearing force, and bending moment as a function of the position on the beam.

Guidelines for writing beam diagram equations

1. Draw the load diagram showing all externally applied loads and reactions.
2. Compute the values for all reactions.
3. Label points along the length of the beam where concentrated loads are applied or where distributed loads begin or end.
4. Draw the shearing force and bending moment diagrams using the techniques shown earlier in this chapter, noting values at the critical points defined in Step 3.
5. Establish conventions for denoting positions on the beam and signs of shearing forces and bending moment. In most cases, we will use the following conventions:
   a. Position on the beam will be denoted by the variable x measured from the left end of the beam.
   b. Downward loads will be negative.
   c. A positive shearing force is one that acts downward within the beam at a given section. An alternative way of determining this is to analyze the net external vertical force acting on that part of the beam to the left of the section of interest. If the next external force is upward, the internal shearing force in the beam is positive. See Figures 6–19 to 6–25.
   d. A positive bending moment is one that acts counterclockwise within the beam at a given section. See Figures 6–26 to 6–31. A positive bending moment will tend to cause a beam to bend in a concave upward shape, typical of a simply supported beam carrying downward loads between the supports.
6. Consider separately each segment of the beam between the points defined in Step 3. The shearing force curve should be continuous within each segment.
7. If the shearing force diagram consists of all straight lines caused by concentrated or uniformly distributed loads, the fundamental principles of analytic geometry can be used to write equations for the shearing force versus position on the beam for each segment. The

resulting equations will be of the form,

$$V_{AB} = \text{Constant} \qquad \text{(zero degree equation)}$$
$$V_{BC} = ax + b \qquad \text{(first degree equation)}$$

The subscripts define the beginning and end of the segment of interest.

8. If the shearing force diagram contains segments that are curved caused by varying distributed loads, first write equations for the load versus position on the beam. Then derive the equations for the shearing force versus position on the beam from

$$V_{AB} = \int w_{AB}\, dx + C$$

where w_{AB} is the equation for load in the segment AB as a function of x, and C is a constant of integration. The resulting shearing force equation will be of the second degree or higher, depending on the complexity of the loading pattern. Compute the value of the constants of integration using known values of V at given locations x.

9. Derive equations for the bending moment as a function of position on the beam for each segment, using the method,

$$M_{AB} = \int V_{AB}\, dx + C$$

This is the mathematical equivalent of the *area rule* for beam diagrams used earlier because the process of integration determines the area under the shearing force curve. Compute the value of the constants of integration using known values of M at given locations x.

10. The result at this point is a set of equations for shearing force and bending moment for each segment of the beam. It would be wise to check the equations for accuracy by substituting key values of x for which the shearing force and bending moment are known into the equations to ensure that the correct values for V and M are computed.

11. Determine the maximum values of shearing force and bending moment if they are not already known by substituting values for x in the appropriate equations where the maximum values are expected. Recall the rule that the maximum bending moment will occur at a point where the shearing force curve crosses the x-axis—that is, where $V = 0$.

This procedure is demonstrated by the following four examples.

Simply supported beam with a concentrated load. The objective is to write the equations for the shearing force and bending moment diagrams for the beam and loading shown in Figure 6–43, using the guidelines given in this section.

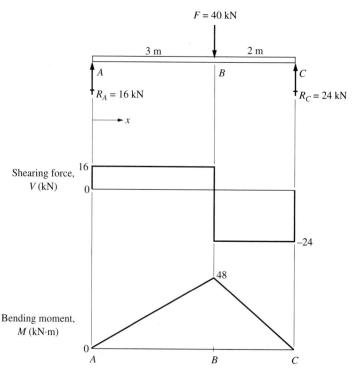

FIGURE 6–43 Simply supported beam with a concentrated load.

Steps 1 to 4 have been completed and shown in Figure 6–43. Points of interest are labeled A at the left support, B at the point of application of the load, and C at the right support. Equations will be developed for the two segments, AB and BC, where AB applies from $x = 0$ to $x = 3$ m and BC applies from $x = 3$ m to $x = 5$ m. Steps 5 and 6 will be as defined in the guidelines.

Step 7 can be applied to write the equations for the shearing force curve as follows:

$$V_{AB} = 16$$
$$V_{BC} = -24$$

The units for shearing force are taken to be kN.

Step 8 does not apply to this example.

Step 9 is now applied to derive the equations for the bending moment in the two segments.

$$M_{AB} = \int V_{AB}\,dx + C = \int 16\,dx + C = 16x + C$$

To evaluate the constant of integration C, we can note that at $x = 0$, $M_{AB} = 0$. Substituting these values into the moment equation gives

$$0 = 16(0) + C$$

Then, $C = 0$. The final equation can now be written as,

$$M_{AB} = 16x$$

As a check, we can see that at $x = 3$ m, the bending moment $M_B = 48$ kN·m, as shown in the bending moment diagram. We can now derive the equation for the bending moment in the segment BC.

$$M_{BC} = \int V_{BC}\,dx + C = \int -24\,dx + C = -24x + C$$

To evaluate C for this segment, we can use the condition at $x = 5$, $M_{BC} = 0$. Then,

$$0 = -24(5) + C$$

Then, $C = 120$. The final equation is

$$M_{BC} = -24x + 120$$

To check this equation, substitute $x = 3$.

$$M_{BC} = -24(3) + 120 = -72 + 120 = 48 \quad \text{(check)}$$

In summary, the equations for the shearing force and bending moment diagrams are:
In the segment AB from $x = 0$ to $x = 3$ m:

$$V_{AB} = 16$$
$$M_{AB} = 16x$$

In the segment BC from $x = 3$ m to $x = 5$ m:

$$V_{BC} = -24$$
$$M_{BC} = -24x + 120$$

The maximum values for the shearing force and bending moment are obvious from the diagrams.

$$V_{\max} = -24 \text{ kN throughout the segment } BC$$
$$M_{\max} = 48 \text{ kN·m at point } B \ (x = 3 \text{ m})$$

This example is concluded.

Simply supported beam with a partial uniformly distributed load. The objective is to write the equations for the shearing force and bending moment diagrams for the beam and loading shown in Figure 6–44, using the guidelines given in this section. Note that this is the same beam and loading that was shown earlier in Figure 6–31.

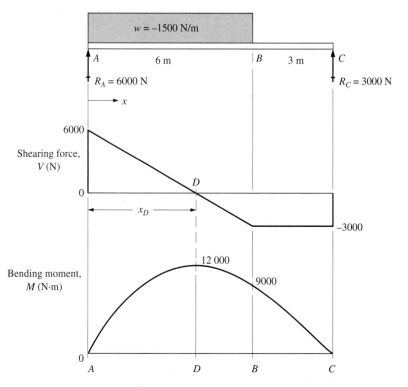

FIGURE 6–44 Simply supported beam with a partial uniformly distributed load.

Steps 1 to 4 have been completed and shown in Figure 6–44. Points of interest are labeled A at the left support, B at the point where the distributed load ends, and C at the right support. Equations will be developed for the two segments, AB and BC, where AB applies from $x = 0$ to $x = 6$ m and BC applies from $x = 6$ m to $x = 9$ m.

Step 5(b) can be used to write an equation for the loading in the segment AB:

$$w_{AB} = -1500 \text{ N/m}$$

Step 7 can be applied to write the equations for the shearing force curve. In the segment AB, the curve is a straight line, so we can write it in the form,

$$V_{AB} = ax + b$$

where a is the slope of the line and b is the intercept of the line with the V axis at $x = 0$. A convenient way to determine the slope is to observe that the slope is equal to the rate of loading for the distributed load. That is, $a = -1500$ N/m. The value of the intercept b can be observed from the shearing force diagram; $b = 6000$ N. Then the final form of the shearing force equation is

$$V_{AB} = -1500x + 6000$$

We can check the equation by substituting $x = 6$ m and computing V_B.

$$V_{AB} = -1500(6) + 6000 = -3000$$

This checks with the known value of shearing force at point B.

Note that we could have used Step 8 to determine the equation for V_{AB}. Note that,

$$w_{AB} = -1500 \text{ N/m}$$

Then,

$$V_{AB} = \int w_{AB}\,dx + C = \int -1500\,dx + C = -1500x + C$$

The value of C can be found by substituting $V_{AB} = 6000$ at $x = 0$.

$$6000 = -1500(0) + C$$

Then, $C = 6000$. Finally,

$$V_{AB} = -1500x + 6000$$

This is identical to the previous result.

In the segment BC the shearing force is a constant value,

$$V_{BC} = -3000$$

Before proceeding to determine the equations for the bending moment diagram, recall that a critical point occurs where the shearing force curve crosses the zero axis. That will be a point of maximum bending moment. Let's call this point D and find the value of x_D where $V = 0$ by setting the equation for V_{AB} equal to zero and solving for x_D.

$$V_{AB} = 0 = -1500x_D + 6000$$
$$x_D = 6000/1500 = 4.0 \text{ m}$$

We will use this value later to find the bending moment at D.

Step 9 of the guidelines is now used to determine the equations for the bending moment diagram. First in the segment AB,

$$M_{AB} = \int V_{AB}\,dx + C = \int (-1500x + 6000)\,dx + C$$
$$M_{AB} = -750x^2 + 6000x + C$$

To evaluate C, note that at $x = 0$, $M_{AB} = 0$. Then, $C = 0$. And,

$$M_{AB} = -750x^2 + 6000x$$

We can check the equation by finding M_B at $x = 6$ m.

$$M_B = -750(6)^2 + 6000(6) = 9000 \quad \text{(check)}$$

Also, we need the value of the maximum moment at D where $x = 4.0$ m.

$$M_D = -750(4)^2 + 6000(4) = 12\,000 \quad \text{(check)}$$

For the segment BC:

$$M_{BC} = \int V_{BC}\,dx + C = \int -3000\,dx + C = -3000x + C$$

But, at $x = 9$ m, $M_{BC} = 0$. Then,

$$0 = -3000(9) + C$$

and $C = 27\,000$. Finally,

$$M_{BC} = -3000x + 27\,000$$

Check this equation at point B where $x = 6$ m.

$$M_B = -3000(6) + 27\,000 = -18\,000 + 27\,000 = 9000 \quad \text{(check)}$$

In summary, the equations for the shearing force and bending moment diagrams are:
In the segment AB from $x = 0$ to $x = 6$ m:

$$V_{AB} = -1500x + 6000$$
$$M_{AB} = -750x^2 + 6000x$$

In the segment BC from $x = 6$ m to $x = 9$ m:

$$V_{BC} = -3000$$
$$M_{BC} = -3000x + 27\,000$$

The maximum values for the shearing force and bending moment are obvious from the diagrams.

$$V_{\text{max}} = 6000 \text{ N at the left end at point } A$$
$$M_{\text{max}} = 12\,000 \text{ N·m at point } D \ (x = 4 \text{ m})$$

This example is concluded.

Cantilever with a varying distributed load. The objective is to write the equations for the shearing force and bending moment diagrams for the beam and loading shown in Figure 6–45, using the guidelines given in this section. Note that this is the same beam and loading shown earlier in Figure 6–36.

There will be only one segment for this example, encompassing the entire length of the beam, because the load, shearing force, and bending moment curves are continuous.

We must first write an equation for the loading that varies linearly from a rate of -200 lb/ft at the left end at A to zero at point B where $x = 8$ ft. You must note that the loading is shown on top of the beam acting downward in the manner we are accustomed to viewing loads. But the downwardly directed load is actually negative. To aid in writing the equation, you could redraw the loading diagram in the form of a *graph* of load versus position x, as shown in Figure 6–46. Then we can write the equation for the straight line,

$$w_{AB} = ax + b$$

The slope, a, can be evaluated by the ratio of the change in w over a given distance x. Using the entire length of the beam gives

$$a = \frac{w_1 - w_2}{x_1 - x_2} = \frac{-200 - 0}{0 - 8} = 25$$

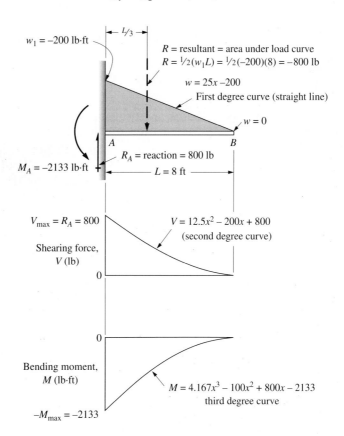

FIGURE 6–45 Cantilever with a varying distributed load.

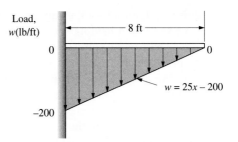

FIGURE 6–46 Alternate display of loading on the beam in Figure 6–45.

The value of $b = -200$ can be observed from the diagram in Figure 6–46. Then the final equation for the loading is

$$w_{AB} = 25x - 200$$

We can check this equation by evaluating w at $x = 8$ ft at the end of the beam.

$$w_{AB} = 25(8) - 200 = 0 \quad \text{(check)}$$

Now we can derive the equation for the shearing force diagram.

$$V_{AB} = \int w_{AB}\,dx + C = \int (25x - 200)\,dx + C = 12.5x^2 - 200x + C$$

Use the condition that at $x = 0$, $V_{AB} = 800$ to evaluate C.

$$800 = 12.5(0)^2 - 200(0) + C$$

Then, $C = 800$. And the final equation for the shearing force is

$$V_{AB} = 12.5x^2 - 200x + 800$$

We can check this equation by evaluating V at $x = 8$ ft at the end of the beam.

$$V_{AB} = 12.5(8)^2 - 200(8) + 800 = 0 \quad \text{(check)}$$

Now we can derive the equation for the bending moment diagram.

$$M_{AB} = \int V_{AB}\,dx + C = \int (12.5x^2 - 200x + 800)\,dx + C$$
$$M_{AB} = 4.167x^3 - 100x^2 + 800x + C$$

Using the condition that at $x = 0$, $M_{AB} = -2133$, we can evaluate C.

$$-2133 = 4.167(0)^3 - 100(0)^2 + 800(0) + C$$

Then, $C = -2133$. The final equation for bending moment is

$$M_{AB} = 4.167x^3 - 100x^2 + 800x - 2133$$

We can check this equation by evaluating M at $x = 8$ ft at the end of the beam.

$$M_{AB} = 4.167(8)^3 - 100(8)^2 + 800(8) - 2133 = 0 \quad \text{(check)}$$

In summary, the equations for the load, shearing force, and bending moment diagrams for the beam shown in Figure 6–45 are

$$w_{AB} = 25x - 200 \qquad \text{(a first degree curve; straight line)}$$
$$V_{AB} = 12.5x^2 - 200x + 800 \qquad \text{(a second degree curve)}$$
$$M_{AB} = 4.167x^3 - 100x^2 + 800x - 2133 \qquad \text{(a third degree curve)}$$

This example is concluded.

Simply supported beam with a varying distributed load. The objective is to write the equations for the shearing force and bending moment diagrams for the beam and loading shown in Figure 6–47, using the guidelines given in this section. A pictorial of one manner of creating this loading pattern is shown in Figure 6–6.

Because of the symmetry of the loading, the two reactions will be of equal magnitude. Each will be equal to the area under one-half of the load diagram. Breaking that into a rectangle 0.2 kN/m high by 2.30 m wide and a triangle 1.0 kN/m high and 2.30 m wide, we can compute:

$$R_A = R_C = (0.2)(2.30) + 0.5(1.0)(2.30) = 0.46 + 1.15 = 1.61 \text{ kN}$$

The general shapes of the shearing force and bending moment diagrams are sketched in Figure 6–47. We can reason that the shearing force curve crosses the zero axis at the middle of the beam at $x = 2.30$ m. Therefore, the maximum bending moment will occur at that point also. In principle, the magnitude of the maximum bending moment is equal to the area under the shearing force curve between points A and B. But the calculation of that area is difficult because the curve is of the second degree and it does not start at its vertex. Then the formulas in Appendix A–1 cannot be used directly. This is one reason for developing the equations for the shearing force and bending moment diagrams.

Let's first write the equation for the load on the left half of the beam from A to B. The rate of loading starts at -0.20 kN/m (downward) and increases in magnitude to -1.20 kN/m. Again, as was done in the preceding example, it may help to draw the load diagram as a graph, as shown in Figure 6–48. Then we can write the equation for the straight line in the form,

$$w_{AB} = ax + b$$

The slope, a, can be evaluated by the ratio of the change in w over a given distance x. Using half the length of the beam gives

$$a = \frac{w_1 - w_2}{x_1 - x_2} = \frac{-0.20 - (-1.20)}{0 - 2.30} = -0.4348$$

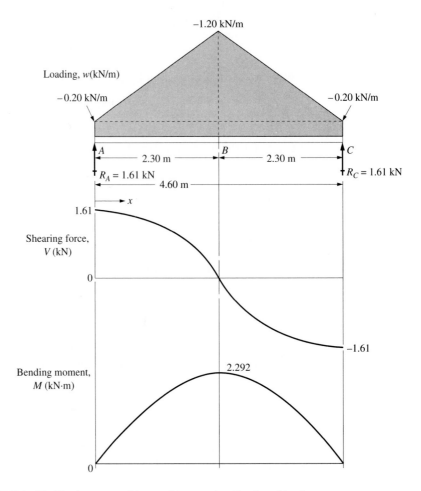

FIGURE 6–47 Simply supported beam with a varying distributed load.

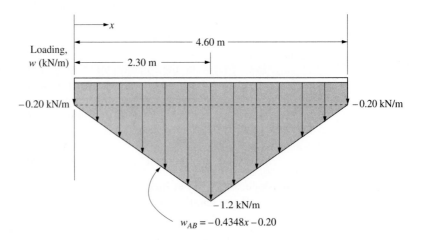

FIGURE 6–48 Alternate display of loading on the beam in Figure 6–47.

The value of $b = -0.20$ can be observed from the diagram in Figure 6–47. Then the final equation for the loading is

$$w_{AB} = ax + b = -0.4348x - 0.20$$

We can check this equation at $x = 2.30$ at the middle of the beam.

$$w_{AB} = -0.4348x - 0.20 = 0.4348(2.30) - 0.20 = 1.20 \text{ kN/m} \quad \text{(check)}$$

Now we can derive the equation for the shearing force diagram for the segment AB.

$$V_{AB} = \int w_{AB}\,dx + C = \int (-0.4348x - 0.20)\,dx + C = -0.2174x^2 - 0.20x + C$$

Use the condition that at $x = 0$, $V_{AB} = 1.61$ to evaluate C. Then, $C = 1.61$ and the final form of the equation is

$$V_{AB} = -0.2174x^2 - 0.20x + 1.61$$

We can check this equation at the middle of the beam by substituting $x = 2.30$ m.

$$V_{AB} = -0.2174(2.30)^2 - 0.20(2.30) + 1.61 = 0 \quad \text{(check)}$$

Now we can derive the equation for the bending moment diagram.

$$M_{AB} = \int V_{AB}\,dx + C = \int (-0.2174x^2 - 0.20x + 1.61)\,dx + C$$
$$M_{AB} = -0.07246x^3 - 0.10x^2 + 1.61x + C$$

Using the condition that at $x = 0$, $M_{AB} = 0$, we can evaluate $C = 0$. And,

$$M_{AB} = -0.07246x^3 - 0.10x^2 + 1.61x$$

Checking at $x = 2.30$ m, gives $M_B = 2.292$ kN·m. The equations for the right side of the diagrams could be derived similarly. But because of the symmetry of the diagrams, the completion of the curves is the mirror image of those alredy computed. In summary, the equations for the left half of the load, shearing force, and bending moment diagrams are

$$w_{AB} = -0.4348x - 0.20$$
$$V_{AB} = -0.2174x^2 - 0.20x + 1.61$$
$$M_{AB} = -0.07246x^3 - 0.10x^2 + 1.61x$$

The maximum shearing force is 1.61 kN at each support and the maximum bending moment is 2.292 kN·m at the middle of the beam.

This example is concluded.

PROBLEMS

Figures P6–1 to P6–76 show a variety of beam types and loading conditions. For the beam in each figure, any or all of the following problem statements can be applied:

1. Compute the reactions at the supports using the techniques shown in Section 6–3.

2. Draw the complete shearing force and bending moment diagrams using the techniques shown in Sections 6–4 to 6–7.

3. Determine the magnitude and location of the maximum absolute value of the shearing force and bending moment.

4. Use the free-body diagram approach shown in Sections 6–4 and 6–5 to determine the internal shearing force and bending moment at any specified point in a beam.

5. Write equations for all segments of the shearing force and bending moment diagrams using the guidelines presented in Section 6–9.

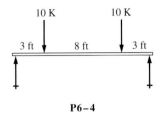

P6–4

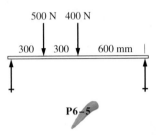

P6–5

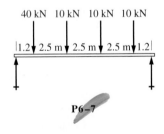

P6–6

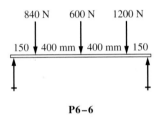

P6–7

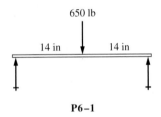

P6–1

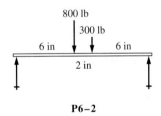

P6–2

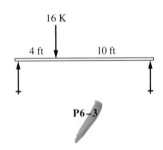

P6–3

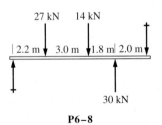

P6–8

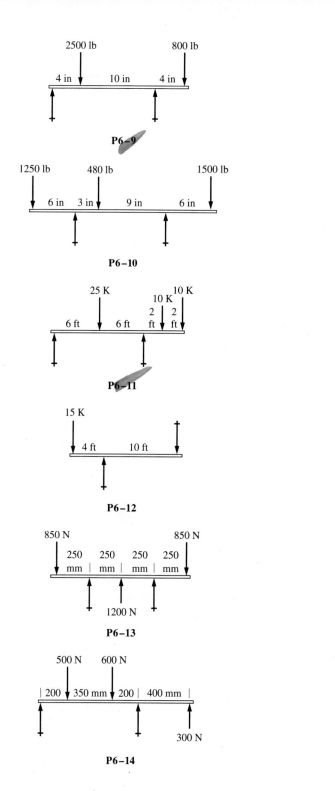

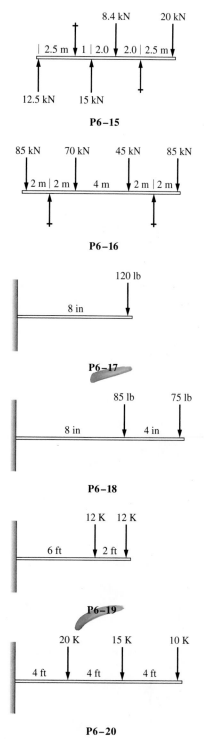

P6–9

P6–10

P6–11

P6–12

P6–13

P6–14

P6–15

P6–16

P6–17

P6–18

P6–19

P6–20

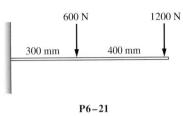

P6–21

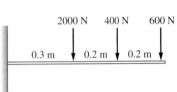

P6–22

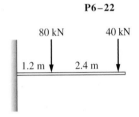

P6–23

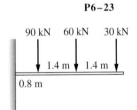

P6–24

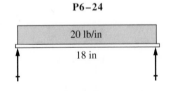

P6–25

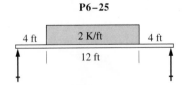

P6–26

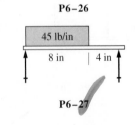

P6–27

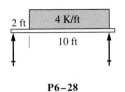

P6–28

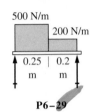

P6–29

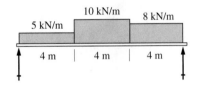

P6–30

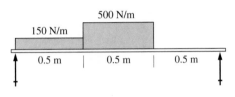

P6–31

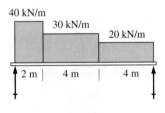

P6–32

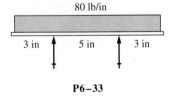

P6–33

a

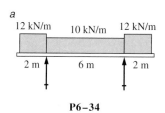

P6–34

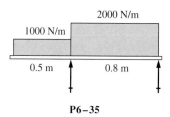

P6–35

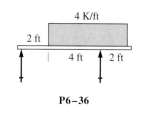

P6–36

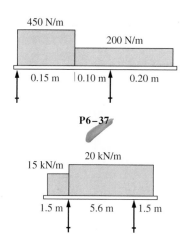

P6–37

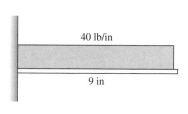

P6–38

40 lb/in

9 in

P6–39

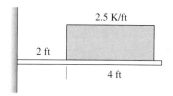

P6–40

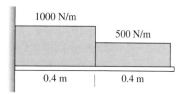

P6–41

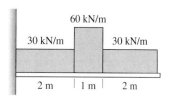

P6–42

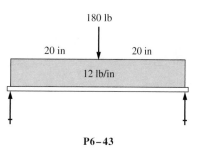

P6–43

P6–44

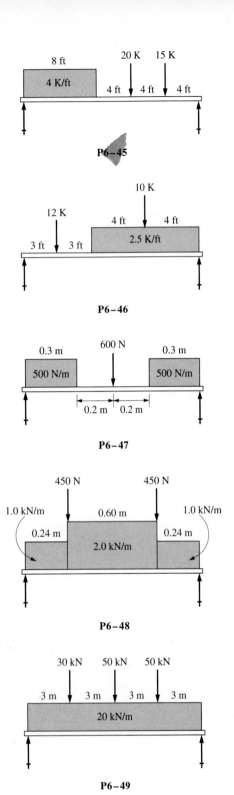

P6–45

P6–46

P6–47

P6–48

P6–49

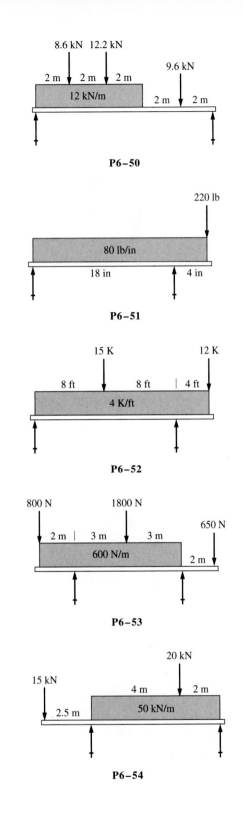

P6–50

P6–51

P6–52

P6–53

P6–54

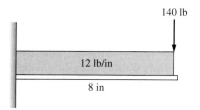

P6–55

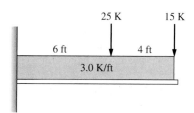

P6–56

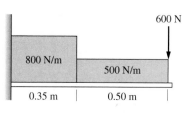

P6–57

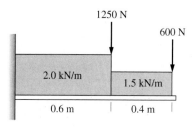

P6–58

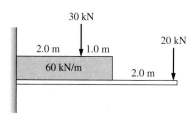

P6–59

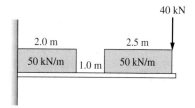

P6–60

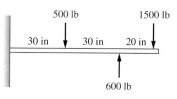

P6–61

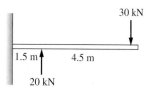

P6–62

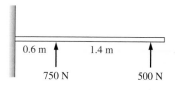

P6–63

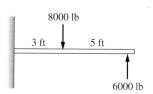

P6–64

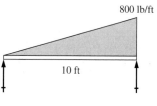

P6–65

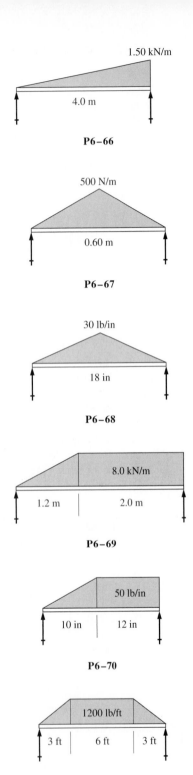

P6–66

P6–67

P6–68

P6–69

P6–70

P6–71

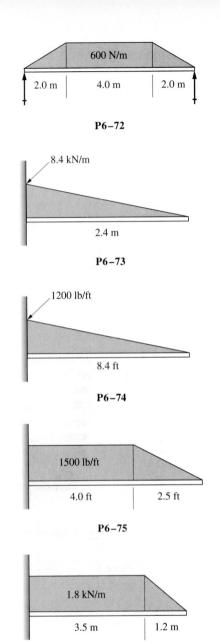

P6–72

P6–73

P6–74

P6–75

P6–76

Problems for Figures P6–77 to P6–84.

Each figure shows a mechanical device in which one or more forces are applied parallel to and away from the axis of the main, beam-like part. The devices are supported by bearings at the locations marked with an × which can provide reaction forces in any direction perpendicular to the axis of the beam. One of the bearings has the capability of resisting horizontally directed forces. For each figure the objectives are:

1. Break the compound beam into parts consisting of each of the straight components.

2. Show the complete free-body diagram of each component part including all external loads and internal forces and bending moments required to keep the part in equilibrium.

3. For the main, horizontal part only, draw the complete shearing force and bending moment diagrams. Refer to Section 6–8 for examples.

Dimensions in mm

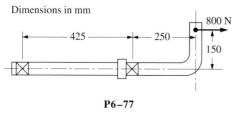

P6–77

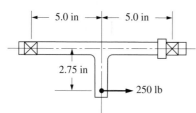

P6–78

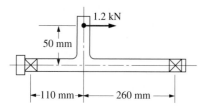

P6–79

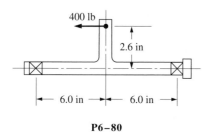

P6–80

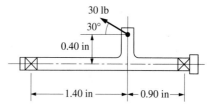

P6–81

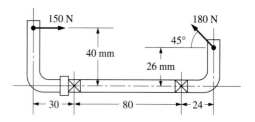

P6–82

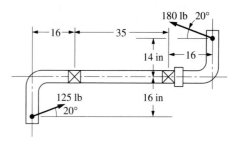

P6–83

P6–84

7

Centroids and Moments of Inertia of Areas

7–1 OBJECTIVES OF THIS CHAPTER

In Chapter 6 you learned to determine the value of shearing forces and bending moments in all parts of beams as part of the requirements for computing shearing stresses and bending stresses in later chapters. This chapter continues this pattern by presenting the properties of the shape of the cross section of the beam, also required for complete analysis of the stresses and deformations of beams.

The properties of the cross-sectional area of the beam that are of interest here are the *centroid* and the *moment of inertia with respect to the centroidal axis.* Some readers will already have mastered these topics through a study of *statics.* For them, this chapter should present a worthwhile review and a tailoring of the subject to the applications of interest in strength of materials. For those who have not studied centroids and moments of inertia, the concepts and techniques presented here will enable you to solve the beam analysis problems throughout this book and in many real design situations.

After completing this chapter, you should be able to:

1. Define *centroid.*

2. Locate the centroid of simple shapes by inspection.

3. Compute the location of the centroid for complex shapes by treating them as composites of two or more simple shapes.

4. Define *moment of inertia* as it applies to the cross-sectional area of beams.

5. Use formulas to compute the moment of inertia for simple shapes with respect to the centroidal axes of the area.

6. Compute the moment of inertia of complex shapes by treating them as composites of two or more simple shapes.

7. Properly use the *transfer-of-axis theorem* in computing the moment of inertia of complex shapes.

8. Analyze composite beam shapes made from two or more standard structural shapes to determine the resulting centroid location and moment of inertia.

9. Recognize what types of shapes are efficient in terms of providing a large moment of inertia relative to the amount of area of the cross section.

7–2 THE CONCEPT OF CENTROID—SIMPLE SHAPES

The *centroid* of an area is the point about which the area could be balanced if it was supported from that point. The word is derived from the word *center*, and it can be thought of as the geometrical center of an area. For three-dimensional bodies, the term *center of gravity*, or *center of mass*, is used to define a similar point.

For simple areas, such as the circle, the square, the rectangle, and the triangle, the location of the centroid is easy to visualize. Figure 7–1 shows the locations,

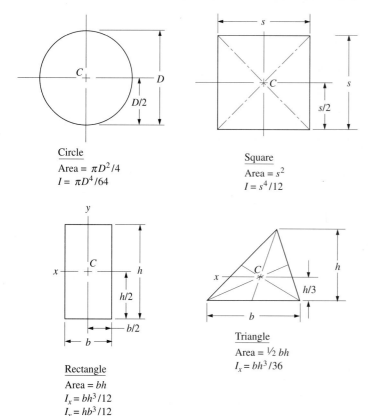

Circle
Area = $\pi D^2 / 4$
$I = \pi D^4 / 64$

Square
Area = s^2
$I = s^4 / 12$

Rectangle
Area = bh
$I_x = bh^3 / 12$
$I_y = hb^3 / 12$

Triangle
Area = $\frac{1}{2} bh$
$I_x = bh^3 / 36$

FIGURE 7–1 Properties of simple areas. The centroid is denoted as *C*.

denoted by C. If these shapes were carefully made and the location for the centroid carefully found, the shapes could be balanced on a pencil point at the centroid. Of course, a steady hand is required. How's yours?

Appendix A–1 is a more complete source of data for centroids and other properties of areas for a variety of shapes.

7–3 CENTROID OF COMPLEX SHAPES

Most complex shapes can be considered to be made up by combining several simple shapes. This can be used to facilitate the location of the centroid, as will be demonstrated later.

Another concept that aids in the location of centroids is that if the area has an axis of symmetry, the centroid will be on that axis. Some complex shapes have two axes of symmetry, and therefore the centroid is at the intersection of these two axes. Figure 7–2 shows examples where this occurs.

Where two axes of symmetry do not occur, the *method of composite areas* can be used to locate the centroid. For example, consider the shape shown in Figure 7–3. It has a vertical axis of symmetry but not a horizontal axis of symmetry. Such areas can be considered to be a composite of two or more simple areas for which the centroid can be found by applying the following principle:

> The product of the total area times the distance to the centroid of the total area is equal to the sum of the products of the area of each component part times the distance to its centroid, with the distances measured from the same reference axis.

This principle uses the concept of the *moment of an area,* that is, the product of the area times the distance from a reference axis to the centroid of the area. The principle states:

> The moment of the total area with respect to a particular axis is equal to the sum of the moments of all the component parts with respect to the same axis.

This can be stated mathematically as

$$A_T \overline{Y} = \sum (A_i y_i) \qquad (7\text{–}1)$$

where A_T = total area of the composite shape
$\overline{Y}$ = distance to the centroid of the composite shape measured from some reference axis
A_i = area of one component part of the shape
y_i = distance to the centroid of the component part from the reference axis

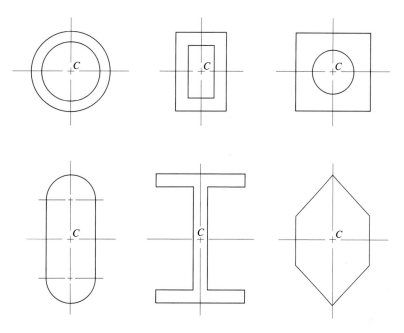

FIGURE 7–2 Composite shapes having two axes of symmetry. The centroid is denoted as *C*.

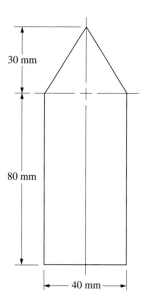

FIGURE 7–3 Shape for Example Problem 7–1.

The subscript i indicates that there may be several component parts, and the product A_iy_i for each must be formed and then summed together, as called for in Equation (7–1). Since our objective is to compute $\overline{Y}$, Equation (7–1) can be solved:

$$\overline{Y} = \frac{\sum (A_iy_i)}{A_T} \tag{7–2}$$

A tabular form of writing the data helps keep track of the parts of the calculations called for in Equation (7–2). An example will illustrate the method.

Example Problem 7–1   Find the location of the centroid of the area shown in Figure 7–3.

Solution   **Objective**   Compute the location of the centroid.

Given   Shape shown in Figure 7–3

Analysis   Because the shape has a vertical axis of symmetry, the centroid lies on that line. The vertical distance from the bottom of the shape to the centroid will be computed using Equation (7–2). The total area is divided into a rectangle (part 1) and a triangle (part 2), as shown in Figure 7–4. Each part is a simple shape for which the centroid is found using the data from Figure 7–1. The distances to the centroids from the bottom of the shape are shown in Figure 7–4 as y_1 and y_2.

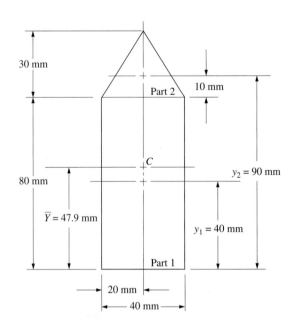

FIGURE 7–4   Data used in Example Problem 7–1.

The following table facilitates the calculations for data required in Equation (7–2).

Part	A_i	y_i	$A_i y_i$
1	3200 mm^2	40 mm	128 000 mm^3
2	600 mm^2	90 mm	54 000 mm^3
	A_T = 3800 mm^2		$\Sigma (A_i y_i)$ = 182 000 mm^3

Now $\overline{Y}$ can be computed.

$$\overline{Y} = \frac{\Sigma (A_i y_i)}{A_T} = \frac{182\,000 \text{ mm}^3}{3800 \text{ mm}^2} = 47.9 \text{ mm}$$

This locates the centroid as shown in Figure 7–4.

Comment In summary, the centroid is on the vertical axis of symmetry at a distance of 47.9 mm up from the bottom of the shape.

The composite area method works also for sections where parts are removed as well as added. In this case the removed area is considered negative, illustrated as follows.

Example Problem 7–2 Find the location of the centroid of the area shown in Figure 7–5.

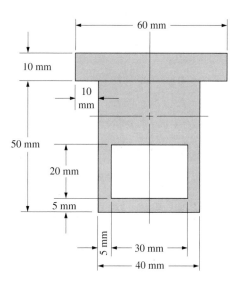

FIGURE 7–5 Shape for Example Problem 7–2.

Solution   **Objective**   Compute the location of the centroid.

Given   Shape shown in Figure 7–5

Analysis   Because the shape has a vertical axis of symmetry, the centroid lies on that line. The vertical distance from the bottom of the shape to the centroid will be computed using Equation (7–2). The total area is divided into three rectangles, as shown in Figure 7–6. Part 1 is the total large rectangle, 50 mm

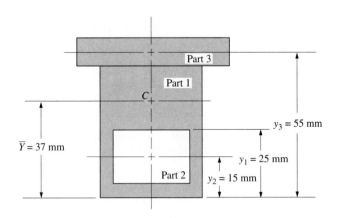

FIGURE 7–6   Data used in Example Problem 7–2.

high and 40 mm wide. Part 2 is the 20 mm × 30 mm rectangle that is *removed* from the composite area and the area, A_2, will be considered to be negative. Part 3 is the 10 mm × 60 mm rectangle on top. The distances to the centroids from the bottom of the shape are shown in Figure 7–6 as y_1, y_2, and y_3.

Results   The following table facilitates the calculations for data required in Equation (7–2).

Part	A_i	y_i	$A_i y_i$
1	2000 mm^2	25 mm	50 000 mm^3
2	−600 mm^2	15 mm	−9000 mm^3
3	600 mm^2	55 mm	33 000 mm^3
	$A_T = 2000$ mm^2		$\Sigma (A_i y_i) = 74\,000$ mm^3

Then

$$\overline{Y} = \frac{\Sigma (A_i y_i)}{A_T} = \frac{74\,000 \text{ mm}^3}{2000 \text{ mm}^2} = 37.0 \text{ mm}$$

Comment In summary, the centroid is on the vertical axis of symmetry at a distance of 37.0 mm up from the bottom of the shape.

7–4 THE CONCEPT OF MOMENT OF INERTIA

In the study of strength of materials, the property of *moment of inertia* is an indication of the stiffness of a beam, that is, the resistance to deflection of the beam when carrying loads that tend to cause it to bend. The deflection of a beam is inversely proportional to the moment of inertia as described in Chapter 12. The use of the moment of inertia in the calculation of stress due to bending is discussed in Chapter 8 of this book. Stresses due to vertical shearing forces also depend on moment of inertia and are discussed in Chapter 9.

Of interest is the moment of inertia of the shape of the *cross section* of the beam. For example, consider the overhanging beam shown in Chapter 6 in Figure 6–11. Its cross section is in the form of an "I," as sketched in Figure 7–7. Because of this shape, such a beam is often referred to as an "I-beam."

Another example is shown in Figure 6–12 where the horizontal boom of the crane assembly is a cantilever whose cross section is a hollow rectangle, as sketched in Figure 7–8. Notice in the original figure, the vertical dimension of the hollow rectangle decreases for sections farther away from the left end where the boom attaches to the support post. Chapter 8 describes why.

Both examples shown in Figures 7–7 and 7–8 represent shapes that are relatively efficient in the use of material to produce large values of their moment of inertia. In most important cases in the study of strength of materials, the moment of inertia of a shape, denoted by the symbol I, is a function of the placement of the area with respect to the *centroidal axis* of the shape, the axis that passes through the centroid of the shape. It is most desirable, from the standpoint of the efficient use of material, to place

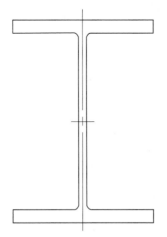

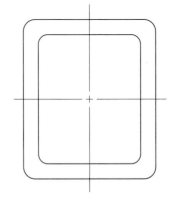

FIGURE 7–7 Typical cross-sectional shape of an I-beam.

FIGURE 7–8 Typical cross-sectional shape of a hollow rectangular tube.

as much of the material as far away from the centroidal axis as practical. This observation is based on the definition of moment of inertia given here.

> *The moment of inertia of an area with respect to a particular axis is defined as the sum of the products obtained by multiplying each infinitesimally small element of the area by the square of its distance from the axis.*

Thus you should be able to reason that if much of the area is placed far away from the centroidal axis, the moment of inertia would tend to be large.

The mathematical formula for the moment of inertia, *I*, follows from the definition. An approximate method involves the process of *summation,* indicated by Σ.

$$I = \sum y^2(\Delta A) \qquad (7-3)$$

This would require that the total area be divided into many very small parts represented by ΔA, and that the distance *y* to the centroid of each part from the axis of interest be determined. Then, the product of $y^2(\Delta A)$ would be computed for each small part, followed by summing all such products. This is a very tedious process, and, fortunately, one that is not often used.

A refinement of the summation method indicated by Equation (7–3) is the process of *integration* which is the mathematical technique of summing infinitesimal quantities over an entire area. The true mathematical definition of moment of inertia requires the use of integration as follows:

$$I = \int y^2\, dA \qquad (7-4)$$

Here, the term *dA* is an area of infinitesimally small size and *y*, as before, is the distance to the centroid of *dA*. We will demonstrate the use of Equation (7–4) in a later section. But in many practical problems, it is not necessary to perform the integration process.

There are several methods of determining the magnitude of the moment of inertia.

1. For simple shapes it is convenient to use standard formulas that have been derived from the basic definition given above. Figure 7–1 shows such formulas for four shapes and Appendix A–1 gives several more. Reference 2 includes a table of formulas for *I* for 42 different shapes.

2. For standard commercially available shapes such as wide-flange beams (W shapes), channels (C shapes), angles (L shapes), and pipe, values of moment of inertia are tabulated in published references such as Reference 1. See also Appendixes A–4 to A–12.

3. For more complex shapes for which no standard formulas are available, it is often practical to divide the shape into component parts that are themselves simple shapes. Examples are shown in Figures 7–4 to 7–8. The details of calculating the moment of inertia of such shapes, called *composite shapes,* de-

pend on the nature of the shapes and will be demonstrated later in this chapter. Important concepts are stated here.

> **a.** If all component parts of a composite shape have the same centroidal axis, the total moment of inertia for the shape can be found by adding or subtracting the moments of inertia of the component parts with respect to the centroidal axis. See Section 7–5.
>
> **b.** If all component parts of a composite shape do **not** have the same centroidal axis, the use of a process called the **transfer of axis theorem** is required. See Section 7–6.

4. The fundamental definition of moment of inertia, Equation (7–4), can be used if the geometry of the shape can be represented in mathematical terms that can be integrated. See Section 7–7.

5. Many computer aided design software systems include automatic calculation of the location of the centroid and the moment of inertia of any closed shape drawn in the system.

6. For the special case of a shape that can be represented as a composite of rectangles having sides perpendicular or parallel to the centroidal axis, a special tabulation technique can be applied that is described in the last section of this chapter. This technique lends itself well to solution by using a programmable calculator or a simple computer program.

7–5 MOMENT OF INERTIA OF COMPOSITE SHAPES WHOSE PARTS HAVE THE SAME CENTROIDAL AXIS

A composite shape is one made up of two or more parts that are themselves simple shapes for which formulas are available to calculate the moment of inertia, I. A special case is when all parts have the same centroidal axis. Then the moment of inertia for the composite shape is found by combining the values of I for all parts according to the following rule:

> If the component parts of a composite area all have the same centroidal axis, the total moment of inertia can be found by adding or subtracting the moments of inertia of the component parts with respect to the centroidal axis. The value of I is added if the part is a positive solid area. If the part is a void, the value of I is subtracted.

Figure 7–9 shows an example of such a shape, composed of a vertical central stem, 30 mm wide and 80 mm high, and two side parts, 30 mm wide and 40 mm high. Notice that all have their own centroidal axis coincident with the centroidal axis x-x for the composite section. The rule just stated can then be used to compute the total value of I for the cross by adding the values of I for each of the three parts. See Example Problem 7–3.

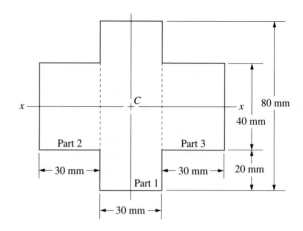

FIGURE 7–9 Shape for Example Problem 7–3.

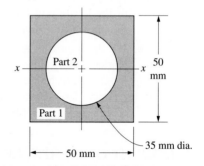

FIGURE 7–10 Shape for Example Problem 7–4.

Figure 7–10 shows an example where there is a 35 mm diameter circular hole removed from a square whose sides measure 50 mm. The circle and the square have the same centroidal axis *x*-*x*. The rule can then be used to compute the value of *I* for the square and then to subtract the value of *I* for the circle to obtain the total value for *I* of the composite shape. See Example Problem 7–4.

Example Problem 7–3	Compute the moment of inertia of the cross-shape shown in Figure 7–9 with respect to its centroidal axis.
Solution **Objective**	Compute the centroidal moment of inertia.
Given	Shape shown in Figure 7–9.
Analysis	The centroid of the cross-shape is at the intersection of the horizontal and vertical axes of symmetry. Dividing the cross into the three parts shown in the figure results in each part having the same centroidal axis, *x*-*x*, as the

entire composite section. Therefore, we can compute the value of I for each part and sum them to obtain the total value, I_T. That is,

$$I_T = I_1 + I_2 + I_3$$

Results Referring to Figure 7–1 for the formula for I for a rectangle gives

$$I_1 = \frac{bh^3}{12} = \frac{30(80)^3}{12} = 1.28 \times 10^6 \text{ mm}^4$$

$$I_2 = I_3 = \frac{30(40)^3}{12} = 0.16 \times 10^6 \text{ mm}^4$$

Then

$$I_T = 1.28 \times 10^6 \text{ mm}^4 + 2(0.16 \times 10^6 \text{ mm}^4) = 1.60 \times 10^6 \text{ mm}^4$$

Example Problem 7–4 Compute the moment of inertia of the shape shown in Figure 7–10 with respect to its centroidal axis.

Solution **Objective** Compute the centroidal moment of inertia.

Given Shape shown in Figure 7–10.

Analysis The centroid of the composite shape is at the intersection of the horizontal and vertical axes of symmetry. This coincides with the centroid of both the square and the circle. The composite shape can be considered to be the square with the circle removed. Therefore, we can compute the total value of I_T by computing the value of I_1 for the square and subtracting I_2 for the circle. That is,

$$I_T = I_1 - I_2$$

Results $$I_1 = \frac{s^4}{12} = \frac{(50)^4}{12} = 520.8 \times 10^3 \text{ mm}^4$$

$$I_2 = \frac{\pi D^4}{64} = \frac{\pi (35)^4}{64} = 73.7 \times 10^3 \text{ mm}^4$$

For the composite section,

$$I_x = I_1 - I_2 = 447.1 \times 10^3 \text{ mm}^4$$

7–6 MOMENT OF INERTIA FOR COMPOSITE SHAPES—GENERAL CASE—USE OF THE TRANSFER-OF-AXIS-THEOREM

When a composite section is composed of parts whose centroidal axes do not lie on the centroidal axis of the entire section, the process of simply summing the values of I for the parts *cannot* be used. It is necessary to employ the *transfer-of-axis-theorem*.

The general statement of the transfer-of-axis-theorem is,

> The moment of inertia of a shape with respect to a certain axis is equal to the sum of the moment of inertia of the shape with respect to its own centroidal axis plus an amount called the **transfer term** computed from Ad^2, where A is the area of the shape and d is the distance from the centroid of the shape to the axis of interest.

This theorem can be applied to compute the total moment of inertia for a general composite shape by using the following procedure. In this case, the axis of interest is the centroidal axis of the composite shape that must be found using the method given in Section 7–3.

General procedure for computing the moment of inertia for a composite shape

1. Divide the composite shape into component parts which are simple shapes for which formulas are available to compute the moment of inertia of the part with respect to its own centroidal axis. Identify the parts as *1, 2, 3*, and so forth.
2. Locate the distance from the centroid of each component part to some convenient reference axis, typically the bottom of the composite section. Call these distances y_1, y_2, y_3, and so forth.
3. Locate the centroid of the composite section using the method given in Section 7–3. Call the distance from the reference axis used in step 2 to the centroid, $\overline{Y}$.
4. Compute the moment of inertia of each part with respect to its own centroidal axis, calling these values I_1, I_2, I_3, and so forth.
5. Determine the distance from the centroid of the composite shape to the centroid of each part, calling these values d_1, d_2, d_3, and so forth. Note that $d_1 = \overline{Y} - y_1, d_2 = \overline{Y} - y_2, d_3 = \overline{Y} - y_3$, and so forth. Use the absolute value of each distance.
6. Compute the *transfer term* for each part from $A_i d_i^2$ where A_i is the area of the part and d_i is the distance found in step 5.
7. Compute the total moment of inertia of the composite section with respect to its centroidal axis from

$$I_T = I_1 + A_1 d_1^2 + I_2 + A_2 d_2^2 + I_3 + A_3 d_3^2 + \ldots \qquad (7\text{–}5)$$

Equation (7–5) is called the transfer-of-axis theorem because it defines how to transfer the moment of inertia of an area from one axis to any parallel axis. As applied here, the two axes are the centroidal axis of the component part and the centroidal axis of the composite section. For each part of a composite section, the sum $I + Ad^2$ is a measure of its contribution to the total moment of inertia.

Implementation of the *General procedure for computing the moment of inertia for a composite shape* can be facilitated by preparing a table that is an extension of that used in Section 7–3 to find the location of the centroid of the shape. The general design of this table follows.

Part	A_i	y_i	$A_i y_i$	I_i	$d_i = \overline{Y} - y_i$	$A_i d_i^2$	$I_i + A_i d_i^2$
1							
2							
3							
$A_T = \Sigma A_i =$		$\Sigma (A_i y_i) =$			$I_T = \Sigma (I_i + A_i d_i^2) =$		

Distance to centroid $= \overline{Y} = \dfrac{\Sigma (A_i y_i)}{A_T} =$

The use of the table and the general guidelines are demonstrated in Example Problems 7–5, 7–6, and 7–7. The benefit from using this type of table becomes greater as the number of components gets greater. Also, the use of a computer spreadsheet to make the appropriate calculations is very convenient.

Example Problem 7–5   Compute the moment of inertia of the tee shape in Figure 7–11 with respect to its centroidal axis.

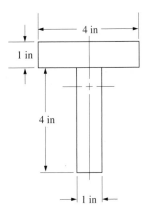

FIGURE 7–11 Shape for Example Problem 7–5.

Solution **Objective** Compute the moment of inertia.

Given Shape shown in Figure 7–11.

Analysis Use the **General Procedure** listed in this section. As step 1, divide the tee shape into two parts, as shown in Figure 7–12. Part 1 is the vertical stem and part 2 is the horizontal flange.

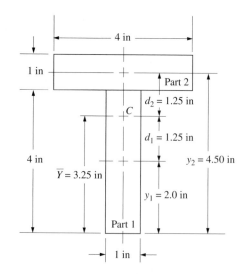

FIGURE 7–12 Data used in Example
Problem 7–5.

Results The following table summarizes the complete set of data used to compute the total moment of inertia with respect to the centroid of the tee shape. Some of the data are shown also in Figure 7–12. Comments are given here to show the manner of arriving at certain data.

Part	A_i	y_i	$A_i y_i$	I_i	d_i	$A_i d_i^2$	$I_i + A_i d_i^2$
1	4.0 in^2	2.0 in	8.0 in^3	5.33 in^4	1.25 in	6.25 in^4	11.58 in^4
2	4.0 in^2	4.5 in	18.0 in^3	0.33 in^4	1.25 in	6.25 in^4	6.58 in^4
$A_T = 8.0$ in^2		$\Sigma(A_i y_i) = 26.0$ in^3					$I_T = 18.16$ in^4

$$\overline{Y} = \frac{\Sigma(A_i y_i)}{A_T} = \frac{26.0 \text{ in}^3}{8.0 \text{ in}^2} = 3.25 \text{ in}$$

Steps 2, 3. The first three columns of the table give data for computing the location of the centroid using the technique shown in Section 7–3. Distances are measured upward from the bottom of the tee. $\overline{Y} = 3.25$ in.

Step 4. Both parts are simple rectangles. Then, the values of I are

$$I_1 = bh^3/12 = (1.0)(4.0)^3/12 = 5.33 \text{ in}^4$$
$$I_2 = bh^3/12 = (4.0)(1.0)^3/12 = 0.33 \text{ in}^4$$

Step 5. Distance from the overall centroid to the centroid of each part,

$$d_1 = \overline{Y} - y_1 = 3.25 \text{ in} - 2.0 \text{ in} = 1.25 \text{ in}$$
$$d_2 = y_2 - \overline{Y} = 4.5 \text{ in} - 3.25 \text{ in} = 1.25 \text{ in}$$

Step 6. Transfer term for each part,

$$A_1 d_1^2 = (4.0 \text{ in}^2)(1.25 \text{ in})^2 = 6.25 \text{ in}^4$$
$$A_2 d_2^2 = (4.0 \text{ in}^2)(1.25 \text{ in})^2 = 6.25 \text{ in}^4$$

It is just coincidence that the transfer terms for each part are the same in this problem.

Step 7. Total moment of inertia,

$$I_T = I_1 + A_1 d_1^2 + I_2 + A_2 d_2^2$$
$$= 5.33 \text{ in}^4 + 6.25 \text{ in}^4 + 0.33 \text{ in}^4 + 6.25 \text{ in}^4$$
$$= 18.16 \text{ in}^4$$

Comment Notice that the transfer terms contribute approximately 2/3 of the total value to the moment of inertia.

7–7 MATHEMATICAL DEFINITION OF MOMENT OF INERTIA

As stated in the Section 7–4, the moment of inertia, I, is defined as the sum of the products obtained by multiplying each element of the area by the square of its distance from the reference axis. The mathematical formula for moment of inertia follows from that definition and is now given. Note that the process of summing over an entire area is accomplished by integration.

$$I = \int y^2 \, dA \tag{7–4}$$

Refer to Figure 7–13 for an illustration of the terms in this formula for the special case of a rectangle, for which we want to compute the moment of inertia with respect to its centroidal axis. The small element of area is shown as a thin strip parallel to the centroidal axis where the width of the strip is the total width of the rectangle, b, and the thickness of the strip is a small value, dy. Then the area of the element is

$$dA = b \cdot dy$$

The distance, y, is the distance from the centroidal axis to the centroid of the elemental area as shown. Substituting these values into Equation (7–4) allows the derivation of the formula for the moment of inertia of the rectangle with respect to its centroidal axis. Note that to integrate over the entire area requires the limits for the integral to be from $-h/2$ to $+h/2$.

$$I = \int_{-h/2}^{+h/2} y^2 \, dA = \int_{-h/2}^{+h/2} y^2 (b \cdot dy)$$

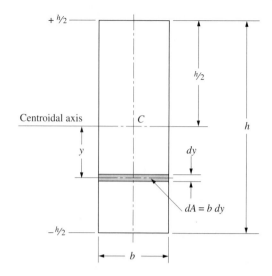

FIGURE 7–13 Data used in derivation of moment of inertia for a rectangle.

Because b is a constant, it can be taken outside the integral, giving

$$I = b \int_{-h/2}^{+h/2} y^2 \, dy = b \left[\frac{y^3}{3} \right]_{-h/2}^{+h/2}$$

Inserting the limits for the integral gives

$$I = b \left[\frac{h^3}{24} - \frac{(-h)^3}{24} \right] = b \left[\frac{2h^3}{24} \right] = \frac{bh^3}{12}$$

This is the formula reported in the tables. Similar procedures can be used to develop the formulas for other shapes.

7–8 COMPOSITE SECTIONS MADE FROM COMMERCIALLY AVAILABLE SHAPES

In Section 1–16 commercially available structural shapes were described for wood, steel, and aluminum. Properties of representative sizes of these shapes are listed in the following appendix tables:

Appendix A–4 for wood beams

Appendix A–5 for structural steel angles

Appendix A–6 for structural steel channels

Appendix A–7 for structural steel wide-flange shapes

Appendix A–8 for structural steel American Standard beams

Appendix A–9 for structural tubing—square and rectangular

Appendix A–10 for aluminum standard channels

Appendix A–11 for aluminum standard I-beams

Appendix A–12 for standard schedule 40 steel pipe

In addition to being very good by themselves for use as beams, these shapes are often combined to produce special composite shapes with enhanced properties.

When used separately, the properties for designing can be read directly from the tables for area, moment of inertia, and pertinent dimensions. When combined into composite shapes, the area and the moment of inertia of the component shapes with respect to their own centroidal axes are needed and can be read from the tables. Also, the tables give the location of the centroid for the shape which is often needed to determine distances needed to compute the transfer-of-axis term, Ad^2, in the moment of inertia calculation. The following example problems illustrate these processes.

Example Problem 7–6

Compute the moment of inertia of the composite I-beam shape shown in Figure 7–14 with respect to its centroidal axis. The shape is formed by welding a 0.50 in thick by 6.00 in wide plate to both the top and bottom flanges to increase the stiffness of the standard aluminum I-beam.

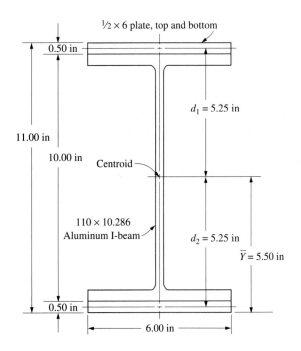

FIGURE 7–14 Data used in Example Problem 7–6.

Solution **Objective** Compute the moment of inertia.

Given Shape shown in Figure 7–14. For the I10 × 10.286 beam shape:
$I = 155.79$ in⁴; $A = 8.747$ in² (from Appendix A–11)

Analysis Use the **General Procedure** listed earlier in this chapter. As **step 1,** divide the beam shape into three parts. Part 1 is the I-beam; part 2 is the bottom plate; part 3 is the top plate. As **steps 2 and 3,** the centroid is coincident with the centroid of the I-beam because the composite shape is symmetrical. Thus, $\overline{Y} = 5.50$ in, or one-half of the total height of the composite shape. No separate calculation of $\overline{Y}$ is needed.

Results The following table summarizes the complete set of data used in steps 4–7 to compute the total moment of inertia with respect to the centroid of the beam shape. Some of the data are shown also in Figure 7–14. Comments are given here to show the manner of arriving at certain data.

Part	A_i	y_i	$A_i y_i$	I_i	$d_i = \overline{Y} - y_i$	$A_i d_i^2$	$I_i + A_i d_i^2$
1	8.747	5.50	–	155.79	0	0	155.79
2	3.00	0.25	–	0.063	5.25	82.69	82.75
3	3.00	10.75	–	0.063	5.25	82.69	82.75

$A_T = \Sigma A_i = 14.747 \text{ in}^2$ $\Sigma (A_i y_i) = $ – $\qquad I_T = \Sigma (I_i + A_i d_i^2) = 321.29 \text{ in}^4$

Distance to centroid $= \overline{Y} = \dfrac{\Sigma (A_i y_i)}{A_T} = 5.50$ in (by inspection)

Step 4. For each rectangular plate

$$I_2 = I_3 = bh^3/12 = (6.0)(0.5)^3/12 = 0.063 \text{ in}^4$$

Step 5. Distance from the overall centroid to the centroid of each part,

$$d_1 = 0.0 \text{ in because the centroids are coincident}$$
$$d_2 = 5.50 - 0.25 = 5.25 \text{ in}$$
$$d_3 = 10.75 - 5.50 = 5.25 \text{ in}$$

Step 6. Transfer term for each part,

$$A_1 d_1^2 = 0.0 \text{ because } d_1 = 0.0$$
$$A_2 d_2^2 = A_3 d_3^2 = (3.00)(5.25)^2 = 82.69 \text{ in}^4$$

Step 7. Total moment of inertia,

$$I_T = I_1 + I_2 + A_2 d_2^2 + I_3 + A_3 d_3^2$$
$$I_T = 155.79 + 0.063 + (3.0)(5.25)^2 + 0.063 + (3.0)(5.25)^2$$
$$= 321.29 \text{ in}^4$$

Comment Notice that the two added plates more than double the total value to the moment of inertia as compared with the original I-beam shape. Also, virtually all of the added value is due to the transfer terms and not to the basic moment of inertia of the plates themselves.

Example Problem 7–7

Compute the moment of inertia of the fabricated I-beam shape shown in Figure 7–15 with respect to its centroidal axis. The shape is formed by welding four standard L4 × 4 × 1/2 steel angles to a 1/2 × 16 vertical plate.

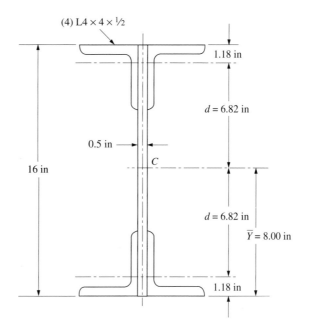

FIGURE 7–15 Data used in Example Problem 7–7.

Solution

Objective Compute the moment of inertia.

Given Shape shown in Figure 7–15. For each angle:
$I = 5.56$ in^4; $A = 3.75$ in^2 (from Appendix A–5)

Analysis Use the **General Procedure** listed above in this section. As **step 1,** we can consider the vertical plate to be part 1. Because the angles are all the same and placed an equal distance from the centroid of the composite shape, we can compute key values for one angle and multiply the results by 4. The location of the angles places the flat faces even with the top and bottom of the vertical plate. The location of the centroid of each angle is then 1.18 in from the top or bottom, based on the location of the centroid of the angles themselves, as listed in Appendix A–5. As **steps 2 and 3,** the centroid is coincident with the centroid of the I-beam because the composite shape is symmetrical. Thus, $\overline{Y} = 8.00$ in, or one-half of the total height of the composite shape. No separate calculation of $\overline{Y}$ is needed.

Results The following table summarizes the complete set of data used in steps 4–7 to compute the total moment of inertia with respect to the centroid of the beam shape. Some of the data are shown also in Figure 7–15. Comments are given here to show the manner of arriving at certain data. The second line of the table, shown in italics, gives data for one angle for reference only.

Line 3 gives the data for all four angles. Then the final results are found by summing lines 1 and 3.

Part	A_i	y_i	$A_i y_i$	I_i	$d_i = \overline{Y} - y_i$	$A_i d_i^2$	$I_i + A_i d_i^2$
1	8.00	8.00	–	170.67	0	0	170.67
(2)	3.75	1.18	–	5.56	6.82	174.42	179.98
4 × (2)	15.00	–	–	22.24	–	697.68	719.93
$A_T = \Sigma A_i = 23.00$ in²		$\Sigma (A_i y_i) =$	–			$I_T = \Sigma (I_i + A_i d_i^2) = 890.60$ in⁴	

Distance to centroid $= \overline{Y} = \dfrac{\Sigma (A_i y_i)}{A_T} = 8.00$ in (by inspection)

Step 4. For the vertical rectangular plate,

$$I_2 = bh^3/12 = (0.5)(16)^3/12 = 170.67 \text{ in}^4$$

Step 5. Distance from the overall centroid to the centroid of each part,

$$d_1 = 0.0 \text{ in because the centroids are coincident}$$
$$d_2 = 8.00 - 1.18 = 6.82 \text{ in}$$

Step 6. Transfer term for each part,

$$A_1 d_1^2 = 0.0 \text{ because } d_1 = 0.0$$
$$A_2 d_2^2 = A_3 d_3^2 = (3.75)(6.82)^2 = 174.42 \text{ in}^4$$

Step 7. Total moment of inertia,

$$I_T = 170.67 + 4[5.56 + 3.75(6.82)^2]$$
$$= 170.67 + 719.93 = 890.60 \text{ in}^4$$

Comment Approximately 80% of the total value of moment of inertia is contributed by the four angles.

7–9 MOMENT OF INERTIA FOR SHAPES WITH ALL RECTANGULAR PARTS

A method is shown here for computing the moment of inertia of special shapes that can be divided into parts, all of which are rectangles with their sides perpendicular and parallel to the axis of interest. An example would be the tee section analyzed in Example Problem 7–5 and shown in Figure 7–11. The method is somewhat simpler than the method described in Section 7–6 that used the transfer-of-axis-theorem, although both methods are based on the same fundamental principles.

The method involves the following steps:

1. Divide the composite section into a convenient number of parts so that each part is a rectangle with its sides perpendicular and parallel to the horizontal axis.

2. For each part, identify the following dimensions:

b = width

y_1 = distance from the base of the composite section to the bottom of that part

y_2 = distance from the base of the composite section to the top of that part

3. Compute the area of each part from the equation,

$$A = b(y_2 - y_1)$$

4. Compute the moment of the area from the equation,

$$M = b(y_2^2 - y_1^2)/2$$

5. Compute the location of the centroid relative to the base of the composite section from

$$\overline{Y} = M/A$$

6. Compute the moment of inertia with respect to the base of the composite section from

$$I_b = b(y_2^3 - y_1^3)/3$$

7. Compute the moment of inertia with respect to the centroid of the composite section from

$$I_c = I_b - A_T\overline{Y}^2$$

where A_T = total area = sum of the areas of all the parts.

This process lends itself very well to automated computation using a programmable calculator, a computer program, or a spreadsheet. As an illustration, Figure 7–16 shows the spreadsheet calculation of the centroidal moment of inertia for the tee section shown in Figure 7–11 and for which the calculation of the moment of inertia was done in Example Problem 7–5 using the transfer-of-axis theorem. The results are, of course, identical. See also Figure 7–17 for data.

Note that there are some blank lines in the spreadsheet because allowance was made for up to six parts for the composite section whereas this one has only two. The spreadsheet could be expanded to include any number of parts.

Moment of inertia for shape with all rectangular parts | Problem ID: Ex Prob 7-5

For each part: b = width; y_1 = distance to bottom of part; y_2 = distance to top of part

	Dimensions			Area, A	Moment, M	I with respect to base, I_b
	b	y_1	y_2	$b(y_2 - y_1)$	$b(y_2^2 - y_1^2)/2$	$b(y_2^3 - y_1^3)/3$
Units:	in	in	in	in^2	in^3	in^4
Part 1	1.000	0.000	4.000	4.000	8.000	21.333
Part 2	4.000	4.000	5.000	4.000	18.000	81.333
Part 3				0.000	0.000	0.000
Part 4				0.000	0.000	0.000
Part 5				0.000	0.000	0.000
Part 6				0.000	0.000	0.000
			TOTALS:	8.000	26.000	102.667

Distance from base to centroid = $\overline{Y}$:

$\overline{Y} = M/A =$ 3.25 in

Moment of inertia with respect to centroid = I_c:

$I_c = I_b - A_T \overline{Y}^2 =$ 18.167 in^4

FIGURE 7–16 Solution for Example Problem 7–5 using a spreadsheet and the solution procedure for moment of inertia in Section 7–9.

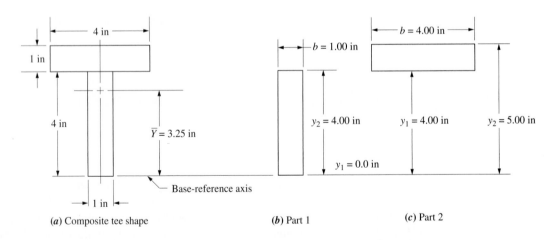

(*a*) Composite tee shape (*b*) Part 1 (*c*) Part 2

FIGURE 7–17 Tee section for illustration of the method of computing moment of inertia described in Section 7–9.

REFERENCES

1. American Institute of Steel Construction, *Manual of Steel Construction,* 9th ed., Chicago, IL, 1989.
2. Oberg, Erik, et al., *Machinery's Handbook,* 24th ed., Industrial Press, New York, 1992.

PROBLEMS

For each of the shapes in Figures P7–1 through P7–40, determine the location of the horizontal centroidal axis and the magnitude of the moment of inertia of the shape with respect to that axis using the transfer-of-axis theorem described in Section 7–6. Figures P7–1 through P7–20 are special shapes that might be produced by plastic or aluminum extrusion, by machining from solid bar stock, or by welding separate components. Figures P7–21 through P7–24 are composite beam sections that can be made by fastening standard wood shapes together by nails, screws, or adhesive. Figures P7–25 through P7–40 are composite beam sections fabricated from standard steel or aluminum structural shapes.

The method described in Section 7–9 can be applied to compute the moment of inertia for those shapes consisting of two or more rectangular parts whose sides are perpendicular and parallel to the horizontal axis. Included are the shapes in Figures P7–1 through P7–15 and the composite shapes made from standard wood beam shapes in Figures P7–21 through P7–24.

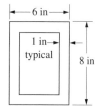

P7–3

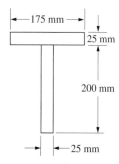

P7–4

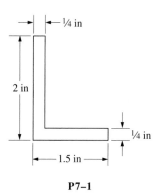

P7–1

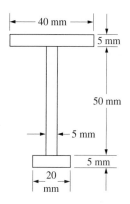

P7–5

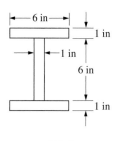

P7–2

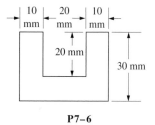

P7–6

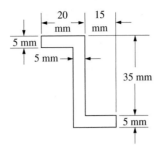

P7–7

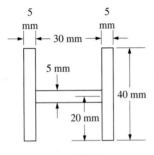

P7–8

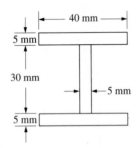

P7–9

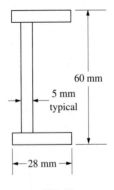

P7–10

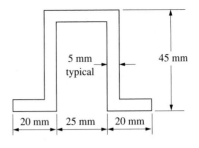

P7–11

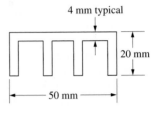

P7–12

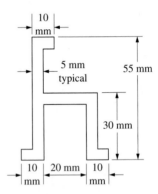

P7–13

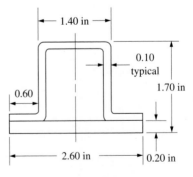

P7–14

268

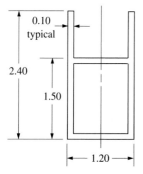

Dimensions in inches

P7–15

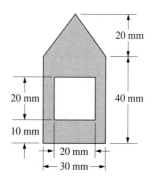

P7–18

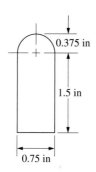

P7–16

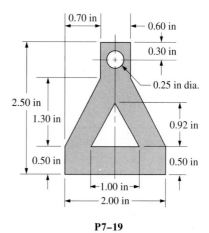

P7–19

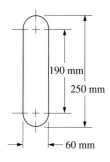

P7–17

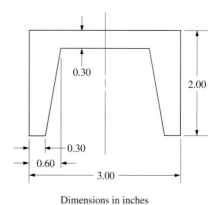

Dimensions in inches

P7–20

269

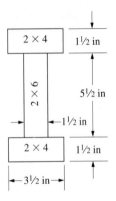

P7–21

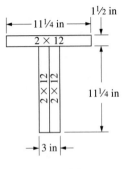

P7–24

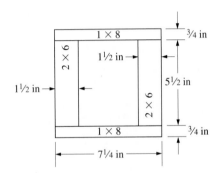

P7–22

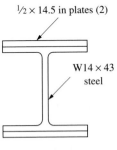

P7–25

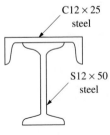

P7–26

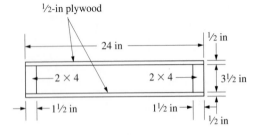

P7–23

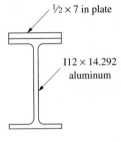

P7–27

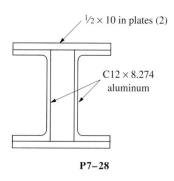

$\frac{1}{2} \times 10$ in plates (2)

C12 × 8.274
aluminum

P7–28

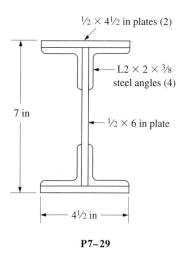

$\frac{1}{2} \times 4\frac{1}{2}$ in plates (2)

L2 × 2 × $\frac{3}{8}$
steel angles (4)

$\frac{1}{2} \times 6$ in plate

7 in

$4\frac{1}{2}$ in

P7–29

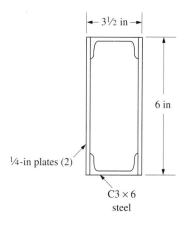

$3\frac{1}{2}$ in

6 in

$\frac{1}{4}$-in plates (2)

C3 × 6
steel

P7–30

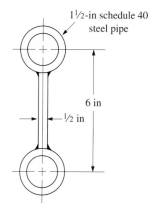

$1\frac{1}{2}$-in schedule 40
steel pipe

6 in

$\frac{1}{2}$ in

P7–31

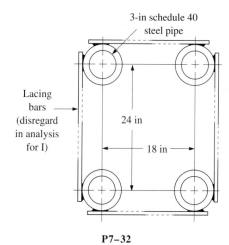

3-in schedule 40
steel pipe

Lacing
bars
(disregard
in analysis
for I)

24 in

18 in

P7–32

12 in

$\frac{1}{2} \times 12$ plate

L4 × 3 × $\frac{1}{4}$

$4\frac{3}{4}$ in

$\frac{1}{4} \times 10$ plate

10 in

P7–33

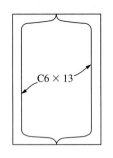

P7–34

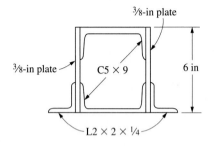

$\frac{3}{8}$-in plate

$\frac{3}{8}$-in plate

C5 × 9

6 in

L2 × 2 × ¼

P7–35

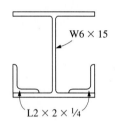

W6 × 15

L2 × 2 × ¼

P7–36

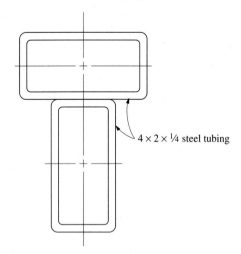

4 × 2 × ¼ steel tubing

P7–37

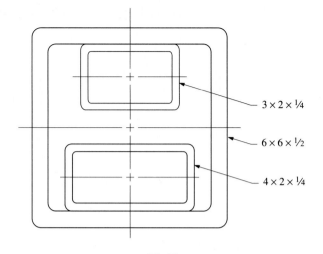

3 × 2 × ¼

6 × 6 × ½

4 × 2 × ¼

P7–38

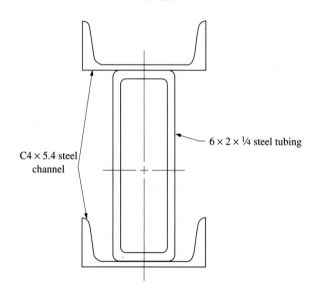

6 × 2 × ¼ steel tubing

C4 × 5.4 steel channel

P7–39

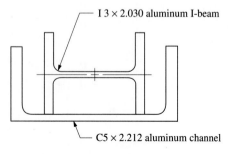

I 3 × 2.030 aluminum I-beam

C5 × 2.212 aluminum channel

P7–40

COMPUTER PROGRAMMING ASSIGNMENTS

1. For a generalized I-shape having equal top and bottom flanges similar to that shown in Figure P7–2, write a computer program to compute the location of the horizontal centroidal axis, the total area, and the moment of inertia with respect to the horizontal centroidal axis for any set of actual dimensions to be input by the program operator.

2. For the generalized T-shape similar to that shown in Figure P7–4, write a computer program to compute the location of the horizontal centroidal axis, the total area, and the moment of inertia with respect to the horizontal centroidal axis for any set of actual dimensions to be input by the program operator.

3. For the generalized I-shape similar to that shown in Figure P7–5, write a computer program to compute the location of the horizontal centroidal axis, the total area, and the moment of inertia with respect to the horizontal centroidal axis for any set of actual dimensions to be input by the program operator.

4. For any generalized shape that can be subdivided into some number of rectangular components with horizontal axes, write a computer program to compute the location of the horizontal centroidal axis, the total area, and the moment of inertia with respect to the horizontal centroidal axis for any set of actual dimensions to be input by the program operator. Use the transfer-of-axis theorem.

5. For the generalized hat-section shape similar to that shown in Figure P7–11, write a computer program to compute the location of the horizontal centroidal axis, the total area, and the moment of inertia with respect to the horizontal centroidal axis for any set of actual dimensions to be input by the program operator.

6. Given a set of standard dimension lumber, compute the area and moment of inertia with respect to the horizontal centroidal axis for the generalized box shape similar to that shown in Figure P7–22. The data for the lumber should be input by the operator of the program.

7. Build a data file containing the dimensions of a set of standard dimension lumber. Then permit the operator of the program to select sizes for the top and bottom plates and the two vertical members for the box shape shown in Figure P7–22. Then compute the area and moment of inertia with respect to the horizontal centroidal axis for the shape being designed.

8. Write a computer program to compute the area and moment of inertia with respect to the horizontal centroidal axis for a standard W or S beam shape with identical plates attached to both the top and bottom flanges similar to that shown in Figure 7–14. The data for the beam shape and the plates are to be input by the operator of the program.

9. Build a data base of standard W or S beam shapes. Then write a computer program to compute the area and moment of inertia with respect to the horizontal centroidal axis for a selected beam shape with identical plates attached to both the top and bottom flanges, as shown in Figure 7–14. The data for the plates are to be input by the operator of the program.

10. Given a standard W or S beam shape and its properties, write a computer program to compute the required thickness for plates to be attached to the top and bottom flanges to produce a specified moment of inertia of the composite section, as shown in Figure 7–14. Make the width of the plates equal to the width of the flange. For the resulting section, compute the total area.

11. Using the computer program written for Assignment 1 for the generalized I shape, perform an analysis of the area (A), the moment of inertia (I), and the ratio of I to A, as the thickness of the web is varied over a specified range. Keep all other dimensions for the shape the same. Note that the ratio of I to A is basically the same as the ratio of the stiffness of a beam having this shape to its weight because the deflection of a beam is inversely proportional to the moment of inertia and the weight of the beam is proportional to its cross-sectional area.

12. Repeat Assignment 11 but vary the height of the section while keeping all other dimensions the same.

13. Repeat Assignment 11 but vary the thickness of the flange while keeping all other dimensions the same.

14. Repeat Assignment 11 but vary the width of the flange while keeping all other dimensions the same.

15. Write a computer program to compute the moment of inertia with respect to the horizontal centroidal axis for any composite shape that can be divided into parts, all of which are rectangular having sides perpendicular and parallel to the horizontal axis using the method described in Section 7–9. Produce output from the program for any of the shapes in Figures P7–1 through P7–15 and Figures P7–21 through P7–24.

8

Stress Due to Bending

8-1 OBJECTIVES OF THIS CHAPTER

Beams were defined in Chapter 6 to be load-carrying members on which the loads act perpendicular to the long axis of the beam. Methods were presented there to determine the magnitude of the bending moment at any point on the beam. The bending moment, acting internal to the beam, causes the beam to bend and develops stresses in the fibers of the beam. The magnitude of the stress thus developed is dependent on the moment of inertia of the cross section, found using the methods reviewed in Chapter 7.

This chapter uses the information from previous chapters to enable the calculation of *stress due to bending* in beams. Specific objectives are:

1. To learn the statement of the *flexure formula* and to apply it properly to compute the maximum stress due to bending at the outer fibers of the beam.

2. To be able to compute the stress at any point within the cross section of the beam and to describe the variation of stress with position in the beam.

3. To understand the conditions on the use of the flexure formula.

4. To recognize that it is necessary to ensure that the beam does not twist under the influence of the bending loads.

5. To define the *neutral axis* and to understand that it is coincident with the centroidal axis of the cross section of the beam.

6. To understand the derivation of the flexure formula and the importance of the moment of inertia to bending stress.

7. To determine the appropriate design stress for use in designing beams.

8. To design beams to carry a given loading safely.

9. To define the *section modulus* of the cross section of the beam.

10. To select standard structural shapes for use as beams.

11. To recognize when it is necessary to use stress concentration factors in the analysis of stress due to bending and to apply appropriate factors properly.

12. To define the *flexural center* and describe its proper use in the analysis of stress due to bending.

8–2 THE FLEXURE FORMULA

Beams must be designed to be safe. When loads are applied perpendicular to the long axis of a beam, bending moments are developed inside the beam, causing it to bend. By observing a thin beam, the characteristically curved shape shown in Figure 8–1 is evident. The fibers of the beam near its top surface are shortened and placed in compression. Conversely, the fibers near the bottom surface are stretched and placed in tension.

Taking a short segment of the beam from Figure 8–1 we show in Figure 8–2 how the shape would change under the influence of the bending moments inside the beam. In part (a) the segment is in its initially straight form when it is not carrying a load.

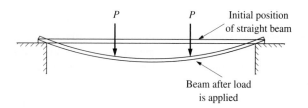

FIGURE 8–1   Example of a beam.

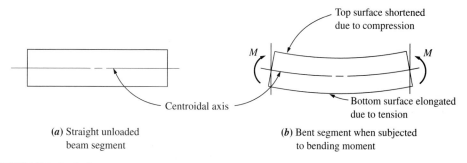

FIGURE 8–2   Influence of bending moment on beam segment.

Part (b) shows the same segment as it is deformed by the application of the bending moments. Lines that were initially horizontal become curved. The ends of the segment, which were initially straight and vertical, remain straight. But now they are inclined, having rotated about the centroidal axis of the cross section of the beam. The result is that the material along the top surface has been placed under compression and consequently shortened. Also, the material along the bottom surface has been placed under tension and has elongated.

In fact, all of the material above the centroidal axis is in compression. But the maximum shortening (compressive strain) occurs at the top. Because stress is proportional to strain, then it can be reasoned that the maximum compressive stress occurs at the top surface. Similarly, all of the material below the centroidal axis is in tension. But the maximum elongation (tensile strain) occurs at the bottom, producing the maximum tensile stress.

We can also reason that, if the upper part of the beam is in compression and the lower part is in tension, then there must be some place in the beam where there is no strain at all. That place is called the *neutral axis* and it will be shown later that it is coincident with the *centroidal axis* of the beam. In summary, we can conclude that,

> ***In a beam subjected to a bending moment of the type shown in Figure 8–2, material above the centroidal axis will be in compression with the maximum compressive stress occurring at the top surface. Material below the centroidal axis will be in tension with the maximum tensile stress occurring at the bottom surface. Along the centroidal axis itself, there is zero strain and zero stress due to bending. This is called the neutral axis.***

In designing or analyzing beams, it is usually the objective to determine the maximum tensile and compressive stress. It can be concluded from the discussion above that these maximums are dependent on the distance from the neutral axis (centroidal axis) to the top and bottom surfaces. We will call that distance, c.

The stress due to bending is also proportional to the magnitude of the bending moment applied to the section of interest. The shape and dimensions of the cross section of the beam determine its ability to withstand the applied bending moment. It will be shown later that the bending stress is inversely proportional to the moment of inertia of the cross section with respect to its horizontal centroidal axis.

We now state the *flexural formula* which can be used to compute the maximum stress due to bending.

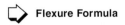 **Flexure Formula**

$$\sigma_{max} = \frac{Mc}{I} \qquad (8\text{–}1)$$

where σ_{max} = maximum stress at the outermost fiber of the beam

M = bending moment at the section of interest

c = distance from the centroidal axis of the beam to the outermost fiber

I = moment of inertia of the cross section with respect to its centroidal axis

The following steps are typically taken to apply Equation (8–1).

Guidelines for applying the flexure formula

1. Determine the maximum bending moment on the beam by drawing the shear and bending moment diagrams.
2. Locate the centroid of the cross section of the beam.
3. Compute the moment of inertia of the cross section with respect to its centroidal axis.
4. Compute the distance c from the centroidal axis to the top or bottom of the beam, whichever is greater.
5. Compute the stress from the flexure formula,

$$\sigma_{max} = \frac{Mc}{I}$$

The flexure formula is discussed in greater detail later. A sample problem will now be shown to illustrate the application of the formula.

Example Problem 8–1

For the beam shown in Figure 8–3, compute the maximum stress due to bending. The cross section of the beam is a rectangle 100 mm high and 25 mm wide. The load at the middle of the beam is 1500 N, and the length of the beam is 3.40 m.

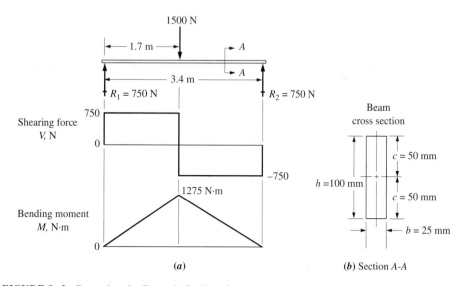

FIGURE 8–3 Beam data for Example Problem 8–1.

Solution   **Objective**   Compute the maximum stress due to bending.

Given   Beam and loading shown in Figure 8–3.

Analysis The guidelines defined in this section will be used.

Results *Step 1.* The shearing force and bending moment diagrams have been drawn and included in Figure 8–3. The maximum bending moment is 1275 N·m at the middle of the beam.

Step 2. The centroid of the rectangular cross section is at the intersection of the two axes of symmetry, 50 mm from either the top or the bottom surface of the beam.

Step 3. The moment of inertia for the rectangular shape with respect to the centroidal axis is

$$I = \frac{bh^3}{12} = \frac{25(100)^3}{12} = 2.08 \times 10^6 \text{ mm}^4$$

Step 4. The distance $c = 50$ mm from the centroidal axis to either the top or the bottom surface.

Step 5. The maximum stress due to bending occurs at the top or the bottom of the beam at the point of maximum bending moment. Applying Equation (8–1) gives

$$\sigma_{max} = \frac{Mc}{I} = \frac{(1275 \text{ N·m})(50 \text{ mm})}{2.08 \times 10^6 \text{ mm}^4} \times \frac{10^3 \text{ mm}}{\text{m}}$$
$$= 30.6 \text{ N/mm}^2 = 30.6 \text{ MPa}$$

8–3 CONDITIONS ON THE USE OF THE FLEXURE FORMULA

The proper application of the flexure formula requires the understanding of the conditions under which it is valid, listed as follows:

1. The beam must be straight or very nearly so.
2. The cross section of the beam must be uniform.
3. All loads and support reactions must act perpendicular to the axis of the beam.
4. The beam must not twist while the loads are being applied.
5. The beam must be relatively long and narrow in proportion to its depth.
6. The material from which the beam is made must be homogeneous, and it must have an equal modulus of elasticity in tension and compression.
7. The stress resulting from the loading must not exceed the proportional limit of the material.
8. No part of the beam may fail from instability, that is, from the buckling or crippling of thin sections.

Although the list of conditions appears to be long, the flexure formula still applies to a wide variety of real cases. Beams violating some of the conditions can be analyzed by using a modified formula or by using a combined stress approach. For example, for condition 2, a change in cross section will cause stress concentrations which can be handled as described in Section 8–9. The combined bending and axial stress or bending and torsional stress produced by violating condition 3 is discussed in Chapter 11. If the other conditions are violated, special analyses are required which are not covered in this book.

Condition 4 is important, and attention must be paid to the shape of the cross section to ensure that twisting does not occur. In general, if the beam has a vertical axis of symmetry and if the loads are applied through that axis, no twisting will result. Figure 8–4 shows some typical shapes used for beams that satisfy condition 4. Conversely, Figure 8–5 shows several that do not satisfy condition 4; in each of these cases, the beam would tend to twist as well as bend as the load is applied in the manner shown. Of course, these sections can support some load, but the actual stress condition in them is different from that which would be predicted from the flexure formula. More about these kinds of beams is presented in Section 8–10.

Condition 8 is important because long, thin members and, sometimes, thin sections of members tend to buckle at stress levels well below the yield strength of the material. Such failures are called *instability* and are to be avoided. Frequently, cross braces or local stiffeners are added to beams to relieve the problem of instability. An example can be seen in the wood joist floor construction of many homes and commercial buildings. The relatively slender wood joists are braced near the midpoints to avoid buckling.

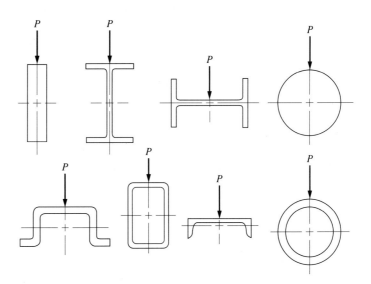

FIGURE 8–4 Example beam shapes with loads acting through an axis of symmetry.

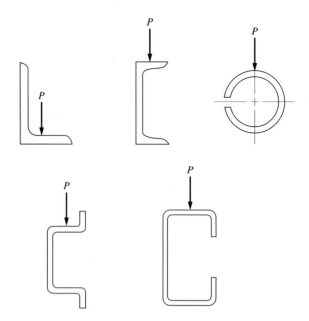

FIGURE 8–5 Example beam shapes with loads not acting through an axis of symmetry, resulting in twisting of the beam.

8–4 STRESS DISTRIBUTION ON A CROSS SECTION OF A BEAM

Refer again to Figure 8–2 showing the manner in which a segment of a beam deforms under the influence of a bending moment. The segment assumes the characteristic "bent" shape as the upper fibers are shortened and the lower fibers are elongated. The neutral axis, coincident with the centroidal axis of the cross section of the beam, bends but it is not strained. Therefore, at the neutral axis the stress due to bending is zero.

Figure 8–2 also shows that the ends of the beam segment that were initially straight and vertical, remain straight. But as the bending moment is applied they rotate. The linear distance from a point on the initial vertical end line to the corresponding point on the rotated end line is an indication of the amount of strain produced at that point in the cross section. It can be reasoned, therefore, that there is a linear variation of strain with position in the cross section as a function of the distance away from the neutral axis. Moving from the neutral axis toward the top of the section results in greater compressive strain while moving downward toward the bottom results in greater tensile strain. For materials following Hooke's law, stress is proportional to strain. The resulting stress distribution, then, is as shown in Figure 8–6.

If we desire to represent the stress at some point within the cross section, we can express it in terms of the maximum stress by noting the linear variation of stress with distance away from the neutral axis. Calling that distance y, we can write an equation for the stress, σ, at any point as,

$$\sigma = \sigma_{max}\frac{y}{c} \qquad (8–2)$$

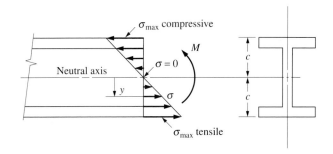

FIGURE 8–6 Stress distribution on a symmetrical section.

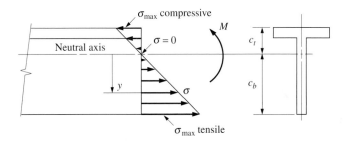

FIGURE 8–7 Stress distribution on a non-symmetrical section.

The general form of the stress distribution shown in Figure 8–6 would occur in any beam section having a centroidal axis equidistant from the top and bottom surfaces. For such cases, the magnitude of the maximum compressive stress would equal the maximum tensile stress.

If the centroidal axis of the section is not the same distance from both the top and bottom surfaces, the stress distribution shown in Figure 8–7 would occur. Still the stress at the neutral axis would be zero. Still the stress would vary linearly with distance from the neutral axis. But now the maximum stress at the bottom of the section is greater than that at the top because it is farther from the neutral axis. Using the distances c_b and c_t as indicated in Figure 8–7, the stresses would be

$$\sigma_{max} = \frac{Mc_b}{I} \quad \text{(tension at the bottom)}$$

$$\sigma_{max} = \frac{Mc_t}{I} \quad \text{(compression at the top)}$$

8–5 DERIVATION OF THE FLEXURE FORMULA

A better understanding of the basis for the flexure formula can be had by following the analysis used to derive it. The principles of static equilibrium are used here to show two concepts that were introduced earlier in this chapter but that were stated without proof. One is that the *neutral axis* is coincident with the *centroidal axis* of the cross section. The second is the flexure formula itself and the significance of the moment of inertia of the cross section.

Refer to Figure 8–6, which shows the distribution of stress over the cross section of a beam. The shape of the cross section is not relevant to the analysis and the I-shape is shown merely for example. The figure shows a portion of a beam, cut at some arbitrary section, with an internal bending moment acting on the section. The stresses, some tensile and some compressive, would tend to produce forces on the cut section in the axial direction. Equilibrium requires that the net sum of these forces must be zero. In general, force equals stress times area. Because the stress varies with position on the cross section, it is necessary to look at the force on any small elemental area and then sum these forces over the entire area using the process of integration. These concepts can be shown analytically as:

$$\text{Equilibrium condition: } \sum F = 0$$

$$\text{Force on any element of area } dA = \sigma \, dA$$

Total force on the cross-sectional area:

$$\sum F = \int_A \sigma \, dA = 0 \tag{8–3}$$

Now we can express the stress σ at any point in terms of the maximum stress by using Equation (8–2):

$$\sigma = \sigma_{max} \frac{y}{c}$$

where y is the distance from the neutral axis to the point where the stress is equal to σ. Substituting this into Equation (8–3) gives

$$\sum F = \int_A \sigma \, dA = \int_A \sigma_{max} \frac{y}{c} \, dA = 0$$

But because σ_{max} and c are constants, they can be taken outside the integral sign.

$$\sum F = \frac{\sigma_{max}}{c} \int_A y \, dA = 0$$

Neither σ_{max} nor c is zero, so the other factor, $\int_A y \, dA$, must be zero. But by definition and as illustrated in Chapter 7,

$$\int_A y \, dA = \overline{Y}(A)$$

where $\overline{Y}$ is the distance to the centroid of the area from the reference axis and A is the total area. Again, A cannot be zero, so, finally, it must be true that $\overline{Y} = 0$. Because the reference axis is the neutral axis, this shows that the neutral axis is coincident with the centroidal axis of the cross section.

The derivation of the flexure formula is based on the principle of equilibrium, which requires that the sum of the moments about any point must be zero. Figure 8–6 shows that a bending moment M acts at the cut section. This must be balanced by the net moment created by the stress on the cross section. But moment is the product of force times the distance from the reference axis to the line of action of the force. As used above,

$$\sum F = \int_A \sigma \, dA = \int_A \sigma_{max} \frac{y}{c} \, dA$$

Multiplying this by distance y gives the resultant moment of the force that must be equal to the internal bending moment M. That is,

$$M = \sum F(y) = \int_A \underbrace{\sigma_{max}}_{\text{stress}} \underbrace{\frac{y}{c}}_{} \underbrace{dA}_{\text{area}} \overbrace{(y)}^{\text{force}} \underbrace{}_{\text{moment arm}}$$

Simplifying, we obtain

$$M = \frac{\sigma_{max}}{c} \int_A y^2 \, dA$$

By definition, and as illustrated in Chapter 7, the last term in this equation is the moment of inertia I of the cross section with respect to its centroidal axis.

$$I = \int_A y^2 \, dA$$

Then

$$M = \frac{\sigma_{max}}{c} I$$

Solving for σ_{max} yields

$$\sigma_{max} = \frac{Mc}{I}$$

This is the form of the flexure formula shown earlier as Equation (8–1).

8–6 APPLICATIONS—BEAM ANALYSIS

Example Problem
8–2 The tee section shown in Figure 8–8 is from a simply supported beam which carries a bending moment of 100 000 lb·in due to a load on the top surface. Earlier, in Example Problem 7–5, this section was analyzed to determine that $I = 18.16\ \text{in}^4$. The centroid of

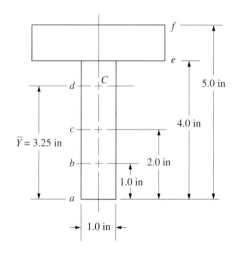

FIGURE 8–8 Tee section for beam in Example Problem 8–2.

the section is 3.25 in up from the bottom of the beam. Compute the stress due to bending in the beam at the six axes a to f indicated in the figure. Then plot a graph of stress versus position in the cross section.

Solution **Objective** Compute the bending stress at six axes a–f. Plot a graph of stress versus position in the cross section.

Given $M = 100\,000$ lb·in. Tee shape of cross section shown in Fig. 8–8.
$I = 18.16\ \text{in}^4$. $\overline{Y} = 3.25$ in from bottom of beam.

Analysis Equation (8–1) will be used to compute σ_{max}, that occurs at the bottom of the beam (axis a) because that is the location of the outermost fiber of the beam, farthest from the centroidal axis. Then the stress at other axes will be computed using Equation (8–2). See Figure 8–9 for values of y.

Results **At Axis a.** In Equation (8–1), use $c = \overline{Y} = 3.25$ in.

$$\sigma_{max} = \sigma_a = \frac{Mc}{I} = \frac{(100\,000\ \text{lb·in})\,(3.25\ \text{in})}{18.16\ \text{in}^4}$$

$$\sigma_a = 17\,900\ \text{psi}\quad\text{(tension)}$$

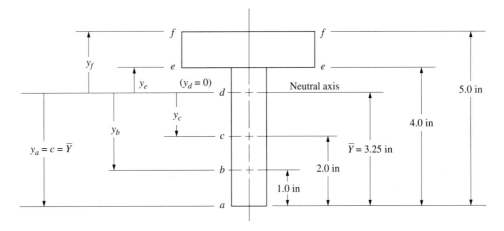

FIGURE 8–9 Data for Example Problem 8-2.

At Axis b.

$$\sigma_b = \sigma_{max}\frac{y}{c} = \sigma_a\frac{y}{c}$$

$$y = 3.25 \text{ in} - 1.0 \text{ in} = 2.25 \text{ in}$$

$$\sigma_b = 17\,900 \text{ psi} \times \frac{2.25}{3.25} = 12\,400 \text{ psi} \quad \text{(tension)}$$

At Axis c.

$$y = 3.25 \text{ in} - 2.0 \text{ in} = 1.25 \text{ in}$$

$$\sigma_c = 17\,900 \text{ psi} \times \frac{1.25}{3.25} = 6900 \text{ psi} \quad \text{(tension)}$$

At Axis d. At the centroid $y = 0$ and $\sigma_d = 0$.
At Axis e.

$$y = 4.0 \text{ in} - 3.25 \text{ in} = 0.75 \text{ in}$$

$$\sigma_e = 17\,900 \text{ psi} \times \frac{0.75}{3.25} = 4100 \text{ psi} \quad \text{(compression)}$$

At Axis f.

$$y = 5.0 \text{ in} - 3.25 \text{ in} = 1.75 \text{ in}$$

$$\sigma_f = 17\,900 \text{ psi} \times \frac{1.75}{3.25} = 9600 \text{ psi} \quad \text{(compression)}$$

The graph of these data is shown in Figure 8–10.

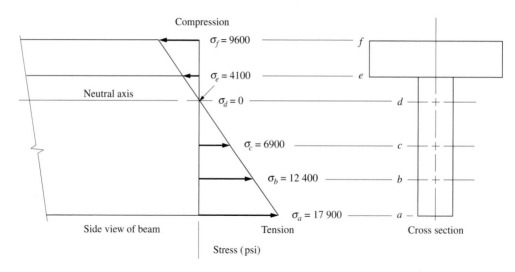

Compression

$\sigma_f = 9600$ ——————— f

$\sigma_e = 4100$ ——————— e

Neutral axis

$\sigma_d = 0$ ——————— d

$\sigma_c = 6900$ ——————— c

$\sigma_b = 12\ 400$ ——————— b

$\sigma_a = 17\ 900$ ——————— a

Side view of beam   Tension   Cross section

Stress (psi)

FIGURE 8–10 Stress distribution on the tee section for Example Problem 8–2.

Comment Notice the linear variation of stress with distance from the neutral axis and that the stresses above the neutral axis are compressive while those below are tensile.

Example Problem 8–3 Figure 8–11 shows the bending moment diagram for a 25-ft long beam in a large machine structure. It has been proposed that the beam be made from a standard W14 × 43 steel shape. Compute the maximum stress due to bending in the beam.

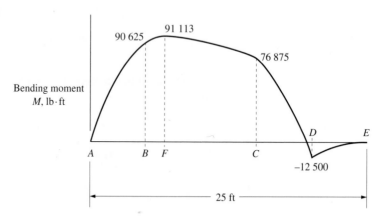

Bending moment
M, lb·ft

90 625

91 113

76 875

A B F C D E

−12 500

25 ft

FIGURE 8–11 Bending moment diagram for beam in Example Problem 8–3.

Solution **Objective** Compute the maximum stress due to bending.

Given Bending moment diagram shown in Figure 8–11. W14 × 43 beam shape.

Analysis Use Equation (8–1). In Figure 8–11, identify the maximum bending moment on the beam to be 91 113 lb·ft at point *F*. Find the values of *I* and *c* from the table of properties for W-beam shapes in Appendix A–7.

$$I = 428 \text{ in}^4$$
$$c = \text{depth}/2 = 13.66 \text{ in}/2 = 6.83 \text{ in}$$

Results $\sigma_{max} = \dfrac{Mc}{I} = 91\,113 \text{ lb·ft} \times \dfrac{12 \text{ in}}{\text{ft}} \times \dfrac{6.83 \text{ in}}{428 \text{ in}^4} = 17\,450 \text{ psi}$

Comment This maximum stress would occur as a tensile stress on the bottom surface of the beam and as a compressive stress on the top surface at the position *F*.

8–7 APPLICATIONS—BEAM DESIGN AND DESIGN STRESSES

To design a beam, its material, length, placement of loads, placement of supports, and the size and shape of its cross section must be specified. Normally, the length, placement of loads, and placement of supports are given by the requirements of the application. Then the material specification and the size and shape of the cross section are determined by the designer.

The primary duty of the designer is to ensure the safety of the design. This requires a stress analysis of the beam and a decision concerning the allowable or design stress to which the chosen material may be subjected. The examples presented here will concentrate on these items. Also of interest to the designer are cost, appearance, physical size, weight, compatibility of the design with other components of the machine or structure, and the availability of the material or beam shape.

Two basic approaches will be shown for beam design. One involves the specification of the *material* from which the beam will be made and the general *shape* of the beam (circular, rectangular, W-beam, etc.), with the subsequent determination of the required dimensions of the cross section of the beam. The second requires specifying the *dimensions* and *shape* of the beam and then computing the required strength of a material from which to make the beam. Then the actual material is specified.

Design stress for metals—General guidelines. When specifying design stresses it is important to keep in mind that both tensile and compressive stresses are produced in beams. If the material is reasonably homogeneous and isotropic having the same strength in either tension or compression, then design is based on the highest stress developed in the beam. When a material has different strengths in tension and compression, as is the case for cast iron or wood, then both the maximum tensile and the maximum compressive stresses must be checked.

The approach used most often in this book to determine design stresses is similar to that first described in Sections 3–3 through 3–6 and it would be wise to review those sections at this time. Table 8–1 lists the design stress guidelines we will use for

TABLE 8–1 Design stress guidelines—bending stresses

Manner of loading	Ductile material	Brittle material
Static	$\sigma_d = s_y/2$	$\sigma_d = s_u/6$
Repeated	$\sigma_d = s_u/8$	$\sigma_d = s_u/10$
Impact or shock	$\sigma_d = s_u/12$	$\sigma_d = s_u/15$

beams in machines and special structures under conditions where loads and material properties are well known. Larger design factors may be used where greater uncertainty exists. We will use Table 8–1 for problems in this book involving metals, unless stated otherwise.

Design stresses from selected codes. Table 8–2 gives a summary of specifications for design bending stresses for structural steel as defined by the American Institute of Steel Construction (AISC) for structural steel and by the Aluminum Association (AA) for aluminum alloys. These data pertain to beams under static loads such as those found in building-type structures.

Additional analysis is required for the parts of beams under compressive stresses because of the possibility of local buckling, especially in shapes having thin sections or extended flanges. Long beams must also be checked for the possibility of twisting. Lateral supports for the compression flanges of long beams are often required to resist the tendency for the beam to twist. See References 1 and 2 for more detailed discussions of these specifications.

Design stresses for nonmetals. When problems involve nonmetals such as wood, plastics, and composites, the concept of yield strength is not typically used. Furthermore, the strengths listed in tables are often based on statistical averages of many tests. Variations in material composition and structure can lead to variations in strength properties. Whenever possible, the actual material to be used in a structure should be tested to determine its strength.

Appendix A–18 lists *allowable* stress values for three species of wood according to the listed grades for applications in building structures and similar static load uses. If load conditions are very well known, a beam can be loaded up to the listed bending stress values. To the extent that there are uncertainties in the conditions of loading, design factors can be applied to the listed values resulting in lower design stresses. No

TABLE 8–2 Design stresses from selected codes—bending stresses–static loads on building-like structures

Structural steel (AISC):

$$\sigma_d = s_y/1.5 = 0.66\, s_y$$

Aluminum (Aluminum Association):

$$\sigma_d = s_y/1.65 = 0.61\, s_y \qquad \text{or} \qquad \sigma_d = s_u/1.95 = 0.51\, s_u$$

whichever is lower

Chapter 8 ▪ Stress Due to Bending

firm guidelines are given here and testing is advised. We will use the listed allowable stresses unless otherwise stated.

The properties of plastics listed in Appendix A–19 can be considered typical for the listed types. But it should be noted that there are many variables involved in producing plastics and it is important to obtain more complete data from manufacturers or by testing the actual material to be used. Also, plastics differ dramatically from one to another in their ability to withstand repeated loading, shock, and impact. In this chapter, we will take the *flexural strength* from Appendix A–19 to be the representative strength of the listed plastics when used in beams. We will assume that failure is imminent at these stress levels. For general, static load cases, we will apply a design factor of $N = 2$ to those values to determine the design stress.

Composites have many advantages when applied to the design of beams because the placement of material can be optimized to provide efficient, lightweight beams. But the resulting structure is typically not homogeneous, so the properties are highly anisotropic. Therefore, the flexure formula as stated in Equations (8–1) and (8–2) cannot be relied upon to give accurate values of stress. General approaches to using composites in beams will be discussed later in this chapter.

8–8 SECTION MODULUS AND DESIGN PROCEDURES

The stress analysis will require the use of the flexure formula

$$\sigma_{max} = \frac{Mc}{I}$$

But a modified form is desirable for cases in which the determination of the dimensions of a section is to be done. Notice that both the moment of inertia I and the distance c are geometrical properties of the cross-sectional area of the beam. Therefore, the quotient I/c is also a geometrical property. For convenience, we can define a new term, *section modulus,* denoted by the letter S.

 Section Modulus

$$S = \frac{I}{c} \tag{8–4}$$

The flexure formula then becomes

$$\sigma_{max} = \frac{M}{S} \tag{8–5}$$

This is the most convenient form for use in design. Example problems will demonstrate the use of the section modulus. It should be noted that some designers use the symbol Z in place of S for section modulus. Appendix A–1 gives formulas for S for some shapes.

Design procedures. Here we show two approaches to design problems. The first applies when the loading pattern and material are known and the shape and dimensions of the cross section of the beam are to be determined. The second procedure applies when the loading pattern, the shape of the beam cross section, and its dimensions have been specified and the objective is to specify a suitable material for the beam to ensure safety.

A. Design procedure to determine the required dimensions for a beam. Given: Loading pattern and material from which the beam is to be made.

   1. Determine the maximum bending moment in the beam, typically by drawing the complete shearing force and bending moment diagrams.

   2. Determine the applicable approach for specifying the design stress from Section 8–7.

   3. Compute the value of the design stress.

   4. Using the flexure formula expressed in terms of the section modulus, Equation (8–5), solve for the section modulus, S. Then let the maximum stress equal the design stress and compute the required minimum value of the section modulus to limit the actual stress to no more than the design stress.

   5. For a specially designed beam shape, determine the required minimum dimensions of the shape to achieve the required section modulus. Then specify the next larger convenient dimensions using the tables of preferred sizes in Appendix A–2.

   6. To select a standard structural shape such as those listed in Appendixes A–4 through A–12, consult the appropriate table of data and specify one having at least the value of the section modulus, S, computed in step 4. Typically, it is recommended that the *lightest* suitable beam shape be specified because the cost of the beam made from a given material is generally directly related to its weight. Reference 1 includes extensive tables for beam shapes with their section modulus values ordered by the weight of the section to facilitate the selection of the lightest beam. Where space limitations exist, the actual dimensions of the shape must be considered.

B. Design procedure for specifying a material for a given beam. Given: Loading pattern, shape, and dimensions for the beam.

   1. Determine the maximum bending moment in the beam, typically by drawing the complete shearing force and bending moment diagrams.

   2. Compute the section modulus for the beam cross section.

   3. Compute the maximum bending stress from the flexure formula, Equation (8–5).

   4. Determine the applicable approach for specifying the design stress from Section 8–7 and specify a suitable design factor.

   5. Set the computed maximum stress from step 3 equal to the formula for the design stress.

6. Solve for the required minimum value of the strength of the material, either s_y or s_u.
7. Select the type of material from which the beam is to be made such as steel, aluminum, cast iron, titanium, or copper.
8. Consult the tables of data for material properties such as those in Appendixes A–13 through A–19 and identify a set of candidate materials having at least the required strength.
9. Considering any factor appropriate to the application, such as ductility, cost, corrosion, potential, ease of fabrication, or weight, specify the material to be used. For metals, it is essential to specify the condition of the material in addition to the alloy.

Example problems are now shown that illustrate these approaches.

**Example Problem
8–4**
A beam is to be designed to carry the static loads shown in Figure 8–12. The cross section of the beam will be rectangular and made from ASTM A36 structural steel plate having a thickness of 1.25 in. Specify a suitable height for the cross section.

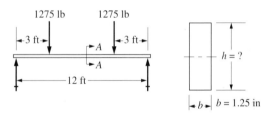

Cross section of beam – Section A-A

FIGURE 8–12   Loading and cross section for beam in Example Problem 8–4.

Solution   **Objective**   Specify the height of the rectangular cross section.

Given   Loading pattern shown in Figure 8–12. ASTM A36 structural steel. Width of the beam to be 1.25 in. Static loads.

Analysis   We will use the design procedure A from this section.

Results   *Step 1.* Figure 8–13 shows the completed shearing force and bending moment diagrams. The maximum bending moment is 45 900 lb·in between the loads, in the middle of the beam span from $x = 3.0$ ft to 9 ft.

Step 2. From Table 8–1, for static load on a ductile material,

$$\sigma_d = s_y/2$$

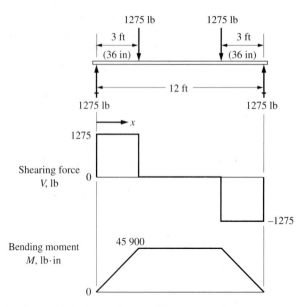

FIGURE 8–13 Load, shearing force, and bending moment diagrams for Example Problem 8–4.

Step 3. From Appendix A–15 $s_y = 36\,000$ psi for A36 steel. For a static load, a design factor of $N = 2$ based on yield strength is reasonable. Then

$$\sigma_d = \frac{s_y}{N} = \frac{36\,000 \text{ psi}}{2} = 18\,000 \text{ psi}$$

Step 4. The required S is

$$S = \frac{M}{\sigma_d} = \frac{45\,900 \text{ lb·in}}{18\,000 \text{ lb/in}^2} = 2.55 \text{ in}^3$$

Step 5. The formula for the section modulus for a rectangular section with a height h and a thickness b is

$$S = \frac{I}{c} = \frac{bh^3}{12(h/2)} = \frac{bh^2}{6}$$

For the beam in this design problem, b will be 1.25 in. Then solving for h gives

$$S = \frac{bh^2}{6}$$

$$h = \sqrt{\frac{6S}{b}} = \sqrt{\frac{6(2.55 \text{ in}^3)}{1.25 \text{ in}}}$$

$$= 3.50 \text{ in}$$

Comment The computed minimum value for *h* is a convenient size. Specify *h* = 3.50 in. The beam will then be rectangular in shape, with dimensions of 1.25 in × 3.50 in. Note that because the beam is rather long, 12 ft, there may be a tendency for it to deform laterally because of elastic instability. Lateral bracing may be required.

Example Problem 8–5 The roof of an industrial building is to be supported by wide-flange beams spaced 4 ft on centers across a 20-ft span, as sketched in Figure 8–14. The roof will be a poured concrete slab, 4 in thick. The design live load on the roof is 200 lb/ft². Specify a suitable wide-flange beam that will limit the stress in the beam to 22 000 psi.

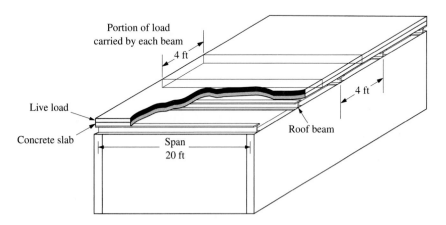

FIGURE 8–14 Roof structure for building in Example Problem 8–5.

Solution **Objective** Specify a suitable wide-flange beam.

Given Loading pattern in Figure 8–15. Design stress = σ_d = 22 000 psi

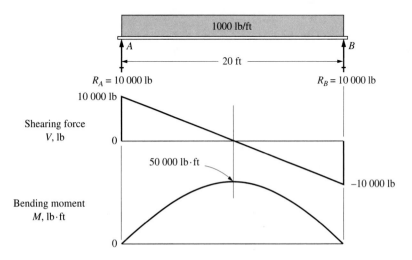

FIGURE 8–15 Load, shearing force, and bending moment diagrams for Example Problem 8–5.

Analysis   Design procedure *A* from this section will be used.

Results   ***Step 1.*** We must first determine the load on each beam of the roof structure. Dividing the load evenly among adjacent beams would result
in each beam carrying a 4-ft-wide portion of the roof load. In addition to the 200-lb/ft² live load, the weight of the concrete slab offers
a sizeable load. In Section 2–10 we find that the concrete weighs
150 lb/ft³. Then each square foot of the roof, 4.0 in thick, would
weigh 50 lb. This is called the *dead load*. Then the total loading due
to the roof is 250 lb/ft². Now, notice that each foot of length of
the beam carries 4 ft² of the roof. Therefore, the load on the beam
is a uniformly distributed load of 1000 lb/ft. Figure 8–15 shows the
loaded beam and the shearing force and bending moment diagrams. The maximum bending moment is 50 000 lb·ft.

Steps 2 and 3 are completed by specifying $\sigma_d = 22\,000$ psi.

Step 4. In order to select a wide-flange beam, the required section modulus
must be calculated.

$$\sigma = \frac{M}{S}$$

$$S = \frac{M}{\sigma_d} = \frac{50\,000 \text{ lb·ft}}{22\,000 \text{ lb/in}^2} \times \frac{12 \text{ in}}{\text{ft}} = 27.3 \text{ in}^3$$

Step 5 does not apply in this problem.

Step 6. A beam must be found from Appendix A–7 which has a value of *S*
greater than 27.3 in³. In considering alternatives, of which there are
many, you should search for the lightest beam that will be safe,
since the cost of the beam is based on its weight. Some possible
beams are

W14 × 26: $S = 35.3$ in³, 26 lb/ft

W12 × 30: $S = 38.6$ in³, 30 lb/ft

Of these, the W14 × 26 would be preferred because it is the
lightest.

Example Problem   A support beam for a conveyor system carries the loads shown in Figure 8–16. Support
8–6   points are at points *A* and *C*. The 20 kN load at *B* and the 10 kN load at *D* are to be ap-

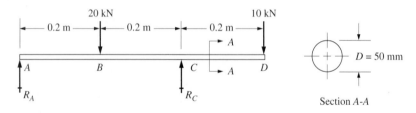

FIGURE 8–16   Beam loading and cross section for beam in Example Problem 8–6.

Chapter 8 ▪ Stress Due to Bending

plied repeatedly many thousands of times. It has been proposed to use a 50 mm diameter circular steel bar for the beam. Specify a suitable steel for the beam.

Solution   **Objective**   Specify a suitable steel.

Given   Loading pattern in Figure 8–16. Loads are repeated.
Beam is to be circular, $D = 50$ mm.

Analysis   Use the type *B* procedure given in this section.

Results   **Step 1.** Figure 8–17 shows the completed shearing force and bending moment diagrams. The maximum bending moment is 2.00 kN·m at the support at *C*.

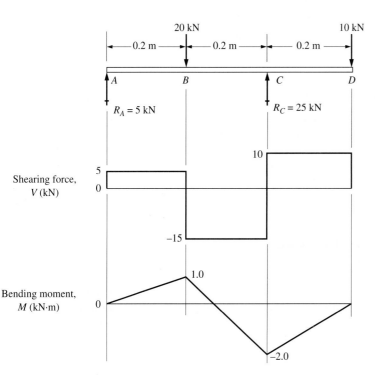

FIGURE 8–17   Shearing force and bending moment diagrams for Example Problem 8–6.

Step 2. Appendix A–1 gives the formula for *S* for a round bar.

$$S = \pi D^3/32 = \pi(50 \text{ mm})^3/32 = 12\,272 \text{ mm}^3$$

Step 3. Using Equation (8–5),

$$\sigma_{max} = \frac{M}{S} = \frac{2.0 \text{ kN·m}}{12\,272 \text{ mm}^3} \times \frac{10^3 \text{ mm}}{\text{m}} \times \frac{10^3 \text{ N}}{\text{kN}}$$

$$\sigma_{max} = 163 \text{ N/mm}^2 = 163 \text{ MPa}$$

Step 4. We can use Table 8–1 to determine an appropriate formula for design stress. The steel selected should be highly ductile because of the repeated load. Then we will use $\sigma_d = s_u/8$.

Step 5. Let $\sigma_{max} = 163$ MPa $= \sigma_d = s_u/8$

Step 6. Solving for s_u gives

$$s_u = 8(\sigma_{max}) = 8(163 \text{ MPa}) = 1304 \text{ MPa}$$

Step 7. It was decided to use steel.

Step 8. Appendix A–13 lists several common steel alloys. From that table, we can select candidate materials that have good ductility and an ultimate strength of at least 1304 MPa. Four are listed here.

AISI 1080 OQT 700; $s_u = 1303$ MPa; 12% elongation

AISI 1141 OQT 700; $s_u = 1331$ MPa; 9% elongation

AISI 4140 OQT 700; $s_u = 1593$ MPa; 12% elongation

AISI 5160 OQT 900; $s_u = 1351$ MPa; 12% elongation

Step 9. For applications to beams carrying repeated loads, it is typical to use a medium carbon steel. Either the AISI 4140 or the AISI 5160 could be used. With 12% elongation, ductility should be adequate.

Comment Note in Appendix A–13 that AISI 4140 OQT 900 has an ultimate strength of 1289 MPa and 15% elongation. The strength is within 2% of the computed value. It may be suitable to specify this material to gain better ductility. A slight reduction in the design factor would result. But, because the values in Table 8–1 are somewhat conservative, this would normally be justified.

8–9 STRESS CONCENTRATIONS

The conditions specified for valid use of the flexure formula in Section 8–3 included the statement that the beam must have a uniform cross section. Changes in cross section result in higher local stresses than would be predicted from the direct application of the flexure formula. Similar observations were made in earlier chapters concerning direct axial stresses and torsional shear stresses. The use of *stress concentration factors* will allow the analysis of beams that do include changes in cross section.

In the design of round shafts for carrying power-transmitting elements, the use of steps in the diameter is encountered frequently. Examples were shown in Chapter 5, where torsional shear stresses were discussed. Figure 8–18 shows such a shaft. Considering the shaft as a beam subjected to bending moments, there would be stress concentrations at the shoulder (2), the keyseat (3), and the groove (4).

At sections where stress concentrations occur, the stress due to bending would be calculated from a modified form of the flexure formula.

Flexure Formula with Stress Concentration

$$\sigma_{max} = \frac{McK_t}{I} = \frac{MK_t}{S} \tag{8–6}$$

The stress concentration factor K_t is found experimentally, with the values reported in graphs such as those in Appendix A–21, cases 4, 5, 8, 9, 10, and 11.

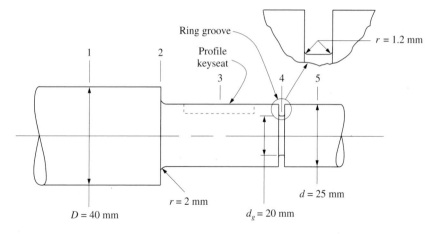

FIGURE 8–18 Portion of a shaft with several changes in cross section producing stress concentrations.

Figure 8–18 shows a portion of a round shaft where a gear is mounted. A bending moment of 30 N·m is applied at this location. Compute the stress due to bending at sections 2, 3, and 4.

Solution   **Objective**   Compute the stress due to bending at sections 2, 3, and 4.

Given   Beam geometry in Figure 8–18. $M = 30$ N·m.

Analysis   Stress concentrations must be considered because of the several changes in geometry in the area of interest. Equation (8–6) will be used to compute the maximum stress at each section. Appendix 21 is the source of data for stress concentration factors, K_t.

Results   **Section 2.** The step in the shaft causes a stress concentration to occur. Then the stress is

$$\sigma_2 = \frac{MK_t}{S}$$

The smaller of the diameters at section 2 is used to compute S. From Appendix A–1,

$$S = \frac{\pi d^3}{32} = \frac{\pi (25 \text{ mm})^3}{32} = 1534 \text{ mm}^3$$

The value of K_t depends on the ratios r/d and D/d. (See Appendix 21–9)

$$\frac{r}{d} = \frac{2 \text{ mm}}{25 \text{ mm}} = 0.08$$

$$\frac{D}{d} = \frac{40 \text{ mm}}{25 \text{ mm}} = 1.60$$

$$\sigma_2 = \frac{MK_t}{S} = \frac{(30 \text{ N·m})(1.87)}{1534 \text{ mm}^3} \times \frac{10^3 \text{ mm}}{\text{m}} = 36.6 \text{ N/mm}^2$$

$$= 36.6 \text{ MPa}$$

Section 3. The keyseat causes a stress concentration factor of 2.0 as listed in Appendix A–21–11. Then

$$\sigma_3 = \frac{MK_t}{S} = \frac{(30 \text{ N·m})(2.0)}{1534 \text{ mm}^3} \times \frac{10^3 \text{ mm}}{\text{m}} = 39.1 \text{ N/mm}^2$$

$$= 39.1 \text{ MPa}$$

Section 4. The groove requires the use of Appendix A–21–9 again to find K_t. Note that the nominal stress is based on the root diameter of the groove, d_g. For the groove,

$$\frac{r}{d_g} = \frac{1.2 \text{ mm}}{20 \text{ mm}} = 0.06$$

$$\frac{d}{d_g} = \frac{25 \text{ mm}}{20 \text{ mm}} = 1.25$$

Then, $K_t = 1.93$. The section modulus at the root of the groove is

$$S = \frac{\pi d_g^3}{32} = \frac{\pi (20 \text{ mm})^3}{32} = 785 \text{ mm}^3$$

Now the stress at section 4 is

$$\sigma_4 = \frac{MK_t}{S} = \frac{(30 \text{ N·m})(1.93)}{785 \text{ mm}^3} \times \frac{10^3 \text{ mm}}{\text{m}} = 73.8 \text{ N/mm}^2$$

$$= 73.8 \text{ MPa}$$

Comment   The stress at section 4 at the ring groove is by far the largest because of the small diameter and the relatively high value of K_t. Design of the shaft to determine a suitable material must use this level of stress as the actual maximum stress.

Example Problem 8–8   Figure 8–19 shows a cantilever bracket carrying a partial uniformly distributed load over the left 10 in and a concentrated load at its right end. The geometry varies at sections A, B, and C as shown. The bracket is made from aluminum 2014-T6 and it is desired to have a minimum design factor of 8 based on the ultimate strength. Evaluate the acceptability of the given design. If any section is unsafe, propose a redesign that will result in a satisfactory stress level. Consider stress concentrations at sections B and C. The attachment at A is blended smoothly such that it can be assumed that $K_t = 1.0$.

Solution   Objective   Evaluate the beam shown in Figure 8–19 to ensure that the minimum design factor is 8 based on ultimate strength. If not, redesign the beam.

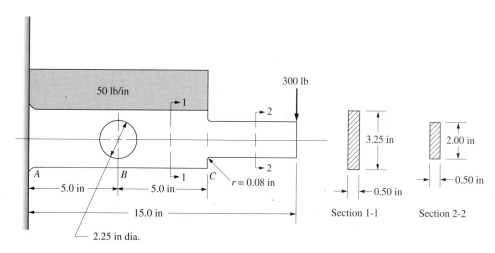

FIGURE 8–19 Beam and loading for Example Problem 8–8.

Given Loading and beam geometry in Figure 8–19. Aluminum 2014-T6.

Analysis 1. The shearing force and bending moment diagrams will be drawn.

2. The design stress will be computed from $\sigma_d = s_u/8$.

3. The stress will be computed at sections *A*, *B*, and *C*, considering stress concentrations at *B* and *C*. These are the three likely points of failure because of either bending moment or stress concentration. Everywhere else will have a lower bending stress. At each section, the stress will be computed from $\sigma = MK_t/S$, and the values of *M*, K_t, and *S* must be determined at each section.

4. The computed stresses will be compared with the design stress.

5. For any section having a stress higher than the design stress, a redesign will be proposed and the stress will be recomputed to verify that it is safe as redesigned.

Results *Step 1.* Figure 8–20 shows the completed shearing force and bending moment diagrams. Note the values of the bending moment at sections *B* and *C* have been computed. You should verify the given values.

Step 2. From Appendix A–17, we find $s_u = 70$ ksi $= 70\,000$ psi. Then,

$$\sigma_d = s_u/8 = (70\,000)/8 = 8750 \text{ psi}$$

Steps 3 and 4. At each section, $\sigma = MK_t/S = K_t\sigma_{\text{nom}}$.

Section A: $K_t = 1.0$ (given). $M_A = 7000$ lb·in.
Dimensions: $b = 0.50$ in; $h = 3.25$ in; rectangle.

$$S = bh^2/6 = (0.50 \text{ in})(3.25 \text{ in})^2/6 = 0.880 \text{ in}^3$$
$$\sigma = \frac{(7000 \text{ lb·in})(1.0)}{0.880 \text{ in}^3} = 7953 \text{ psi} < 8750 \text{ psi} \quad \textbf{OK}$$

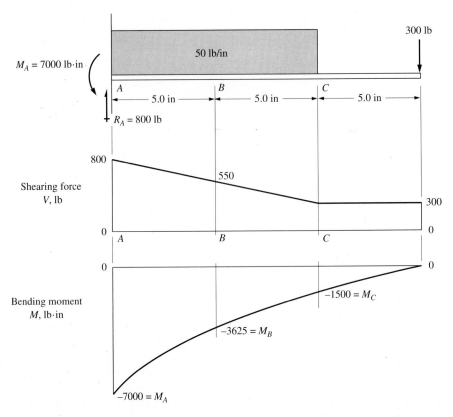

FIGURE 8–20 Shearing force and bending moment diagrams for Example Problem 8–8.

Section B: $M_B = 3625$ lb·in. Find K_t from Appendix A–21–4.
Dimensions: $t = b = 0.50$ in; $w = h = 3.25$ in; $d = 2.25$ in
$d/w = 2.25/3.25 = 0.692$; then $K_t = 1.40$ (curve C).

$$\sigma = K_t \sigma_{nom} = \frac{K_t(6Mw)}{(w^3 - d^3)t} = \frac{(1.40)(6)(3625)(3.25)}{[(3.25)^3 - (2.25)^3](0.50)}$$

$$\sigma = 8629 \text{ psi} < 8750 \text{ psi} \quad \textbf{OK}$$

Section C: $M_C = 1500$ lb·in. Find K_t from Appendix A–21–10.
Dimensions: $t = 0.50$ in; $H = 3.25$ in; $h = 2.00$ in; $r = 0.08$ in
$H/h = 3.25/2.0 = 1.625$; $r/h = 0.08/2.00 = 0.04$.
Then, $K_t = 2.40$.

Section modulus $= S = th^2/6 = (0.50)(2.00)^2/6 = 0.333$ in³

$$\sigma = \frac{MK_t}{S} = \frac{(1500 \text{ lb·in})(2.40)}{0.333 \text{ in}^3} = 10\,800 \text{ psi} \quad \textbf{too high}$$

Step 5. Proposed redesign at C: Because the stress concentration factor at section *C* is quite high, increase the fillet radius. The maximum

Chapter 8 ▪ Stress Due to Bending

allowable stress concentration factor is found by solving the stress equation for K_t and letting $\sigma = \sigma_d = 8750$ psi. Then,

$$K_t = \frac{S\sigma_d}{M} = \frac{(0.333 \text{ in}^3)(8750 \text{ lb/in}^2)}{1500 \text{ lb·in}} = 1.94$$

From Appendix A−21−10, the minimum value of $r/h = 0.08$ to limit K_t to 1.94. Then, $r_{min} = 0.08(h) = 0.08(2.00) = 0.16$. Let $r = 0.20$ in; $r/h = 0.20/2.00 = 0.10$; $K_t = 1.80$. Then,

$$\sigma = \frac{MK_t}{S} = \frac{(1500 \text{ lb·in})(1.80)}{0.333 \text{ in}^3} = 8100 \text{ psi} \quad \textbf{OK}$$

Comment   This problem is a good illustration of the necessity of analyzing any point within a beam where high stress may occur because of high bending moment, high stress concentration, small section modulus, or some combination of these. It also demonstrates one method of redesigning a beam to ensure safety.

8−10 FLEXURAL CENTER (SHEAR CENTER)

The flexure formula is valid for computing the stress in a beam provided the applied loads pass through a point called the *flexural center,* or sometimes, the *shear center.* If a section has an axis of symmetry and if the loads pass through that axis, then they also pass through the flexural center. The beam sections shown in Figure 8−4 are of this type.

For sections loaded away from an axis of symmetry, the position of the flexural center, indicated by Q, must be found. Such sections were identified in Figure 8−5.

In order to result in pure bending, the loads must pass through Q, as shown in Figure 8−21. If they don't, then a condition of *unsymmetrical bending* occurs and other analyses would have to be performed which are not discussed in this book. The sections of the type shown in Figure 8−21 are used frequently in structures. Some lend themselves nicely to production by extrusion and are therefore very economical. But because of the possibility of producing unsymmetrical bending, care must be taken in their application.

Example Problem 8−9   Determine the location of the flexural center for the two sections shown in Figure 8−22.

Solution   **Objective**   Locate the shear center, Q, for the two shapes.

Given   Shapes in Figure 8−22: channel in 8−22(a); hat section in 8−22(b).

Analysis   The general location of the shear center for each shape is shown in Figure 8−21 along with the means of computing the value of e that locates Q relative to specified features of the shapes.

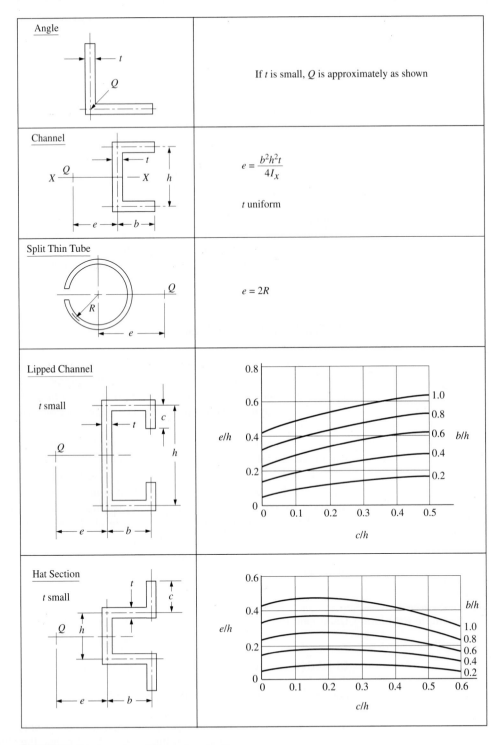

FIGURE 8–21 Location of flexural center Q.

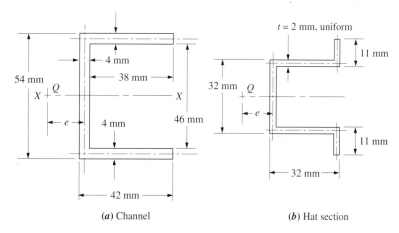

(a) Channel   (b) Hat section

FIGURE 8–22   Beam sections for which the flexural centers are computed in Example Problem 8–9.

Results   ***Channel Section (a).*** From Figure 8–21, the distance e to the flexural center is

$$e = \frac{b^2 h^2 t}{4 I_x}$$

Note that the dimensions b and h are measured to the middle of the flange or web. Then $b = 40$ mm and $h = 50$ mm. Because of symmetry about the X axis, I_x can be found by the difference between the value of I for the large outside rectangle (54 mm by 42 mm) and the smaller rectangle removed (46 mm by 38 mm).

$$I_x = \frac{(42)(54)^3}{12} - \frac{(38)(46)^3}{12} = 0.243 \times 10^6 \text{ mm}^4$$

Then

$$e = \frac{(40)^2(50)^2(4)}{4(0.243 \times 10^6)} \text{ mm} = 16.5 \text{ mm}$$

This dimension is drawn to scale in Figure 8–22(a).

Hat Section (b).   Here the distance e is a function of the ratios c/h and b/h.

$$\frac{c}{h} = \frac{10}{30} = 0.3$$

$$\frac{b}{h} = \frac{30}{30} = 1.0$$

Then from Figure 8–21, $e/h = 0.45$. Solving for e yields

$$e = 0.45h = 0.45(30 \text{ mm}) = 13.5 \text{ mm}$$

This dimension is drawn to scale in Figure 8–22(b).

Comment   Now, can you devise a design for using either section as a beam and provide for the application of the load through the flexural center Q to produce pure bending?

8-11 PREFERRED SHAPES FOR BEAM CROSS SECTIONS

Recall the discussion earlier in this chapter of the stress distribution in the cross section of a beam characterized by the equations,

$$\sigma_{max} = Mc/I = M/S \quad \text{at the outermost fiber of the beam}$$
$$\sigma = \sigma_{max}(y/c) \quad \text{at any point at a distance } y \text{ from neutral axis}$$

Figures 8–6 and 8–7 illustrate the stress distribution. It must be understood that these equations apply strictly only to beams made from homogeneous, isotropic materials; that is, those having equal properties in all directions.

Because the larger stresses occur near the top and bottom of the cross section, the material there provides more of the resistance to the externally applied bending moment than material nearer to the neutral axis. It follows that it is desirable to place the larger part of the material away from the neutral axis to obtain efficient use of the material. In this discussion, efficiency refers to maximizing the moment of inertia and section modulus of the shape for a given amount of material, as indicated by the area of the cross section.

Figure 8–23 shows several examples of efficient shapes for beam cross sections. These illustrations are based on the assumption that the most significant stress is a bending stress caused by loads acting perpendicular to the neutral axis on top of the beam. Examples are shown in Figures 8–1, 8–3, 8–12, and 8–15. In such cases, it is said that the bending is positive about the horizontal neutral axis. It is also assumed that the material is equally strong in compression and tension.

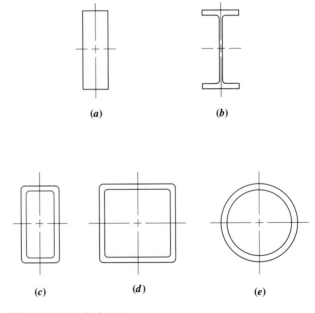

FIGURE 8–23   Efficient shapes for beams.

Starting with the simple rectangular shape in Figure 8–23(a), it is preferred to orient the long dimension vertical as shown because the moment of inertia is proportional to the *cube* of the height of the rectangle, where the height is the dimension perpendicular to the neutral axis. For example, consider the case of a rectangle having dimensions of 40 mm by 125 mm and compare the resulting values of *I* and *S*.

	125 mm dimension vertical	40 mm dimension vertical
$I = bh^3/12$	$I_1 = (40)(125)^3/12 = 6.51 \times 10^6 \text{ mm}^4$	$I_2 = (125)(40)^3/12 = 0.667 \times 10^6 \text{ mm}^4$
$S = bh^2/6$	$S_1 = (40)(125)^2/6 = 1.04 \times 10^5 \text{ mm}^3$	$S_2 = (125)(40)^2/6 = 0.333 \times 10^5 \text{ mm}^3$

Comparing these results gives

$$\frac{I_1}{I_2} = \frac{6.51 \times 10^6 \text{ mm}^4}{0.667 \times 10^6 \text{ mm}^4} = 9.76 \qquad \frac{S_1}{S_2} = \frac{1.04 \times 10^5 \text{ mm}^3}{0.333 \times 10^5 \text{ mm}^3} = 3.12$$

The comparison of the values of the section modulus, *S*, is the most pertinent for comparing stresses in beams because it contains both the moment of inertia, *I*, and the distance, *c*, to the outermost fiber of the beam cross section. While the section with the long dimension vertical has a moment of inertia almost ten times that with the long dimension horizontal, it is over three times taller, resulting in the improvement in the section modulus of approximately three times. Still, that is a significant improvement.

A related factor in the comparison of beam shapes is that the *deflection* of a beam is inversely proportional to the moment of inertia, *I*, as will be shown in Chapter 12. Therefore, the taller rectangular beam in the above example would be expected to deflect only 1/9.76 times as much as the shorter one, only about 10% as much.

The shape shown in Figure 8–23(b) is the familiar "I-beam." Placing most of the material in the horizontal flanges at the top and bottom of the section puts them in the regions of the highest stresses for maximum resistance to the bending moment. The relatively thin vertical web serves to hold the flanges in position and provides resistance to shearing forces, as described in Chapter 9. It would be well to study the proportions of standard steel and aluminum I-shaped sections listed in Appendixes A–7, A–8, and A–11 to get a feel for reasonable flange and web thicknesses. The thickness of that flange which is in compression is critical with regard to buckling when the beam is relatively long. References 1 and 2 give data on proper proportions.

The tall, rectangular tube shown in Figure 8–23(c) is very similar to the I-shape in its resistance to bending moments caused by vertical loads. The two vertical sides serve a similar function as the web of the I-shape. In fact, the moment of inertia with respect to the horizontal centroidal axis for the tube in (c) would be identical to that for the I-shape in (b) if the thickness of the top and bottom horizontal parts were equal and if the vertical sides of the tube were each 1/2 the thickness of the web of the I-shape. The tube is superior to the I-shape when combinations of loads are encountered that cause bending about the vertical axis in addition to the horizontal axis, because the placement of the vertical sides away from the *Y-Y* axis increases the moment of inertia with respect to that axis. The tube is also superior when any torsion is applied, as discussed in Chapter 5. When torsion or bending about the vertical axis is significant, it may be preferred to use the square tube shape shown in Figure 8–23(d).

Circular tubes as shown in Figure 8–23(e) make very efficient beams for the same reasons listed for square tubes above. And they are superior to square tubes when both bending and torsion are present in significant amounts. An obvious example of where a circular tube is preferred is the case of a rotating shaft carrying both bending and torsional loads such as the drive shaft and the axles of a car or truck.

Shapes made from thin materials. Economical production of beams having moderate dimensions can be done using rollforming or pressforming of relatively thin flat sheet materials. Aluminum and many plastics are extruded to produce shapes having a uniform cross section, often with thin walls and extended flanges. Examples are shown in Figures P7–10 through P7–20. Such shapes can be specially adapted to the use of the beam. See if you can identify beam-like members with special shapes around you. In your home you might find such beams used as closet door rails, curtain rods, structures for metal furniture, patio covers or awnings, ladders, parts for plastic toys, tools in the workshop, or parts for appliances or lawn maintenance tools. In your car, look at the windshield wiper arms, suspension members, gear shift levers, linkages or brackets in the engine compartment, and the bumpers. Aircraft structures contain numerous examples of thin-walled shapes designed to take advantage of their very light weight.

Figure 8–24 shows three examples of extruded or roll-formed shapes found around the home. Part (a) shows a closet door rail where the track for the roller that supports the door is produced as an integral part of the aluminum extrusion. The extruded side rail of an aluminum extension ladder is sketched in part (b). Part (c) shows a portion of the roll-formed decking for a patio cover made from aluminum sheet only 0.025 in (0.64 mm) thick. The shape is specially designed to link together to form a continuous panel to cover a wide area. Some design features of these sections should be noted. Extended flanges are reinforced with bulb-like projections to provide local stiffness that resists wrinkling or buckling of the flanges. Broad flat areas are stiffened by ribs or roll-formed corrugations, also to inhibit local buckling. References 1 and 2 provide guidelines for the design of such features.

Beams made from anisotropic materials. The design of beams to be made from materials having different strengths in tension and compression requires special care. Most types of cast iron, for example, have a much higher compressive strength than tensile strength. Appendix A–16 lists the properties of malleable iron ASTM A220, grade 80002 as,

Ultimate tensile strength: $\qquad$ $s_u = 655$ MPa (95 ksi)
Ultimate compressive strength: $\quad$ $s_{uc} = 1650$ MPa (240 ksi)

An efficient beam shape that would account for this difference is the modified I-shape shown in Figure 8–25. Because the typical, positive bending moment places the bottom flange in tension, providing a larger bottom flange moves the neutral axis down and tends to decrease the resulting tensile stress in the bottom flange relative to the compressive stress in the top flange. Example Problem 8–10 illustrates the result with the design factor based on tensile strength being nearly equal to that based on compressive strength.

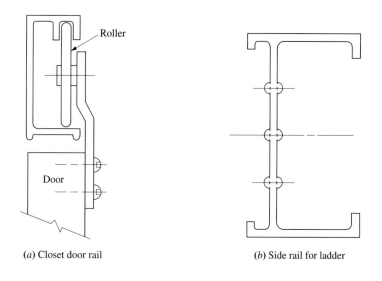

(a) Closet door rail

(b) Side rail for ladder

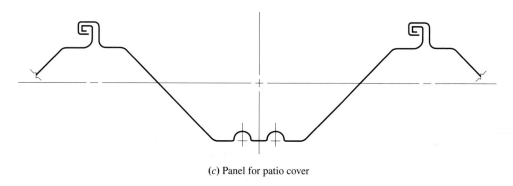

(c) Panel for patio cover

FIGURE 8–24 Examples of thin-walled beam sections.

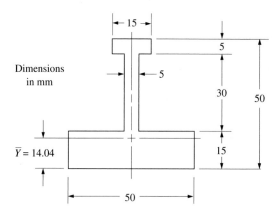

FIGURE 8–25 Nonsymmetrical shape for cross section of beam in Example Problem 8–10.

Figure 8–25 shows the cross section of a beam that is to be made from malleable iron, ASTM A220, grade 80002. The beam is subjected to a maximum bending moment of 1025 N·m, acting in a manner to place the bottom of the beam in tension and the top in compression. Compute the resulting design factor for the beam based on the ultimate strength of the iron. The moment of inertia for the cross section is 1.80×10^5 mm^4.

Solution **Objective** Compute the design factor based on the ultimate strength.

Given Beam shape shown in Figure 8–25. $I = 1.80 \times 10^5$ mm^4. $M = 1025$ N·m. Material is malleable iron, ASTM A220, grade 80002.

Analysis Because the beam cross section is not symmetrical, the value of the maximum tensile stress at the bottom of the beam, σ_{tb}, will be lower than the maximum compressive stress at the top, σ_{ct}. We will compute:

$$\sigma_{tb} = Mc_b/I \quad \text{and} \quad \sigma_{ct} = Mc_t/I$$

where $c_b = \overline{Y} = 14.04$ mm and $c_t = 50 - 14.04 = 35.96$ mm. The tensile stress at the bottom will be compared with the ultimate tensile strength to determine the design factor based on tension, N_t, from

$$\sigma_{tb} = s_u/N_t \quad \text{or} \quad N_t = s_u/\sigma_{tb}$$

where $s_u = 655$ MPa from Appendix A–16. Then the compressive stress at the top will be compared with the ultimate compressive strength to determine the design factor based on compression, N_c, from

$$\sigma_{ct} = s_{uc}/N_c \quad \text{or} \quad N_c = s_{uc}/\sigma_{ct}$$

where $s_{uc} = 1650$ MPa from Appendix A–16. The lower of the two values of N will be the final design factor for the beam.

Results At the bottom of the beam,

$$\sigma_{tb} = \frac{Mc_b}{I} = \frac{(1025 \text{ N·m})(14.04 \text{ mm})}{1.80 \times 10^5 \text{ mm}^4} \cdot \frac{(1000 \text{ mm})}{\text{m}} = 79.95 \text{ MPa}$$

$$N_t = s_u/\sigma_{tb} = 655 \text{ MPa}/79.95 \text{ MPa} = 8.19$$

At the top of the beam,

$$\sigma_{ct} = \frac{Mc_t}{I} = \frac{(1025 \text{ N·m})(35.96 \text{ mm})}{1.80 \times 10^5 \text{ mm}^4} \cdot \frac{(1000 \text{ mm})}{\text{m}} = 204.8 \text{ MPa}$$

$$N_c = s_{uc}/\sigma_{ct} = 1650 \text{ MPa}/204.8 \text{ MPa} = 8.06$$

Comment The compressive stress at the top of the beam is the limiting value in this problem because the smaller design factor exists there. But note that the two values of the design factor were quite close to being equal, indicating that the shape of the cross section has been reasonably well optimized for the different strengths in tension and compression.

8–12 DESIGN OF BEAMS TO BE MADE FROM COMPOSITE MATERIALS

Composite materials, discussed in Chapter 2, have superior properties when applied to beam design because of the ability to tailor the constituents for the composite and their placement in the beam. Composite processing often allows for unique shapes to be designed that optimize the geometry of the structure with regard to the magnitude and direction of loads to be carried. Combining these features with the inherent advantages of composites in terms of high strength-to-weight and stiffness-to-weight ratios make them very desirable for use in beams.

The discussion of Section 8–10 applies equally well to the design of composite beams. The designer should select a shape for the beam cross section that is, itself, efficient in resisting bending moments. In addition, the designer can call for the placement of a higher concentration of the stronger, stiffer fibers in the regions where the higher stresses would be expected: namely at the outermost fibers of the beam farthest from the neutral axis. More plies of a fabric-type filler can be placed in the high-stress regions.

An effective technique for composite beam design is to employ a very light core material for a structure made from a rigid foam or a honeycomb material, covered by relatively thin layers of the strong, stiff fibers in a polymer matrix. If it is known that the bending moments will always act in the same direction, the fibers of the composite can be aligned with the direction of the tensile and compressive stresses in the beam. If the bending moments are expected to act in a variety of directions, a more disperse placement of fibers can be specified or fabric plies can be placed at a variety of angles, as suggested in Figure 2–13.

Care must be exercised in designing and testing composite beam structures because of the multiple possible modes of failure. The structure may fail in the high tensile stress region by failure of the fibers or the matrix or by the disengagement of the fibers from the matrix. But perhaps a more likely failure mode for a laminated composite is interlaminar shear failure in regions of high shear stress near the neutral axis, as discussed in Chapter 9. Failure could also occur in the compressive stress region by local buckling of the shape or by delamination.

When the beam has been designed with the assumption of bending in a certain plane, it is essential that loads are properly applied and that the shape itself promotes pure bending rather than a combination of bending and torsion. The discussion of *flexural center,* Section 8–9, should be reviewed.

The shape and dimensions of the beam cross section can be varied to correspond to the magnitude of the bending moment at various positions in a beam. For example, a cantilever carrying a concentrated load at its end experiences the highest bending moment at the support point and the magnitude of the bending moment decreases linearly out to the end of the beam. Then the cross section can be deeper at the support and progressively smaller toward the end. A simply supported beam with a load at the center has its highest bending moment at the center, decreasing toward each support. Then the beam can be thicker at the center and thinner toward the ends.

Beams with broad, flat or curved surfaces, such as the wings of an aircraft, must be designed for stiffness of the broad panels as well as for adequate strength. The skin of the panel may have to be supported by internal ribs to break it into smaller areas.

Penetrations in a composite beam must be carefully designed to ensure smooth transfer of loads from one part of the beam to another. If feasible, the placement of penetrations should be in regions of low stress. Similarly, fasteners must be carefully designed to ensure proper engagement in the fibrous composite material. Thickened bosses may be provided when fasteners are to be located. It may be possible to minimize the number of fasteners by clever shaping of the structure such as by molding in brackets integral with the main structure.

In summary, the designer of composite beams must carefully analyze the stress distribution in the beam and attempt to optimize the placement of material to optimize the shape and dimensions of the beam. The designer must visualize the path of load transfer from its point of application to the ultimate point of support.

REFERENCES

1. Aluminum Association, *Specifications for Aluminum Structures,* Washington, D.C., 1986.

2. American Institute of Steel Construction, *Manual of Steel Construction,* 9th ed., Chicago, IL, 1989.

3. ASM INTERNATIONAL, *Composites, Engineered Materials Handbook, Volume 1,* Metals Park, OH, 1987.

4. *Design Guide for Advanced Composites Applications,* Advanstar Communications, Inc., Duluth, MN, 1993.

5. Mallick, P.K., *Fiber-Reinforced Composites: Materials, Manufacturing, and Design,* Marcel Dekker, New York, 1988.

6. Weeton, John W., Dean M. Peters, and Karyn L. Thomas, eds., *Engineers' Guide to Composite Materials,* ASM INTERNATIONAL, Metals Park, OH, 1987.

PROBLEMS

Analysis of Bending Stresses

8–1.M A square bar 30 mm on a side is used as a simply supported beam subjected to a bending moment of 425 N·m. Compute the maximum stress due to bending in the bar.

8–2.M Compute the maximum stress due to bending in a round rod 20 mm in diameter if it is subjected to a bending moment of 120 N·m.

8–3.E A bending moment of 5800 lb·in is applied to a beam having a rectangular cross section with dimensions of 0.75 in × 1.50 in. Compute the maximum bending stress in the beam (a) if the vertical side is 1.50 in, and (b) if the vertical side is 0.75 in.

8–4.E A wood beam carries a bending moment of 15 500 lb·in. It has a rectangular cross section 1.50 in wide by 7.25 in high. Compute the maximum stress due to bending in the beam.

8–5.E The loading shown in Figure P6–4 is to be carried by a W12 × 16 steel beam. Compute the stress due to bending.

8–6.E An American Standard beam, S12 × 35, carries the load shown in Figure P6–11. Compute the stress due to bending.

8–7.E The 24-in long beam shown in Figure P6–10 is an aluminum channel, C4 × 2.331, positioned with the legs down so that the flat 4-in surface can carry the applied loads. Compute the maximum tensile and maximum compressive stresses in the channel.

8–8.E The 650-lb load at the center of the 28-in-long bar shown in Figure P6–1 is carried by a standard steel pipe, $1\frac{1}{2}$-in schedule 40. Compute the stress in the pipe due to bending.

8–9.M The loading shown in Figure P6–7 is to be carried by the fabricated beam shown in Figure P7–28. Compute the stress due to bending in the beam.

8–10.C An aluminum I-beam, I9 × 8.361, carries the load shown in Figure P6–8. Compute the stress due to bending in the beam.

8–11.E A part of a truck frame is composed of two channel-shaped members, as shown in Figure 8–26. If the moment at the section is 60 000 lb·ft, compute the bending stress in the frame. Assume that the two channels act as a single beam.

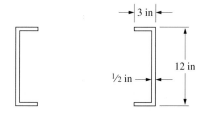

FIGURE 8–26 Truck frame members for Problem 8–11.

Design of Beams

8–12.M Compute the required diameter of a round bar used as a beam to carry a bending moment of 240 N·m with a stress no greater than 125 MPa.

8–13.M A rectangular bar is to be used as a beam subjected to a bending moment of 145 N·m. If its height is to be three times its width, compute the required dimensions of the bar to limit the stress to 55 MPa.

8–14.M The tee section shown in Figure P7–4 is to carry a bending moment of 28.0 kN·m. It is to be made of steel plates welded together. If the load on the beam is a dead load, would AISI 1020 hot-rolled steel be satisfactory for the plates?

8–15.M The modified I-section shown in Figure P7–5 is to be extruded aluminum. Specify a suitable aluminum alloy if the beam is to carry a repeated load resulting in a bending moment of 275 N·m.

8–16.E A standard steel pipe is to be used as a chinning bar for personal exercise. The bar is to be 42 in long and simply supported at its ends. Specify a suitable size pipe if the bending stress is to be limited to 10 000 psi and a 280-lb man hangs by one hand in the middle.

8–17.E A pipeline is to be supported above ground on horizontal beams, 14 ft long. Consider each beam to be simply supported at its ends. Each beam carries the combined weight of 50 ft of 48-in-diameter pipe and the oil flowing through it, about 42 000 lb. Assuming the load acts at the center of the beam, specify the required section modulus of the beam to limit the bending stress to 20 000 psi. Then specify a suitable wide flange or American Standard beam.

8–18.E A wood platform is to be made of standard plywood and finished lumber using the cross section shown in Figure P7–23. Would the platform be safe if four men, weighing 250 lb each, were to stand 2 ft apart, as shown in Figure 8–27? Consider only bending stresses (see Chapter 9 for shear stresses).

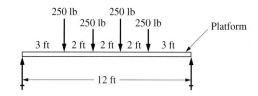

FIGURE 8–27 Platform loading for Problem 8–18.

8–19.E A diving board has a hollow rectangular cross section 30 in wide and 3.0 in thick and is supported as shown in Figure 8–28. Compute the maximum stress due to bending in the board if a 300-lb person stands at the end. Would the

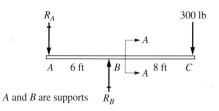

(*a*) Loads on diving board

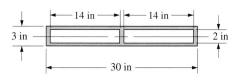

(*b*) Section *A-A* through board

FIGURE 8–28 Diving board for Problem 8–19.

board be safe if it was made of extruded 6061-T4 aluminum and the person landed at the end of the board with an impact?

8–20.M The loading shown in Figure P6–6 is to be carried by an extruded aluminum hat-section beam having the cross section shown in Figure P7–11. Compute the maximum stress due to bending in the beam. If it is made of extruded 6061-T4 aluminum and the loads are dead loads, would the beam be safe?

8–21.M The extruded shape shown in Figure P7–12 is to be used to carry the loads shown in Figure P6–5, which is a part of a business machine frame. The loads are due to a motor mounted on the frame and can be considered dead loads. Specify a suitable aluminum alloy for the beam.

8–22.M A beam is being designed to support the loads shown in Figure 8–29. The four shapes pro-

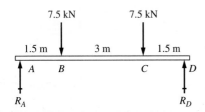

FIGURE 8–29 Beam for Problem 8–22.

posed are (a) a round bar, (b) a square bar, (c) a rectangular bar with the height made four times the thickness, and (d) the lightest American Standard beam. Determine the required dimensions of each proposed shape to limit the maximum stress due to bending to 80 MPa. Then compare the magnitude of the cross-sectional areas of the four shapes. Since the weight of the beam is proportional to its area, the one with the smallest area will be the lightest.

8–23.E A children's play gym includes a cross beam carrying four swings, as shown in Figure 8–30. Assume that each swing carries 300 lb. It is desired to use a standard steel pipe for the beam, keeping the stress due to bending below 10 000 psi. Specify the suitable size pipe for the beam.

8–24.E A 60 in long beam simply supported at its ends is to carry two 4800 lb loads, each placed 14 in from an end. Specify the lightest suitable steel tube for the beam, either square or rectangular, to produce a design factor of 4 based on yield

strength. The tube is to be cold formed from ASTM A500, grade A steel.

8–25.E Repeat Problem 8–24, but specify the lightest standard aluminum I-beam from Appendix A–11. The beam will be extruded using alloy 6061-T6.

8–26.E Repeat Problem 8–24, but specify the lightest wide-flange steel shape from Appendix A–7. The beam will be made from ASTM A36 structural steel.

8–27.E Repeat Problem 8–24, but specify the lightest structural steel channel from Appendix A–6. The channel is to be installed with the legs down so that the loads can be applied to the flat back of the web of the channel. The channel will be made from ASTM A36 structural steel.

8–28.E Repeat Problem 8–24, but specify the lightest standard schedule 40 steel pipe from Appendix A–12. The pipe is to be made from ASTM A501 hot-formed steel.

8–29.E Repeat Problem 8–24, but design the beam using any material and shape of your choosing to achieve a safe beam that is lighter than any of the results from Problems 8–24 through 8–28.

8–30.E The shape shown in Figure P7–15 is to be made from extruded plastic and used as a simply supported beam, 12 ft long, to carry two electric cables weighing a total of 6.5 lb/ft of length. Specify a suitable plastic for the extrusion to provide a design factor of 4.0 based on flexural strength.

8–31.C The loading shown in Figure P6–34 represents the load on a floor beam of a commercial building. Determine the maximum bending moment on the beam, and then specify a wide-flange shape that will limit the stress to 150 MPa.

8–32.M Figure P6–35 represents the loading on a motor shaft; the two supports are bearings in the motor housing. The larger load between the supports is due to the rotor plus dynamic forces. The smaller, overhung load is due to externally applied loads. Using AISI 1141 OQT 1300 steel for the shaft, specify a suitable diameter based on bending stress only. Use a design factor of 8 based on ultimate strength.

8–33 to Using the indicated loading, specify the lightest
8–42. standard wide-flange beam shape (W shape) that will limit the stress due to bending to the

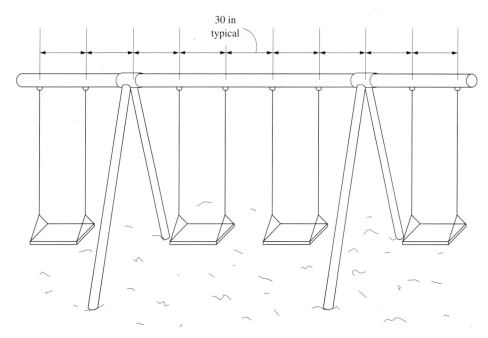

30 in
typical

FIGURE 8–30 Swing set for Problem 8–23.

allowable design stress from the AISC specification. All loads are static and the beams are made from ASTM A36 structural steel.

8–33.E Use the loading in Figure P6–3.

8–34.C Use the loading in Figure P6–7.

8–35.C Use the loading in Figure P6–8.

8–36.C Use the loading in Figure P6–11.

8–37.C Use the loading in Figure P6–16.

8–38.C Use the loading in Figure P6–36.

8–39.C Use the loading in Figure P6–40.

8–40.C Use the loading in Figure P6–52.

8–41.C Use the loading in Figure P6–63.

8–42.E Use the loading in Figure P6–64.

8–43 to Repeat Problems 8–33 through 8–42 but
8–52. specify the lightest American Standard Beam (S shape).

8–53 to Repeat Problems 8–33 through 8–42 but use
8–62. ASTM A572 Grade 60 high-strength low-alloy structural steel.

8–63.E A floor joist for a building is to be made from a standard wooden beam selected from Appendix A–4. If the beam is to be simply supported at its ends and carry a uniformly distributed load of 125 lb/ft over the entire 10-ft length, specify a suitable beam size. The beam will be made from No. 2 grade southern pine. Consider only bending stress.

8–64.E A bench for football players is to carry the load shown in Figure 8–31 approximating the case when 10 players, each weighing 300 lb, sit close together, each taking 18 in of the length of the bench. If the cross section of the bench is made as shown in Figure 8–31, would it be safe for bending stress? The wood is No. 2 grade hemlock.

8–65.E A bench is to be designed for football players. It is to carry the load shown in Figure 8–31 approximating the case when 10 players, each weighing 300 lb, sit close together, each taking 18 in of the length of the bench. The bench is to be T-shaped, made from No. 2 grade hemlock, as shown with a 2 × 12 top board. Specify the required vertical member of the tee if the bench is to be safe for bending stress.

8–66.E Repeat Problem 8–65, but use the cross-section shape shown in Figure 8–32.

8–67.E Repeat Problem 8–65, but use any cross-section shape of your choosing made from standard

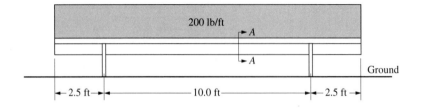

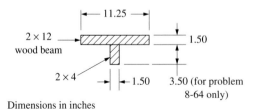

Dimensions in inches

Section *A-A*

FIGURE 8–31 Bench and load for Problems 8–64, 8–65, 8–66, and 8–67.

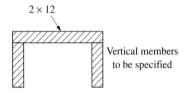

FIGURE 8–32 Shape for bench cross section for Problem 8–66.

wooden beams from Appendix A–4. Try to achieve a lighter design than in either Problem 8–65 or 8–66. Note that a lighter design would have a smaller cross-section area.

8–68.E A wood deck is being designed to carry a uniformly distributed load over its entire area of 100 lb/ft^2. Joists are to be used as shown in Figure 8–33, set 16 in on center. If the deck is to be 8 ft by 12 ft in size, determine the required size for the joists. Use standard wooden beam sections from Appendix A–4 and No. 2 hemlock.

8–69.E Repeat Problem 8–68, but run the joists across the 12-ft length rather than the 8-ft width.

8–70.E Repeat Problem 8–68, but set the support beams in 18 in from the ends of the joists instead of at the ends.

8–71.E Repeat Problem 8–69, but set the support beams in 18 in from the ends of the joists instead of at the ends.

8–72.E For the deck design shown in Figure 8–33 specify a suitable size for the cross beams that support the joists.

8–73.E Design a bridge to span a small stream. Assume that rigid supports are available on each bank, 10.0 ft apart. The bridge is to be 3.0 ft wide and carry a uniformly distributed load of 60 lb/ft^2 over its entire area. Design only the deck boards and beams. Use two or more beams of any size from Appendix A–4 or others of your own design.

8–74.E Would the bridge you designed in Problem 8–73 be safe if a horse and rider weighing 2200 lb walked slowly across it?

8–75.E Millwrights in a factory need to suspend a machine weighing 10 500 lb from a beam having a span of 12.0 ft so that a truck can back under it. Assume that the beam is simply supported at its ends. The load is applied by two cables, each 3.0 ft from a support. Design a suitable beam. Consider standard wooden or steel beams or one of your own design.

8–76.E In an amateur theater production, a pirate is to "walk the plank." If the pirate weighs 220 lb, would the design shown in Figure 8–34 be safe? If not, design one to be safe.

8–77.M A branch of a tree has the approximate dimensions shown in Figure 8–35. Assuming the bending strength of the wood to be similar to that of No. 3 grade hemlock, would it be safe for a person having a mass of 135 kg to sit in the swing?

8–78.E Would it be safe to use a standard 2 × 4 made from No. 2 grade southern pine as a lever as shown in Figure 8–36 to lift one side of a machine? If not, what would you suggest be used?

Chapter 8 ▪ Stress Due to Bending

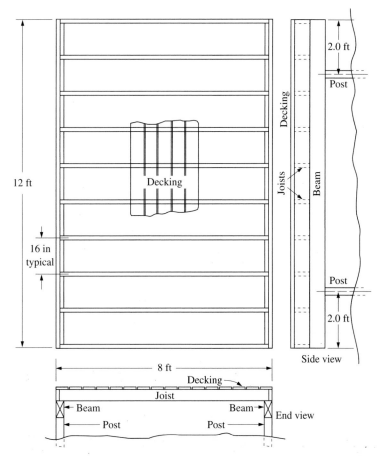

2.0 ft

Post

Decking

Joists

Beam

Post

2.0 ft

Side view

Decking

12 ft

Decking

16 in
typical

8 ft

Decking

Joist

Beam

Beam

Post

Post

End view

FIGURE 8–33 Deck design for Problem 8–68.

8–79.M Figure P7–6 shows the cross section of an extruded plastic beam made from nylon 6/6. Specify the largest uniformly distributed load in N/mm the beam could carry if it is simply supported with a span of 0.80 m. The maximum stress due to bending is not to exceed one-half of the flexural strength of the nylon.

8–80.M The I-beam shape in Figure P7–9 is to carry two identical concentrated loads of 2.25 kN each, symmetrically placed on a simply supported beam with a span of 0.60 m. Each load is 0.2 m from an end. Which of the plastics from Appendix A–19 would carry these loads with a stress due to bending no more than one-third of their flexural strength?

8–81.M A bridge for a toy construction set is to be made from polypropylene with the flexural strength

listed in Appendix A–19. The cross section of one beam is shown in Figure P7–5. What maximum concentrated load could be applied to the middle of the beam if the span was 1.25 m and the ends are simply supported? Do not exceed one-half of the flexural strength of the plastic.

8–82.M A structural element in a computer printer is to carry the load shown in Figure 8–37 with the uniformly distributed load representing electronic components mounted on a printed circuit board and the concentrated loads applied from a power supply. It is proposed to make the beam from polycarbonate with the cross-section shape shown in Figure P7–12. Compute the stress in the beam and compare it with the flexural strength for polycarbonate from Appendix A–19.

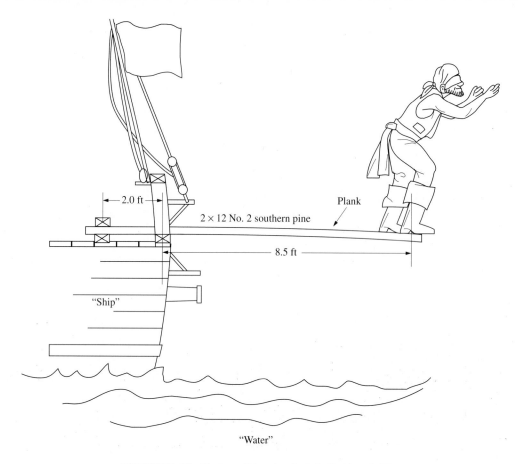

2.0 ft

2 × 12 No. 2 southern pine

Plank

8.5 ft

"Ship"

"Water"

FIGURE 8–34 Pirate walking the plank in Problem 8–76.

8–83.E It is proposed to make the steps for a child's slide from molded polystyrene with the cross section shown in Figure P7–20. The steps are to be 14.0 in wide and will be simply supported at the ends. What maximum weight can be applied at the center of the step if the stress due to bending must not exceed one-third of the flexural strength of the polystyrene listed in Appendix A–19?

8–84.E The shape shown in Figure P7–15 is used as a beam for a carport. The span of the beam will be 8.0 ft. Compute the maximum uniformly distributed load that can be carried by the beam if it is extruded from 6061-T4 aluminum. Use a design factor of 2 based on yield strength.

8–85.E Figure P7–14 shows the cross section of an aluminum beam made by fastening a flat plate to the bottom of a roll-formed hat section. If the

beam is used as a cantilever, 24 in long, compute the maximum allowable concentrated load that can be applied at the end if the maximum stress is to be no more than one-eighth of the ultimate strength of 2014-T4 aluminum.

8–86.E Repeat Problem 8–85 but use only the hat section without the cover plate.

8–87.E Figure P7–19 shows the cross section of a beam that is to be extruded from aluminum 6061-T6 alloy. If the beam is to be used as a cantilever, 42 in long, compute the maximum allowable uniformly distributed load it could carry while limiting the stress due to bending to one-sixth of the ultimate strength of the aluminum.

8–88.E Repeat Problem 8–87, but use Figure P7–20 for the cross section and casting alloy 356.0-T6.

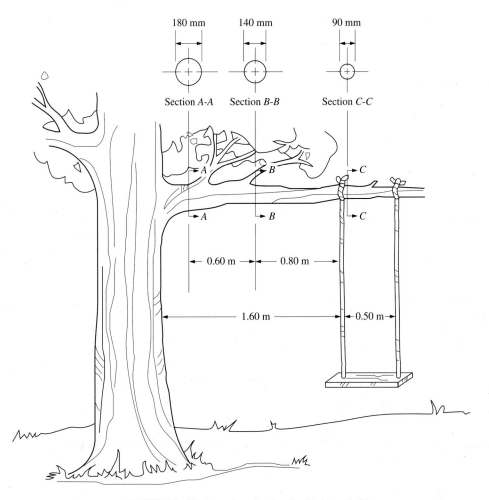

180 mm   140 mm   90 mm

Section *A-A*   Section *B-B*   Section *C-C*

A   *B*   *C*

A   *B*   *C*

0.60 m   0.80 m

1.60 m   0.50 m

FIGURE 8–35   Branch and swing for Problem 8–77.

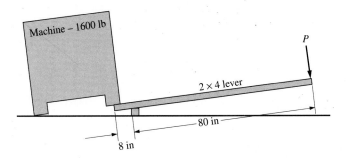

Machine – 1600 lb

P

2 × 4 lever

80 in

8 in

FIGURE 8–36   2 × 4 used as a lever in Problem 8–78.

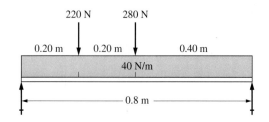

FIGURE 8–37 Beam in a computer printer for Problem 8–82.

Beams with Stress Concentrations and Varying Cross Sections

8–89.E In Figure 8–38 the 4-in pipe mates smoothly with its support, so that no stress concentration exists at D. At C the $3\frac{1}{2}$-in pipe is placed inside the 4-in pipe with a spacer ring to provide a good fit. Then a $\frac{1}{4}$-in smoothly radiused fillet weld is used to secure the section together. Accounting for the stress concentration at the joint, determine how far out point C must be to limit the stress to 20 000 psi. Use Appendix A–21–9 for the stress concentration factor. Is the 4-in pipe safe at D?

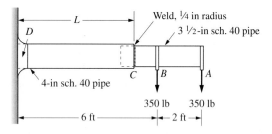

FIGURE 8–38 Data for Problem 8–89.

8–90.M Figure 8–39 shows a round shaft from a gear transmission. Gears are mounted at points A, C, and E. Supporting bearings at B and D. The forces transmitted from the gears to the shaft are shown, all acting downward. Compute the maximum stress due to bending in the shaft, accounting for stress concentrations.

8–91.M The forces shown on the shaft in Figure 8–40 are due to gears mounted at B and C. Compute the maximum stress due to bending in the shaft.

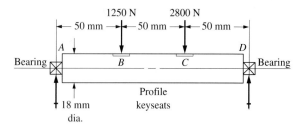

FIGURE 8–40 Data for Problem 8–91.

8–92.E Figure 8–41 shows a machine shaft supported by two bearings at its ends. The two forces are exerted on the shaft by gears. Considering only bending stresses, compute the maximum stress in the shaft and tell where it occurs.

8–93.E Figure 8–42 shows a lever made from a rectangular bar of steel. Compute the stress due to bending at the fulcrum, 20 in from the pivot and at each of the holes in the bar. The diameter of each hole is 0.75 in.

8–94.E Repeat Problem 8–93 but use the diameter of the holes as 1.38 in.

8–95.E In Figure 8–42, the holes in the bar are provided to permit the length of the lever to be changed relative to the pivot. Compute the maximum bending stress in the lever as the pivot is moved to each hole. Use the diameter of the holes as 1.25 in.

8–96.M The bracket shown in Figure 8–43 carries the opposing forces created by a spring. If the force, F, is 2500 N, compute the bending stress at a section such as A-A, away from the holes.

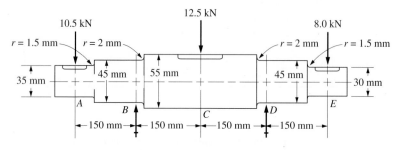

FIGURE 8–39 Data for Problem 8–90.

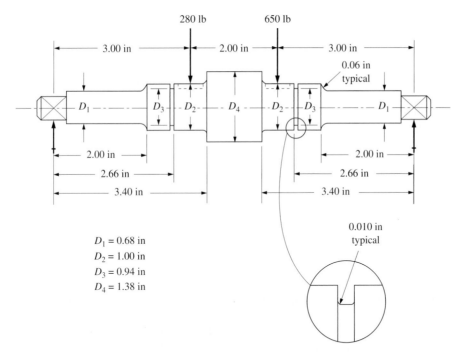

280 lb 650 lb

3.00 in   2.00 in   3.00 in

D_1 D_3 D_2 D_4 D_2 D_3 D_1

0.06 in
typical

2.00 in   2.00 in

2.66 in   2.66 in

3.40 in   3.40 in

$D_1 = 0.68$ in
$D_2 = 1.00$ in
$D_3 = 0.94$ in
$D_4 = 1.38$ in

0.010 in
typical

FIGURE 8–41   Shaft for Problem 8–92.

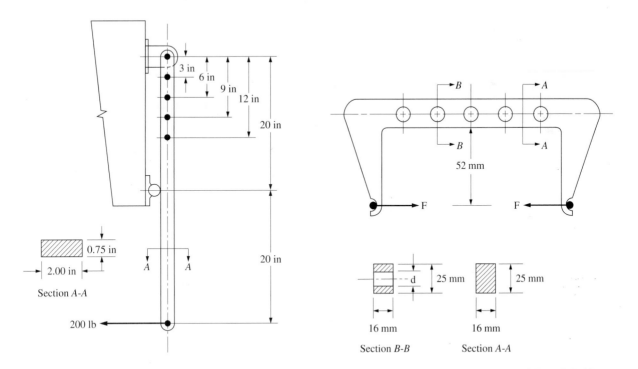

3 in
6 in
9 in
12 in
20 in

20 in

0.75 in

2.00 in

Section *A-A*

A   *A*

200 lb

FIGURE 8–42   Lever for Problems 8–93 through 8–95.

B   *A*

B   *A*

52 mm

F   F

d   25 mm   25 mm

16 mm   16 mm

Section *B-B*   Section *A-A*

FIGURE 8–43   Bracket for Problems 8–96 through 8–99.

8–97.M If the force, F, in Figure 8–43 is 2500 N, compute the bending stress at a section through the holes, such as B-B. Use $d = 12$ mm for the diameter of the holes.

8–98.M Repeat Problem 8–97 but use $d = 15$ mm for the diameter of the holes.

8–99.M For the resulting stress computed in Problem 8–98, specify a suitable steel for the bracket if the force is repeated many thousands of times.

8–100.M Figure 8–44 shows a stepped flat bar in bending. If the bar is made from AISI 1040 cold-drawn steel, compute the maximum repeated force, F, that can safely be applied to the bar.

8–101.M Repeat Problem 8–100, but use $r = 2.0$ mm for the fillet radius.

8–102.M For the stepped flat bar shown in Figure 8–44 change the 75-mm dimension that locates the step to a value that makes the bending stress at the step equal to that at the point of application of the load.

8–103.M For the stepped flat bar shown in Figure 8–44 change the size of the fillet radius to make the

bending stress at the fillet equal to that at the point of application of the load.

8–104.M Repeat Problem 8–100, but change the depth of the bar from 60 mm to 75 mm.

8–105.M For the stepped flat bar in Figure 8–44, would it be possible to drill a hole in the middle of the 60-mm depth of the bar between the two forces without increasing the maximum bending stress in the bar? If so, what is the maximum size hole that can be put in?

8–106.M Figure 8–45 shows a stepped flat bar carrying three concentrated loads. Let $P = 200$ N, $L_1 = 180$ mm, $L_2 = 80$ mm, and $L_3 = 40$ mm. Compute the maximum stress due to bending and state where it occurs. The bar is braced against lateral bending and twisting. Note that the length dimensions in the figure are not drawn to scale.

8–107.M For the data of Problem 8–106, specify a suitable material for the bar to produce a design factor of 8 based on ultimate strength.

8–108.M Repeat Problem 8–107 except use $r = 1.50$ mm for the fillet radius.

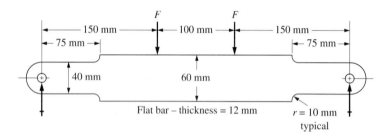

FIGURE 8–44 Stepped flat bar for Problems 8–100 through 8–105.

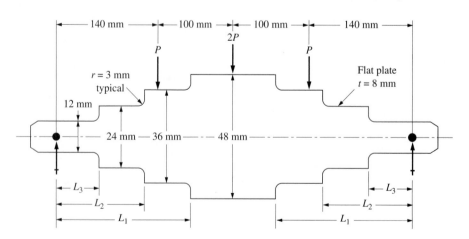

FIGURE 8–45 Stepped flat bar for Problems 8–106 through 8–110.

8–109.M For the stepped flat bar shown in Figure 8–45, let $P = 400$ N. The bar is to be made from titanium, Ti-6Al-4V, and a design factor of 8 based on ultimate strength is desired. Specify the maximum permissible lengths, L_1, L_2, and L_3, that would be safe.

8–110.M The stepped flat bar in Figure 8–45 is to be made from AISI 4140 OQT 1100 steel. Use $L_1 = 180$ mm, $L_2 = 80$ mm, and $L_3 = 40$ mm. Compute the maximum allowable force P that could be applied to the bar if a design factor of 8 based on ultimate strength is desired.

8–111.M Figure 8–46 shows a flat bar that has a uniform thickness of 20 mm. The depth tapers from $h_1 = 40$ mm to $h_2 = 20$ mm in order to save weight. Compute the stress due to bending in the bar at points spaced 40 mm apart from the support to the load. Then create a graph of stress versus distance from the support. The bar is symmetrical with respect to its middle. Let $P = 5.0$ kN.

8–112.M For the bar shown in Figure 8–46 let $h_1 = 60$ mm and $h_2 = 20$ mm. The bar is to be made from polycarbonate plastic. Compute the maximum permissible load P that will produce a design factor of 4 based on the flexural strength of the plastic. The bar is symmetrical with respect to its middle.

8–113.M In Figure 8–46 the load $P = 1.20$ kN and the bar is to be made from AISI 5160 OQT 1300 steel. Compute the required dimensions h_1 and h_2 that will produce a design factor of 8 based on ultimate strength. The bar is symmetrical with respect to its middle.

8–114.E A rack is being designed to support large sections of pipe, as shown in Figure 8–47. Each pipe exerts a force of 2500 lb on the support

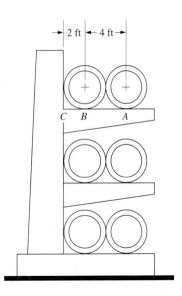

FIGURE 8–47 Pipe storage rack for Problem 8–114.

arm. The height of the arm is to be tapered as suggested in the figure, but the thickness will be a constant 1.50 in. Determine the required height of the arm at sections B and C, considering only bending stress. Use AISI 1040 hot-rolled steel for the arms and a design factor of 4 based on yield strength.

8–115.E Specify the lightest wide-flange beam shape (W shape) that can carry a uniformly distributed static load of 2.5 kip/ft over the entire length of a simply supported span, 12.0 ft long. Use the AISC specification and ASTM A36 structural steel.

8–116.E A proposal is being evaluated to save weight for the beam application of Problem 8–115. The result for that problem required a W14 × 26 beam

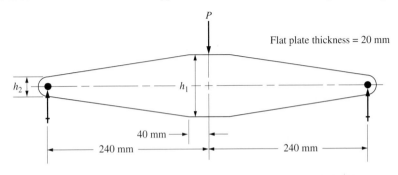

FIGURE 8–46 Tapered flat bar for Problems 8–111 through 8–113.

that would weigh 312 lb for the 12-ft length. A W12 × 16 beam would only weigh 192 lb but does not have sufficient section modulus S. To increase S, it is proposed to add steel plates, 0.25 in thick and 3.50 in wide, to both the top and the bottom flange over a part of the middle of the beam. Perform the following analyses:

(a) Compute the section modulus of the portion of the W12 × 16 beam with the cover plates.

(b) If the result of part (a) is satisfactory to limit the stress to an acceptable level, compute the required length over which the plates would have to be applied to the nearest 0.5 ft.

(c) Compute the resulting weight of the composite beam and compare it to the original W14 × 26 beam.

8–117.E Figure P7–26 shows a composite beam made by adding a channel to an American Standard beam shape. If the beam is simply supported and carries a uniformly distributed load over a span of 15.0 ft, compute the allowable load for the composite beam and for the S shape by itself. The load is static and the AISC specification is to be used for A36 structural steel.

Flexural Center

8–118.M Compute the location of the flexural center of a channel-shaped member shown in Figure 8–48 measured from the left of the vertical web.

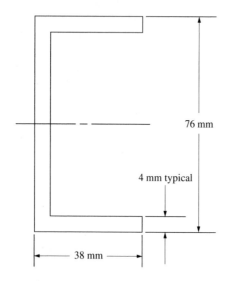

76 mm

4 mm typical

38 mm

FIGURE 8–48 Channel shape for Problem 8–118.

8–119.M A company plans to make a series of three channel-shaped beams by roll-forming them from flat sheet aluminum. Each channel is to have the same outside dimensions as shown in Figure 8–48, but they will have different material thicknesses, 0.50, 1.60, and 3.00 mm. For each design, compute the moment of inertia with respect to the horizontal centroidal axis and the location of the flexural center, measured from the left face of the vertical web.

8–120.E Compute the location of the flexural center for the hat section shown in Figure 8–49 measured from the left face of the vertical web.

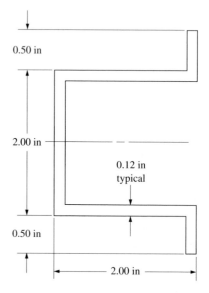

0.50 in

2.00 in

0.12 in typical

0.50 in

2.00 in

FIGURE 8–49 Hat section for Problem 8–120.

8–121.E A company plans to make a series of three hat sections by roll-forming them from flat sheet aluminum. Each hat section is to have the same outside dimensions as shown in Figure 8–49, but they will have different material thicknesses, 0.020, 0.063, and 0.125 in. For each design, compute the location of the flexural center, measured from the left face of the vertical web.

8–122.M Compute the location of the flexural center for the lipped channel shown in Figure 8–50, measured from the left face of the vertical web.

8–123.M A company plans to make a series of three lipped channels by roll-forming them from flat sheet aluminum. Each channel section is to have

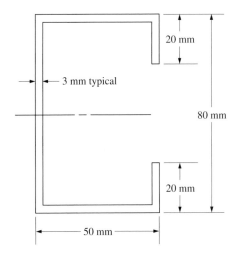

FIGURE 8–50 Lipped channel for Problem 8–122.

the same outside dimensions as shown in Figure 8–50, but they will have different material thicknesses, 0.50, 1.60, and 3.00 mm. For each design, compute the location of the flexural center, measured from the left face of the vertical web.

8–124.M Compute the location of the flexural center of a split, thin tube if it has an outside diameter of 50 mm and a wall thickness of 4 mm.

8–125.E For an aluminum channel C2 × 0.577 with its web oriented vertically, compute the location of its flexural center. Neglect the effect of the fillets between the flanges and the web.

8–126.M If the hat section shown in Figure P7–11 were turned 90 deg from the position shown, compute the location of its flexural center.

Beams Made from Anisotropic Materials

8–127.E The beam section shown in Figure P7–15 is to be extruded from 6061-T6 aluminum. The allowable tensile strength is 19 ksi. Because of the relatively thin extended legs on the top, the allowable compressive strength is only 14 ksi. The beam is to span 6.5 ft and will be simply supported at its ends. Compute the maximum allowable uniformly distributed load on the beam.

8–128.E Repeat Problem 8–127 but turn the section upside down. With the legs pointed downward, they are in tension and can withstand 19 ksi. The part of the section in compression at the top is now well supported and can withstand 21 ksi.

8–129.M The shape in Figure P7–6 is to carry a single concentrated load at the center of a 1200-mm span. The allowable strength in tension is 100 MPa, while the allowable strength in compression anywhere is 70 MPa. Compute the allowable load.

8–130.M Repeat Problem 8–129 with the section turned upside down.

8–131.M Repeat Problem 8–129 with the shape shown in Figure P7–8.

8–132.M Repeat Problem 8–129 with the shape shown in Figure P7–9.

8–133.M The T-shaped beam cross section shown in Figure P7–4 is to be made from gray cast iron, ASTM A48 Grade 40. It is to be loaded with two equal loads P, 1.0 m from the ends of the 2.80 m long beam. Specify the largest static load P that the beam could carry. Use N = 4.

8–134.M The modified I-beam shape shown in Figure P7–5 is to carry a uniformly distributed static load over its entire 1.20 m length. Specify the maximum allowable load if the beam is made from malleable iron, ASTM A220, class 80002. Use N = 4.

8–135.M Repeat Problem 8–134 but turn the beam upside down.

8–136.M A wide beam is made as shown in Figure 8–51 from ductile iron, ASTM A536, Grade 120-90-2. Compute the maximum load P that can be

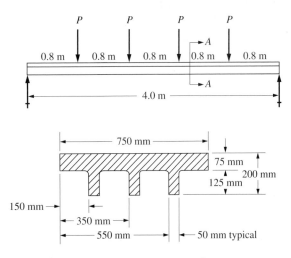

Section A-A – beam cross section

FIGURE 8–51 Wide beam for Problem 8–136.

carried with a resulting design factor of 10 based on either tensile or compressive ultimate strength.

8–137.M Repeat Problem 8–136 but increase the depth of the vertical ribs by a factor of 2.0.

8–138.M Problems 8–133 through 8–137 illustrate that a beam shape made as a modified I-shape more nearly optimizes the use of the available strength of a material having different strengths in tension and compression. Design an I-shape that has a nearly uniform design factor of 6 based on ultimate strength in either tension or compression when made from gray iron, Grade 20, and which carries a uniformly distributed load of 20 kN/m over its 1.20 m length. (Note: You may want to use the computer program written for Assignment 3 at the end of Chapter 7 to facilitate the computations. A trial and error solution may be used.)

COMPUTER PROGRAMMING ASSIGNMENTS

1. Write a program to compute the maximum bending stress for a simply supported beam carrying a single concentrated load at its center. Allow the operator to input the load, span, and beam section properties. The output should include the maximum bending moment and the maximum bending stress and indicate where the maximum stress occurs.

Enhancements for Assignment 1

(a) For the computed stress, compute the required strength of the material for the beam to produce a given design factor.

(b) In addition to (a), include a table of properties for a selected material such as the data for steel in Appendix A–13. Then search the table for a suitable steel from which the beam can be made.

2. Repeat Assignment 1 except use a uniformly distributed load.

3. Repeat Assignment 1 except the beam is a cantilever with a single concentrated load at its end.

4. Write a program to compute the maximum bending moment for a simply supported beam carrying a single concentrated load at its center. Allow the operator to input the load and span. Then compute the required section modulus for the cross section of the beam to limit the maximum bending stress to a given level or to achieve a given design factor for a given material. The output should include the maximum bending moment and the required section modulus.

Enhancements for Assignment 4

(a) After computing the required section modulus, have the program complete the design of the beam cross section for a given general shape, such as rectangular with a given ratio of thickness to depth (see Problem 8–13), or circular.

(b) Include a table of properties for standard beam sections such as any of those in Appendixes A–4 through A–12 and have the program search for a suitable beam section to provide the required section modulus.

5. Repeat Assignment 4 but use a uniformly distributed load.

6. Repeat Assignment 4 but use the load described in Problem 8–22.

7. Repeat Assignment 4 but use any loading pattern assigned by the instructor.

8. Write a computer program to facilitate the solution of Problem 8–138, including the computation of section properties for the modified I-shape using the techniques of Chapter 7. Use the loading pattern of Figure P6–3 but allow the user to specify the span of the beam, the magnitude of the load, and the placement of the load.

9. Write a computer program to facilitate the solution of problems of the type given in Problem 8–116. Make the program general, permitting the user to input the loading on the beam, the desired beam section properties, and the dimensions of the plates to be added to the basic beam section.

10. Write a computer program to perform the computations called for in Problem 8–111, but make the program more general, permitting the user to input values for the load, span, beam cross-section dimensions, and the interval for computing the bending stress. Have the program produce the graph of stress versus position on the beam.

11. Write a computer program to compute the location for the flexural center for the generalized channel shape in Figure 8–48. Permit the user to input data for all dimensions.

12. Write a computer program to compute the location for the flexural center for the generalized hat section shown in Figure 8–49. Permit the user to input data for all dimensions. Curve-fitting techniques and interpolation may be used to interpret the graph in Figure 8–21.

13. Write a computer program to compute the location for the flexural center for the generalized lipped channel shown in Figure 8–50. Permit the user to input data for all dimensions. Curve-fitting techniques and interpolation may be used to interpret the graph in Figure 8–21.

9

Shearing Stresses in Beams

9–1 OBJECTIVES OF THIS CHAPTER

Continuing the analysis of beams, this chapter is concerned with the stresses created within a beam due to the presence of shearing forces. As shown in Figure 9–1, shearing forces are visualized to act within the beam on its cross section and to be directed transverse, that is perpendicular, to the axis of the beam. Thus they would tend to create *transverse shearing stresses,* sometimes called *vertical shearing stresses.*

But if a small stress element subjected to such shearing stresses is isolated, as shown in Figure 9–2, it can be seen that horizontal shearing stresses must also exist in order to cause the element to be in equilibrium. Thus, both vertical and horizontal shearing stresses, having the same magnitude at a given point, are created by shearing stresses in beams.

After completing this chapter, you should be able to:

1. Describe the conditions under which shearing stresses are created in beams.
2. Compute the magnitude of shearing stresses in beams by using the general shear formula.
3. Define and evaluate the *statical moment* required in the analysis of shearing stresses.
4. Specify where the maximum shearing stress occurs on the cross section of a beam.
5. Compute the shearing stress at any point within the cross section of a beam.

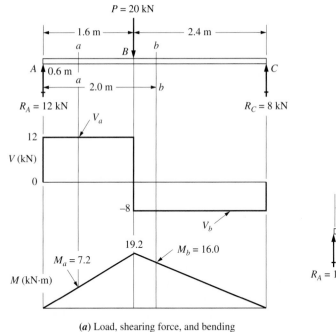

$P = 20$ kN

1.6 m   2.4 m

a

B b

A 0.6 m

a

2.0 m b

$R_A = 12$ kN $R_C = 8$ kN

12

V_a

V (kN)

0

−8

V_b

19.2

$M_b = 16.0$

$M_a = 7.2$

M (kN·m)

(*a*) Load, shearing force, and bending
moment diagrams

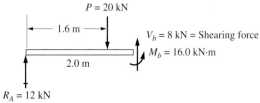

$M_a = 7.2$ kN·m

0.6 m $V_a = 12$ kN = Shearing force

$R_A = 12$ kN

(*b*) Free-body diagram at *a-a*

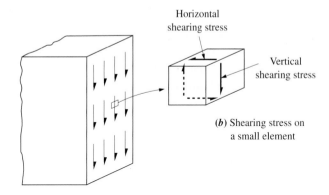

$P = 20$ kN

1.6 m $V_b = 8$ kN = Shearing force

$M_b = 16.0$ kN·m

2.0 m

$R_A = 12$ kN

(*c*) Free-body diagram at *b-b*

FIGURE 9–1 Shearing forces in beams.

Horizontal
shearing stress

Vertical
shearing stress

(*b*) Shearing stress on
a small element

(*a*) Shearing stress on a cut section of a beam

FIGURE 9–2 Shearing stress in a beam.

6. Describe the general distribution of shearing stress as a function of position within the cross section of a beam.

7. Understand the basis for the development of the general shearing stress formula.

8. Describe four design applications where shearing stresses are likely to be critical in beams.

9. Develop and use special shear formulas for computing the maximum shearing stress in beams having rectangular or solid circular cross sections.

10. Understand the development of approximate relationships for estimating the maximum shearing stress in beams having cross sections with tall thin webs or those with thin-walled hollow tubular shapes.

11. Specify a suitable design shearing stress and apply it to evaluate the acceptability of a given beam design.

12. Define *shear flow* and compute its value.

13. Use the shear flow to evaluate the design of fabricated beam sections held together by nails, bolts, rivets, adhesives, welding, or other means of fastening.

9–2 VISUALIZATION OF SHEARING STRESS IN BEAMS

The existence of horizontal shearing stress in beams can also be observed by considering a beam made from several flat strips, as illustrated in Figure 9–3. A demonstrator can be made from cardboard, sheet metal, plastic, or other materials.

One thin flat strip would make a very poor beam if it was simply supported near its ends and loaded with a concentrated load at the middle of the span. The beam would deflect a large amount and it would tend to break at a very small load.

Laying several strips on top of each other would produce a beam with increased strength and decreased deflection for a given load, but to only a very small extent. As shown in Figure 9–3(b), the strips would slide over one another at the surfaces of contact and the beam would still be relatively flexible and weak.

A stronger and stiffer beam can be made by fastening the strips together in such a way that the sliding between strips is prevented. This can be done by using an adhesive, welding, brazing, or mechanical fasteners such as rivets, screws, bolts, pins, nails, or even staples. In this way, the tendency for one strip to slide on the next is prevented and the fastening means is subjected to a shearing force directed horizontally, parallel

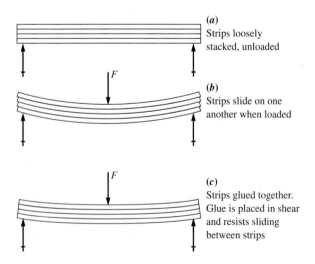

(a)
Strips loosely stacked, unloaded

(b)
Strips slide on one another when loaded

(c)
Strips glued together. Glue is placed in shear and resists sliding between strips

FIGURE 9–3 Illustration of the presence of shearing stress in a beam.

to the neutral axis of the beam. This is a visualization of the *horizontal shearing stress* in a beam. See Figure 9–3(c).

A similar condition exists in a solid beam. Here the tendency for horizontal sliding of one part of a beam relative to the part above or below it is resisted by *the material of the beam*. Therefore, a shearing stress is developed on any horizontal plane. Again, as shown in Figure 9–2, shearing stresses must exist simultaneously in the vertical plane to maintain equilibrium of any stress element.

9–3 IMPORTANCE OF SHEARING STRESSES IN BEAMS

Several situations exist in practical design in which the mode of failure is likely to be shearing of a part of a beam or of a means of fastening a composite beam together. Five such situations are described here.

Wooden beams. Wood is inherently weak in shear along the planes parallel to the grain of the wood. Consider the beam shown in Figure 9–4, which is similar to the joists used in floor and roof structures for wood-frame construction. The grain runs generally parallel to the long axis in commercially available lumber. When subjected to transverse loads, the initial failure in a wooden beam is likely to be by separation along the grain of the wood, due to excessive horizontal shearing stress. Note in Appendix A–18 that the allowable shearing stress in common species of wood ranges from only 70 to 95 psi (0.48 to 0.66 MPa), very low values.

Thin-webbed beams. An efficient beam cross section would be one with relatively thick horizontal flanges on the top and bottom with a thin vertical web connecting them together. This generally describes the familiar "I-beam," the wide-flange beam, or the American Standard beam, as sketched in Figure 9–5. Actual dimensions for such beam sections are given in Appendixes A–7, A–8, and A–11.

But if the web is excessively thin, it would not have sufficient stiffness and stability to hold its shape and it would fail due to shearing stress in the thin web. The American Institute of Steel Construction (AISC) defines the allowable shearing stress in the webs of steel beams. See Reference 1. See also the *web shear formula,* defined later in this chapter, Section 9–7.

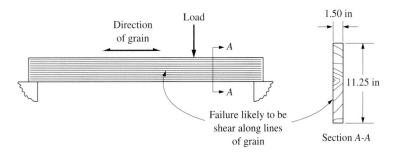

FIGURE 9–4 Shear failure in a wood beam.

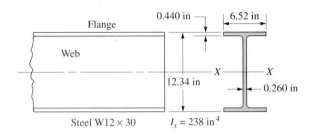

Flange

Web

0.440 in

6.52 in

X ——— X

12.34 in

0.260 in

Steel W12 × 30

$I_x = 238\ \text{in}^4$

FIGURE 9–5 Example of thin-webbed beam shape.

Short beams. In very short beams, the bending moment, and therefore the bending stress, is likely to be small. In such beams, the shearing stress may be the limiting stress.

Fastening means in fabricated beams. As shown in Figure 9–3, the fasteners in a composite beam section are subjected to shearing stresses. The concept of *shear flow*, developed later, can be used to evaluate the safety of such beams or to specify the required type, number, and spacing of fasteners to use. Also, beams made of composite materials are examples of fabricated beams. Separation of the layers of the composite, called *interlaminar shear*, is a potential mode of failure.

Stressed skin structures. Aircraft and aerospace structures and some ground-based vehicles and industrial equipment are made using a *stressed skin* design. Sometimes called *monocoque* structures, they are designed to carry much of the load in the thin skins of the structure. The method of shear flow is typically used to evaluate such structures, but this application is not developed in this book.

9–4 THE GENERAL SHEAR FORMULA

Presented here is the general shear formula from which you can compute the magnitude of the shearing stress at any point within the cross section of a beam carrying a vertical shearing force. In Section 9–7, the formula itself is developed. It may be desirable to study the development of the formula along with this section.

The general shear formula is stated as follows:

 **General Shear Formula**

$$\tau = \frac{VQ}{It} \tag{9–1}$$

where V = *vertical shearing force* at the section of interest. The value of V can be found from the shearing force diagram developed as described in Chapter 6. Generally, the maximum absolute value of V, positive or negative, is used.

I = *moment of inertia* of the entire cross section of the beam with respect to its centroidal axis. This is the same value of I used in the flexure formula ($\sigma = Mc/I$) to compute the bending stress.

t = *thickness* of the cross section taken at the axis where the shearing stress is to be computed.

Q = *statical moment.*

The *statical moment* is defined as the moment, with respect to the overall centroidal axis, of the area of that part of the cross section that lies away from the axis where the shearing stress is to be computed. By definition,

Statical Moment

$$Q = A_p \bar{y} \qquad (9\text{--}2)$$

where A_p = area of that *part* of the cross section that lies away from the axis where the shearing stress is to be computed.

$\bar{y}$ = distance to the centroid of A_p from the centroidal axis of the entire cross section.

Note that the statical moment is the *moment of an area;* that is, area times distance. Therefore, it will have the units of length cubed, such as in^3, m^3, or mm^3.

Careful evaluation of the statical moment Q is critical to proper use of the general shear formula. It is helpful to draw a sketch of the beam cross section and then to highlight the partial area, A_p. Then show the location of the centroid of the partial area on the sketch. Figure 9–6 shows an example for which this has been done. In this example, the objective is to calculate the shearing stress at the axis labeled *a-a*. The crosshatched area is A_p, shown as that part away from the axis *a-a*.

The following three example problems illustrate the method of computing Q. In each, this is the procedure used.

Method of computing the statical moment, Q

1. Locate the centroidal axis for the entire cross section.
2. Draw in the axis where the shearing stress is to be calculated.
3. Identify the partial area A_p away from the axis of interest and shade it for emphasis.

If the partial area A_p is a simple area for which the centroid is readily found by simple calculations, use steps 4–7 to compute Q. Otherwise, use steps 8–11.

4. Compute the magnitude of A_p.
5. Locate the centroid of the partial area.
6. Compute the distance $\bar{y}$ from the centroidal axis of the full section to the centroid of the partial area.
7. Compute $Q = A_p \bar{y}$.

For cases in which the partial area is itself a composite area made up of several component parts, steps 8–11 are used.

8. Divide A_p into component parts which are simple areas and label them A_1, A_2, A_3, and so on. Compute their values.
9. Locate the centroid of each component area.
10. Determine the distances from the centroidal axis of the full section to the centroid of each component area, calling them y_1, y_2, y_3, and so on.
11. Compute $Q = A_p \bar{y}$ from

$$Q = A_p \bar{y} = A_1 y_1 + A_2 y_2 + A_3 y_3 + \cdots \qquad (9\text{--}3)$$

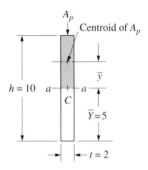

All dimensions in inches

FIGURE 9–6 Example of A_p and $\bar{y}$ for use in computing Q.

Example Problem 9–1   For the rectangular section in Figure 9–6, compute the statical moment Q as it would be used in the general shear formula to compute the vertical shearing stress at the section marked *a-a*.

Solution

Objective   Compute the value of Q.

Given   Shape and dimensions of cross section in Figure 9–6.

Analysis   Use the method defined in this section.

Results   **Step 1.** The neutral axis for this section is at its midheight, $h/2$, from the bottom. For this problem, $h/2 = 5.00$ in.

Step 2. The axis of interest is *a-a*, coincident with the netural axis for this example.

Step 3. The partial area, A_p, is shown crosshatched in the figure to be the upper half of the rectangle.

Because the partial area is itself a simple rectangle, steps 4–7 are used to compute Q.

Step 4. The partial area is

$$A_p = t(h/2) = (2.0 \text{ in})(5.0 \text{ in}) = 10 \text{ in}^2$$

Step 5. The centroid of the partial area is at its midheight, 2.5 in above *a-a*.

Step 6. Because the centroidal axis is coincident with the axis *a-a*, $\bar{y} = 2.5$ in.

Step 7. Now Q can be computed.

$$Q = A_p\bar{y} = (10.0 \text{ in}^2)(2.5 \text{ in}) = 25.0 \text{ in}^3$$

Example Problem 9–2 For the I-shaped section in Figure 9–7, compute the statical moment Q as it would be used in the general shear formula to compute the vertical shearing stress at the section marked a-a.

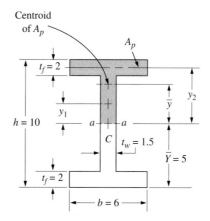

All dimensions in inches

FIGURE 9–7   I-shape for Example Problem 9.2.

Solution   **Objective**   Compute the value of Q.

Given   Shape and dimensions of cross section in Figure 9–7.

Analysis   Use the method defined in this section.

Results   **Step 1.**   The I-shape is symmetrical and, therefore, the neutral axis lies at half the height from its base, 5.0 in.

Step 2.   The axis of interest is a-a, coincident with the neutral axis for this example.

Step 3.   The partial area A_p is shown crosshatched in the figure to be the upper half of the I-shape.

Because the partial area is in the form of a "T," steps 8–11 are used to compute Q.

Step 8.   The T-shape is divided into two parts: the upper half of the vertical web is part 1 and the entire top flange is part 2. The magnitudes of these areas are

$$A_1 = \left(\frac{h}{2} - t_f\right)(t_w) = (5.0 \text{ in} - 2.0 \text{ in})(1.5 \text{ in}) = 4.5 \text{ in}^2$$

$$A_2 = bt_f = (6.0 \text{ in})(2.0 \text{ in}) = 12.0 \text{ in}^2$$

Step 9.   Each part is a rectangle for which the centroid is at its midheight, as shown in the figure.

Step 10. The required distances are

$$y_1 = \frac{1}{2}\left(\frac{h}{2} - t_f\right) = \frac{1}{2}(5.0 \text{ in} - 2.0 \text{ in}) = 1.5 \text{ in}$$

$$y_2 = \left(\frac{h}{2} - \frac{t_f}{2}\right) = (5.0 \text{ in} - 1.0 \text{ in}) = 4.0 \text{ in}$$

Step 11. Using Equation (9–3) gives us

$$Q = A_1 y_1 + A_2 y_2$$
$$Q = (4.5 \text{ in}^2)(1.5 \text{ in}) + (12.0 \text{ in}^2)(4.0 \text{ in}) = 54.75 \text{ in}^3$$

Example Problem 9–3 For the T-shaped section in Figure 9–8, compute the statical moment Q as it would be used in the general shear formula to compute the vertical shearing stress at the section marked a-a at the very top of the web just below where it joins the flange.

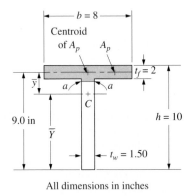

All dimensions in inches

FIGURE 9–8 T-shape for Example Problem 9.3.

Solution **Objective** Compute the value of Q.

Given Shape and dimensions of cross section in Figure 9–8.

Analysis Use the method defined in this section.

Results **Step 1.** Locate the centroid.

$$\overline{Y} = \frac{A_w y_w + A_f y_f}{A_w + A_f}$$

where the subscript w refers to the vertical web and the subscript f refers to the top flange. Then

$$\overline{Y} = \frac{(12)(4) + (16)(9)}{12 + 16} = 6.86 \text{ in}$$

Step 2. The axis of interest, *a-a*, is at the very top of the web, just below the flange.

Step 3. The partial area above *a-a* is the entire flange.

Step 4. $A_p = (8 \text{ in})(2 \text{ in}) = 16 \text{ in}^2$

Step 5. The centroid of A_p is 1.0 in down from the top of the flange, which is 9.0 in above the base of the tee.

Step 6. $\overline{y} = 9.0 \text{ in} - \overline{Y} = 9.0 \text{ in} - 6.86 \text{ in} = 2.14 \text{ in}$

Step 7. $Q = A_p\overline{y} = (16 \text{ in}^2)(2.14 \text{ in}) = 34.2 \text{ in}^3$

Comment   It should be noted that the value of *Q* would be the same if the axis of interest *a-a* were to be taken at the very bottom of the flange just above the web. But the thickness of the section, *t*, would be equal to the entire width of the flange, whereas for the axis *a-a* used in this problem, the thickness of the web is used. The resulting shearing stresses would be markedly different. This will be shown later.

Use of the general shear formula.   Example problems are presented here to illustrate the use of the general shear formula [Equation (9–1)] to compute the vertical shearing stress in a beam. The following procedure is typical of that used in solving such problems.

Guidelines for computing shearing stresses in beams

The overall objective is to compute the shearing stress at any specified position on the beam at any specified axis within the cross section using the general shear formula,

$$\tau = \frac{VQ}{It} \qquad (9\text{--}1)$$

1. Determine the vertical shearing force *V* at the section of interest. This may require preparation of the complete shearing force diagram using the procedures of Chapter 6.
2. Locate the centroid of the entire cross section and draw the neutral axis through the centroid.
3. Compute the moment of inertia of the section with respect to the neutral axis.
4. Identify the axis for which the shearing stress is to be computed and determine the thickness *t* at that axis. Include all parts of the section that are cut by the axis of interest when computing *t*.
5. Compute the statical moment of the partial area from the axis of interest with respect to the neutral axis. Use the procedure developed in this section.
6. Compute the shearing stress using Equation (9–1).

Compute the shearing stress at the axis *a-a* for a beam with the rectangular cross section shown in Figure 9–6. The shearing force, *V*, on the section of interest is 1200 lb.

Solution   **Objective**   Compute the shearing stress at the axis *a-a*.

Given   Cross section shape and dimensions in Figure 9–6. $V = 1200$ lb.

Analysis   Use the *Guidelines for computing shearing stresses in beams.*

Results   **Step 1.** $V = 1200$ lb (given)

Step 2. For the rectangular shape, the centroid is at the midheight, as shown in Figure 9–6, coincident with axis *a-a*. $\overline{Y} = 5.00$ in.

Step 3. $I = bh^3/12 = (2.0)(10.0)^3/12 = 166.7$ in^4

Step 4. Thickness $= t = 2.0$ in at axis *a-a*.

Step 5. Normally we would compute $Q = A_p\overline{y}$ using the method shown earlier in this chapter. But the value of Q for the section in Figure 9–6 was computed in Example Problem 9–1. Use $Q = 25.0$ in^3.

Step 6. Using Equation (9–1),

$$\tau = \frac{VQ}{It} = \frac{(1200 \text{ lb})(25.0 \text{ in}^3)}{(166.7 \text{ in}^4)(2.0 \text{ in})} = 90.0 \text{ psi}$$

Example Problem
9–5

Compute the shearing stress at the axes *a-a* and *b-b* for a beam with the T-shaped cross section shown in Figure 9–8. Axis *a-a* is at the very top of the vertical web, just below the flange. Axis *b-b* is at the very bottom of the flange. The shearing force, *V*, on the section of interest is 1200 lb.

Solution   **Objective**   Compute the shearing stress at the axes *a-a* and *b-b*.

Given   Cross section shape and dimensions in Figure 9–8. $V = 1200$ lb.

Analysis   Use the *Guidelines for computing shearing stresses in beams.*

Results   For the axis *a-a*:

Step 1. $V = 1200$ lb (given)

Step 2. This particular T-shape was analyzed in Example Problem 9–3. Use $\overline{Y} = 6.86$ in.

Step 3. We will use the methods of Chapter 7 to compute *I*. Let the web be part 1 and the flange be part 2. For each part, $I = bh^3/12$ and $d = \overline{Y} - \overline{y}$.

Part	I	A	d	Ad^2	$I + Ad^2$
1	64.00	12.0	2.86	98.15	162.15
2	5.33	16.0	2.14	73.27	78.60
				Total $I =$	240.75 in^4

Step 4. Thickness = t = 1.5 in at axis a-a in the web.

Step 5. Normally we would compute $Q = A_p\overline{y}$ using the method shown earlier in this chapter. But the value of Q for the section in Figure 9−8 was computed in Example Problem 9−3. Use Q = 34.2 in³.

Step 6. Using Equation (9−1),

$$\tau = \frac{VQ}{It} = \frac{(1200 \text{ lb})(34.2 \text{ in}^3)}{(240.75 \text{ in}^4)(1.5 \text{ in})} = 114 \text{ psi}$$

For the axis b-b: Some of the data will be the same as at a-a.

Step 1. V = 1200 lb (given)

Step 2. Again, use $\overline{Y}$ = 6.86 in.

Step 3. I = 240.75 in⁴

Step 4. Thickness = t = 8.0 in at axis b-b in the flange.

Step 5. Again, use Q = 34.2 in³. The value is the same as at axis a-a because both A_p and $\overline{y}$ are the same.

Step 6. Using Equation (9−1),

$$\tau = \frac{VQ}{It} = \frac{(1200 \text{ lb})(34.2 \text{ in}^3)}{(240.75 \text{ in}^4)(8.0 \text{ in})} = 21.3 \text{ psi}$$

Comment   Note the dramatic reduction in the value of the shearing stress when moving from the web to the flange.

9−5   DISTRIBUTION OF SHEARING STRESS IN BEAMS

Most applications require that the maximum shearing stress be determined to evaluate the acceptability of the stress relative to some criterion of design. For most sections used for beams, the maximum shearing stress occurs at the neutral axis, coincident with the centroidal axis, about which bending occurs. The following rule can be used to decide when to apply this observation.

> Provided that the thickness at the centroidal axis is not greater than at some other axis, the maximum shearing stress in the cross section of a beam occurs at the centroidal axis.

Thus the computation of the shearing stress only at the centroidal axis would give the maximum shearing stress in the section, making the computations at other axes unnecessary.

The logic behind this rule can be seen by examining Equation (9−1), the general shear formula. To compute the shearing stress at any axis, the values of the shearing force V and the moment of inertia I are the same. Because the thickness, t, is in the denominator, the smallest thickness would tend to produce the largest shearing stress,

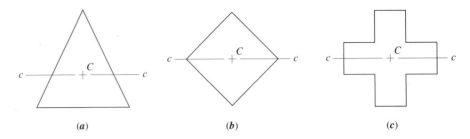

(a)   (b)   (c)

FIGURE 9–9 Beam cross sections for which the maximum shearing stress may not occur at the centroidal axis, c-c.

as implied in the statement of the rule. But the value of the statical moment Q also varies at different axes, decreasing as the axis of interest moves toward the outside of the section. Recall that Q is the product of the partial area A_p and the distance $\bar{y}$ to the centroid of A_p. For axes away from the centroidal axis, the area decreases at a faster rate than $\bar{y}$ increases, resulting in the value of Q decreasing. Thus the maximum value of Q will be that for stress computed at the centroidal axis. It follows that the maximum shearing stress will always occur at the centroidal axis *unless the thickness at some other axis is smaller than that at the centroidal axis.*

The shapes shown in Figures 9–6, 9–7, and 9–8 are all examples that conform to the rule that the maximum shearing stress occurs at the neutral axis because each has its smallest thickness at the neutral axis. Figure 9–9 shows three examples where the rule *does not apply.* In each, at some axes away from the neutral axis, the thickness is smaller than that at the neutral axis. In such cases, the maximum shearing stress *may* occur at some other axis. An example problem to follow illustrates this observation by analyzing the triangular section.

The solid and hollow circular sections are important examples of where the maximum shearing stress does occur at the neutral axis even though the thickness decreases at other axes. It can be shown that the ratio Q/t continuously decreases from axes away from the neutral axis at the diameter.

The following example problems illustrate the shearing stress distribution in beams of different shapes. Note the comments at the end of each problem for some general conclusions.

Example Problem 9–6   Compute the distribution of shearing stress with position in the cross section for a beam with the rectangular shape shown in Figure 9–6. The actual dimensions are 2.0 in by 10.0 in. Plot the results. The shearing force, V, on the section of interest is 1200 lb.

Solution   **Objective**   Compute the shearing stress at several axes and plot τ versus position.

Given   Cross section shape and dimensions in Figure 9–6. $V = 1200$ lb.

Analysis   Use the *Guidelines for computing shearing stresses in beams.* Because the shape is symmetrical with respect to the centroidal axis, we choose to compute the shearing stresses in the upper part at the axes a-a, b-b, c-c, and d-d, as shown in Figure 9–10. Then, the values of stresses in the lower part

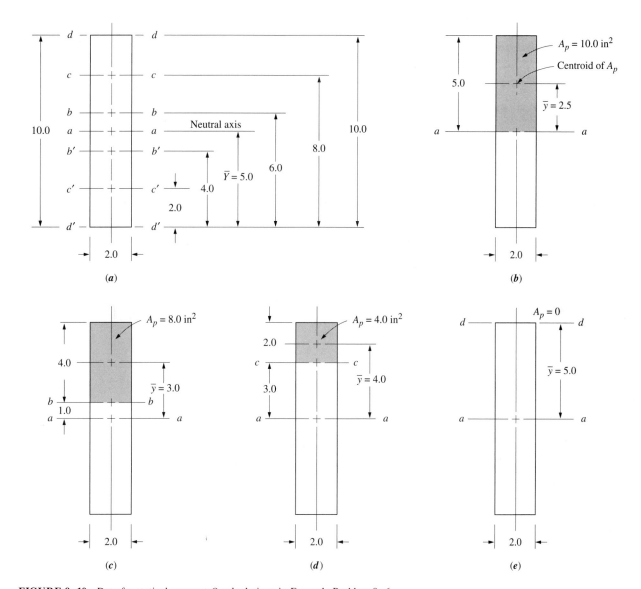

FIGURE 9–10 Data for statical moment Q calculations in Example Problem 9–6.

at sections b'-b', c'-c', and d'-d' will be the same as the corresponding points above.

Results **Step 1.** $V = 1200$ lb (given)

Step 2. For the rectangular shape, the centroid is at the midheight, as shown in Figure 9–6, coincident with axis a-a. $\overline{Y} = 5.00$ in.

Step 3. $I = bh^3/12 = (2.0)(10.0)^3/12 = 166.7$ in⁴

Step 4. Thickness $= t = 2.0$ in at all axes.

Step 5. We will compute $Q = A_p y$ for each axis using the method shown earlier in this chapter. Recall that the value of Q for this section at the centroidal axis was computed in Example Problem 9–1 where we found $Q = 25.0$ in³. A similar calculation is summarized in the table following Step 6, using data from Figure 9–10.

Step 6. Using Equation (9–1), the calculation for shearing stress at the neutral axis *a-a* is shown here. The calculation would be the same at the other axes with only the value of Q changing. See the table below.

$$\tau = \frac{VQ}{It} = \frac{(1200 \text{ lb})(25.0 \text{ in}^3)}{(166.7 \text{ in}^4)(2.0 \text{ in})} = 90.0 \text{ psi}$$

Axis	V	I	t	A_p	y	$Q = A_p y$	$\tau = VQ/It$
a-a	1200	166.7	2.0	10.0	2.5	25.0	90.0 psi
b-b	1200	166.7	2.0	8.0	3.0	24.0	86.4 psi
c-c	1200	166.7	2.0	4.0	4.0	57.6	57.6 psi
d-d	1200	166.7	2.0	0.0	5.0	0.0	0.0 psi

The results of shearing stress versus position are shown in Figure 9–11 alongside the rectangular section itself.

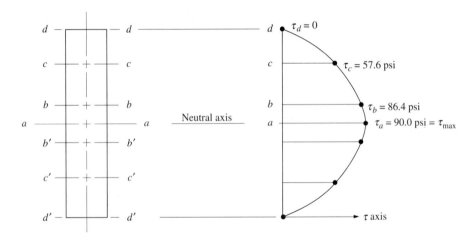

FIGURE 9–11   Distribution of shearing stress on rectangular section for Example Problem 9–6.

Comments   Note that the maximum shearing stress does occur at the neutral axis as predicted. The variation of shearing stress with position is parabolic, ending with zero stress at the top and bottom surfaces.

Example Problem 9–7   For the triangular beam cross section shown in Figure 9–12, compute the shearing stress that occurs at the axes *a* through *g*, each 50 mm apart. Plot the variation of stress with position on the section. The shearing force is 50 kN.

Solution   **Objective**   Compute the shearing stress at seven axes and plot τ versus position.

Given   Cross section shape and dimensions in Figure 9–12. $V = 50$ kN.

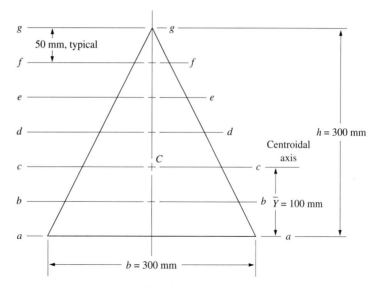

FIGURE 9–12 Triangular cross section for a beam for which the maximum shearing stress does not occur at the centroidal axis.

Analysis   Use the *Guidelines for computing shearing stresses in beams.*

Results   In the general shear formula, the values of V and I will be the same for all computations. V is given to be 50 kN and

$$I = \frac{bh^3}{36} = \frac{(300)(300)^3}{36} = 225 \times 10^6 \text{ mm}^4$$

Table 9–1 shows the remaining computations. Obviously, the value for Q for axes *a-a* and *g-g* is zero because the area outside each axis is zero. Note that because of the unique shape of the given triangle, the thickness t at any axis is equal to the height of the triangle above the axis.

TABLE 9–1

Axis	A_p (mm^2)	$\bar{y}$ (mm)	$Q = A_p\bar{y}$ (mm^3)	t (mm)	τ (MPa)
a-a	0	100	0	300	0
b-b	13 750	75.8	1.042×10^6	250	0.92
c-c	20 000	66.7	1.333×10^6	200	1.48
d-d	11 250	100.0	1.125×10^6	150	1.67
e-e	5 000	133.3	0.667×10^6	100	1.48
f-f	1 250	166.7	0.208×10^6	50	0.92
g-g	0	200	0	0	0

Figure 9–13 shows a plot of these stresses. The maximum shearing stress occurs at half the height of the section, and the stress at the centroid (at $h/3$) is lower. This illustrates the general statement made earlier that for sections whose minimum thickness does not occur at the centroidal axis,

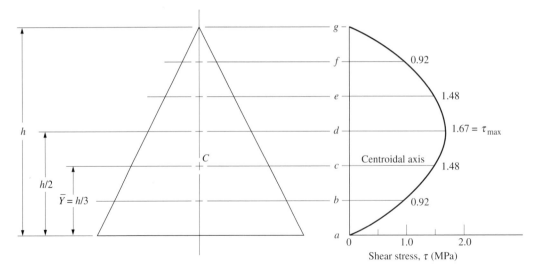

FIGURE 9–13 Shearing stress distribution in the triangular cross section for Example Problem 9–7.

the maximum shearing stress may occur at some axis other than the centroidal axis.

Comment One further note can be made about the computations shown for the triangular section. For the axis *b-b*, the partial area A_p was taken as that area *below b-b*. The resulting section is the trapezoid between *b-b* and the bottom of the beam. For all other axes, the partial area A_p was taken as the triangular area *above* the axis. The area below the axis could have been used, but the computations would have been more difficult. When computing Q, it does not matter whether the area above or below the axis of interest is used for computing A_p and $\bar{y}$.

By reviewing the results of the several example problems worked thus far in this chapter, the following conclusions can be drawn.

Summary of observations about the distribution of shearing stress in the cross section of a beam

1. The shearing stress at the outside of the section away from the centroidal axis is zero.
2. The maximum shearing stress in the cross section occurs at the centroidal axis provided that the thickness there is no greater than at some other axis.
3. Within a part of the cross section where the thickness is constant, the shearing stress varies in a curved fashion, decreasing as the distance from the centroidal axis increases. The curve is actually a part of a parabola.
4. At an axis where the thickness changes abruptly, as where the web of a tee or an I-shape joins the flange, the shearing stress also changes abruptly, being much smaller in the flange than in the thinner web. See Figure 9–14.

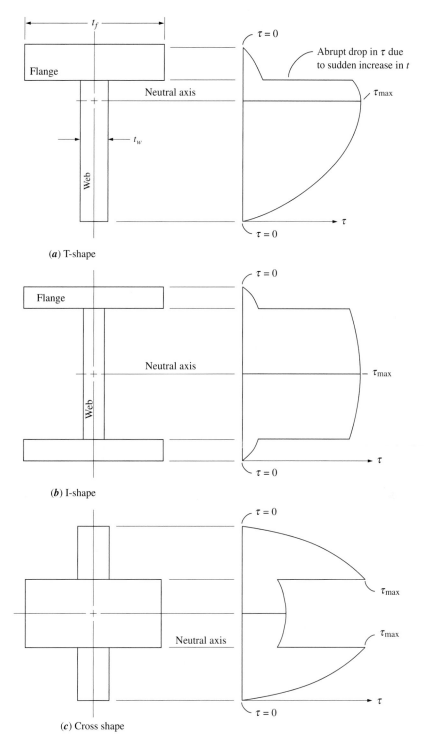

(a) T-shape

(b) I-shape

(c) Cross shape

FIGURE 9–14 Stress distribution for shapes with abrupt changes in thickness.

9–6 DEVELOPMENT OF THE GENERAL SHEAR FORMULA

This section presents the background information on the general shear formula. Figure 9–15 shows a beam carrying two transverse loads and the corresponding shearing force and bending moment diagrams to help you visualize certain relationships.

The *moment-area* principle of beam diagrams states that *the change in bending moment between two points on a beam is equal to the area under the shearing force curve between those two points.* For example, consider two points in segment *A-B* of the beam in Figure 9–15, marked x_1 and x_2, a small distance dx apart. The moment at x_1 is M_1 and the moment at x_2 is M_2. Then the moment-area rule states that

$$M_2 - M_1 = V(dx) = dM$$

This can also be stated,

$$V = \frac{dM}{dx} \tag{9-4}$$

That is, the differential change in bending moment for a differential change in position on the beam is equal to the shearing force occurring at that position.

Equation (9–4) can also be developed by looking at a free-body diagram of the small segment of the beam between x_1 and x_2, as shown in Figure 9–16(a). As this is a cut section from the beam, the internal shearing forces and bending moments

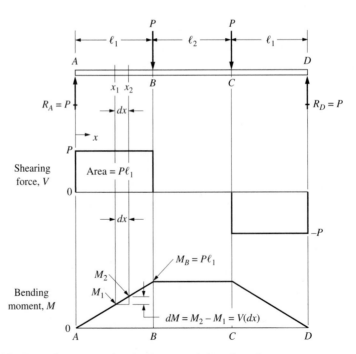

FIGURE 9–15   Beam diagrams used to develop general shear formula.

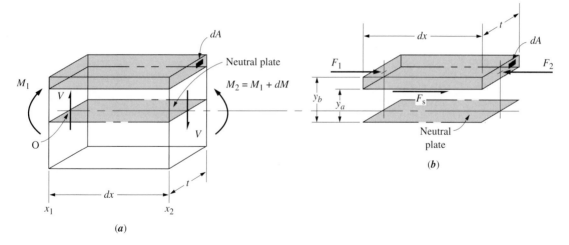

FIGURE 9–16 Forces on a portion of a cut segment of a beam. (a) Free-body diagram of beam segment. (b) Isolated portion of segment.

are shown acting on the cut faces. Because the beam itself is in equilibrium, this segment is also. Then the sum of moments about a point in the left face at O must be zero. This gives

$$\sum M_O = 0 = M_1 - M_2 + V(dx) = -dM + V(dx)$$

Or, as shown before,

$$V = \frac{dM}{dx}$$

Any *part* of the beam segment in Figure 9–16(a) must also be in equilibrium. The shaded portion isolated in Figure 9–16(b) is acted on by forces parallel to the axis of the beam. On the left side, F_1 is due to the bending stress acting at that section on the area. On the right side, F_2 is due to the bending stress acting at that section on the area. In general, the values of F_1 and F_2 will be different and there must be a third force acting on the bottom face of the shaded portion of the segment to maintain equilibrium. This is the shearing force, F_s, which causes the shearing stress in the beam. Figure 9–16(b) shows F_s acting on the area $t(dx)$. Then the shearing stress is

$$\tau = \frac{F_s}{t(dx)} \tag{9–5}$$

By summing forces in the horizontal direction, we find

$$F_s = F_2 - F_1 \tag{9–6}$$

We will now develop the equations for the forces F_1 and F_2. Each force is the product of the bending stress times the area over which it acts. But the bending stress varies with position in the cross section. From the flexure formula, the bending stress at any position y relative to the neutral axis is

$$\sigma = \frac{My}{I}$$

Then the total force acting on the shaded area of the left face of the beam segment is

$$F_1 = \int_A \sigma \, dA = \int_{y_a}^{y_b} \frac{M_1 y}{I} \, dA \qquad (9\text{--}7)$$

where dA is a small area within the shaded area. The values of M_1 and I are constant and can be taken outside the integral sign. Equation (9–7) then becomes

$$F_1 = \frac{M_1}{I} \int_{y_a}^{y_b} y \, dA \qquad (9\text{--}8)$$

Now the last part of Equation (9–8) corresponds to the definition of the centroid of the shaded area. That is,

$$\int_{y_a}^{y_b} y \, dA = \bar{y} A_p \qquad (9\text{--}9)$$

where A_p is the area of the shaded portion of the left face of the segment and $\bar{y}$ is the distance from the neutral axis to the centroid of A_p. This product of $\bar{y} A_p$ is called the *static moment* Q in the general shear formula. Making this substitution in Equation (9–8) gives

$$F_1 = \frac{M_1}{I} \int_{y_a}^{y_b} y \, dA = \frac{M_1}{I} \bar{y} A_p = \frac{M_1 Q}{I} \qquad (9\text{--}10)$$

Similar reasoning can be used to develop the relationship for the force F_2 on the right face of the segment.

$$F_2 = \frac{M_2 Q}{I} \qquad (9\text{--}11)$$

Substitutions can now be made for F_1 and F_2 in Equation (9–6) to complete the development of the shearing force.

$$F_s = F_2 - F_1 = \frac{M_2 Q}{I} - \frac{M_1 Q}{I} = \frac{Q}{I}(M_2 - M_1) \qquad (9\text{--}12)$$

Earlier we defined $(M_2 - M_1) = dM$. Then

$$F_s = \frac{Q(dM)}{I} \qquad (9\text{--}13)$$

Then, in Equation (9–5),

$$\tau = \frac{F_s}{t(dx)} = \frac{Q(dM)}{It(dx)}$$

But, from Equation (9–4), $V = dM/dx$. Then

$$\tau = \frac{VQ}{It}$$

This is the form of the general shear formula [Equation (9–1)] used in this chapter.

9–7 SPECIAL SHEAR FORMULAS

As demonstrated in several example problems, the general shear formula can be used to compute the shearing stress at any axis on any cross section of the beam. However, frequently it is desired to know only the *maximum shearing stress*. For many common shapes used for beams, it is possible to develop special simplified formulas that will give the maximum shearing stress quickly. The rectangle, circle, thin-walled hollow tube, and thin-webbed shapes can be analyzed this way. The formulas are developed in this section.

For all of these section shapes, the maximum shearing stress occurs at the neutral axis. The rectangle and thin-webbed shapes conform to the rule stated in Section 9–5 because the thickness at the neutral axis is no greater than at other axes in the section. The circle and the thin-walled tube do not conform to the rule. However, it can be shown that the ratio Q/t in the general shear formula decreases continuously as the axis of interest moves away from the neutral axis, resulting in the decrease in the shearing stress.

Rectangular shape.   Figure 9–17 shows a typical rectangular cross section having a thickness t and a height h. The three geometrical terms in the general shear formula can be expressed in terms of t and h.

$$I = \frac{th^3}{12}$$

$$t = t$$

$$Q = A_p \overline{y} \quad \text{(for area above centroidal axis)}$$

$$Q = \frac{th}{2} \cdot \frac{h}{4} = \frac{th^2}{8}$$

Putting these terms in the general shear formula gives

$$\tau_{\max} = \frac{VQ}{It} = V \cdot \frac{th^2}{8} \cdot \frac{12}{th^3} \cdot \frac{1}{t} = \frac{3}{2} \frac{V}{th}$$

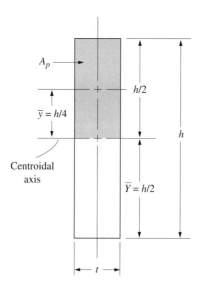

FIGURE 9–17   Rectangular shape.

But since th is the total area of the section,

**Special Shear
Formula for
Rectangle**

$$\tau_{max} = \frac{3V}{2A} \qquad (9\text{–}14)$$

Equation (9–14) can be used to compute exactly the maximum shearing stress in a rectangular beam at its centroidal axis.

Note that $\tau = V/A$ represents the *average* shearing stress on the section. Thus the maximum shearing stress on a rectangular cross section is 1.5 times higher than the average.

**Example Problem
9–8**

Compute the maximum shearing stress that would occur in the rectangular cross section of a beam like that shown in Figure 9–17. The shearing force is 1000 lb, $t = 2.0$ in, and $h = 8.0$ in.

Solution

Using Equation (9–14) yields

$$\tau_{max} = \frac{3V}{2A} = \frac{3(1000 \text{ lb})}{2(2 \text{ in})(8 \text{ in})} = 93.8 \text{ psi}$$

Circular shape.   The special shear formula for the circular shape is developed in a similar manner to that used for the rectangular shape. Equations for Q, I, and t are written in terms of the primary size variable for the circular shape, its diameter. Then the general shear formula is simplified (refer to Figure 9–18).

$$t = D$$

$$I = \frac{\pi D^4}{64}$$

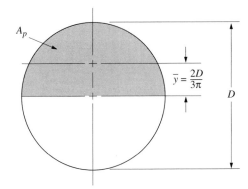

FIGURE 9–18 Circular shape.

$$Q = A_p \bar{y} \quad \text{(for the semicircle above the centroid)}$$

$$Q = \frac{\pi D^2}{8} \cdot \frac{2D}{3\pi} = \frac{D^3}{12}$$

Then the maximum shearing stress is

$$\tau_{max} = \frac{VQ}{It} = V \times \frac{D^3}{12} \times \frac{64}{\pi D^4} \times \frac{1}{D} = \frac{64V}{12\pi D^2}$$

To refine the equation, factor out a 4 from the numerator and then note that the total area of the circular section is $A = \pi D^2/4$.

$$\tau_{max} = \frac{16(4)V}{12\pi D^2} = \frac{16V}{12A}$$

$$\tau_{max} = \frac{4V}{3A} \tag{9–15}$$

Special Shear Formula for Circle

This shows that the maximum shearing stress is 1.33 times higher than the average on the circular section.

Example Problem 9–9

Compute the maximum shearing stress that would occur in a circular shaft, 50 mm in diameter, if it is subjected to a vertical shearing force of 110 kN.

Solution

Equation (9–15) will give the maximum shearing stress at the horizontal diameter of the shaft.

$$\tau_{max} = \frac{4V}{3A}$$

But

$$A = \frac{\pi D^2}{4} = \frac{\pi (50 \text{ mm})^2}{4} = 1963 \text{ mm}^2$$

Then

$$\tau_{max} = \frac{4(110 \times 10^3 \text{ N})}{3(1963 \text{ mm}^2)} = 74.7 \text{ MPa}$$

Hollow thin-walled tubular shape. Removing material from the center of a circular cross section tends to increase the local value of the shearing stress, especially near the diameter where the maximum shearing stress occurs. Although not giving a complete development here, it is observed that the maximum shearing stress in a thin-walled tube is approximately twice the average. That is,

Special Shear Formula for Thin-Walled Tube

$$\tau_{max} \approx 2\frac{V}{A} \tag{9-16}$$

where A is the total cross-sectional area of the tube.

Example Problem 9-10

Compute the approximate maximum shearing stress that would occur in a 3-in schedule 40 steel pipe if it is used as a beam and subjected to a shearing force of 6200 lb.

Solution

Equation (9-16) should be used. From Appendix A-12 we find that the cross-sectional area of the 3-in schedule 40 steel pipe is 2.228 in². Then an estimate of the maximum shearing stress in the pipe, occurring near the horizontal diameter, is

$$\tau_{max} \approx 2\frac{V}{A} = \frac{2(6200 \text{ lb})}{2.228 \text{ in}^2} = 5566 \text{ psi}$$

Thin-webbed shapes. Structural shapes such as W- and S-beams have relatively thin webs. The distribution of shearing stress in such beams is typically like that shown in Figure 9-19. The maximum shearing stress is at the centroidal axis. It decreases slightly in the rest of the web and then drastically in the flanges. Thus most of the resistance to the vertical shearing force is provided by the web. Also, the average shearing stress in the web would be just slightly smaller than the maximum stress. For these reasons, the *web shear formula* is often used to get a quick estimate of the shearing stress in thin-webbed shapes.

Web Shear Formula for Thin-Webbed Shapes

$$\tau_{max} \approx \frac{V}{A_{web}} = \frac{V}{th} \tag{9-17}$$

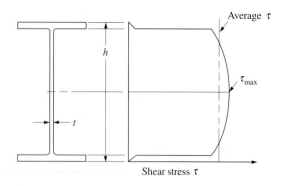

FIGURE 9-19 Distribution of shearing stress in a thin-webbed shape.

The thickness of the web is t. The simplest approach would be to use the full height of the beam for h. This would result in a shearing stress approximately 15% lower than the actual maximum shearing stress at the centroidal axis for typical beam shapes. Using just the web height between the flanges would produce a closer approximation of the maximum shearing stress, probably less than 10% lower than the actual value. In problems using the web shear formula, we use the full height of the cross section unless otherwise stated.

In summary, for thin-webbed shapes, compute the shearing stress from the web shear formula using the full height of the beam for h and the actual thickness of the web for t. Then, to obtain a more accurate estimate of the maximum shearing stress, increase this value by about 15%.

Example Problem 9–11

Using the web shear formula, compute the shearing stress in a W12 × 16 beam if it is subjected to a shearing force of 25 000 lb.

Solution

In Appendix A–7 for W-beams, it is found that the web thickness is 0.220 in and the overall depth (height) of the beam is 11.99 in. Then, using Equation (9–17), we have

$$\tau_{max} \approx \frac{V}{th} = \frac{25\,000 \text{ lb}}{(0.220 \text{ in})(11.99 \text{ in})} = 9478 \text{ psi}$$

9–8 DESIGN SHEAR STRESS

The design shear stress depends greatly upon the material from which the beam is to be made and on the form of the member subjected to the shearing stress. A limited amount of data is presented in this book and the reader is advised to check more complete references, such as References 1, 2, and 3.

For wood beams, data are given in Appendix A–18 for allowable horizontal shear stress. Note that the values are quite low, typically less than 100 psi (0.69 MPa). Shear failure is frequently the limiting factor for wood beams.

For shear stress in the webs of rolled steel beam shapes, the AISC generally recommends

$$\tau = 0.40\, s_y \tag{9–18}$$

But there are extensive discussions in Reference 1 for special cases of short beams, beams with unusually tall, thin webs, and beams with stiffeners applied either in the vertical or horizontal directions. Careful consideration of these factors is advised.

The Aluminum Association also provides extensive data for various conditions of loading and beam geometry. For example, Reference 2 gives actual data for allowable shear stress of the more popular aluminum alloys for several applications. It is not practical to summarize such data in this book.

As a general guideline, we will use the same design shear stress for ductile metals under static loads as listed in Chapter 3, Table 3–4. That is, a design factor of $N = 2$

based on the yield strength of the material in shear, s_{ys}, is suggested. And, an approximation for the value of s_{ys} is one-half of the yield strength in tension, s_y. In summary,

$$\tau_d = \frac{s_{ys}}{N} = \frac{0.5s_y}{N} = \frac{s_y}{2N} \qquad (9\text{–}19)$$

For $N = 2$,

$$\tau_d = \frac{s_y}{4} = 0.25s_y$$

9–9 SHEAR FLOW

Built-up sections used for beams, such as those shown in Figures 9–20 and 9–21, must be analyzed to determine the proper size and spacing of fasteners. The discussion in preceding sections showed that horizontal shearing forces exist at the planes joined by the nails, bolts, and rivets. Thus the fasteners are subjected to shear. Usually, the size and material of the fastener will permit the specification of an allowable shearing force on each. Then the beam must be analyzed to determine a suitable spacing for the fasteners that will ensure that all parts of the beam act together.

The term *shear flow* is useful for analyzing built-up sections. Called q, the shear flow is found by multiplying the shearing stress at a section by the thickness at that section. That is,

$$q = \tau t \qquad (9\text{–}20)$$

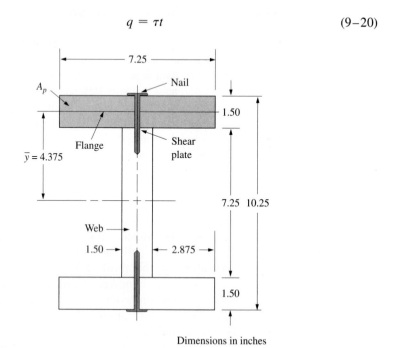

Dimensions in inches

FIGURE 9–20 Beam shape for Example Problem 9–12.

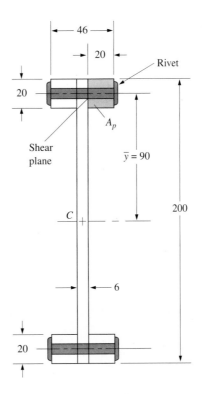

Dimensions in mm

FIGURE 9–21 Built-up beam shape for Example Problem 9–13.

But from the general shear formula,

$$\tau = \frac{VQ}{It}$$

Then

$$q = \tau t = \frac{VQ}{I} \qquad (9\text{–}21)$$

The units for q are *force per unit length,* such as N/m, N/mm, or lb/in. The shear flow is a measure of how much shearing force must be resisted at a particular section per unit length. Knowing the shearing force capacity of a fastener then allows the determination of a safe spacing for the fasteners.

For example, if a particular style of nail can safely withstand 150 lb of shearing force, we will define

$$F_{sd} = 150 \text{ lb}$$

Then, if in a particular place on a beam fabricated by nailing boards together and the shear flow is computed to be $q = 28.5$ lb/in, the maximum spacing, s, of the nail is

$$s = \frac{F_{sd}}{q} = \frac{150 \text{ lb}}{28.5 \text{ lb/in}} = 5.26 \text{ in} \qquad (9-22)$$

Guidelines for specifying the spacing of fasteners

The objective is to specify suitable spacing for fasteners to hold parts of a composite beam shape together while resisting an applied vertical shearing force.

Required data: Applied shearing force, V.
  Geometry of the cross section of the beam.
  Allowable shearing force on each fastener, F_{sd}.

1. Compute the moment of inertia, I, for the entire cross section with respect to its centroidal axis.
2. Compute the value of the statical moment, Q, for that part of the cross section outside the fasteners. Use $Q = A_p y$ as defined in Section 9–4.
3. Compute the shear flow, q, from

$$q = \frac{VQ}{I}$$

The result will be the amount of shearing force that must be resisted per unit length along the beam.

4. Compute the maximum allowable spacing of the fasteners, s, from

$$s = F_{sd}/q$$

5. Specify a convenient spacing between fasteners less than the maximum allowable.

Example Problem 9–12

Determine the proper spacing of the nails used to secure the flange boards to the web of the built-up I-beam shown in Figure 9–20. All boards are standard 2 × 8 wood shapes. The nails to be used can safely resist 250 lb of shearing force each. The load on the beam is shown in Figure 9–15 with $P = 500$ lb.

Solution   **Objective**   Specify a suitable spacing for the nails.

Given   Loading in Figure 9–15. $P = 500$ lb. $F_{sd} = 250$ lb/nail.
Beam shape and dimensions in Figure 9–20.

Analysis   Use the *Guidelines for specifying the spacing of fasteners.*

Results The maximum shearing force on the beam is 500 lb, occurring between each support and the applied loads.

Step 1. The moment of inertia can be computed by subtracting the two open-space rectangles at the sides of the web from the full rectangle surrounding the I-shape.

$$I = \frac{7.25(10.25)^3}{12} - \frac{2(2.875)(7.25)^3}{12} = 468.0 \text{ in}^4$$

Step 2. At the place where the nails join the boards, Q is evaluated for the area of the top (or bottom) flange board.

$$Q = A_p \bar{y} = (1.5 \text{ in})(7.25 \text{ in})(4.375 \text{ in}) = 47.6 \text{ in}^3$$

Step 3. Then the shear flow is

$$q = \frac{VQ}{I} = \frac{(500 \text{ lb})(47.6 \text{ in}^3)}{468 \text{ in}^4} = 50.9 \text{ lb/in}$$

This means that 50.9 lb of force must be resisted along each inch of length of the beam at the point between the flange and the web boards.

Step 4. Since each nail can withstand 250 lb, the maximum spacing is

$$s = \frac{F_{sd}}{q} = \frac{250 \text{ lb}}{50.9 \text{ lb/in}} = 4.92 \text{ in}$$

Step 5. A spacing of 4.5 in would be reasonable.

The principle of shear flow also applies for sections like that shown in Figure 9–21, in which a beam section is fabricated by riveting square bars to a vertical web plate to form an I-shape. The shear flow occurs from the web plate to the flange bars. Thus, when evaluating the statical moment Q, the partial area, A_p, is taken to be the area of one of the flange bars.

Example Problem 9–13 A fabricated beam is made by riveting square aluminum bars to a vertical plate, as shown in Figure 9–21. The bars are 20 mm square. The plate is 6 mm thick and 200 mm high. The rivets can withstand 800 N of shearing force across one cross section. Determine the required spacing of the rivets if a shearing force of 5 kN is applied.

Solution **Objective** Specify a suitable spacing for the rivets.

Given Shearing force = 5 kN. F_{sd} = 800 N/rivet.
Beam shape and dimensions in Figure 9–21.

Analysis Use the *Guidelines for specifying the spacing of fasteners.*

Results **Step 1.** *I* is the moment of inertia of the entire cross section,

$$I = \frac{6(200)^3}{12} + 4\left[\frac{20^4}{12} + (20)(20)(90)^2\right]$$
$$= 17.0 \times 10^6 \text{ mm}^4$$

Step 2. Q is the product of $A_p\bar{y}$ for the area *outside* the section where the shear is to be calculated. In this case, the partial area A_p is the 20-mm square area *to the side* of the web. For the beam in Figure 9–19,

$$Q = A_p\bar{y} \quad \text{(for one square bar)}$$
$$= (20)(20)(90) \text{ mm}^3 = 36\,000 \text{ mm}^3$$

Step 3. Then for $V = 5$ kN,

$$q = \frac{VQ}{I} = \frac{(5 \times 10^3 \text{ N})(36 \times 10^3 \text{ mm}^3)}{17.0 \times 10^6 \text{ mm}^4} = 10.6 \text{ N/mm}$$

Thus a shearing force of 10.6 N is to be resisted for each millimeter of length of the beam.

Step 4. Since each rivet can withstand 800 N of shearing force, the maximum spacing is

$$s = \frac{F_{sd}}{q} = \frac{800 \text{ N}}{10.6 \text{ N/mm}} = 75.5 \text{ mm}$$

Step 5. Specify a spacing of $s = 75$ mm.

REFERENCES

1. Aluminum Association, *Specifications for Aluminum Structures,* Washington, D.C., 1986.

2. American Institute of Steel Construction, *Manual of Steel Construction,* 9th ed., Chicago, 1989.

3. U.S. Department of Agriculture Forest Products Laboratory, *Handbook of Wood and Wood-Based Materials for Engineers, Architects, and Builders,* Hemisphere Publishing Corp., New York, 1989.

PROBLEMS

For Problems 9–1 through 9–20, compute the shearing stress at the horizontal neutral axis for a beam having the cross-sectional shape shown in the given figure for the given shearing force. Use the general shear formula.

9–1.M Use a rectangular shape having a width of 50 mm and a height of 200 mm. $V = 7500$ N.

9–2.M Use a rectangular shape having a width of 38 mm and a height of 180 mm. $V = 5000$ N.

9–3.E Use a rectangular shape having a width of 1.5 in and a height of 7.25 in. $V = 12\,500$ lb.

9–4.E Use a rectangular shape having a width of 3.5 in and a height of 11.25 in. $V = 20\,000$ lb.

9–5.M Use a circular shape having a diameter of 50 mm. $V = 4500$ N.

9–6.M Use a circular shape having a diameter of 38 mm. $V = 2500$ N.

9–7.E Use a circular shape having a diameter of 2.00 in. $V = 7500$ lb.

9–8.E Use a circular shape having a diameter of 0.63 in. $V = 850$ lb.

9–9.E Use the shape shown in Figure P7–16. $V = 1500$ lb.

9–10.E Use the shape shown in Figure P7–2. $V = 850$ lb.

9–11.E Use the shape shown in Figure P7–3. $V = 850$ lb.

9–12.M Use the shape shown in Figure P7–4. $V = 112$ kN.

9–13.M Use the shape shown in Figure P7–17. $V = 71.2$ kN.

9–14.M Use the shape shown in Figure P7–18. $V = 1780$ N.

9–15.M Use the shape shown in Figure P7–5. $V = 675$ N.

9–16.M Use the shape shown in Figure P7–6. $V = 2.5$ kN.

9–17.M Use the shape shown in Figure P7–8. $V = 10.5$ kN.

9–18.E Use the shape shown in Figure P7–14. $V = 1200$ lb.

9–19.E Use the shape shown in Figure P7–15. $V = 775$ lb.

9–20.E Use the shape shown in Figure P7–33. $V = 2500$ lb.

For Problems 9–21 through 9–30, assume that the indicated shape is the cross section of a beam made from wood having an allowable shearing stress of 70 psi, which is that of No. 2 grade southern pine listed in Appendix A–18. Compute the maximum allowable shearing force for each shape. Use the general shear formula.

9–21.E Use a standard 2×4 wooden beam with the long dimension vertical.

9–22.E Use a standard 2×4 wooden beam with the long dimension horizontal.

9–23.E Use a standard 2×12 wooden beam with the long dimension vertical.

9–24.E Use a standard 2×12 wooden beam with the long dimension horizontal.

9–25.E Use a standard 10×12 wooden beam with the long dimension vertical.

9–26.E Use a standard 10×12 wooden beam with the long dimension horizontal.

9–27.E Use the shape shown in Figure P7–21.

9–28.E Use the shape shown in Figure P7–22.

9–29.E Use the shape shown in Figure P7–23.

9–30.E Use the shape shown in Figure P7–24.

9–31.E For a beam having the I-shape cross section shown in Figure P7–2, compute the shearing stress on horizontal axes 0.50 in apart from the bottom to the top. At the ends of the web where it joins the flanges, compute the stress in both the web and the flange. Use a shearing force of 500 lb. Then plot the results.

9–32.E For a beam having the box-shape cross section shown in Figure P7–3, compute the shearing stress on horizontal axes 0.50 in apart from the bottom to the top. At the ends of the vertical sides where they join the flanges, compute the stress in both the web and the flange. Use a shearing force of 500 lb. Then plot the results.

9–33.E For a standard W14 $\times$ 43 steel beam, compute the shearing stress at the neutral axis when subjected to a shearing force of 33 500 lb. Use the general shear formula. Neglect the fillets at the intersection of the web with the flanges.

9–34.E For the same conditions listed in Problem 9–33, compute the shearing stress at several axes and plot the variation of stress with position in the beam.

9–35.E For a standard W14 $\times$ 43 steel beam, compute the shearing stress from the web shear formula when it carries a shearing force of 33 500 lb. Compare this value with that computed in Problem 9–33 and plot it on the graph produced for Problem 9–34.

9–36.E For an Aluminum Association Standard I8 $\times$ 6.181 beam, compute the shearing stress at the neutral axis when subjected to a shearing force of 13 500 lb. Use the general shear formula. Neglect the fillets at the intersection of the web with the flanges.

9–37.E For the same conditions listed in Problem 9–36, compute the shearing stress at several axes and plot the variation of stress with position in the beam.

9–38.E For an aluminum I8 × 6.181 beam, compute the shearing stress from the web shear formula when the beam carries a shearing force of 13 500 lb. Compare this value with that computed in Problem 9–36 and plot it on the graph produced for Problem 9–37.

Note: In problems calling for design stresses, use the following:

For structural steel:

$$\text{In bending:} \quad \sigma_d = 0.66s_y$$
$$\text{In shear:} \quad \tau_d = 0.4s_y$$

For any other metal:

$$\text{In bending:} \quad \sigma_d = \frac{s_y}{N}$$

$$\text{In shear:} \quad \tau_d = 0.5\frac{s_y}{N}$$

For wood:

Use allowable stresses in Appendix A–18.

9–39.E The loading shown in Figure P6–4 is to be carried by a W10 × 15 steel beam. Compute the shearing stress using the web shear formula. Also compute the maximum bending stress. Then compare the stresses to the design stresses for ASTM A36 structural steel.

9–40.E Specify a suitable wide-flange beam to be made from ASTM A36 structural steel to carry the load shown in Figure P6–4 based on the design stress in bending. Then, for the beam selected, compute the shearing stress from the web shear formula and compare it with the design shear stress.

9–41.E Specify a suitable wide-flange beam to be made from ASTM A36 structural steel to carry the load shown in Figure P6–52 based on the design stress in bending. Then, for the beam selected, compute the shearing stress from the web shear formula and compare it with the design shear stress.

9–42.C Specify a suitable wide-flange beam to be made from ASTM A36 structural steel to carry the load shown in Figure P6–54 based on the design stress in bending. Then, for the beam selected, compute the shearing stress from the web shear formula and compare it with the design shear stress.

9–43.E Specify a suitable standard steel pipe from Appendix A–12 to be made from AISI 1020 hot-rolled steel to carry the load shown in Figure P6–51 based on the design stress in bending with a design factor of 3. Then, for the pipe selected, compute the shearing stress from the special shear formula for hollow tubes and compute the resulting design factor from the design shear stress formula.

9–44.E An Aluminum Association standard channel (Appendix A–10) is to be specified to carry the load shown in Figure P6–9 to produce a design factor of 4 in bending. The legs of the channel are to point down. The channel is made from 6061-T6 aluminum. For the channel selected, compute the maximum shearing stress.

9–45.E A wooden joist in the floor of a building is to carry a uniformly distributed load of 200 lb/ft over a length of 12.0 ft. Specify a suitable standard wooden beam shape for the joist, made from No. 2 grade hemlock, to be safe in both bending and shear (see Appendixes A–4 and A–18).

9–46.C A wooden beam in an outdoor structure is to carry the load shown in Figure P6–53. If it is to be made from No. 3 grade Douglas fir, specify a suitable standard wooden beam to be safe in both bending and shear (see Appendixes A–4 and A–18).

9–47.E The box beam shown in Figure P7–22 is to be made from No. 1 grade southern pine. It is to be 14 ft long and carry two equal concentrated loads, each 3 ft from an end. The beam is simply supported at its ends. Specify the maximum allowable load for the beam to be safe in both bending and shear.

9–48.C An aluminum I-beam, 19 × 8.361, carries the load shown in Figure P6–8. Compute the shearing stress in the beam using the web shear formula.

9–49.C Compute the bending stress for the beam in Problem 9–48.

9–50.E A 2 × 8 wooden floor joist in a home is simply supported, 12 ft long and carries a uniformly distributed load of 80 lb/ft. Compute the shearing stress in the joist. Would it be safe if it is made from No. 2 southern pine wood?

9–51.E A steel beam is made as a rectangle, 0.50 in wide by 4.00 in high.
   (a) Compute the shearing stress in the beam if it carries the load shown in Figure P6–10.
   (b) Compute the stress due to bending.

(c) Specify a suitable steel for the beam to produce a design factor of 3 for either bending or shear.

9–52.M An aluminum beam is made as a rectangle, 16 mm wide by 60 mm high.
  (a) Compute the shearing stress in the beam if it carries the load shown in Figure P6–6.
  (b) Compute the stress due to bending.
  (c) Specify a suitable aluminum for the beam to produce a design factor of 3 for either bending or shear.

9–53.M It is planned to use a rectangular bar to carry the load shown in Figure P6–47. Its thickness is to be 12 mm and it is to be made from aluminum 6061-T6. Determine the required height of the rectangle to produce a design factor of 4 in bending based on yield strength. Then compute the shearing stress in the bar and the resulting design factor for shear.

9–54.M A round shaft, 40 mm in diameter, carries the load shown in Figure P6–48.
  (a) Compute the maximum shearing stress in the shaft.
  (b) Compute the maximum stress due to bending.
  (c) Specify a suitable steel for the shaft to produce a design factor of 4 based on yield strength in either shear or bending.

9–55.M Compute the required diameter of a round bar to carry the load shown in Figure P6–47 while limiting the stress due to bending to 120 MPa. Then compute the resulting shearing stress in the bar and compare it to the bending stress.

9–56.E Compute the maximum allowable vertical shearing force on a wood dowel having a diameter of 1.50 in, if the maximum allowable shearing stress is 70 psi.

9–57.E A standard steel pipe is to be selected from Appendix A–12 to be used as a chinning bar in a gym. It is to be simply supported at the ends of its 36-in length. Men weighing up to 400 lb are expected to hang on the bar with either one or two hands at any place along its length. The pipe is to be made from AISI 1020 hot-rolled steel. Specify a suitable pipe to provide a design factor of 6 based on yield strength in either bending or shear.

9–58.E A standard steel pipe is to be simply supported at its ends and carry a single concentrated load of 2800 lb at its center. The pipe is to be made from AISI 1020 hot-rolled steel. The minimum design

factor is to be 4 based on yield strength for either bending or shear. Specify a suitable pipe size from Appendix A–12 if the length of the pipe is
  (a) 1.5 in
  (b) 3.0 in
  (c) 4.5 in
  (d) 6.0 in

Shear Flow Problems

9–59.E The shape in Figure P7–14 is to be made by gluing the flat plate to the hat section. If the beam made from this section is subjected to a shearing force of 1200 lb, compute the shear flow at the joint. What must be the shearing strength of the glue in psi?

9–60.E The shape in Figure P7–26 is to be made by using a metal-bonding glue between the S-beam and the web of the channel. Compute the shear flow at the joint and the required shearing strength of the glue for a shearing force of 2500 lb.

9–61.E The shape in Figure P7–33 is to be fabricated by riveting the bottom plate to the angles and then welding the top plate to the angles. When used as a beam, there are four potential failure modes: bending stress, shearing stress in the angles, shear in the welds, and shear of the rivets. The shape is to be used as the seat of a bench carrying a uniformly distributed load over a span of 10.0 ft. Compute the maximum allowable distributed load for the following design limits.
  (a) The material of all components is aluminum 6061-T4 and a design factor of 4 is required for either bending or shear.
  (b) The allowable shear flow on each weld is 1800 lb/in.
  (c) The rivets are spaced 4.0 in apart along the entire length of the beam. Each rivet can withstand 600 lb of shear.

9–62.E An alternative design for the bench described in Problem 9–61 is to use the built-up wood T-shape shown in Figure P7–24. The wood is to be No. 3 grade southern pine. One nail is to be driven into each vertical 2 × 12. Each nail can withstand 160 lb in shear and the nails are spaced 6.0 in apart along the length of the beam. Compute the maximum allowable distributed load on the beam.

9–63.E The shape shown in Figure P7–21 is glued together and the allowable shearing strength of the glue is 800 psi. The components are No. 2 grade Douglas fir. If the beam is to be simply supported and carry a single concentrated load at its center, compute the maximum allowable load. The length is 10 ft.

9–64.E The I-section shown in Figure P7–21 is fabricated from three wooden boards by nailing through the top and bottom flanges into the web. Each nail can withstand 180 lb of shearing force. If the beam having this section carries a vertical shearing force of 300 lb, what spacing would be required between nails?

9–65.E The built-up section shown in Figure P7–22 is nailed together by driving one nail through each side of the top and bottom boards into the $1\frac{1}{2}$-in-thick sides. If each nail can withstand 150 lb of shearing force, determine the required nail spacing when the beam carries a vertical shearing force of 600 lb.

9–66.E The platform whose cross section is shown in Figure P7–23 is glued together. How much force per unit length of the platform must the glue withstand if it carries a vertical shearing force of 500 lb?

9–67.C The built-up section shown in Figure P7–25 is fastened together by passing two $\frac{3}{8}$-in rivets through the top and bottom plates into the flanges of the beam. Each rivet will withstand 2650 lb in shear. Determine the required spacing of the rivets along the length of the beam if it carries a shearing force of 175 kN.

9–68.E A fabricated beam having the cross section shown in Figure P7–26 carries a shearing force of 50 kN. The channel is riveted to the S-beam with two $\frac{1}{4}$-in-diameter rivets which can withstand 1750 lb each in shear. Determine the required rivet spacing.

10

The General Case of Combined Stress and Mohr's Circle

10–1 OBJECTIVES OF THIS CHAPTER

In the earlier chapters of this book we focused on the computation of simple stresses, those cases in which only one type of stress was of interest. We studied direct stresses due to tension, compression, bearing, and shear; torsional shear stress; stress due to bending; and shearing stresses in beams. Also presented were many practical problems for which the computation of the simple stress was the appropriate method of analysis.

But a much wider array of real, practical problems involve *combined stresses,* situations in which two or more different components of stress act at the same point in a load-carrying member. In this chapter we develop the general approaches used to properly combine stresses. In Chapter 11 we develop several practical special cases involving combined stress.

After completing this chapter, you should be able to:

1. Recognize cases for which combined stresses occur.

2. Represent the stress condition on a stress element.

3. Understand the development of the equations for combined stresses, from which you can compute the following:
   a. The maximum and minimum principal stresses
   b. The orientation of the principal stress element
   c. The maximum shear stress on an element
   d. The orientation of the maximum shear stress element

e. The normal stress that acts along with the maximum shear stress

f. The normal and shear stress that occurs on the element oriented in any direction

4. Construct Mohr's circle for biaxial stress.

5. Interpret the information available from Mohr's circle for the stress condition at a point in any orientation.

6. Use the data from Mohr's circle to draw the principal stress element and the maximum shear stress element.

10–2 THE STRESS ELEMENT

In general, combined stress refers to cases in which two or more types of stress are exerted on a given point at the same time. The component stresses can be either *normal* (i.e., tension or compression) or *shear* stresses.

When a load-carrying member is subjected to two or more different kinds of stresses, the first task is to compute the stress due to each component. Then a decision is made about which point within the member has the highest *combination* of stresses and the combined stress analysis is completed for that point. For some special cases, it is desired to know the stress condition at a given point regardless of whether or not it is the point of maximum stress. Examples would be near welds in a fabricated structure, along the grain of a wood member, or near a point of attachment of one member to another.

After the point of interest is identified, the stress condition at that point is determined from the classical stress analysis relationships presented in this book if possible. At times, because of the complexity of the geometry of the member or the loading pattern, it is not possible to complete a reliable stress analysis entirely by computations. In such cases experimental stress analysis can be employed in which strain gauges, photoelastic models, or strain-sensitive coatings give data experimentally. Also, with the aid of finite element stress analysis techniques, the stress condition can be determined using computer-based analysis.

Then, after using one of these methods, you should know the information required to construct the *initial stress element,* as shown in Figure 10–1. The element is assumed to be infinitesimally small and aligned with known directions on the member being analyzed. The complete element, as shown, could have a normal stress (either tensile or compressive) acting on each pair of faces in mutually perpendicular directions, usually named the x and y axes. As the name *normal stress* implies, these stresses act normal (perpendicular) to the faces. As shown, σ_x is aligned with the x-axis and is a tensile stress tending to pull the element apart. Recall that tensile stresses are considered positive. Then σ_y is shown to be compressive, tending to crush the element. Compressive stresses are considered negative.

In addition, there may be shear stresses acting along the sides of the element as if each side were being cut away from the adjacent material. Recall from earlier discussion of shearing stresses that a set of four shears, all equal in magnitude, exist on any

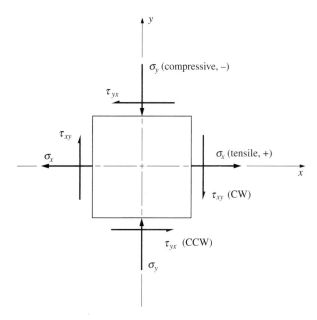

FIGURE 10–1 Complete stress element.

element in equilibrium. On any two opposite faces the shear stresses will act in opposite directions, thus creating a *couple* that tends to rotate the element. Then there must exist a pair of shear stresses on the adjacent faces producing an oppositely directed couple for the element to be in equilibrium. We will refer to each *pair* of shears using a double subscript notation. For example, τ_{xy} refers to the shear stress acting perpendicular to the *x*-axis and parallel to the *y*-axis. Conversely, τ_{yx} acts perpendicular to the *y*-axis and parallel to the *x*-axis. Rather than establish a convention for signs of the shear stresses, we will refer to them as *clockwise* (CW) or *counterclockwise* (CCW), according to how they tend to rotate the stress element.

10–3 STRESS DISTRIBUTION CREATED BY BASIC STRESSES

Figures 10–2, 10–3, and 10–4 show direct tension, direct compression, and bending stresses in which normal stresses were developed. It is important to know the distribution of stress within the member as shown in the figures. Also shown are stress elements subjected to these kinds of stresses. Listed below are the principal formulas for computing the value of these stresses.

Direct tension:	$\sigma = \dfrac{P}{A}$	Uniform over the area
(see Figure 10–2)		(Chapters 1, 3)
Direct compression:	$\sigma = \dfrac{-P}{A}$	Uniform over the area
(see Figure 10–3)		(Chapters 1, 3)

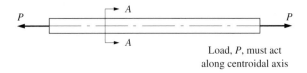

Load, P, must act along centroidal axis

(a) Load condition – Direct tension

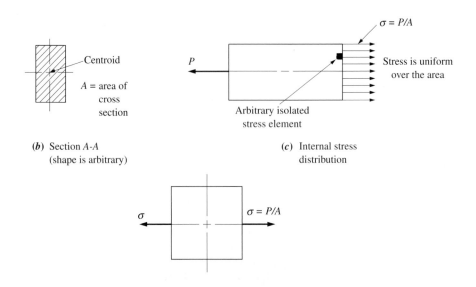

Centroid

A = area of cross section

$\sigma = P/A$

P

Stress is uniform over the area

Arbitrary isolated stress element

(b) Section *A-A* (shape is arbitrary)

(c) Internal stress distribution

σ

$\sigma = P/A$

(d) Stress element – normal tensile stress

FIGURE 10–2 Distribution of normal stress for direct tension.

Stress due to bending:
(see Figure 10–4)

$$\sigma_{max} = \pm \frac{Mc}{I}$$

Maximum stress at outer surfaces
(Chapter 8)

$$\sigma = \pm \frac{My}{I}$$

Bending stress at any point *y*
(Chapter 8)

Figures 10–5, 10–6, and 10–7 show three cases in which shearing stresses are produced along with the stress distributions and the stress elements subjected to these types of stress. Listed below are the principal formulas used to compute shearing stresses.

Direct shear:
(see Figure 10–5)

$$\tau = \frac{P}{A_s}$$

Uniform over the area
(Chapters 1, 3)

Torsional shear:
(see Figure 10–6)

$$\tau_{max} = \frac{Tc}{J}$$

Maximum at outer surface
(Chapter 5)

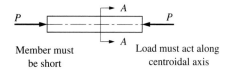

Member must   Load must act along
be short   centroidal axis

(a) Load condition – Direct compressive stress

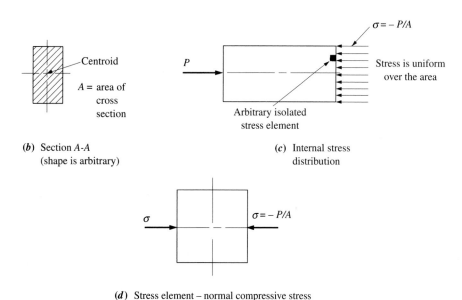

(b) Section A-A   (c) Internal stress
    (shape is arbitrary)   distribution

(d) Stress element – normal compressive stress

FIGURE 10–3 Distribution of normal stress for direct compression.

$$\tau = \frac{Tr}{J}$$ Torsional shear at any radius
                                                 (Chapter 5)

Shearing stresses in beams: $$\tau = \frac{VQ}{It}$$ (Chapter 9)
(see Figure 10–7)

10–4   CREATING THE INITIAL STRESS ELEMENT

One major objective of this chapter is to develop relationships from which the *maximum principal stresses* and the *maximum shear stress* can be determined. Before this can be done, it is necessary to know the state of stress at a point of interest in some orientation. In this section we demonstrate the determination of the initial stress condition by direct calculation from the formulas for basic stresses.

    Figure 10–8 shows an L-shaped lever attached to a rigid surface with a downward load *P* applied to its end. The shorter segment at the front of the lever is loaded as a cantilever beam as shown in Figure 10–9(a), with the moment at its left end resisted by the other segment of the lever.

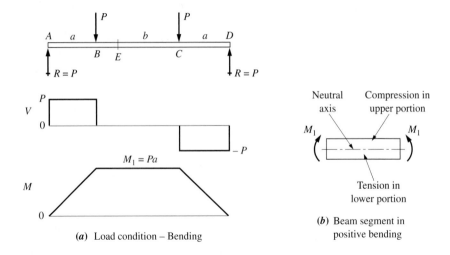

(a) Load condition – Bending

(b) Beam segment in positive bending

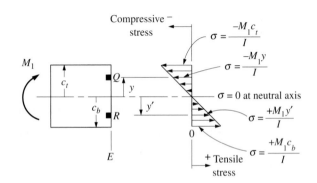

(c) Internal stress distribution

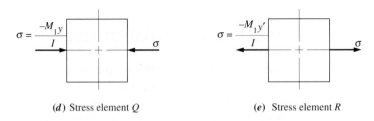

(d) Stress element Q

(e) Stress element R

FIGURE 10–4 Distribution of normal stress for bending.

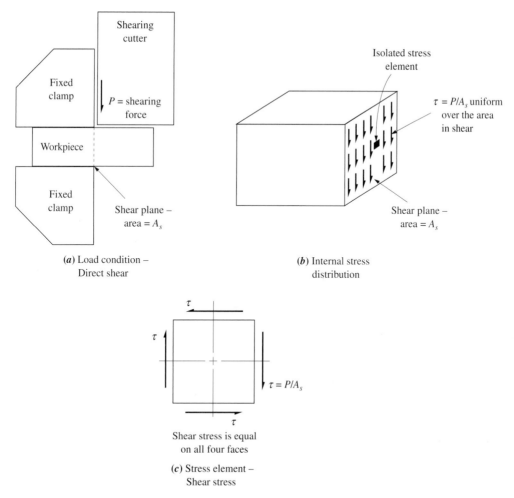

(a) Load condition –
Direct shear

(b) Internal stress
distribution

$\tau = P/A_s$ uniform
over the area
in shear

Shear plane –
area $= A_s$

Shear plane –
area $= A_s$

Isolated stress
element

$\tau = P/A_s$

Shear stress is equal
on all four faces

(c) Stress element –
Shear stress

FIGURE 10–5 Distribution of shearing stress for direct shear.

Figure 10–9(b) shows the longer segment as a free-body diagram. At the front, the force P and the torque T are the reactions to the force and moment at the left end of the shorter segment in (a). Then at the rear, there must be reactions M, P, and T to maintain equilibrium. The bar is subjected to a combined stress condition with the following kinds of stresses:

Bending stress due to the bending moment

Torsional shear stress due to the torque

Shearing stress due to the vertical shearing force

Reviewing Figures 10–4, 10–6, and 10–7, we can conclude that one of the points where the stress is likely to be the highest is on the top of the long segment near the

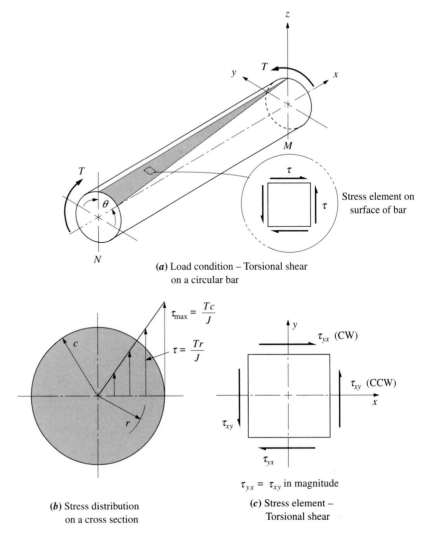

$$\tau_{max} = \frac{Tc}{J}$$

$$\tau = \frac{Tr}{J}$$

τ_{yx} (CW)

τ_{xy} (CCW)

$\tau_{yx} = \tau_{xy}$ in magnitude

(*a*) Load condition – Torsional shear
on a circular bar

(*b*) Stress distribution
on a cross section

(*c*) Stress element –
Torsional shear

FIGURE 10–6   Distribution of torsional shear stress in a solid round shaft.

support. Called element K in Figure 10–9(b), it would see the maximum tensile bend-ing stress and the maximum torsional shear stress. But the vertical shearing stress would be zero because it is at the outer surface away from the neutral axis.

Element K is then subjected to a combined stress, as shown in Figure 10–10. The tensile normal stress, σ_x, is directed parallel to the x-axis along the top surface of the bar. The applied torque tends to produce a shear stress τ_{xy} in the negative y-direction on the face toward the front of the bar and in the positive y-direction on the face toward the rear. Together they create a counterclockwise couple on the element. The element is completed by showing the shear stresses τ_{yx} producing a clockwise couple on the other faces.

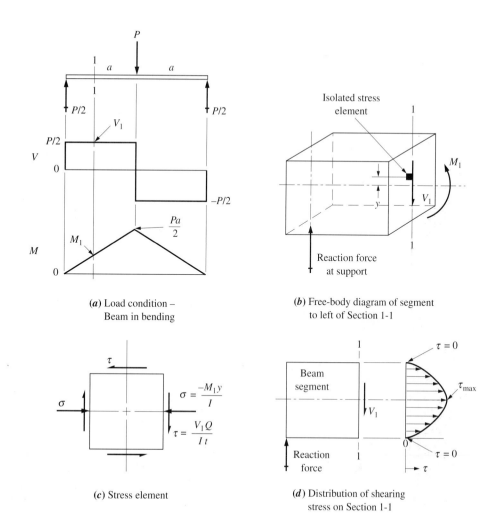

FIGURE 10–7 Distribution of shearing stress in a beam.

(*a*) Load condition –
Beam in bending

(*b*) Free-body diagram of segment
to left of Section 1-1

(*c*) Stress element

(*d*) Distribution of shearing
stress on Section 1-1

In figure (c):
$$\sigma = \frac{-M_1 y}{I}$$
$$\tau = \frac{V_1 Q}{I t}$$

The example problem shown next includes illustrative calculations for the values of the stresses that are shown in Figure 10–10.

**Example Problem
10–1**

Figure 10–8 shows an L-shaped lever carrying a downward force at its end. Compute the stress condition that exists on a point on top of the lever near the support. Let $P = 1500$ N, $a = 150$ mm, $b = 300$ mm, and $D = 30$ mm. Show the stress condition on a stress element.

Solution

Objective Compute the stress condition and draw the stress element.

Given Geometry and loading in Figure 10–8. $P = 1500$ N
Dimensions: $a = 150$ mm, $b = 300$ mm, $D = 30$ mm.

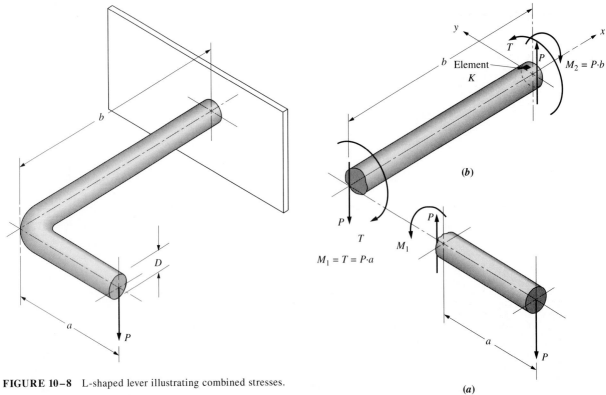

FIGURE 10–8 L-shaped lever illustrating combined stresses.

FIGURE 10–9 Free-body diagrams of segments of an L-shaped lever. (a) Shorter segment of lever. (b) Longer segment of lever.

$M_1 = T = P \cdot a$

$M_2 = P \cdot b$

(a)

(b)

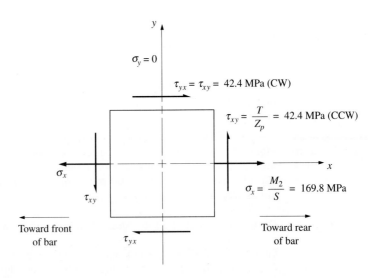

$\sigma_y = 0$

$\tau_{yx} = \tau_{xy} = 42.4$ MPa (CW)

$\tau_{xy} = \dfrac{T}{Z_p} = 42.4$ MPa (CCW)

$\sigma_x = \dfrac{M_2}{S} = 169.8$ MPa

Toward front of bar

Toward rear of bar

FIGURE 10–10 Stress on element K from Figure 10–9 with data from Example Problem 10–1.

Analysis   Figure 10–9 shows the lever separated into two free-body diagrams. The point of interest is labeled "Element K" at the right end of the lever where it joins the support. The element is subjected to bending stress due to the reaction moment at the support, M_2. It is also subjected to torsional shear stress due to the torque, T.

Results   Using the free-body diagrams in Figure 10–9, we can show that

$$M_1 = T = Pa = (1500 \text{ N})(150 \text{ mm}) = 225\,000 \text{ N·mm}$$
$$M_2 = Pb = (1500 \text{ N})(300 \text{ mm}) = 450\,000 \text{ N·mm}$$

Then the bending stress on top of the bar, shown as element K in Figure 10–9(b), is

$$\sigma_x = \frac{M_2 c}{I} = \frac{M_2}{S}$$

But the section modulus S is

$$S = \frac{\pi D^3}{32} = \frac{\pi (30 \text{ mm})^3}{32} = 2651 \text{ mm}^3$$

Then

$$\sigma_x = \frac{450\,000 \text{ N·mm}}{2651 \text{ mm}^3} = 169.8 \text{ N/mm}^2 = 169.8 \text{ MPa}$$

The torsional shear stress is at its maximum value all around the outer surface of the bar, with the value of

$$\tau = \frac{T}{Z_p}$$

But the polar section modulus Z_p is

$$Z_p = \frac{\pi D^3}{16} = \frac{\pi (30 \text{ mm})^3}{16} = 5301 \text{ mm}^3$$

Then

$$\tau = \frac{225\,000 \text{ N·mm}}{5301 \text{ mm}^3} = 42.4 \text{ N/mm}^2 = 42.4 \text{ MPa}$$

Comments   The bending and shear stresses are shown on the stress element K in Figure 10–10. This is likely to be the point of maximum combined stress which will be discussed later in this chapter. At a point on the side of the bar on the y-axis, a larger shearing stress would exist because the maximum torsional shear stress combines with the maximum vertical shear stress due to bending. But the bending stress there is zero. That element should also be analyzed.

10–5 EQUATIONS FOR STRESSES IN ANY DIRECTION

The *initial stress element* discussed in Section 10–4 was oriented in a convenient direction related to the member being analyzed. The methods of this section allow you to compute the stresses in any direction and to compute the maximum normal stresses and the maximum shear stress directly.

Figure 10–11 shows a stress element with orthogonal axes u ands v superimposed on the initial element such that the u-axis is at an angle ϕ relative to the given x-axis. In general, there will be a normal stress σ_u and a shear stress τ_{uv} acting on the inclined surface AC. The following development will result in equations for those stresses.

Before going on, note that Figure 10–11(a) shows only two dimensions of an element that is really a three-dimensional cube. Part (b) of the figure shows the entire cube with each side having the dimension h.

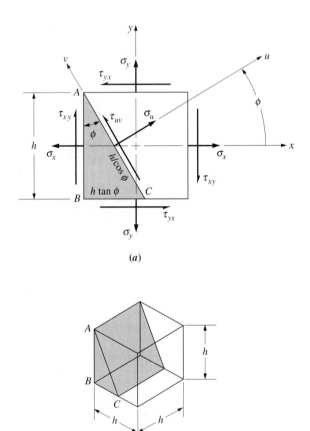

(a)

(b)

FIGURE 10–11 Initial stress element with u and v axes included. (a) Stress element with inclined surface. (b) 3-dimensional element showing wedge.

Normal stress in the *u*-direction, σ_u. Visualize a wedge-shaped part of the initial element as shown in Figure 10–12. Note how the angle ϕ is located. The side *AB* is the original left side of the initial element having the height *h*. The bottom of the wedge, side *BC*, is only a part of the bottom of the initial element where the length is determined by the angle ϕ itself.

$$BC = h \cdot \tan \phi$$

Also, the length of the sloped side of the wedge, *AC*, is

$$AC = \frac{h}{\cos \phi}$$

These lengths are important because now we are going to consider all the forces that act on the wedge. Because force is the product of stress times area, we need to know the area on which each stress acts. Starting with σ_x, it acts on the entire left face of the wedge having an area equal to h^2. Remember that the depth of the wedge perpendicular to the paper is also *h*. Then,

$$\text{force due to } \sigma_x = \sigma_x h^2$$

Using similar logic,

$$\text{force due to } \sigma_y = \sigma_y h^2 \tan \phi$$
$$\text{force due to } \tau_{xy} = \tau_{xy} h^2$$
$$\text{force due to } \tau_{yx} = \tau_{yx} h^2 \tan \phi$$

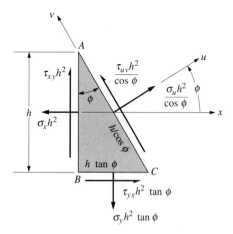

FIGURE 10–12 Free-body diagram of wedge showing forces acting on each face.

The stresses acting on the inclined face of the wedge must also be considered:

$$\text{force due to } \sigma_u = \frac{\sigma_u h^2}{\cos \phi}$$

$$\text{force due to } \tau_{uv} = \frac{\tau_{uv} h^2}{\cos \phi}$$

Now using the principle of equilibrium, we can sum forces in the u-direction. From the resulting equation, we can solve for σ_u. The process is facilitated by resolving all forces into components perpendicular and parallel to the inclined face of the wedge. Figure 10–13 shows this for each force except for those due to σ_u and τ_{uv} which are already aligned with the u and v axes. Then,

$$\sum F_u = 0 = \frac{\sigma_u h^2}{\cos \phi} - \sigma_x h^2 \cos \phi - \sigma_y h^2 \tan \phi \sin \phi + \tau_{xy} h^2 \sin \phi$$
$$+ \tau_{yx} h^2 \tan \phi \cos \phi$$

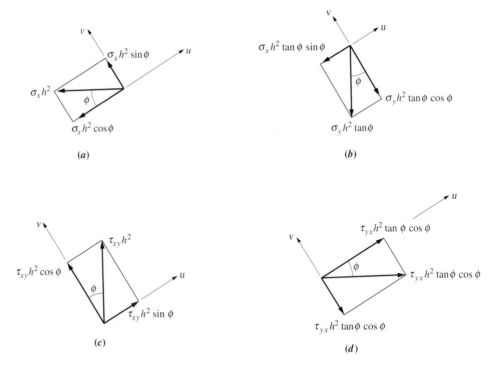

FIGURE 10–13 Resolution of forces along u and v directions. (a) Components of force due to σ_x. (b) Components of force due to σ_y. (c) Components of force due to τ_{xy}. (d) Components of force due to τ_{yx}.

To begin solving for σ_u, all terms include h^2, which can be canceled. Also, we have noted that $\tau_{xy} = \tau_{yx}$ and we can let $\tan \phi = \sin \phi / \cos \phi$. The equilibrium equation then becomes

$$0 = \frac{\sigma_u}{\cos \phi} - \sigma_x \cos \phi - \frac{\sigma_y \sin \phi \sin \phi}{\cos \phi} + \tau_{xy} \sin \phi + \frac{\tau_{xy} \sin \phi \cos \phi}{\cos \phi}$$

Now multiply by $\cos \phi$ to obtain

$$0 = \sigma_u - \sigma_x \cos^2 \phi - \sigma_y \sin^2 \phi + \tau_{xy} \sin \phi \cos \phi + \tau_{xy} \sin \phi \cos \phi$$

Combine the last two terms and solve for σ_u.

$$\sigma_u = \sigma_x \cos^2 \phi + \sigma_y \sin^2 \phi - 2\tau_{xy} \sin \phi \cos \phi$$

This is a usable formula for computing σ_u, but a more convenient form can be obtained by using the following trigonometric identities:

$$\cos^2 \phi = \tfrac{1}{2} + \tfrac{1}{2} \cos 2\phi$$
$$\sin^2 \phi = \tfrac{1}{2} - \tfrac{1}{2} \cos 2\phi$$
$$\sin \phi \cos \phi = \tfrac{1}{2} \sin 2\phi$$

Making the substitutions gives

$$\sigma_u = \tfrac{1}{2}\sigma_x + \tfrac{1}{2}\sigma_x \cos 2\phi + \tfrac{1}{2}\sigma_y - \tfrac{1}{2}\sigma_y \cos 2\phi - \tau_{xy} \sin 2\phi$$

Combining terms, we obtain

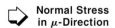

Normal Stress in μ-Direction

$$\sigma_u = \tfrac{1}{2}(\sigma_x + \sigma_y) + \tfrac{1}{2}(\sigma_x - \sigma_y) \cos 2\phi - \tau_{xy} \sin 2\phi \qquad (10\text{–}1)$$

Equation (10–1) can be used to compute the normal stress in any direction provided that the stress condition in some direction, indicated by the x and y axes, is known.

Shearing stress, τ_{uv}, acting parallel to the cut plane. Now the equation for the shearing stress, τ_{uv}, acting parallel to the cut plane and perpendicular to σ_u will be developed. Again referring to Figures 10–12 and 10–13, we can sum the forces on the wedge-shaped element acting in the v-direction.

$$\sum F_v = 0 = \frac{\tau_{uv}h^2}{\cos \phi} + \sigma_x h^2 \sin \phi - \sigma_y h^2 \tan \phi \cos \phi$$
$$+ \tau_{xy} h^2 \cos \phi - \tau_{yx} h^2 \tan \phi \sin \phi$$

Using the same techniques as before, this equation can be simplified and solved for τ_{uv}, resulting in

Shearing Stress, τ_{uv} on Face of Element

$$\tau_{uv} = -\tfrac{1}{2}(\sigma_x - \sigma_y) \sin 2\phi - \tau_{xy} \cos 2\phi \qquad (10\text{--}2)$$

Equation (10–2) can be used to compute the shearing stress that acts on the face of the element at any angular orientation.

10–6 PRINCIPAL STRESSES

In design and stress analysis, the maximum stresses are frequently desired in order to ensure the safety of the load-carrying member. Equation (10–1) can be used to compute the maximum normal stress if we know at what angle ϕ it occurred.

From the study of calculus, we know that the value of the angle ϕ at which the maximum or minimum normal stress occurs can be found by differentiating the function and setting the result equal to zero, then solving for ϕ. Differentiating Equation (10–1) gives

$$\frac{d\sigma_u}{d\phi} = 0 = 0 + \tfrac{1}{2}(\sigma_x - \sigma_y)(-\sin 2\phi)(2) - \tau_{xy} \cos 2\phi(2)$$

Dividing by $\cos 2\phi$ and simplifying gives

$$0 = -(\sigma_x - \sigma_y) \tan 2\phi - 2\tau_{xy}$$

Solving for $\tan 2\phi$ gives

$$\tan 2\phi = \frac{-2\tau_{xy}}{\sigma_x - \sigma_y} = \frac{-\tau_{xy}}{\tfrac{1}{2}(\sigma_x - \sigma_y)} \qquad (10\text{--}3)$$

The angle ϕ is then

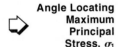
Angle Locating Maximum Principal Stress, σ_1

$$\phi = \tfrac{1}{2} \tan^{-1} \frac{-\tau_{xy}}{\tfrac{1}{2}(\sigma_x - \sigma_y)} \qquad (10\text{--}4)$$

If we substitute the value of ϕ defined by Equations (10–3) and (10–4) into Equation (10–1), we can develop an equation for the maximum normal stress on the element. In addition, we can develop the equation for the minimum normal stress. These two stresses are called the *principal stresses* with σ_1 used to denote the *maximum principal stress* and σ_2 denoting the *minimum principal stress*.

Note from Equation (10–1) that we need values for $\sin 2\phi$ and $\cos 2\phi$. Figure 10–14 is a graphical aid to obtaining expressions for these functions. The right triangle has the opposite and adjacent sides defined by the terms of the tangent function in Equation (10–3).

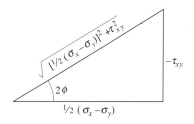

The triangle is defined by Equation (10-3):

$$\tan 2\phi = \frac{-\tau_{xy}}{\frac{1}{2}(\sigma_x - \sigma_y)}$$

$$\sin 2\phi = \frac{-\tau_{xy}}{\sqrt{[\frac{1}{2}(\sigma_x - \sigma_y)]^2 + \tau_{xy}^2}}$$

$$\cos 2\phi = \frac{\frac{1}{2}(\sigma_x - \sigma_y)}{\sqrt{[\frac{1}{2}(\sigma_x - \sigma_y)]^2 + \tau_{xy}^2}}$$

FIGURE 10–14 Development of $\sin 2\phi$ and $\cos 2\phi$ for principal stress formulas.

Substituting into Equation (10–1) and simplifying gives

Maximum Principal Stress, σ_1

$$\sigma_{\max} = \sigma_1 = \tfrac{1}{2}(\sigma_x + \sigma_y) + \sqrt{[\tfrac{1}{2}(\sigma_x - \sigma_y)]^2 + \tau_{xy}^2} \qquad (10-5)$$

Because the square root has two possible values, + and −, we can also find the expression for the minimum principal stress, σ_2.

Minimum Principal Stress, σ_2

$$\sigma_{\min} = \sigma_2 = \tfrac{1}{2}(\sigma_x + \sigma_y) - \sqrt{[\tfrac{1}{2}(\sigma_x - \sigma_y)]^2 + \tau_{xy}^2} \qquad (10-6)$$

Conceivably, there would be a shear stress existing along with these normal stresses. But it can be shown that by substituting the value of ϕ from Equation (10–4) into the shear stress Equation (10–2) that the result will be zero. In conclusion,

> On the element on which the principal stresses act, the shear stress is zero.

10–7 MAXIMUM SHEAR STRESS

The same technique can be used to find the maximum shear stress by working with Equation (10–2). Differentiating with respect to ϕ and setting the result equal to zero gives

$$\frac{d\tau_{uv}}{d\phi} = 0 = -\tfrac{1}{2}(\sigma_x - \sigma_y)(\cos 2\phi)(2) - \tau_{xy}(-\sin 2\phi)(2)$$

Dividing by cos 2ϕ and simplifying gives

$$\tan 2\phi = \frac{(\sigma_x - \sigma_y)}{2\tau_{xy}} = \frac{\frac{1}{2}(\sigma_x - \sigma_y)}{\tau_{xy}} \qquad (10\text{--}7)$$

Solving for ϕ yields

$$\phi = \tfrac{1}{2} \tan^{-1} \frac{\frac{1}{2}(\sigma_x - \sigma_y)}{\tau_{xy}} \qquad (10\text{--}8)$$

Obviously, the value for ϕ from Equation (10–8) is different from that of Equation (10–4). In fact, we will see that the two values are *always 45 deg apart*.

Figure 10–15 shows the right triangle, from which we can find sin 2ϕ and cos 2ϕ as we did in Figure 10–14. Substituting these values into Equation (10–2) gives the maximum shear stress.

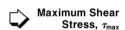

$$\tau_{max} = \pm\sqrt{[\tfrac{1}{2}(\sigma_x - \sigma_y)]^2 + \tau_{xy}^2} \qquad (10\text{--}9)$$

Here we should check to see if there is a normal stress existing on the element having the maximum shear stress. Substituting the value of ϕ from Equation (10–8) into the general normal stress Equation (10–1) gives

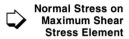

$$\sigma_{avg} = \tfrac{1}{2}(\sigma_x + \sigma_y) \qquad (10\text{--}10)$$

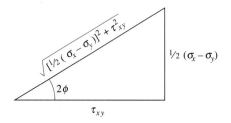

The triangle is defined by Equation (10-7):

$$\tan 2\phi = \frac{\frac{1}{2}(\sigma_x - \sigma_y)}{\tau_{xy}}$$

$$\sin 2\phi = \frac{\frac{1}{2}(\sigma_x - \sigma_y)}{\sqrt{[\frac{1}{2}(\sigma_x - \sigma_y)]^2 + \tau_{xy}^2}}$$

$$\cos 2\phi = \frac{\tau_{xy}}{\sqrt{[\frac{1}{2}(\sigma_x - \sigma_y)]^2 + \tau_{xy}^2}}$$

FIGURE 10–15 Development of sin 2ϕ and cos 2ϕ for maximum shear stress formula.

This is the formula for the *average* of the initial normal stresses, σ_x and σ_y. Thus we can conclude:

> On the element on which the maximum shear stress occurs, there will also be a normal stress equal to the average of the initial normal stresses.

10–8 MOHR'S CIRCLE FOR STRESS

The use of Equations (10–1) through (10–10) often presents difficulties because of the many possible combinations of signs for the terms σ_x, σ_y, τ_{xy}, and ϕ. Also, two roots of the square root and the fact that the inverse tangent function can result in angles in any of the four quadrants presents difficulties. Fortunately, there is a graphical aid, called *Mohr's circle*, available that can help to overcome these problems. The use of Mohr's circle should give you a better understanding of the general case of stress at a point.

It can be shown that the two equations, (10–1) and (10–2), for the normal and shear stress at a point in any direction can be combined and arranged in the form of the equation for a circle. First presented by Otto Mohr in 1895, the circle allows a rapid and exact computation of:

1. The maximum and minimum principal stresses [Equations (10–5) and (10–6)]
2. The maximum shear stress [Equation (10–9)]
3. The angles of orientation of the principal stress element and the maximum shear stress element [Equations (10–4) and (10–8)]
4. The normal stress that exists along with the maximum shear stress on the maximum shear stress element [Equation (10–10)]
5. The stress condition at any angular orientation of the stress element [Equations (10–1) and (10–2)]

Mohr's circle is drawn on a set of perpendicular axes with shear stress, τ, plotted vertically and normal stresses, σ, horizontally, as shown in Figure 10–16. The following convention is used in this book:

Sign Conventions:
1. Positive normal stresses (tensile) are to the right.
2. Negative normal stresses (compressive) are to the left.
3. Shear stresses that tend to rotate the stress element clockwise (CW) are plotted upward on the τ-axis.
4. Shear stresses that tend to rotate the stress element counterclockwise (CCW) are plotted downward.

The procedure listed next can be used to draw Mohr's circle. Steps 1–7 are shown in Figure 10–16. The complete stress element as shown in Figure 10–1 is the basis for this example.

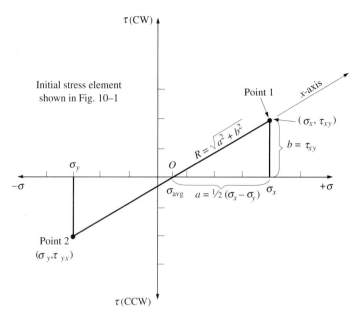

FIGURE 10–16 Steps 1 through 7 of Mohr's circle construction procedure.

Procedure for drawing Mohr's circle

1. Identify the stress condition at the point of interest and represent it as the *initial stress element* in the manner shown in Figure 10–1.
2. The combination σ_x and τ_{xy} is plotted as *point 1* on the σ–τ plane.
3. The combination σ_y and τ_{yx} is then plotted as *point 2*. Note that τ_{xy} and τ_{yx} always act in opposite directions. Therefore, one point will be plotted above the σ-axis and one will be below.
4. Draw a straight line between the two points.
5. This line crosses the σ-axis at the center of Mohr's circle, which is also the value of the *average normal stress* applied to the initial stress element. The location of the center can be observed from the data used to plot the points or computed from Equation (10–10), repeated here:

$$\sigma_{avg} = \tfrac{1}{2}(\sigma_x + \sigma_y)$$

For convenience, label the center O.
6. Identify the line from O through point 1 (σ_x, τ_{xy}) as the *x*-axis. This line corresponds to the original *x*-axis and is essential to correlating the data from Mohr's circle to the original *x* and *y* directions.
7. The points O, σ_x, and point 1 form an important right triangle because the distance from O to point 1, the hypotenuse of the triangle,

is equal to the radius of the circle, R. Calling the other two sides a and b, the following calculations can be made:

$$a = \tfrac{1}{2}(\sigma_x - \sigma_y)$$
$$b = \tau_{xy}$$
$$R = \sqrt{a^2 + b^2} = \sqrt{[\tfrac{1}{2}(\sigma_x - \sigma_y)]^2 + \tau_{xy}^2}$$

Note that the equation for R is identical to Equation (10–9) for the maximum shear stress on the element. Thus

The length of the radius of Mohr's circle is equal to the magnitude of the maximum shear stress.

Steps 8–11 are shown in Figure 10–17.

8. Draw the complete circle with the center at O and the radius R.
9. Draw the vertical diameter of the circle. The point at the top of the circle has the coordinates $(\sigma_{avg}, \tau_{max})$, where the shear stress has the clockwise (CW) direction. The point at the bottom of the circle represents $(\sigma_{avg}, \tau_{max})$ where the shear stress is counterclockwise (CCW).

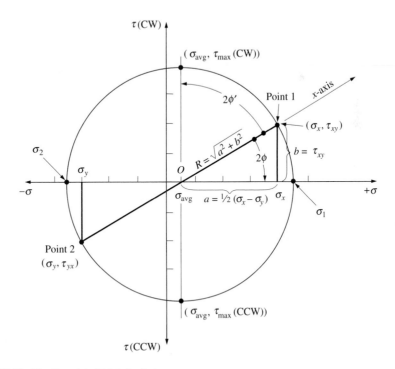

FIGURE 10–17 Completed Mohr's circle.

10. Identify the points on the σ-axis at the ends of the horizontal diameter as σ_1 at the right (the maximum principal stress) and σ_2 at the left (the minimum principal stress). Note that the shear stress is zero at these points.

11. Determine the values for σ_1 and σ_2 from

$$\sigma_1 = \text{"}O\text{"} + R \qquad (10\text{--}11)$$
$$\sigma_2 = \text{"}O\text{"} - R \qquad (10\text{--}12)$$

where "O" represents the coordinate of the center of the circle, σ_{avg}, and R is the radius of the circle. Thus Equations (10–11) and (10–12) are identical to Equations (10–5) and (10–6) for the principal stresses.

The following steps determine the angles of orientation of the principal stress element and the maximum shear stress element. An important concept to remember is that *angles obtained from Mohr's circle are double the true angles*. The reason for this is that the equations on which it is based, Equations (10–1) and (10–2), are functions of 2ϕ.

12. The orientation of the principal stress element is determined by finding the angle *from* the x-axis *to* the "σ_1"-axis, labeled 2ϕ in Figure 10–17. From the data on the circle you can see that

$$2\phi = \tan^{-1} \frac{b}{a}$$

The argument of this inverse tangent function is the same as the absolute value of the argument shown in Equation (10–4). Problems

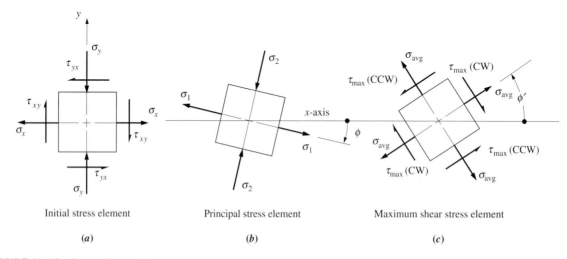

Initial stress element	Principal stress element	Maximum shear stress element
(a)	(b)	(c)

FIGURE 10–18   General form of final results from Mohr's circle analysis.

with signs for the resulting angle are avoided by noting the *direction from the x-axis to the σ_1-axis* on the circle, clockwise for the present example. Then the principal stress element is rotated *in the same direction* from the x-axis by an amount ϕ to locate the face on which the maximum principal stress σ_1 acts.

13. Draw the principal stress element in its proper orientation as determined from step 12 with the two principal stresses σ_1 and σ_2 shown [see Figure 10–18, (a) and (b)].

14. The orientation of the maximum shear stress element is determined by finding the angle *from* the x-axis *to* the τ_{max}-axis, labeled $2\phi'$ in Figure 10–17. In the present example,

$$2\phi' = 90° - 2\phi$$

From trigonometry it can be shown that this is equivalent to finding the inverse tangent of a/b, the reciprocal of the argument used to find 2ϕ. Thus it is a true evaluation of Equation (10–8), derived to find the angle of orientation of the element on which the maximum shear stress occurs.

Again problems with signs for the resulting angle are avoided by noting the *direction from the x-axis to the τ_{max}-axis* on the circle, counterclockwise for the present example. Then the maximum shear stress element is rotated *in the same direction* from the x-axis by an amount ϕ' to locate the face on which the clockwise maximum shear stress acts.

15. Draw the maximum shear stress element in its proper orientation as determined from step 14 with the shear stresses on all four faces and the average normal stress acting on each face [see Figure 10–18(c)]. As a whole, Figure 10–18 is the desired result from a Mohr's circle analysis. Shown are the initial stress element that establishes the x and y axes, the principal stress element drawn in proper rotation relative to the x-axis, and the maximum shear stress element also drawn in the proper rotation relative to the x-axis.

Example Problem 10–2

It has been determined that a point in a load-carrying member is subjected to the following stress condition:

$$\sigma_x = 400 \text{ MPa} \qquad \sigma_y = -300 \text{ MPa} \qquad \tau_{xy} = 200 \text{ MPa (CW)}$$

Perform the following:

(a) Draw the initial stress element.

(b) Draw the complete Mohr's circle, labeling critical points.

(c) Draw the complete principal stress element.

(d) Draw the maximum shear stress element.

Solution The 15-step procedure described before will be used to complete the problem. The initial stress element is drawn in Figure 10–20(a) and Mohr's circle is in Figure 10–19. The numerical results from steps 1–7 are summarized below.

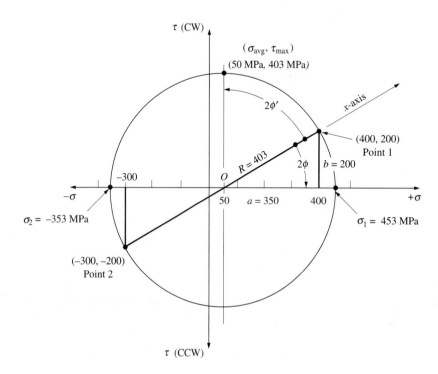

FIGURE 10–19 Complete Mohr's circle for Example Problem 10–2.

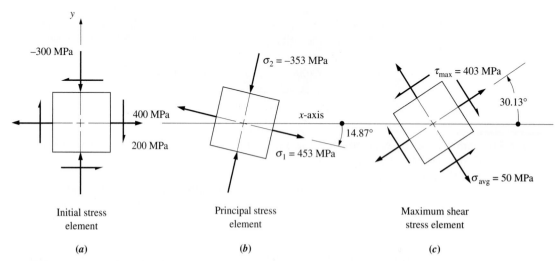

FIGURE 10–20 Results for Example Problem 10–2.

Center of the circle O is at σ_{avg}:

$$\sigma_{avg} = \tfrac{1}{2}(\sigma_x + \sigma_y) = \tfrac{1}{2}[400 + (-300)] = 50 \text{ MPa}$$

The lower side of the triangle,

$$a = \tfrac{1}{2}(\sigma_x - \sigma_y) = \tfrac{1}{2}[400 - (-300)] = 350 \text{ MPa}$$

The vertical side of the triangle,

$$b = \tau_{xy} = 200 \text{ MPa}$$

The radius of the circle,

$$R = \sqrt{a^2 + b^2} = \sqrt{(350)^2 + (200)^2} = 403 \text{ MPa}$$

Step 8 is the drawing of the circle. The significant data points from steps 9–11 are summarized below.

$$\sigma_{avg} = 50 \text{ MPa (same as the location of } O)$$
$$\tau_{max} = 403 \text{ MPa (same as the value of } R)$$
$$\sigma_1 = O + R = 50 + 403 = 453 \text{ MPa}$$
$$\sigma_2 = O - R = 50 - 403 = -353 \text{ MPa}$$

Steps 12–15 are completed in Figures 10–19 and 10–20. The computations of the angles are summarized below.

$$2\phi = \tan^{-1}\frac{b}{a} = \tan^{-1}\frac{200}{350} = 29.74°$$

Note that 2ϕ is CW from the x-axis to σ_1 on the circle.

$$\phi = \frac{29.74°}{2} = 14.87°$$

Thus, in Figure 10–20(b), the principal stress element is drawn rotated 14.87° CW from the original x axis to the face on which σ_1 acts.

$$2\phi' = 90° - 2\phi = 90° - 29.74° = 60.26°$$

Note that $2\phi'$ is CCW from the x-axis to τ_{max} on the circle.

$$\phi' = \frac{60.26°}{2} = 30.13°$$

Thus, in Figure 10–20(c) the maximum shear stress element is drawn rotated 30.13° CCW from the original x-axis to the face on which τ_{max} acts.

Example Problem 10–2 is completed.

Summary of Results for Example Problem 10–2   **Mohr's Circle**

Given $\sigma_x = 400$ MPa $\sigma_y = -300$ MPa $\tau_{xy} = 200$ MPa CW

Results   Figures 10–19 and 10–20.

$\sigma_1 = 453$ MPa $\sigma_2 = -353$ MPa $\phi = 14.87°$ CW

$\tau_{max} = 403$ MPa $\sigma_{avg} = 50$ MPa $\phi' = 30.13°$ CCW

Comment   The x-axis is in the first quadrant.

10–9   EXAMPLES OF THE USE OF MOHR'S CIRCLE

The data for Example Problem 10–2 in the preceding section and Example Problems 10–3 through 10–8 to follow, have been selected to demonstrate a wide variety of results. A major variable is the quadrant in which the x-axis lies and the corresponding definition of the angles of rotation for the principal stress element and the maximum shear stress element.

Example Problems 10–6, 10–7, and 10–8 present the special cases of biaxial stress with no shear, uniaxial tension with no shear, and pure shear. These should help you understand the behavior of load-carrying members subjected to such stresses.

The solution for each Example Problem to follow is Mohr's circle itself along with the stress elements, properly labeled. For each problem, the objectives are to:

(a) Draw the initial stress element
(b) Draw the complete Mohr's circle, labeling critical points
(c) Draw the complete principal stress element
(d) Draw the complete shear stress element.

Example Problem   **Given** $\sigma_x = 60$ ksi $\sigma_y = -40$ ksi $\tau_{xy} = 30$ ksi CCW
10–3
Mohr's Circle   **Results**   Figure 10–21.

$\sigma_1 = 68.3$ ksi $\sigma_2 = -48.3$ ksi $\phi = 15.48°$ CCW

$\tau_{max} = 58.3$ ksi $\sigma_{avg} = 10$ ksi $\phi' = 60.48°$ CCW

Comment   The x-axis is in the second quadrant.

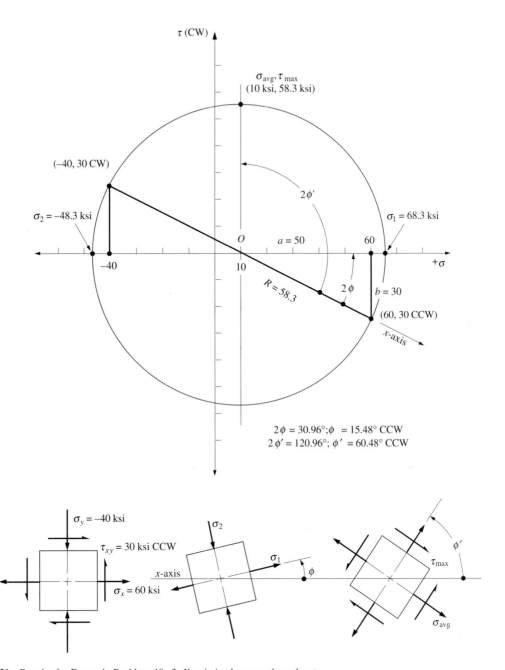

FIGURE 10–21 Results for Example Problem 10–3. *X*-axis in the second quadrant.

Example Problem 10–4
Mohr's Circle

Given $\sigma_x = -120$ MPa $\sigma_y = 180$ MPa $\tau_{xy} = 80$ MPa CCW

Results   Figure 10–22.

$\sigma_1 = 200$ MPa $\sigma_2 = -140$ MPa $\phi = 75.96°$ CCW

$\tau_{max} = 170$ MPa $\sigma_{avg} = 30$ ksi $\phi' = 59.04°$ CW

Comment   The x-axis is in the third quadrant.

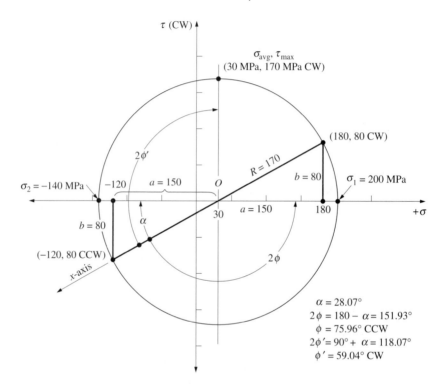

$\alpha = 28.07°$
$2\phi = 180 - \alpha = 151.93°$
$\phi = 75.96°$ CCW
$2\phi' = 90° + \alpha = 118.07°$
$\phi' = 59.04°$ CW

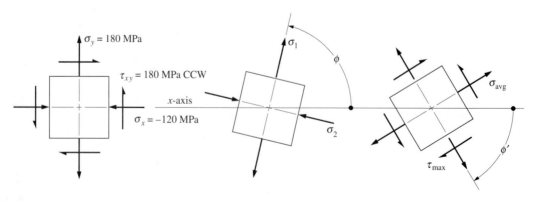

FIGURE 10–22   Results for Example Problem 10–4. X-axis in the third quadrant.

Given $\sigma_x = -30$ ksi $\sigma_y = 180$ ksi $\tau_{xy} = 40$ ksi CW

Results   Figure 10–23.

$\sigma_1 = 42.17$ ksi $\sigma_2 = -52.17$ ksi $\phi = 61.0°$ CW

$\tau_{max} = 47.17$ ksi $\sigma_{avg} = -5$ ksi $\phi' = 16.0°$ CW

Comment   The *x*-axis is in the fourth quadrant.

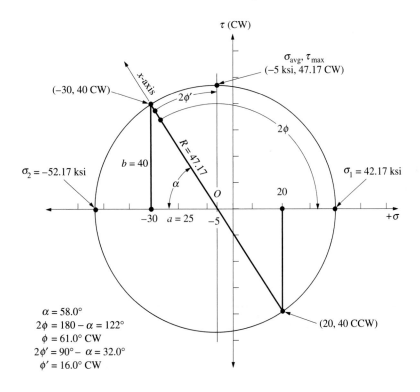

$\alpha = 58.0°$
$2\phi = 180 - \alpha = 122°$
$\phi = 61.0°$ CW
$2\phi' = 90° - \alpha = 32.0°$
$\phi' = 16.0°$ CW

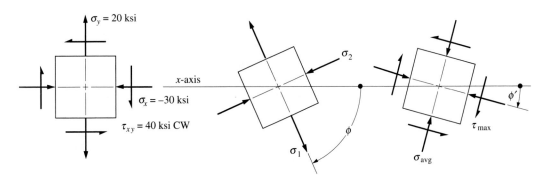

FIGURE 10–23   Results for Example Problem 10–5. *X*-axis in the fourth quadrant.

Given $\sigma_x = 220$ MPa $\sigma_y = -120$ MPa $\tau_{xy} = 0$ MPa

Results   Figure 10–24.

$\sigma_1 = 220$ MPa $\sigma_2 = -120$ MPa $\phi = 0°$

$\tau_{max} = 170$ MPa $\sigma_{avg} = 50$ MPa $\phi' = 45.0°$ CCW

Comment   Special case of biaxial stress with no shear on the given element.

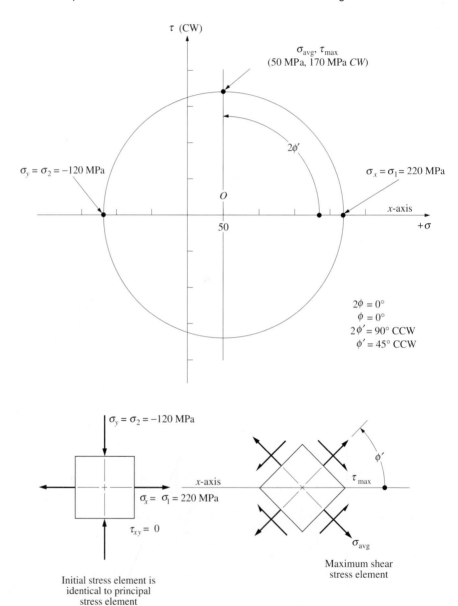

FIGURE 10–24   Results of Example Problem 10–6. Special case of biaxial stress with no shear.

Example Problem 10–7
Mohr's Circle

Given $\sigma_x = 40$ ksi $\sigma_y = 0$ ksi $\tau_{xy} = 0$ ksi

Results   Figure 10–25.

$$\sigma_1 = 40 \text{ ksi} \qquad \sigma_2 = 0 \text{ ksi} \qquad \phi = 0°$$
$$\tau_{max} = 20 \text{ ksi} \qquad \sigma_{avg} = 20 \text{ ksi} \qquad \phi' = 45.0° \text{ CCW}$$

Comment   Special case of uniaxial tension with no shear.

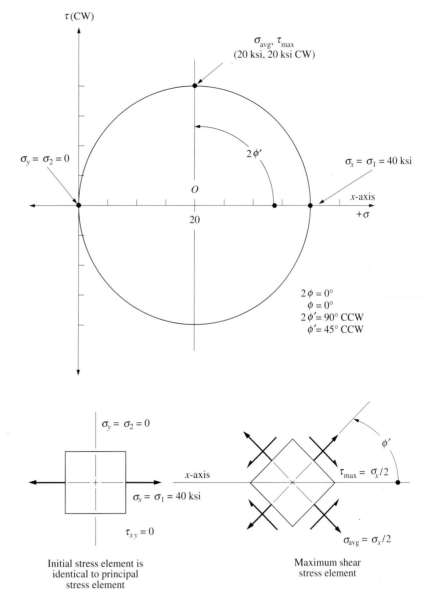

FIGURE 10–25   Results of Example Problem 10–7. Special case of uniaxial tension.

Example Problem 10–8
Mohr's Circle

Given $\sigma_x = 0$ ksi $\sigma_y = 0$ ksi $\tau_{xy} = 40$ ksi CW

Results   Figure 10–26.

$\sigma_1 = 40$ ksi $\sigma_2 = -40$ ksi $\phi = 45°$ CW

$\tau_{max} = 40$ ksi $\sigma_{avg} = 0$ ksi $\phi' = 0°$

Comment   Special case of pure shear.

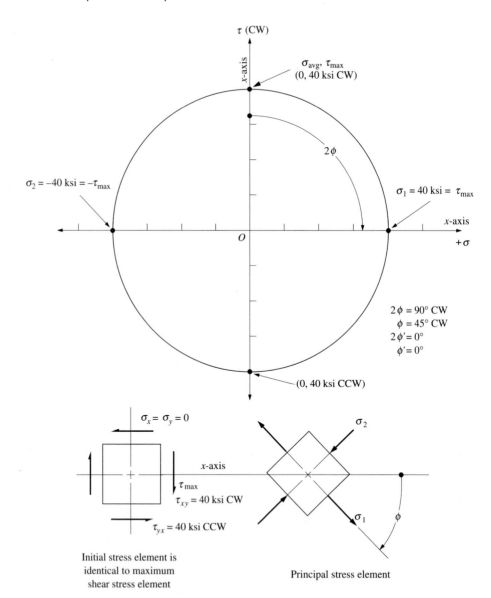

FIGURE 10–26   Results of Example Problem 10–8. Special case of pure shear.

10–10 STRESS CONDITION ON SELECTED PLANES

There are some cases in which it is desirable to know the stress condition on an element at some selected angle of orientation relative to the reference directions. Figures 10–27 and 10–28 show examples.

The wood block in Figure 10–27 shows that the grain of the wood is inclined at an angle of 30° CCW from the given x-axis. Because the wood is very weak in shear parallel to the grain, it is desirable to know the stresses in this direction.

Figure 10–28 shows a member fabricated by welding two components along a seam inclined at an angle relative to the given x-axis. The welding operation could produce a weaker material in the near vicinity of the weld, particularly if the component parts were made from steel that was heat-treated prior to welding. The same is true of many aluminum alloys. For such cases the allowable stresses are somewhat lower along the weld line.

The environmental conditions to which a part is exposed during operation may also affect the material properties. For example, a furnace part may be subjected to local heating by radiant energy along a particular line. The strength of the heated material will be lower than that which remains cool and thus it is desirable to know the stress condition along the angle of the heat-affected zone.

Mohr's circle can be used to determine the stress condition at specified angles of orientation of the stress element. The procedure is outlined here and demonstrated in Example Problem 10–9.

Procedure for Finding the Stress at a Specified Angle	**Given:**	Stress condition on the given element aligned in the x and y directions.
	Objective:	Determine the normal and shear stresses on the element at a specified angle, β, relative to the given x direction.
	Step 1:	Draw the complete Mohr's circle for the element.
	Step 2:	Identify the line representing the x-axis on the circle.
	Step 3:	Measure the angle 2β from the x-axis and draw a line through the center of Mohr's circle extending to the two intersections with the circle. This line represents the axis aligned with the direction of interest.
	Step 4:	Using the geometry of the circle, determine the coordinates (σ and τ) of the first point of intersection. The σ component is the normal stress acting on the element in the direction of β. The τ component is the shear stress acting on the faces of the element. The coordinates of the second point represent the normal and shear stresses acting on the faces of the element of interest that are parallel to the β-axis.
	Step 4:	Draw the element of interest showing the normal and shear stress acting on it.

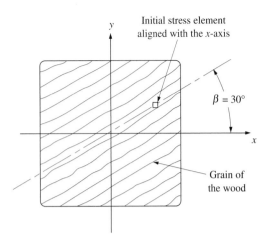

FIGURE 10–27 Cross section of a wood post with the grain running 30° with respect to the x-axis.

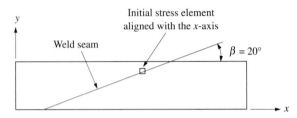

FIGURE 10–28 Flat bar welded along a 20° inclined seam.

Example Problem 10–9

For the flat bar welded along a seam at an angle of 20° CCW from the x-axis, the stress element aligned parallel with the x- and y-axes is subjected to these stresses:

$$\sigma_x = 400 \text{ MPa} \qquad \sigma_y = -300 \text{ MPa} \qquad \tau_{xy} = 200 \text{ MPa CW}$$

Determine the stress condition on the element inclined at the 20° angle, aligned with the weld seam.

Solution **Objective** Draw the stress element aligned with the weld seam at 20° with the x-axis.

Given Note that the given stress element is the same as that used in Example Problem 10–2. The basic Mohr's circle from that problem is shown in Figure 10–19 and it is reproduced in Figure 10–29.

Analysis Use the *Procedure for finding the stress at a specified angle.*

Results *Steps 1 and 2* are shown on the original Mohr's circle.

Step 3. The desired axis is one inclined at 20° CCW from the x-axis. Recalling that angles in Mohr's circle are *double* the true angles, we can draw a line through the center of the circle at an angle of $2\beta = 40°$ CCW from the x-axis. The intersection of this line with the circle, labeled *A* in the figure, locates the point on the circle defining the stress condition of the desired element. The coordinates of this point, (σ_A, τ_A), give the normal and shear stresses acting on one set of faces of the desired stress element.

Step 4. Simple trigonometry and the basic geometry of the circle can be used to determine σ_A and τ_A by projecting lines vertically and horizontally from point *A* to the σ- and τ-axes, respectively. The total angle from the σ-axis to the axis through point *A*, called η (eta) in

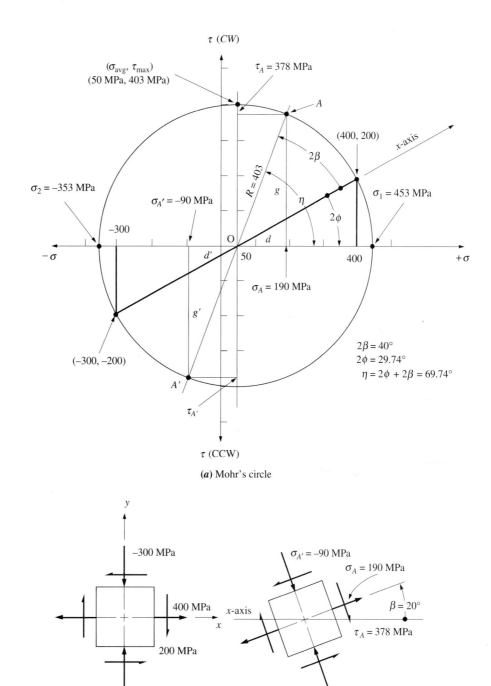

(a) Mohr's circle

(b) Given stress element   (c) Stress element at 20°

FIGURE 10–29 Complete Mohr's circle for Example Problem 10–9 showing stresses on an element inclined at 20° CCW from the x-axis.

the figure, is the sum of 2ϕ and 2β. In Example Problem 10–2, we found $2\phi = 29.74°$. Then

$$\eta = 2\phi + 2\beta = 29.74° + 40° = 69.74°$$

A triangle has been identified in Figure 10–29 with sides labeled d, g, and R. From this triangle, we can compute

$$d = R \cos \eta = (403) \cos 69.74° = 140$$
$$g = R \sin \eta = (403) \sin 69.74° = 378$$

These values enable the computation of

$$\sigma_A = O + d = 50 + 140 = 190 \text{ MPa}$$
$$\tau_A = g = 378 \text{ MPa CW}$$

where O indicates the value of the normal stress at the center of Mohr's circle.

The stresses on the remaining set of faces of the desired stress element are the coordinates of the point A' located 180° from A on the circle and, therefore, 90° from the faces on which (σ_A, τ_A) act. Projecting vertically and horizontally from A' to the σ- and τ-axes locates $\sigma_{A'}$ and $\tau_{A'}$. By similar triangles we can say that $d' = d$ and $g' = g$. Then

$$\sigma_{A'} = O - d' = 50 - 140 = -90 \text{ MPa}$$
$$\tau_{A'} = g' = 378 \text{ MPa CCW}$$

Comment Figure 10–29(c) shows the final element inclined at 20° from the x-axis. This is the stress condition experienced by the material along the weld line.

10–11 SPECIAL CASE IN WHICH BOTH PRINCIPAL STRESSES HAVE THE SAME SIGN

In preceding sections involving Mohr's circle, we used the convention that σ_1 is the maximum principal stress and σ_2 is the minimum principal stress. This is true for cases of plane stress (stresses applied in a single plane) if σ_1 and σ_2 have opposite signs, that is, when one is tensile and one is compressive. Also, in such cases, the shear stress found at the top of the circle (equal to the radius, R) is the true maximum shear stress on the element.

But special care must be exercised when Mohr's circle indicates that σ_1 and σ_2 have the same sign. Even though we are dealing with plane stress, the true stress element is three-dimensional and should be represented as a cube rather than a square, as shown in Figure 10–30. Faces 1, 2, 3, and 4 correspond to the sides of the square ele-

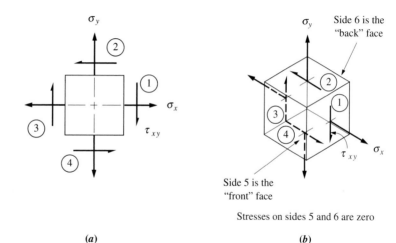

(a)

(b)

FIGURE 10–30 Plane stress shown as two-dimensional and three-dimensional stress elements. (a) Two-dimensional stress element. (b) Three-dimensional stress element.

ment and faces 5 and 6 are the "front" and "back." For plane stress the stresses on faces 5 and 6 are zero.

For the three-dimensional stress element, three principal stresses exist, called σ_1, σ_2, and σ_3, acting on the mutually perpendicular sides of the stress element. Convention calls for the following order:

$$\sigma_1 > \sigma_2 > \sigma_3$$

Thus σ_3 is the true minimum principal stress and σ_1 is the true maximum principal stress. It can also be shown that the true maximum shear stress can be computed from

$$\tau_{max} = \tfrac{1}{2}(\sigma_1 - \sigma_3) \qquad (10\text{–}13)$$

Figure 10–31 illustrates a case in which the three-dimensional stress element must be considered. The initial stress element, shown in part (a), carries the following stresses:

$$\sigma_x = 200 \text{ MPa} \qquad \sigma_y = 120 \text{ MPa} \qquad \tau_{xy} = 40 \text{ MPa CW}$$

Part (b) of the figure shows the conventional Mohr's circle, drawn according to the procedure outlined in Section 10–8. Note that both σ_1 and σ_2 are positive or tensile. Then, considering that the stress on the "front" and "back" faces is zero, this is the true minimum principal stress. We can then say that

$$\sigma_1 = 216.6 \text{ MPa}$$
$$\sigma_2 = 103.4 \text{ MPa}$$
$$\sigma_3 = 0 \text{ MPa}$$

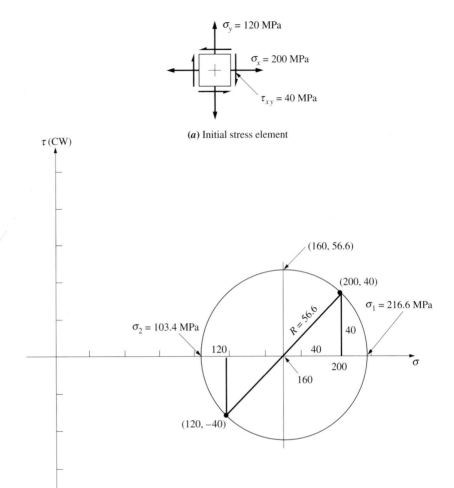

(a) Initial stress element

(b) Mohr's circle

FIGURE 10–31 Mohr's circle for which σ_1 and σ_2 are both positive.

From Equation (10–13), the true maximum shear stress is

$$\tau_{max} = \tfrac{1}{2}(\sigma_1 - \sigma_3) = \tfrac{1}{2}(216.6 - 0) = 108.3 \text{ MPa}$$

These concepts can be visualized graphically by creating a set of three Mohr's circles rather than only one. Figure 10–32 shows the circle obtained from the initial stress element, a second circle encompassing σ_1 and σ_3, and a third encompassing σ_2 and σ_3. Thus each circle represents the plane on which two of the three principal stresses act. The point at the top of each circle indicates the largest shear stress that would occur in that plane. Then the largest circle, drawn for σ_1 and σ_3, produces the true maximum shear stress and its value is consistent with Equation (10–13).

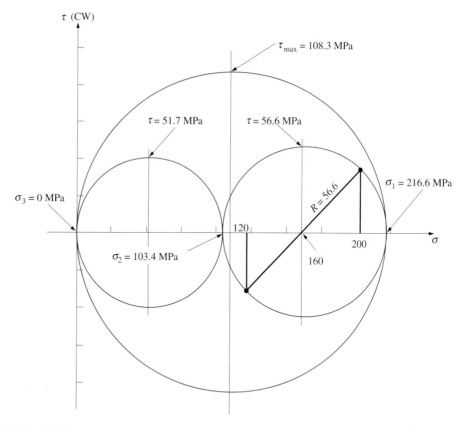

FIGURE 10–32 Three related Mohr's circles showing σ_1, σ_2, σ_3, and τ_{max}.

Figure 10–33 illustrates another case in which the principal stresses from the initial stress element have the same sign, both negative in this case. The initial stresses are

$$\sigma_x = -50 \text{ MPa} \qquad \sigma_y = -180 \text{ MPa} \qquad \tau_{xy} = 30 \text{ MPa CCW}$$

Here, too, the supplementary circles must be drawn. But in this case, the zero stress on the "front" and "back" faces of the element becomes the *maximum* principal stress (σ_1). That is,

$$\sigma_1 = 0 \text{ MPa}$$
$$\sigma_2 = -43.4 \text{ MPa}$$
$$\sigma_3 = -186.6 \text{ MPa}$$

and the maximum shear stress is

$$\tau_{max} = \tfrac{1}{2}(\sigma_1 - \sigma_3) = \tfrac{1}{2}[0 - (-186.6)] = 93.3 \text{ MPa}$$

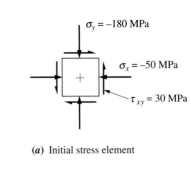

(a) Initial stress element

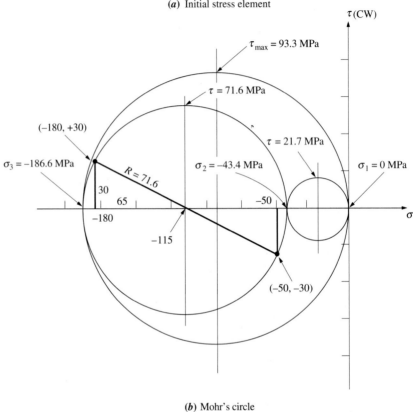

(b) Mohr's circle

FIGURE 10–33 Three related Mohr's circles showing σ_1, σ_2, σ_3, and τ_{max}.

Summary—procedures for dealing with cases in which principal stresses have the same sign

This summary pertains to the situation in which the Mohr's circle analysis of a stress element subjected to plane stress (stresses applied only in two dimensions) yields the result that both principal stresses (σ_1 and σ_2) are of the same sign; that is, both are tensile or both are compressive. In such cases, the following steps should be completed to provide a true picture of the stress condition on the three-dimensional element.

A. Draw the complete Mohr's circle for the plane stress condition, and identify the principal stresses σ_1 and σ_2.

B. If both principal stresses are tensile (positive): [See example in Figure 10-32.]

   1. Consider that the *zero stress* acting in the direction perpendicular to the initial stress element is the true minimum principal stress. Then it is necesssary to define *three* principal stresses such that:

   σ_1 = Maximum principal stress from the first Mohr's circle.

   σ_2 = Minimum principal stress from the first Mohr's circle.

   σ_3 = Zero (true minimum principal stress).

   3. Draw a secondary Mohr's circle whose diameter extends from σ_1 to σ_3 on the σ-axis. The center of the circle will be at the average of σ_1 and σ_3, $(\sigma_1 + \sigma_3)/2$. But, because $\sigma_3 = 0$, the average is $\sigma_1/2$.

   4. The maximum shear stress is at the top of the secondary circle and its value is also $\sigma_1/2$.

C. If both principal stresses are compressive (negative): [See example in Figure 10-33.]

   1. Consider that the *zero stress* acting in the direction perpendicular to the initial stress element is the true maximum principal stress. Then it is necessary to define *three* principal stresses such that:

   σ_1 = Zero (true maximum principal stress).

   σ_2 = Maximum principal stress from the first Mohr's circle.

   σ_3 = Minimum principal stress from the first Mohr's circle.

   3. Draw a secondary Mohr's circle whose diameter extends from σ_1 to σ_3 on the σ-axis. The center of the circle will be at the average of σ_1 and σ_3, $(\sigma_1 + \sigma_3)/2$. But, because $\sigma_1 = 0$, the average is $\sigma_3/2$.

   4. The maximum shear stress is at the top of the secondary circle and its magnitude is also $\sigma_3/2$.

10-12 THE MAXIMUM SHEAR STRESS THEORY OF FAILURE

One of the most widely used principles of design is the *maximum shear stress theory of failure,* which states:

A ductile material can be expected to fail when the maximum shear stress to which the material is subjected exceeds the yield strength of the material in shear.

Of course, to apply this theory, it is necessary to be able to compute the magnitude of the maximum shear stress. If the member is subjected to pure shear, such as torsional shear stress, direct shear stress, or shearing stress in beams in bending, the maximum shear stress can be computed directly from formulas such as those developed in this book. But if a combined stress condition exists, the use of Equation (10–9) or Mohr's circle should be used to determine the maximum shear stress.

One special case of combined stress that occurs often is one in which a normal stress in only one direction is combined with a shear stress. For example, a round bar could be subjected to a direct axial tension while also being twisted. In many types of mechanical power transmissions, shafts are subjected to bending and torsion simultaneously. Certain fasteners may be subjected to tension combined with direct shear.

A simple formula can be developed for such cases using Mohr's circle or Equation (10–9). If only a normal stress in the x-direction, σ_x, combined with a shearing stress, τ_{xy} exists, the maximum shear stress is

$$\tau_{max} = \sqrt{(\sigma_x/2)^2 + \tau_{xy}^2} \tag{10–14}$$

This formula can be developed from Equation (10–9) by letting $\sigma_y = 0$.

Example Problem 10–10 A solid circular bar is 45 mm in diameter and is subjected to an axial tensile force of 120 kN along with a torque of 1150 N·m. Compute the maximum shear stress in the bar.

Solution **Objective** Compute the maximum shear stress in the bar.

Given Diameter = D = 45 mm.
Axial force = F = 120 kN = 120 000 N.
Torque = T = 1150 N·m = 1 150 000 N·mm.

Analysis Use Equation (10–14) to compute τ_{max}.

Results 1. First, the applied normal stress can be found using the direct stress formula.

$$\sigma = F/A$$
$$A = \pi D^2/4 = \pi(45 \text{ mm})^2/4 = 1590 \text{ mm}^2$$
$$\sigma = (120\,000 \text{ N})/(1590 \text{ mm}^2) = 75.5 \text{ N/mm}^2 = 75.5 \text{ MPa}$$

2. Next, the applied shear stress can be found from the torsional shear stress formula.

$$\tau = T/Z_p$$
$$Z_p = \pi D^3/16 = \pi(45 \text{ mm})^3/16 = 17\,892 \text{ mm}^3$$
$$\tau = (1\,150\,000 \text{ N·mm})/(17\,892 \text{ mm}^3) = 64.3 \text{ N/mm}^2 = 64.3 \text{ MPa}$$

Chapter 10 ■ The General Case of Combined Stress and Mohr's Circle

3. Then using Equation (10–14) yields

$$\tau_{max} = \sqrt{\left(\frac{75.5 \text{ MPa}}{2}\right)^2 + (64.3 \text{ MPa})^2} = 74.6 \text{ MPa}$$

Comment This stress should be compared with the design shear stress.

REFERENCES

1. Muvdi, B. B., and J. W. McNabb, *Engineering Mechanics of Materials,* 3rd ed., Springer-Verlag, New York, 1990.

2. Popov, E. P., *Engineering Mechanics of Solids,* Prentice-Hall, Englewood Cliffs, NJ, 1990.

3. Shigley, J. E., and C. R. Mischke, *Mechanical Engineering Design,* 5th ed., McGraw-Hill, New York, 1989.

PROBLEMS

A. For Problems 10–1 to 10–28, determine the principal stresses and the maximum shear stress using Mohr's circle. The data sets below give the stresses on the initial stress element. Perform the following operations.

(a) Draw the complete Mohr's circle, labeling critical points including σ_1, σ_2, τ_{max}, and σ_{avg}.

(b) On Mohr's circle, indicate which line represents the x-axis on the initial stress element.

(c) On Mohr's circle, indicate the angles from the line representing the x-axis to the σ_1-axis and the τ_{max}-axis.

(d) Draw the principal stress element and the maximum shear stress element in their proper orientation relative to the initial stress element.

Problem	σ_x	σ_y	τ_{xy}
10–1	300 MPa	−100 MPa	80 MPa CW
10–2	250 MPa	−50 MPa	40 MPa CW
10–3	80 MPa	−10 MPa	60 MPa CW
10–4	150 MPa	10 MPa	100 MPa CW
10–5	20 ksi	−5 ksi	10 ksi CCW
10–6	38 ksi	−25 ksi	18 ksi CCW
10–7	55 ksi	15 ksi	40 ksi CCW
10–8	32 ksi	−50 ksi	20 ksi CCW

Problem	σ_x	σ_y	τ_{xy}
10–9	−900 kPa	600 kPa	350 kPa CCW
10–10	−580 kPa	130 kPa	75 kPa CCW
10–11	−840 kPa	−35 kPa	650 kPa CCW
10–12	−325 kPa	50 kPa	110 kPa CCW
10–13	−1800 psi	300 psi	800 psi CW
10–14	−6500 psi	1500 psi	1200 psi CW
10–15	−4250 psi	3250 psi	2800 psi CW
10–16	−150 psi	8600 psi	80 psi CW
10–17	260 MPa	0 MPa	190 MPa CCW
10–18	1450 kPa	0 kPa	830 kPa CW
10–19	22 ksi	0 ksi	6.8 ksi CW
10–20	6750 psi	0 psi	3120 psi CCW
10–21	0 ksi	−28 ksi	12 ksi CW
10–22	0 MPa	440 MPa	215 MPa CW
10–23	0 MPa	260 MPa	140 MPa CCW
10–24	0 kPa	−1560 kPa	810 kPa CCW
10–25	225 MPa	−85 MPa	0 MPa
10–26	6250 psi	−875 psi	0 psi
10–27	775 kPa	−145 kPa	0 kPa
10–28	38.6 ksi	−13.4 ksi	0 ksi

B. For problems that result in the principal stresses from the initial Mohr's circle having the same sign, use the procedures from Section 10–11 to draw supplementary circles and find the following:

(a) The three principal stresses: σ_1, σ_2, and σ_3.

(b) The true maximum shear stress.

Problem	σ_x	σ_y	τ_{xy}
10–29	300 MPa	100 MPa	80 MPa CW
10–30	250 MPa	150 MPa	40 MPa CW
10–31	180 MPa	110 MPa	60 MPa CW
10–32	150 MPa	80 MPa	30 MPa CW
10–33	30 ksi	15 ksi	10 ksi CCW
10–34	38 ksi	25 ksi	8 ksi CCW
10–35	55 ksi	15 ksi	5 ksi CCW
10–36	32 ksi	50 ksi	20 ksi CCW
10–37	−840 kPa	−335 kPa	120 kPa CCW
10–38	−325 kPa	−50 kPa	60 kPa CCW
10–39	−1800 psi	−300 psi	80 psi CW
10–40	−6500 psi	−2500 psi	1200 psi CW

C. For the following problems, use the data from the indicated problem for the initial stress element to draw Mohr's circle. Then determine the stress condition on the element at the specified angle of rotation from the given x-axis. Draw the rotated element in its proper re-

lation to the initial stress element and indicate the normal and shear stresses acting on it.

Problem	Problem for the initial stress data	Angle of rotation from the x-axis
10–41	10–1	30 deg CCW
10–42	10–1	30 deg CW
10–43	10–4	70 deg CCW
10–44	10–6	20 deg CW
10–45	10–8	50 deg CCW
10–46	10–10	45 deg CW
10–47	10–13	10 deg CCW
10–48	10–15	25 deg CW
10–49	10–16	80 deg CW
10–50	10–18	65 deg CW

D. For the following problems, use Equation (10–14) to compute the magnitude of the maximum shear stress for the data from the indicated problem.

10–51. Use data from Problem 10–17.

10–52. Use data from Problem 10–18.

10–53. Use data from Problem 10–19.

10–54. Use data from Problem 10–20.

COMPUTER PROGRAMMING ASSIGNMENTS

1. Write a program for a computer, spreadsheet, or a programmable calculator to aid in the construction of Mohr's circle. Input the initial stresses, σ_x, σ_y, and τ_{xy}. Have the program compute the radius of the circle, the maximum and minimum principal stresses, the maximum shear stress, and the average stress. Use the program in conjunction with freehand sketching of the circle for the data for Problems 10–1 through 10–24.

2. Enhance the program in Assignment 1 by computing the angle of orientation of the principal stress element and the angle of orientation of the maximum shear stress element.

3. Enhance the program in Assignment 1 by computing the normal and shear stresses on the element rotated at any specified angle relative to the original x-axis.

4. Enhance the program in Assignment 1 by causing it to detect if the principal stresses from the initial Mohr's circle are of the same sign; and in such cases, print out the three principal stresses in the proper order, σ_1, σ_2, σ_3. Also have the program compute the true maximum shear stress from Equation (10–13).

11

Special Cases of Combined Stresses

11–1 OBJECTIVES OF THIS CHAPTER

This chapter can be studied either after completing Chapter 10, or independently from it. There are several practical cases involving combined stresses that can be solved without resorting to the more rigorous and time-consuming procedures presented in Chapter 10, although the techniques discussed here are based on the principles of that chapter.

When a beam is subjected to both bending and a direct axial stress, either tensile or compressive, simple superposition of the applied stresses can be used to determine the combined stress. Many forms of power transmission equipment involve shafts that are subjected to torsional shear stress along with bending stress. Such shafts can be analyzed by using the maximum shear stress theory of failure and employing the equivalent torque technique of analysis.

After completing this chapter, you should be able to:

1. Compute the combined normal stress resulting from the application of bending stress with either direct tensile or compressive stresses using the principle of superposition.

2. Recognize the importance of visualizing the stress distribution over the cross section of a load-carrying member and considering the stress condition at a point.

3. Recognize the importance of free-body diagrams of components of structures and mechanisms in the analysis of combined stresses.

4. Evaluate the design factor for combined normal stress, including the properties of either isotropic or nonisotropic materials.

5. Optimize the shape and dimensions of a load-carrying member relative to the variation of stress in the member and its strength properties.

6. Analyze members subjected only to combined bending and torsion by computing the resulting maximum shear stress.

7. Use the maximum shear stress theory of failure properly.

8. Apply the equivalent torque technique to analyze members subjected to combined bending and torsion.

9. Consider the stress concentration factors when using the equivalent torque technique.

11–2 COMBINED NORMAL STRESSES

The first combination to be considered is bending with direct tension or compression. In any combined stress problem, it is helpful to visualize the stress distribution caused by the various components of the total stress pattern. You should review Section 10–3 for summaries of the stress distribution for bending and direct tension and compression. Notice that bending results in tensile and compressive stresses, as do both direct tension and direct compression. Since the same kind of stresses are produced, a simple algebraic sum of the stresses produced at any point is all that is required to compute the resultant stress at that point. This process is called *superposition*.

Guidelines for solving problems with combined normal stresses

These guidelines pertain to situations in which two or more loads or components of loads act in a manner that produces normal stresses (tension and/or compression) on the load-carrying member. In general, loads producing bending stress or direct tension or compression are to be included. The objective is to compute the maximum combined stress in the member. Let tensile stresses be positive ($+$) and compressive stresses be negative ($-$).

1. Draw the free-body diagram for the load-carrying member and compute the magnitude of all applied forces.

2. For any force that acts at an angle to the neutral axis of the member, resolve the force into components perpendicular and parallel to the neutral axis.

3. Forces or components acting in a direction coincident with the neutral axis will produce either direct tension or direct compression with uniform stress distribution across the cross section. Compute these stresses.

4. Forces or components acting perpendicular to the neutral axis cause bending stresses. Using the methods of Chapter 6, determine the bending moments caused by these forces, either individually or in combination. Then, for the section subjected to the largest bending moment,

compute the bending stress from the flexure formula, $\sigma = M/S$. The maximum stress will occur at the outermost fibers of the cross section. Note at which points the stress is tensile and which will be compressive.

5. Forces or components acting parallel to the neutral axis but whose line of action is away from that axis also cause bending. The bending moment is the product of the force times the perpendicular distance from the neutral axis to the line of action of the force. Compute the bending stress due to such moments at any section where the combined stress may be the maximum.

6. Considering all normal stresses computed from steps 1–5, use superposition to combine them at any point within any cross section where the combined stress may be the maximum. Superposition is accomplished by algebraically summing all stresses acting at a point, taking care to note if each component stress is tensile ($+$) or compressive ($-$). It may be necessary to evaluate the stress condition at two or more points if it is not obvious where the maximum combined stress occurs. In general, the superposition process can be expressed as,

$$\sigma_{\text{comb}} = \pm \frac{F}{A} \pm \frac{M}{S} \qquad (11\text{--}1)$$

where the term, $\pm F/A$, includes all direct tensile and compressive stresses acting at the point of interest and the term, $\pm M/S$, includes all bending stresses acting at that point. The sign for each stress must be logically determined from the load causing the individual stress.

7. The maximum combined stress on the member can then be compared with the design stress for the material from which the member is to be made to compute the resulting design factor and to assess the safety of the member. For isotropic materials, either tensile or compressive stress could cause failure, whichever is maximum. For nonisotropic materials having different strengths in tension and compression, it is necessary to compute the resulting design factor for both the tensile and the compressive stress to determine which is critical. Also, in general, it will be necessary to consider the stability of those parts of members subjected to compressive stresses by analyzing the tendency for buckling or local crippling of the member. See Chapter 14 for buckling of column-like compression members. The analysis for crippling and buckling of parts of members will require reference to other sources. See the references at the end of Chapters 8, 9, and 10.

An example of a member in which both bending and direct tensile stresses are developed is shown in Figure 11–1. The two horizontal beams support a 10 000-lb load by means of the cable assembly. The beams are rigidly attached to columns, so that they act as cantilever beams. The load at the end of each beam is equal to the tension

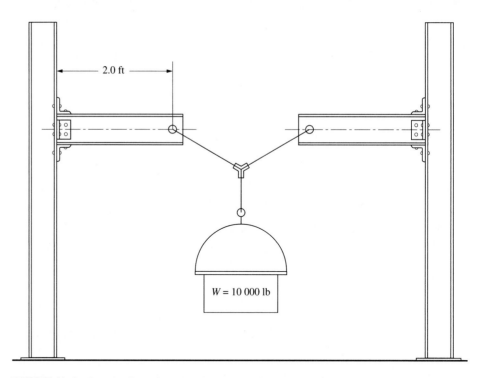

FIGURE 11–1 Two cantilever beams subjected to combined bending and axial tensile stresses.

in the cable. Figure 11–2 shows that the vertical component of the tension in each cable must be 5000 lb. That is,

$$F \cos 60° = 5000 \text{ lb}$$

and the total tension in the cable is

$$F = \frac{5000 \text{ lb}}{\cos 60°} = 10\,000 \text{ lb}$$

This is the force applied to each beam, as shown in Figure 11–3.

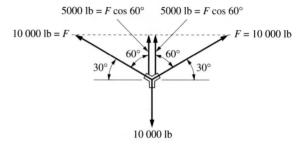

FIGURE 11–2 Force analysis on cable system.

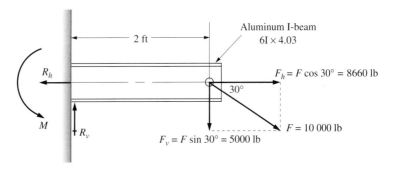

FIGURE 11–3 Forces applied to each beam.

An example problem will now be shown that completes the analysis of the combined stress condition in each of the horizontal cantilever beams in Figure 11–1.

Example Problem 11–1

For the system shown in Figure 11–1, determine the maximum combined tensile or compressive stress in each horizontal cantilever beam when a static load of 10 000 lb is suspended from the cable system between them. The beams are standard Aluminum Association I-beams, 6I × 4.03. Then, if the beams are to be made of 6061-T6 aluminum alloy, compute the resulting design factor.

Solution   **Objective**   Compute the maximum combined stress and the resulting design factor for the horizontal cantilever beams in Figure 11–1.

Given   Load weighs 10 000 lb. Force system detail shown in Figure 11–2. Tension in each cable is 10 000 lb. The 2.0 ft long beams are aluminum 61 × 4.03 made from 6061-T6 aluminum alloy. From Appendix A–11, the properties of the beam are $A = 3.427$ in^2 and $S = 7.33$ in^3.

Analysis   The tension in the cable attached to the end of each cantilever beam will tend to cause direct tensile stress combined with bending stress in the beam. Use the *Guidelines for solving problems with combined normal stresses* outlined in this section.

Results   *Step 1.* Figure 11–3 shows the free-body diagram of one beam with the 10 000 lb force applied by the cable to the end of the beam. The reaction at the left end where the beam is rigidly attached to the column consists of a vertical reaction force, a horizontal reaction force, and a counterclockwise moment.

Step 2. Also shown in Figure 11–3 is the resolution of the 10 000 lb force into vertical and horizontal components where $F_v = 5000$ lb and $F_h = 8660$ lb.

Step 3. The horizontal force, F_h, acts in a direction that is coincident with the neutral axis of the beam. Therefore, it causes direct tensile stress with a magnitude of

$$\sigma_t = \frac{F_h}{A} = \frac{8660 \text{ lb}}{3.427 \text{ in}^2} = 2527 \text{ psi}$$

Step 4. The vertical force, F_v, causes bending in a downward direction such that the top of the beam is in tension and the bottom is in compression. The maximum bending moment will occur at the support at the left end, where

$$M = F_v(2.0 \text{ ft}) = (5000 \text{ lb})(2.0 \text{ ft})(12 \text{ in/ft}) = 120\,000 \text{ lb·in}$$

Then the maximum bending stress caused by this moment is

$$\sigma_b = \frac{M}{S} = \frac{120\,000 \text{ lb·in}}{7.33 \text{ in}^3} = 16\,371 \text{ psi}$$

A stress of this magnitude occurs as a tensile stress at the top surface and as a compressive stress at the bottom surface of the beam at the support.

Step 5. This step does not apply to this problem because there is no horizontal force offset from the netural axis.

Step 6. It can be reasoned that the maximum combined stress occurs at the top surface of the beam at the support because both the direct tensile stress computed in step 3 and the bending stress computed in step 4 are tensile at that point. Therefore, they will add together. Using superposition,

$$\sigma_{\text{top}} = 2527 \text{ psi} + 16\,371 \text{ psi} = 18\,898 \text{ psi tension}$$

For comparison, the combined stress at the bottom surface of the beam is

$$\sigma_{\text{bot}} = 2527 \text{ psi} - 16\,371 \text{ psi} = -13\,844 \text{ psi compression}$$

Figure 11–4 shows a set of diagrams that illustrate the process of superposition. Part (a) is the stress in the beam caused by bending. Part (b) shows the direct tensile stress due to F_v. Part (c) shows the combined stress distribution.

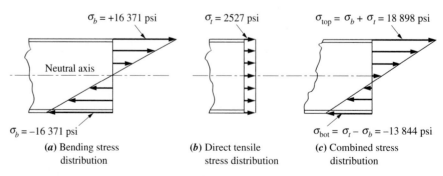

FIGURE 11–4 Diagram of superposition principle applied to the beams of Figure 11–1.

Step 7. Because the load is static, we can compute the design factor based on the yield strength of the 6061-T6 aluminum alloy, where $s_y = 40\,000$ psi (Appendix A–17). Then,

$$N = s_y/\sigma_{top} = (40\,000 \text{ psi})/(18\,898 \text{ psi}) = 2.11$$

Comment This should be adequate for a purely static load. If there is uncertainty about the magnitude of the load or if there is a possibility of the load being applied with some shock or impact, a higher design factor would be preferred.

Example Problem 11–2

A picnic table in a park is made by supporting a circular top on a pipe that is rigidly held in concrete in the ground. Figure 11–5 shows the arrangement. Compute the maximum stress in the pipe if a person having a mass of 135 kg sits on the edge of the table. The pipe is made of an aluminum alloy and has an outside diameter of 170 mm and an inside diameter of 163 mm. If the aluminum is 6061-T4, compute the resulting design factor based on both yield strength and ultimate strength. Then comment on the suitability of the design.

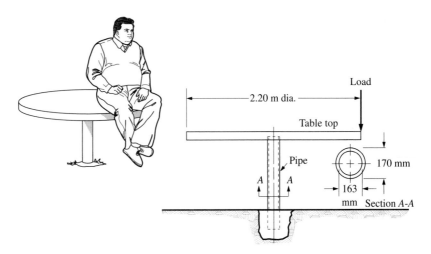

FIGURE 11–5 Picnic table supported by a pipe.

Solution

Objective Compute the maximum stress in the pipe in Figure 11–5 and the design factor based on both yield strength and ultimate strength.

Given Loading and pipe dimensions shown in Figure 11–5. Load is the force due to a 135 kg mass on the edge of the table. Pipe is aluminum, 6061-T4; $s_y = 145$ MPa; $s_u = 241$ MPa.

Analysis The pipe is subjected to combined bending and direct compression as illustrated in Figure 11–6, the free-body diagram of the pipe. The effect of the load is to produce a downward force on the top of the pipe while exerting a moment in the clockwise direction. The moment is the product of the load times the radius of the tabletop. The reaction at the bottom of the pipe, supplied by the concrete, is an upward force combined with a counterclockwise

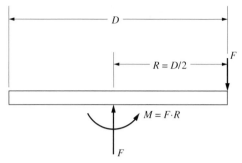

(*a*) Free-body diagram of table top

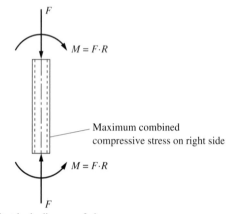

Maximum combined
compressive stress on right side

(*b*) Free-body diagram of pipe

FIGURE 11–6   Free-body diagrams for table top and pipe for
Example Problem 11–2.

moment. Use the *Guidelines for solving problems with combined normal
stresses.*

Results   ***Step 1.*** Figure 11–6 shows the free-body diagram. The force is the gravitational attraction of the 135-kg mass.

$$F = m{\cdot}g = 135 \text{ kg} {\cdot} 9.81 \text{ m/s}^2 = 1324 \text{ N}$$

Step 2. No forces act at an angle to the axis of the pipe.

Step 3. Now the direct axial compressive stress in the pipe is

$$\sigma_a = -\frac{F}{A}$$

But

$$A = \frac{\pi(D_o^2 - D_i^2)}{4} = \frac{\pi(170^2 - 163^2) \text{ mm}^2}{4} = 1831 \text{ mm}^2$$

Then

$$\sigma_a = -\frac{1324 \text{ N}}{1831 \text{ mm}^2} = -0.723 \text{ MPa}$$

This stress is a uniform, compressive stress across any cross section of the pipe.

Step 4. No forces act perpendicular to the axis of the pipe.

Step 5. Since the force acts at a distance of 1.1 m from the axis of the pipe, the moment is

$$M = (1324 \text{ N})(1.1 \text{ m}) = 1456 \text{ N·m}$$

The bending stress computation requires the application of the flexure formula,

$$\sigma_b = \frac{Mc}{I}$$

where $c = \dfrac{D_o}{2} = \dfrac{170 \text{ mm}}{2} = 85 \text{ mm}$

$$I = \frac{\pi}{64}(D_o^4 - D_i^4) = \frac{\pi}{64}(170^4 - 163^4) \text{ mm}^4$$
$$I = 6.347 \times 10^6 \text{ mm}^4$$

Then

$$\sigma_b = \frac{(1456 \text{ N·m})(85 \text{ mm})}{6.347 \times 10^6 \text{ mm}^4} \times \frac{10^3 \text{ mm}}{\text{m}} = 19.5 \text{ MPa}$$

Step 6. The bending stress σ_b produces compressive stress on the right side of the pipe and tension on the left side. Since the direct compression stress adds to the bending stress on the right side, that is where the maximum stress would occur. The combined stress would then be

$$\sigma_c = -\sigma_a - \sigma_b = (-0.723 - 19.5) \text{ MPa} = -20.22 \text{ MPa}$$

Step 7. The design factor based on yield strength is

$$N = \frac{s_y}{\sigma_c} = \frac{145 \text{ MPa}}{20.22 \text{ MPa}} = 7.17$$

The design factor based on ultimate strength is

$$N = \frac{s_u}{\sigma_c} = \frac{241 \text{ MPa}}{20.22 \text{ MPa}} = 11.9$$

Comment These values should be acceptable for this application based on the normal tensile and compressive stresses. Table 3–2 in Chapter 3 suggests $N = 2$ based on yield strength for static loading and $N = 12$ based on ultimate strength for impact. If the person simply sits on the edge of the table, the loading would be considered to be static. But a person jumping on the edge would apply an impact load. The design factor of 11.9 is sufficiently close to the recommended value of 12. However, additional analysis should be performed to evaluate the tendency for the pipe to buckle as a column, as discussed in Chapter 14. Also, Reference 1 defines the procedures for evaluating the tendency for local buckling of a hollow circular pipe subjected to compression.

11–3 COMBINED NORMAL AND SHEAR STRESSES

Rotating shafts in machines transmitting power represent good examples of members loaded in such a way as to produce combined bending and torsion. Figure 11–7 shows a shaft carrying two chain sprockets. Power is delivered to the shaft through the sprocket at C and transmitted down the shaft to the sprocket at B, which in turn delivers it to another shaft. Because it is transmitting power, the shaft between B and C is subject to a torque and torsional shear stress, as you learned in Chapter 5. In order for the sprockets to transmit torque, they must be pulled by one side of the chain. At C, the back side of the chain must pull down with the force F_1 in order to drive the sprocket clockwise. Since the sprocket at B drives a mating sprocket, the front side of the chain would be in tension under the force F_2. The two forces, F_1 and F_2, acting downward cause bending of the shaft. Thus the shaft must be analyzed for both torsional shear stress and bending stress. Then, since both stresses act at the same place on the shaft, their combined effect must be determined. The method of analysis to be used is called the *maximum shear stress theory of failure,* which is described next. Then example problems will be shown.

Maximum shear stress theory of failure. When the tensile or compressive stress caused by bending occurs at the same place that a shearing stress occurs, the two kinds

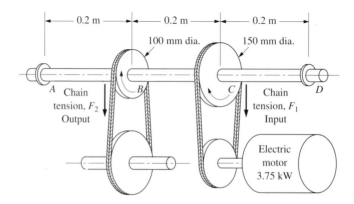

FIGURE 11–7 Power transmission shafts.

of stress combine to produce a larger shearing stress. The maximum stress can be computed from

$$\tau_{\text{max}} = \sqrt{\left(\frac{\sigma}{2}\right)^2 + \tau^2}$$ (11–2)

In Equation (11–2), σ refers to the magnitude of the tensile or compressive stress at the point, and τ is the shear stress at the same point. The result τ_{max} is the maximum shear stress at the point. The basis of Equation (11–2) was shown by using Mohr's circle in Section 10–12.

The maximum shear stress theory of failure states that a member fails when the maximum shear stress exceeds the yield strength of the material in shear. This failure theory shows good correlation with test results for ductile metals such as most steels.

Equivalent torque.   Equation (11–2) can be expressed in a simplified form for the particular case of a circular shaft subjected to bending and torsion. Evaluating the bending stress separately, the maximum tensile or compressive stress would be

$$\sigma = \frac{M}{S}$$

where $S = \dfrac{\pi D^3}{32} =$ section modulus

$D =$ diameter of the shaft

$M =$ bending moment on the section

The maximum stress due to bending occurs at the outside surface of the shaft, as shown in Figure 11–8.

Now consider the torsional shear stress separately. In Chapter 5 the torsional shear stress equation was derived:

$$\tau = \frac{T}{Z_p}$$

where $Z_p = \dfrac{\pi D^3}{16} =$ polar section modulus

$T =$ torque on the section

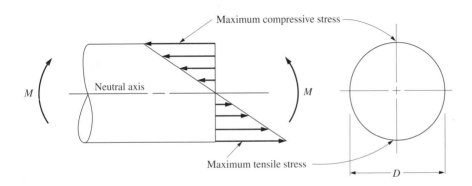

FIGURE 11–8 Bending stress distribution in a round shaft.

The maximum shear stress occurs at the outer surface of the shaft all the way around the circumference, as shown in Figure 11–9.

Thus the maximum tensile stress and the maximum torsional shear stress both occur at the same point in the shaft. Now let's use Equation (11–2) to get an expression for the combined stress in terms of the bending moment M, the torque T, and the shaft diameter D.

$$\tau_{max} = \sqrt{\left(\frac{\sigma}{2}\right)^2 + \tau^2}$$

$$\tau_{max} = \sqrt{\left(\frac{M}{2S}\right)^2 + \left(\frac{T}{Z_p}\right)^2} \tag{11–3}$$

Notice that, from the definitions of S and Z_p given earlier,

$$2S = Z_p$$

Substituting this into Equation (11–3) gives

$$\tau_{max} = \sqrt{\left(\frac{M}{Z_p}\right)^2 + \left(\frac{T}{Z_p}\right)^2}$$

Factoring out Z_p yields

$$\tau_{max} = \frac{1}{Z_p} \sqrt{M^2 + T^2}$$

Sometimes the term $\sqrt{M^2 + T^2}$ is called the *equivalent torque* because it represents the amount of torque that would have to be applied to the shaft by itself to cause the equivalent magnitude of shear stress as the combination of bending and torsion. Calling the equivalent torque T_e,

$$T_e = \sqrt{M^2 + T^2} \tag{11–4}$$

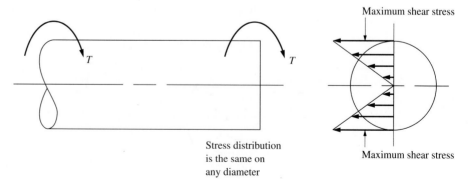

Maximum shear stress

Maximum shear stress

Stress distribution
is the same on
any diameter

FIGURE 11–9 Shear stress distribution in a round shaft.

and

$$\tau_{max} = \frac{T_e}{Z_p} \qquad (11\text{--}5)$$

Equations (11–4) and (11–5) greatly simplify the calculation of the maximum shear stress in a circular shaft subjected to bending and torsion.

In the design of circular shafts subjected to bending and torsion, a design stress can be specified giving the maximum allowable shear stress. This was done in Chapter 5.

$$\tau_d = \frac{s_{ys}}{N}$$

where s_{ys} is the yield strength of the material in shear. Since s_{ys} is seldom known, the approximate value found from $s_{ys} = s_y/2$ can be used. Then

$$\tau_d = \frac{s_y}{2N} \qquad (11\text{--}6)$$

where s_y is the yield strength in tension, as reported in most material property tables, such as those in Appendixes A–13 through A–17. It is recommended that the value of the design factor be *no less than 4*. A rotating shaft subjected to bending is a good example of a repeated and reversed load. With each revolution of the shaft, a particular point on the surface is subjected to the maximum tensile and then the maximum compressive stress. Thus fatigue is the expected mode of failure, and $N = 4$ or greater is recommended, based on yield strength.

Stress concentrations. In shafts, stress concentrations are created by abrupt changes in geometry such as keyseats, shoulders, and grooves. See Appendix A–21 for values of stress concentration factors. The proper application of stress concentration factors to the equivalent torque Equations (11–4) and (11–5) should be considered carefully. If the value of K_t at a section of interest is equal for both bending and torsion, then it can be applied directly to Equation (11–5). That is,

$$\tau_{max} = \frac{T_e K_t}{Z_p} \qquad (11\text{--}7)$$

The form of Equation (11–7) can also be applied as a conservative calculation of τ_{max} by selecting K_t as the largest value for either torsion or bending.

To account for the proper K_t for both torsion and bending, Equation (11–4) can be modified as

$$T_e = \sqrt{(K_{tB} M)^2 + (K_{tT} T)^2} \qquad (11\text{--}8)$$

Then Equation (11–5) can be used directly to compute the maximum shear stress.

Specify a suitable material for the shaft shown in Figure 11–7. The shaft has a uniform diameter of 55 mm and rotates at 120 rpm while transmitting 3.75 kW of power. The chain sprockets at *B* and *C* are keyed to the shaft with sled-runner keyseats. Sprocket *C* receives the power, and sprocket *B* delivers it to another shaft. The bearings at *A* and *D* provide simple supports for the shaft.

Solution

Objective Specify a suitable material for the shaft.

Given Shaft and loading in Figure 11–7.
Power = P = 3.75 kW. Rotational speed = n = 120 rpm.
Shaft diameter = D = 55 mm.
Sled-runner keyseats at *B* and *C*.
Simple supports at *A* and *D*.

Analysis The several steps to be used to solve this problem are outlined below.

1. The torque in the shaft will be computed for the known power and rotational speed from $T = P/n$, as developed in Chapter 5.

2. The tensions in the chains for sprockets *B* and *C* will be computed. These are the forces that produce bending in the shaft.

3. Considering the shaft to be a beam, the shear and bending moment diagrams will be drawn for it.

4. At the section where the maximum bending moment occurs, the equivalent torque T_e will be computed from Equation (11–4).

5. The polar section modulus Z_p and the stress concentration factor K_t will be determined.

6. The maximum shear stress will be computed from Equation (11–7).

7. The required yield strength of the shaft material will be computed by letting $\tau_{max} = \tau_d$ in Equation (11–6) and solving for s_y. Remember, let $N = 4$ or more.

8. A steel that has a sufficient yield strength will be selected from Appendix A–13.

Results **Step 1.** The desirable unit for torque is N·m. Then it is most convenient to observe that the power unit of kilowatts is equivalent to the units of kN·m/s. Also, rotational speed must be expressed in rad/s.

$$n = \frac{120 \text{ rev}}{\text{min}} \times \frac{2\pi \text{ rad}}{\text{rev}} \times \frac{1 \text{ min}}{60 \text{ s}} = 12.57 \text{ rad/s}$$

We can now compute torque.

$$T = \frac{P}{n} = \frac{3.75 \text{ kN·m}}{\text{s}} \times \frac{1}{12.57 \text{ rad/s}} = 0.298 \text{ kN·m}$$

Step 2. The tensions in the chains are indicated in Figure 11–7 by the forces F_1 and F_2. For the shaft to be in equilibrium, the torque on both sprockets must be the same in magnitude but opposite in direction. On either sprocket the torque is the product of the chain force times the *radius* of the pulley. That is,

$$T = F_1 R_1 = F_2 R_2$$

The forces can now be computed.

$$F_1 = \frac{T}{R_1} = \frac{0.298 \text{ kN·m}}{75 \text{ mm}} \times \frac{10^3 \text{ mm}}{\text{m}} = 3.97 \text{ kN}$$

$$F_2 = \frac{T}{R_2} = \frac{0.298 \text{ kN·m}}{50 \text{ mm}} \times \frac{10^3 \text{ mm}}{\text{m}} = 5.96 \text{ kN}$$

Step 3. Figure 11–10 shows the complete shear and bending moment diagrams found by the methods of Chapter 6. The maximum bending moment is 1.06 kN·m at section *B*, where one of the sprockets is located.

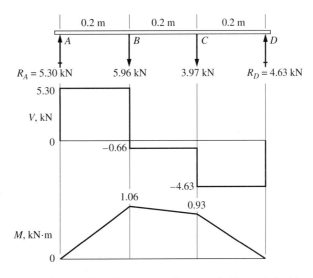

FIGURE 11–10 Shearing force and bending moment diagrams for Example Problem 11–3.

Step 4. At section *B*, the torque in the shaft is 0.298 kN·m and the bending moment is 1.06 kN·m. Then,

$$T_e = \sqrt{M^2 + T^2} = \sqrt{(1.06)^2 + (0.298)^2} = 1.10 \text{ kN·m}$$

Step 5. $Z_p = \dfrac{\pi D^3}{16} = \dfrac{\pi (55 \text{ mm})^3}{16} = 32.67 \times 10^3 \text{ mm}^3$

For the keyseat at section *B* securing the sprocket to the shaft, we will use $K_t = 1.6$, as reported in Figure 5–9.

Step 6. $\tau_{max} = \dfrac{T_e K_t}{Z_p} = \dfrac{1.10 \times 10^3 \text{ N·m}(1.6)}{32.67 \times 10^3 \text{ mm}^3} \times \dfrac{10^3 \text{ mm}}{\text{m}} = 53.9 \text{ MPa}$

Section 11–3 ▪ Combined Normal and Shear Stresses

419

Step 7. Let

$$\tau_{max} = \tau_d = \frac{s_y}{2N}$$

Then

$$s_y = 2N\tau_{max} = (2)(4)(53.9 \text{ MPa}) = 431 \text{ MPa}$$

Step 8. Referring to Appendix A–13, we find that several alloys could be used. For example, AISI 1040, cold drawn, has a yield strength of 490 MPa. Alloy AISI 1141 OQT 1300 has a yield strength of 469 MPa and also a very good ductility, as indicated by the 28% elongation. Either of these would be a reasonable choice.

REFERENCES

1. Aluminum Association, *Specifications for Aluminum Structures,* Washington, DC, 1986.

2. American Institute of Steel Construction, *Manual of Steel Construction,* 9th ed., Chicago, 1989.

3. Mott, Robert L., *Machine Elements in Mechanical Design,* 2nd ed., Merrill, an imprint of Macmillan Publishing Co., New York, 1992.

4. Shigley, J. E., and C. R. Mischke, *Mechanical Engineering Design,* 5th ed., McGraw-Hill Book Company, New York, 1989.

PROBLEMS

Combined Normal Stresses

11–1.E A $2\frac{1}{2}$-in schedule 40 steel pipe is used as a support for a basketball backboard, as shown in Figure 11–11. It is securely fixed into the ground. Compute the stress that would be developed in the pipe if a 230-lb player hung on the base of the rim of the basket.

11–2.M The bracket shown in Figure 11–12 has a rectangular cross section 18 mm wide by 75 mm high. It is securely attached to the wall. Compute the maximum stress in the bracket.

11–3.E The beam shown in Figure 11–13 carries a 6000-lb load attached to a bracket below the beam. Compute the stress at points M and N, where it is attached to the column.

11–4.E For the beam shown in Figure 11–13, compute the stress at points M and N if the 6000-lb load acts vertically downward instead of at an angle.

11–5.E For the beam shown in Figure 11–13, compute the stress at points M and N if the 6000-lb load acts back toward the column at an angle of 40 deg below the horizontal instead of as shown.

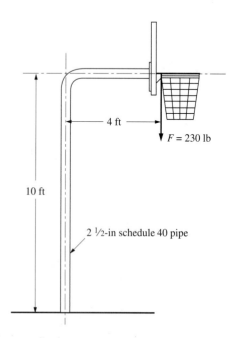

4 ft

$F = 230$ lb

10 ft

2 ½-in schedule 40 pipe

FIGURE 11–11   Basketball backboard for Problem 11–1.

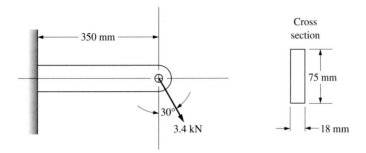

FIGURE 11–12 Bracket for Problem 11–2.

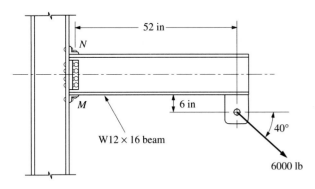

FIGURE 11–13 Beam for Problems 11–3, 11–4, and 11–5.

11–6.M Compute the maximum stress in the top portion of the coping saw frame shown in Figure 11–14 if the tension in the blade is 125 N.

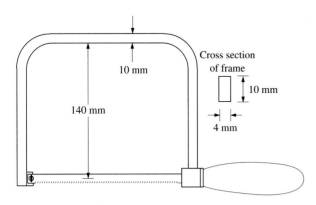

FIGURE 11–14 Coping saw frame for Problem 11–6.

11–7.M Compute the maximum stress in the crane beam shown in Figure 11–15 when a load of 12 kN is applied at the middle of the beam.

11–8.M Figure 11–16 shows a metal-cutting hacksaw. Its frame is made of hollow tubing having an out-

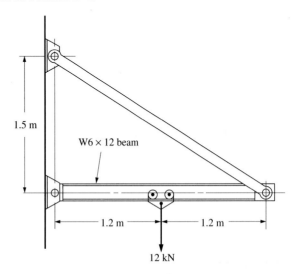

FIGURE 11–15 Crane beam for Problem 11–7.

FIGURE 11–16 Hacksaw frame for Problem 11–8.

side diameter of 12 mm and a wall thickness of 1.0 mm. The blade is pulled taut by the wing nut so that a tensile force of 160 N is applied to the blade. Compute the maximum stress in the top section of the tubular frame.

11–9.M The C-clamp in Figure 11–17 is made of cast zinc, ZA12. Determine the allowable clamping force that the clamp can exert if it is desired to have a design factor of 4 based on ultimate strength in either tension or compression.

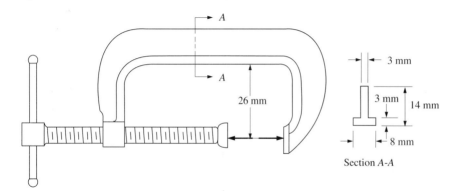

FIGURE 11-17 C-clamp for Problem 11-9.

11-10.M The C-clamp shown in Figure 11-18 is made of cast malleable iron, ASTM A220 grade 45008. Determine the allowable clamping force that the clamp can exert if it is desired to have a design factor of 4 based on ultimate strength in either tension or compression.

11-11.M A tool used for compressing a coil spring to allow its installation in a car is shown in Figure 11-19. A force of 1200 N is applied near the ends of the extended lugs, as shown. Compute the maximum tensile stress in the threaded rod. Assume a stress concentration factor of 3.0 at the thread

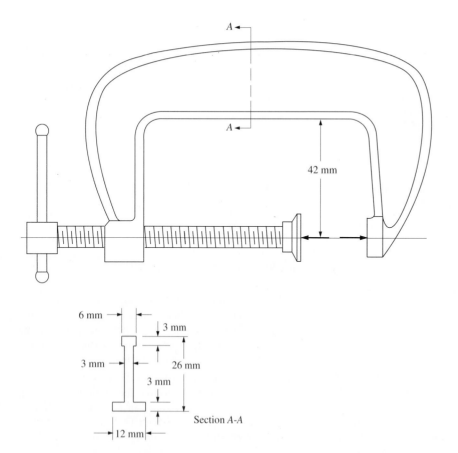

FIGURE 11-18 C-clamp for Problem 11-10.

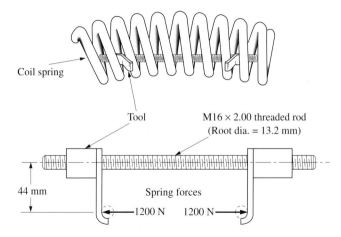

Coil spring

Tool

M16 × 2.00 threaded rod
(Root dia. = 13.2 mm)

44 mm

Spring forces

1200 N   1200 N

FIGURE 11–19 Tool for compressing springs for Problem 11–11.

root for both tensile and bending stresses. Then, using a design factor of 2 based on yield strength, specify a suitable material for the rod.

11–12.M Figure 11–20 shows a portion of the steering mechanism for a car. The steering arm has the detailed design shown. Notice that the arm has a constant thickness of 10 mm so that all cross sections between the end lugs are rectangular. The 225-N force is applied to the arm at a 60-deg angle. Compute the stress in the arm at sections A and B. Then, if the arm is made of ductile iron, ASTM A536, grade 80-55-6, compute the minimum design factor at these two points based on ultimate strength.

11–13.E An S3 × 5.7 American Standard I-beam is subjected to the forces shown in Figure 11–21. The 4600-lb force acts directly in line with the axis of the beam. The 500-lb downward force at A produces the reactions shown at the supports B and C. Compute the maximum tensile and compressive stresses in the beam.

11–14.M The horizontal crane boom shown in Figure 11–22 is made of a hollow rectangular steel tube. Compute the stress in the boom just to the left of point B when a mass of 1000 kg is supported at the end.

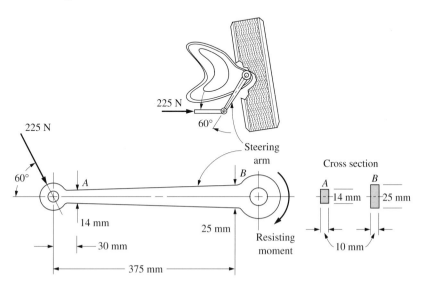

FIGURE 11–20 Steering arm for Problem 11–12.

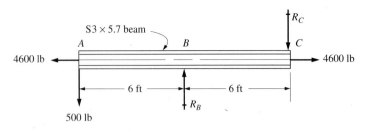

FIGURE 11–21 Beam for Problem 11–13.

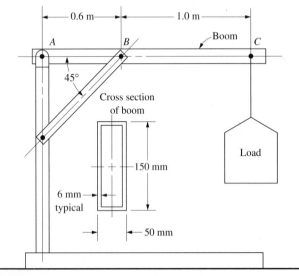

FIGURE 11–22 Crane boom for Problems 11–14 and 11–15.

11–15.M For the crane boom shown in Figure 11–22, compute the load that could be supported if a design factor of 3 based on yield strength is desired. The boom is made of AISI 1040 hot-rolled steel. Analyze only sections where the full box section carries the load, assuming that sections at connections are adequately reinforced.

11–16.E The member *EF* in the truss shown in Figure 11–23 carries an axial tensile load of 54 000 lb in addition to the two 1200-lb loads

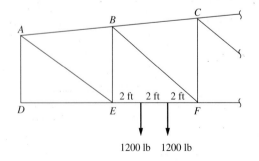

FIGURE 11–23 Truss for Problem 11–16.

shown. It is planned to use two steel angles for the member, arranged back to back. Specify a suitable size for the angles if they are to be made of ASTM A36 structural steel with an allowable stress of 0.6 times the yield strength.

For Problems 11–17 through 11–20, specify a suitable size for the horizontal portion of the member that will maintain the combined tensile or compressive stress to 6000 psi if the data are in the U.S. Customary unit system and 42 MPa if the SI metric system is used. The cross section is to be square. The loading for these members is the same as that used in the specified problems in Chapter 6 for which the free-body diagrams and the shearing force and bending moment diagrams were to be determined.

11–17.M Use Figure P6–77.

11–18.E Use Figure P6–78.

11–19.M Use Figure P6–79.

11–20.E Use Figure P6–80.

Combined Normal and Torsional Shear Stresses

11–21.M A solid circular bar is 40 mm in diameter and is subjected to an axial tensile force of 150 kN along with a torque of 500 N·m. Compute the maximum shear stress in the bar.

11–22.E A solid circular bar has a diameter of 2.25 in and is subjected to an axial tensile force of 47 000 lb along with a torque of 8500 lb·in. Compute the maximum shear stress in the bar.

11–23.E A short solid circular bar has a diameter of 4.00 in and is subjected to an axial compressive force of 40 000 lb along with a torque of 25 000 lb·in. Compute the maximum shear stress in the bar.

11–24.E A short hollow circular post is made from a 12-in schedule 40 steel pipe and it carries an axial compressive load of 250 000 lb along with a torque of 180 000 lb·in. Compute the maximum shear stress in the post.

11–25.E A short hollow circular support is made from a 3-in schedule 40 steel pipe and it carries an axial compressive load of 25 000 lb along with a torque of 15 500 lb·in. Compute the maximum shear stress in the support.

11–26.E A television antenna is mounted on a hollow aluminum tube as sketched in Figure 11–24. During installation, a force of 20 lb is applied to the end of the antenna as shown. Calculate the torsional shear stress in the tube and the stress due to bending. Consider the tube to be simply supported against bending at the clamps, but assume that rotation is not permitted. If the tube is made of 6061-T6 aluminum, would it be safe under this load? The tube has an outside diameter of 1.50 in and a wall thickness of $\frac{1}{16}$ in.

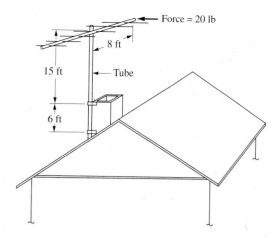

FIGURE 11–24 Antenna mounting tube for Problem 11–26.

11–27.M Figure 11–25 shows a crank to which a force F of 1200 N is applied. Compute the maximum stress in the circular portion of the crank.

11–28.E A standard steel pipe is to be used to support a bar carrying four loads, as shown in Figure 11–26. Specify a suitable pipe that would keep the maximum shear stress to 8000 psi.

Rotating Shafts—Combined Torsional Shear and Bending Stresses

11–29.M A circular shaft carries the load shown in Figure 11–27. The shaft carries a torque of 1500 N·m between sections B and C. Compute the maximum shear stress in the shaft near section B.

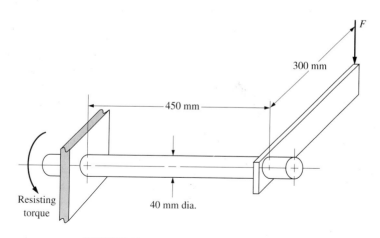

FIGURE 11–25 Crank for Problem 11–27.

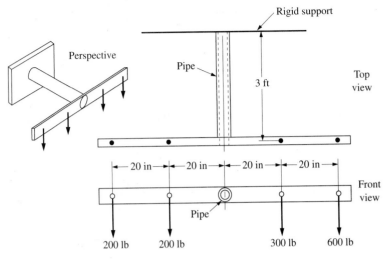

FIGURE 11–26 Bracket for Problem 11–28.

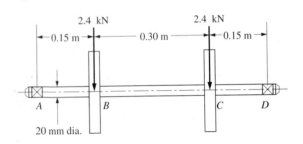

FIGURE 11–27 Shaft for Problem 11–29.

11–30.M A circular shaft carries the load shown in Figure 11–28. The shaft carries a torque of 4500 N·m between sections B and C. Compute the maximum shear stress in the shaft near section B.

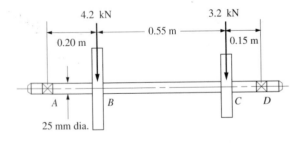

FIGURE 11–28 Shaft for Problem 11–30.

11–31.E A 1.0-in-diameter solid round shaft will carry 25 hp while rotating at 1150 rpm in the arrangement shown in Figure 11–29. The total bending loads at the gears A and C are shown, along with

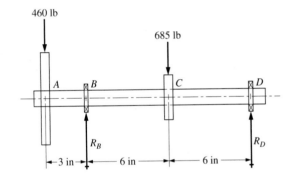

FIGURE 11–29 Shaft for Problem 11–31.

the reactions at the bearings B and D. Use a design factor of 6 for the maximum shear stress theory of failure, and determine a suitable steel for the shaft.

11–32.E Three pulleys are mounted on a rotating shaft, as shown in Figure 11–30. Belt tensions are as shown in the end view. All power is received by the shaft through pulley C from below. Pulleys A and B deliver power to mating pulleys above.

(a) Compute the torque at all points in the shaft.

(b) Compute the bending stress and the torsional shear stress in the shaft at the point where the maximum bending moment occurs if the shaft diameter is 1.75 in.

(c) Compute the maximum shear stress at the point used in step (b). Then specify a suitable material for the shaft.

Chapter 11 ▪ Special Cases of Combined Stresses

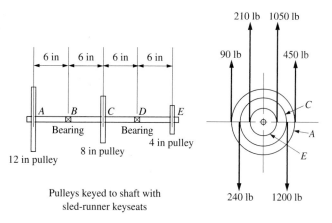

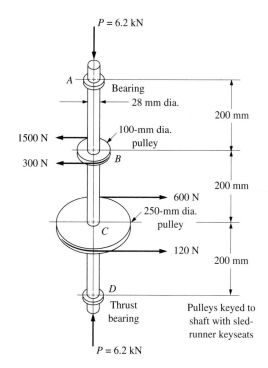

FIGURE 11–30   Shaft for Problem 11–32.

11–33.M The vertical shaft shown in Figure 11–31 carries two belt pulleys. The tensile forces in the belts under operating conditions are shown. Also, the shaft carries an axial compressive load of 6.2 kN. Considering torsion, bending, and axial compressive stresses, compute the maximum shear stress using Equation (11–2).

FIGURE 11–31   Shaft for Problem 11–33.

11–34.M For the shaft in Problem 11–33, specify a suitable steel that would provide a design factor of 4 based on yield strength in shear.

Combined Axial Tension and Direct Shear Stresses

11–35.E A machine screw has Number 8-32 UNC American Standard threads (see Appendix A–3). The screw is subjected to an axial tensile force that produces a direct tensile stress in the threads of 15 000 psi based on the tensile stress area. There is a section under the head without threads that has a diameter equal to the major diameter of the threads. This section is also subjected to a direct shearing force of 120 lb. Compute the maximum shear stress in this section.

11–36.E Repeat Problem 11–35 except the screw threads are $\frac{1}{4}$-20 UNC American Standard and the shearing force is 775 lb.

11–37.E Repeat Problem 11–35 except the screw threads are No. 4-48 UNF American Standard and the shearing force is 50 lb.

11–38.E Repeat Problem 11–35 except the screw threads are $1\frac{1}{4}$-12 UNF and the shearing force is 2500 lb.

11–39.M A machine screw has metric threads with a major diameter of 16 mm and a pitch of 2.0 mm (see Appendix A–3). The screw is subjected to an axial force that produces a direct tensile stress in the threads of 120 MPa based on the tensile stress area. There is a section under the head without threads that has a diameter equal to the major diameter of the threads. This section also is subjected to a direct shearing force of 8.0 kN. Compute the maximum shear stress in this section.

11–40.M A machine screw has metric threads with a major diameter of 48 mm and a pitch of 5.0 mm (see Appendix A–3). The screw is subjected to an axial tensile force that produces a direct tensile stress in the threads of 120 MPa based on the tensile stress area. There is a section under the head without threads that has a diameter equal to the major diameter of the threads. This section also is subjected to a direct shearing force of 80 kN. Compute the maximum shear stress in this section.

Combined Bending and Vertical Shear Stresses

11–41.E A rectangular bar is used as a beam carrying a concentrated load of 5500 lb at the middle of its 60-in length. The cross section is 2.00 in wide and 6.00 in high with the 6.00-in dimension oriented vertically. Compute the maximum shear stress that occurs in the bar near the load at the following points within the section:
(a) At the bottom of the bar.
(b) At the top of the bar.
(c) At the neutral axis.
(d) At a point 1.0 in above the bottom of the bar.
(e) At a point 2.0 in above the bottom of the bar.

11–42.E Repeat Problem 11–41 except the beam is an aluminum I-beam, I6 × 4.692.

11–43.E Repeat Problem 11–41 except the load is a uniformly distributed load of 100 lb/in over the entire length. Consider cross sections near the middle of the beam, near the supports, and 15 in from the left support.

Noncircular Sections—Combined Normal and Torsional Shear Stresses

11–44.M A square bar is 25 mm on a side and carries an axial tensile load of 75 kN along with a torque of 245 N·m. Compute the maximum shear stress in the bar. (*Note:* Refer to Section 5–11 and Figure 5–18.)

11–45.M A rectangular bar is 30 mm by 50 mm in cross section and is subjected to an axial tensile force of 175 kN along with a torque of 525 N·m. Compute the maximum shear stress in the bar. (*Note:* Refer to Section 5–11 and Figure 5–18.)

11–46.M A bar has a cross section in the form of an equilateral triangle, 50 mm on a side. It carries an axial tensile force of 115 kN along with a torque of 775 N·m. Compute the maximum shear stress in the bar. (*Note:* Refer to Section 5–11 and Figure 5–18.)

11–47.E A link in a large mechanism is made from a square structural tube 3 × 3 × $\frac{1}{4}$ (see Appendix A–9). It is originally designed to carry an axial tensile load that produces a design factor of 3 based on the yield strength of ASTM A500 cold-formed structural steel, shaped, grade C.
(a) Determine this load and the maximum shear stress that results in the tube under this load.
(b) In operation, a torque of 950 lb·ft is experienced by the tube in addition to the axial load. Compute the maximum shear stress under this combined loading and compute the resulting design factor based on the yield strength of the steel in shear. (See Section 5–11 and Figure 5–18.)

12

Deflection of Beams

12–1 OBJECTIVES OF THIS CHAPTER

The proper performance of machine parts, the structural rigidity of buildings, vehicles, and machine frames, and the tendency for a part to vibrate are all dependent on deformation in beams. Therefore, the ability to analyze beams for deflections under load is very important.

In this chapter we present the principles on which the computation of the deflection of beams is based, along with four popular methods of deflection analysis: the *formula method,* the *superposition method,* the *successive integration method,* and the *moment-area method.*

Each method has its advantages and disadvantages, and your choice of which method to use depends on the nature of the problem. The formula method is the simplest, but it depends on the availability of a suitable formula to match the application. The superposition method, a modest extension of the formula method, dramatically expands the number of practical problems that can be handled without a significant increase in the complexity of your work. The moment-area method is fairly quick and simple, but it is typically used for computing the deflections of only one or a few points on the beam. Its use requires a high level of understanding of the principle of moments and the techniques of preparing bending moment diagrams. The successive integration method is perhaps the most general, and it can be used to solve virtually any combination of loading and support conditions for statically determinate beams. But its use requires the ability to write the equations for the shearing force and bending moment diagrams and to derive equations for the slope and deflection of the beam using integral calculus. The successive integration method results in equations for the slope and deflection for the entire beam and enables the direct determination of the point of maximum deflection. Published formulas were developed using the successive integration or the moment-area method.

Several computer-assisted beam analysis programs are available to reduce the time and computation required to determine the deflection of beams. Although they can relieve the designer of much work, it is recommended that the principles on which they are based be understood before using them.

After completing this chapter, you should be able to:

1. Understand the need for considering beam deflections.

2. Understand the development of the relationships between the manner of loading and support for a beam and the deflection of the beam.

3. Graphically show the relationships among the load, shearing force, bending moment, slope, and deflection curves for beams.

4. Use standard formulas to compute the deflection of beams at selected points.

5. Use the principle of superposition along with standard formulas to solve problems of greater complexity.

6. Develop formulas for the deflection of beams for certain cases using the successive integration method.

7. Apply the method of successive integration to beams having a variety of loading and support conditions.

8. Use the moment-area method to solve for the slope and deflection for beams.

9. Write computer programs to assist in using the several methods of beam analysis described in this chapter.

The organization of the chapter allows selective coverage. In general, all the information necessary to use each method is included within that part of the chapter. An exception is that the understanding of the formula method is necessary before using the superposition method.

12–2 THE NEED FOR CONSIDERING BEAM DEFLECTIONS

The spindle of a lathe or drill press and the arbor of a milling machine carry cutting tools for machining metals. Deflection of the spindle or arbor would have an adverse effect on the accuracy that the machine could produce. The manner of loading and support of these machine elements indicate that they are beams, and the approach to computing their deflection will be discussed in this chapter.

Precision measuring equipment must also be designed to be very rigid. Deflection caused by the application of measuring forces reduces the precision of the desired measurement.

Power-transmission shafts carrying gears must have sufficient rigidity to ensure that the gear teeth mesh properly. Excessive deflection of the shafts would tend to separate the mating gears, resulting in a movement away from the most desirable point of contact between the gear teeth. Noise generation, decreased power-transmitting capability, and increased wear would result. For straight spur gears, it is recommended that the movement between two gears not exceed 0.005 in (0.13 mm). This limit is the *sum* of the movement of the two shafts carrying the mating gears at the location of the gears.

The floors of buildings must have sufficient rigidity to carry expected loads. Occupants of the building should not notice floor deflections. Machines and other equipment require a stable floor support for proper operation. Beams carrying plastered ceilings must not deflect excessively so as not to crack the plaster. A limit of $\frac{1}{360}$ times the span of the beam carrying a ceiling is often used for deflection.

Frames of vehicles, metal-forming machines, automation devices, and process equipment must also possess sufficient rigidity to ensure satisfactory operation of the equipment carried by the frame. The bead of a lathe, the crown of a punch press, the structure of an automatic assembly device, and the frame of a truck are examples.

Vibration is caused by the forced oscillations of parts of a structure or machine. The tendency to vibrate at a certain frequency and the severity of the vibrations are functions of the flexibility of the parts. Of course, flexibility is a term used to describe how much a part deflects under load. Vibration problems can be solved by either *increasing or decreasing* the stiffness of the part, depending on the circumstances. In either case, an understanding of how to compute deflections of beams is important.

Recommended deflection limits.   It is the designer's responsibility to specify the maximum allowable deflection for a beam in a machine, frame, or structure. Knowledge of the application should give guidance. In the absence of such guidance, the following limits are suggested in References 2 and 3:

General machine part: $y_{max} = 0.0005$ to 0.003 in/in or mm/mm of beam length.

Moderate precision: $y_{max} = 0.00001$ to 0.0005 in/in or mm/mm of beam length.

High precision: $y_{max} = 0.000001$ to 0.00001 in/in or mm/mm of beam length.

12–3   DEFINITION OF TERMS

To describe graphically the condition of a beam carrying a pattern of loading, five diagrams are used, as shown in Figure 12–1. We have used the first three diagrams in earlier work. The *load diagram* is the free-body diagram on which all external loads and support reactions are shown. From that, the *shearing force diagram* was developed which enables the calculation of shearing stresses in the beam at any section. The *bending moment diagram* is a plot of the variation of bending moment with position on the beam with the results used for computing the stress due to bending. The horizontal axis of these plots is the position on the beam, called *x*. It is typical to measure *x* relative to the left end of the beam, but any reference point can be used.

Deflection diagram.   The last two diagrams are related to the deformation of the beam under the influence of the loads. It is convenient to begin discussion with the last

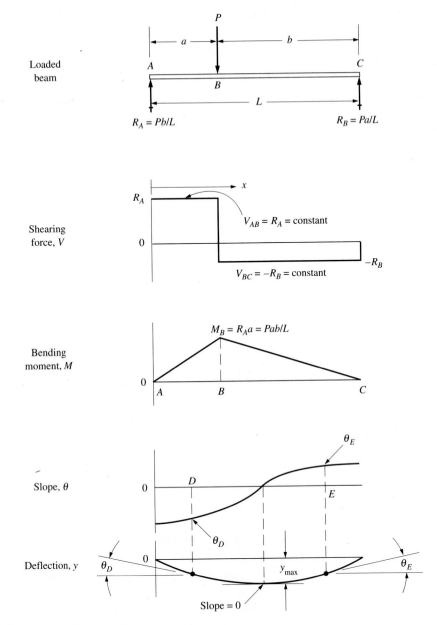

FIGURE 12–1 Five beam diagrams.

diagram, the *deflection diagram,* because this shows the shape of the deflected beam. Actually, it is the plot of the position of the neutral axis of the beam relative to its initial position. The initial position is taken to be the straight line between the two support points. The amount of deflection will be called y, with positive values measured upward. Typical beams carrying downward loads, such as that shown in Figure 12–1, will result in downward (negative) deflections of the beam.

Chapter 12 ■ Deflection of Beams

Slope diagram. A line drawn tangent to the deflection curve at a point of interest would define the slope of the deflection curve at that point. The slope is indicated as the angle, θ, measured in radians, relative to the horizontal, as shown in Figure 12–1. The plot of the slope as a function of position on the beam is the *slope curve,* drawn below the bending moment curve and above the deflection curve. Note on the given beam that the slope of the left portion of the deflection curve is negative and the right portion has a positive slope. The point where the tangent line is itself horizontal is the point of zero slope and defines the location of the maximum deflection. This observation will be used in the discussion of the moment-area method and the successive integration method later in this chapter.

Radius of curvature. Figure 12–2 shows the radius of curvature, R, at a particular point. For practical beams, the curvature is very slight, resulting in a very large value for R. For convenience, the shape of the deflection curve is exaggerated to aid in visualizing the principles and the variables involved in the analysis. Remember from analytic geometry that the radius of curvature at a point is perpendicular to the line drawn tangent to the curve at that point.

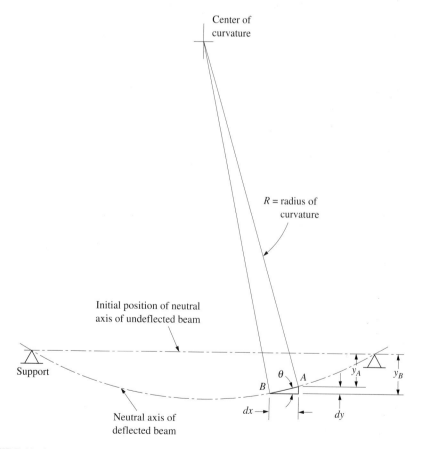

FIGURE 12–2 Illustration of radius of curvature and slope for a beam deflection curve.

The relationship between slope and deflection is also illustrated in Figure 12–2. Over a small distance dx, the deflection changes by a small amount dy. A small portion of the deflection curve itself completes the right triangle from which we can define

$$\tan \theta = \frac{dy}{dx} \qquad (12\text{–}1)$$

The absolute value of θ will be very small because the curvature of the beam is very slight. We can then take advantage of the observation for small angles, $\tan \theta = \theta$. Then

$$\theta = \frac{dy}{dx} \qquad (12\text{–}2)$$

It can then be concluded that:

The slope of the deflection curve at a point is equal to the ratio of the change in the deflection to the change in position on the beam.

Beam stiffness. It will be shown later that the amount of deflection for a beam is inversely proportional to the *beam stiffness,* indicated by the product *EI*, where

E = modulus of elasticity of the material of the beam

I = moment of inertia of the cross section of the beam with respect to the neutral axis

12–4 BEAM DEFLECTIONS USING THE FORMULA METHOD

For many practical configurations of beam loading and support, formulas have been derived that allow the computation of the deflection at any point on the beam. The method of successive integration or the moment-area method, described later, may be used to develop the equations. Appendixes A–22, A–23, and A–24 include many examples of formulas for beam deflections.

The deflection formulas are valid only for the cases where the cross section of the beam is uniform for its entire length. Example problems will demonstrate the application of the formulas.

Appendix A–22 includes ten different conditions of loading on beams that are simply supported, that is, beams having two and only two simple supports. Some are overhanging beams. We have shown earlier that such beams can be analyzed for the values of the reactions using the standard equations of equilibrium. Then the shearing force and bending moment diagrams can be developed using the methods from Chapter 6 from which the stress analysis of the beam can be completed, as discussed in Chapters 8 and 9. It must be understood that both the stress analysis and the deflection analysis of beams should be completed to assess the acceptability of a beam design.

The loading conditions in Appendix A–22 include single concentrated loads, two concentrated loads, a variety of distributed loads, and one case with a concentrated moment. The concentrated moment could be developed in the manner of the examples shown in Section 6–8. The thin phantom line in the diagrams is a sketch of the shape of the deflected beam, somewhat exaggerated. This can help you to visualize where critical points of deflection can be expected.

Note carefully the labeling for loads and dimensions in the diagrams of the beam deflection cases. It is essential that the real beam being analyzed match the general form of a given case and that you identify accurately the variables used in the formulas to the right of the diagrams. For most cases, formulas are given for the maximum deflection to be expected, deflections at the ends of overhangs, and deflections at points of applications of concentrated loads. Some cases include formulas for the deflection at any chosen point.

Take special notice of the general form of the deflection formulas. While some are more complex than others, the following general characteristics can be observed. Understanding these observations can help you make good decisions when designing beams.

1. Deflections are denoted by the variable y and are the change in the position of the neutral axis of the beam from its unloaded condition to the final, loaded condition, measured perpendicular to the original neutral axis.

2. Upward deflections are positive; downward deflections are negative.

3. The variable x, when used, denotes the horizontal position on the beam, measured from one of the supports. In some cases, a second position variable v is indicated, measured from the other support.

4. Deflections are proportional to the load applied to the beam.

5. Deflections are inversely proportional to the *stiffness* of the beam, defined as the product of E, the stiffness of the material from which the beam is made, and I, the moment of inertia of the cross section of the beam.

6. Deflections are proportional to the *cube* of some critical length dimension, typically the span between the supports or the length of an overhang.

Appendix A–23 includes four cases in which cantilever beams carry concentrated loads, distributed loads, or a concentrated moment. The maximum deflection obviously occurs at the free end of the beam. The fixed end constrains the beam against rotation at the support so that the deflection curve has a zero slope there.

Appendix A–24 includes ten cases of *statically indeterminate beams*. This term means that the reactions cannot be computed by the application of the standard equations of equilibrium. Therefore, formulas are given for the reactions and key bending moments along with the deflection formulas. The shapes of the shearing force and bending moment diagrams are also given and they are, in general, quite different from those of statically determinate beams. More is said about statically indeterminate beams in Chapter 13.

Example Problem 12–1	Determine the maximum deflection of a simply supported beam carrying a hydraulic cylinder in a machine used to press bushings into a casting, as shown in Figure 12–3. The force exerted during the pressing operation is 15 kN. The beam is rectangular, 25 mm thick and 100 mm high, and made of steel.

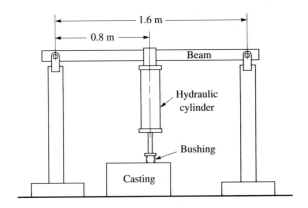

FIGURE 12–3 Beam for Example Problem 12–1.

Solution   **Objective**   Compute the maximum deflection of the given beam.

Given   System in Figure 12–3. Load = P = 15 kN. Span = L = 1.60 m.
Beam cross section: 25 mm wide by 100 mm high. Steel beam.

Analysis   The given beam can be considered to be a simply supported beam with a concentrated force applied in an upward direction at its center. Case *a* in Appendix A–22 applies.

Results   Using the formula from Appendix A–22–a, we find the maximum deflection to be

$$y = \frac{PL^3}{48EI}$$

From Appendix A–13, for steel, E = 207 GPa = 207 × 10^9 N/m^2. For the rectangular beam,

$$I = \frac{(25)(100)^3}{12} = 2.083 \times 10^6 \text{ mm}^4$$

Then

$$y = \frac{PL^3}{48EI} = \frac{(15 \times 10^3 \text{ N})(1.6 \text{ m})^3}{48(207 \times 10^9 \text{ N/m}^2)(2.083 \times 10^6 \text{ mm}^4)} \times \frac{(10^3 \text{ mm})^5}{\text{m}^5}$$

$$y = 2.97 \text{ mm}$$

Comment   This is a relatively high deflection that could adversely affect the accuracy of the bushing placement operation. A stiffer beam shape (one with a higher moment of inertia, I) should be considered. Alternatively, the support system could be modified to decrease the span between the supports, a desirable approach because the deflection is proportional to the cube of the length. Assuming that the overall operation of the system permits the redesign of the span to be one-half of the given span (0.80 m), the deflection would be only 0.37 mm, 1/8 as much as the given design.

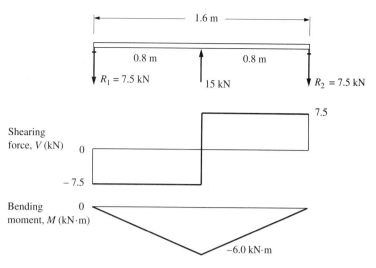

FIGURE 12–4 Beam diagrams for Example Problem 12–1.

The stress in the beam should also be computed to assess the safety of the design. Figure 12–4 shows the load, shearing force, and bending moment diagrams for the original beam design from which we find the maximum bending moment in the beam to be $M = 6.00$ kN·m. The flexure formula can be used to compute the stress.

$$\sigma = \frac{Mc}{I} = \frac{(6.00 \text{ kN·m}) (50 \text{ mm})}{2.083 \times 10^6 \text{ mm}^4} \cdot \frac{10^3 \text{ N}}{\text{kN}} \cdot \frac{10^3 \text{ mm}}{\text{m}} = 144 \text{ MPa}$$

This is a relatively high stress level. To continue the analysis, note that the beam would be subjected to repeated bending stress. Therefore, the recommended design stress is

$$\sigma_d = s_u/8$$

Letting $\sigma_d = \sigma$, we can solve for the required ultimate strength.

$$s_u = 8 \sigma = (8)(144 \text{ MPa}) = 1152 \text{ MPa}$$

Consulting Appendix A–13 for the properties of selected steels, we could specify AISI 4140 OQT 900 steel that has an ultimate strength of 1281 MPa. But this is a fairly expensive, heat-treated steel. A redesign of the type discussed for limiting deflection would reduce the stress and permit the use of a lower-priced steel.

Example Problem 12–2 A round shaft, 45 mm in diameter, carries a 3500-N load, as shown in Figure 12–5. If the shaft is steel, compute the deflection at the load and at the point C, 100 mm from the right end of the shaft. Also compute the maximum deflection.

Solution **Objective** Compute the deflection at points B and C and at the point of maximum deflection.

Given Beam in Figure 12–5. Load $= P = 3500$ N
Beam is round shaft; $D = 45$ mm. Steel beam.

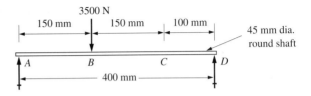

FIGURE 12–5 Shaft for Example Problem 12–2.

Analysis Appendix A–22–b applies because the beam is simply supported and has a single concentrated load place away from the middle of the beam.

Results Formula for deflection at the load, point B:

$$y_B = \frac{-Pa^2b^2}{3EIL}$$

Note that the dimension a is the longer segment between the load and one support; b is the shorter. Express all data in N and mm.

Data: $L = 400$ mm; $a = 250$ mm; $b = 150$ mm

$$I = \frac{\pi D^4}{64} = \frac{\pi (45 \text{ mm})^4}{64} = 0.201 \times 10^6 \text{ mm}^4$$

$$E = \frac{207 \times 10^9 \text{ N}}{\text{m}^2} \times \frac{1 \text{ m}^2}{(10^3 \text{ mm})^2} = 207 \times 10^3 \text{ N/mm}^2$$

Now

$$y_B = \frac{-Pa^2b^2}{3EIL} = \frac{-(3500)(150)^2(250)^2}{3(207 \times 10^3)(0.201 \times 10^6)(400)} = -0.0985 \text{ mm}$$

Formula for deflection at point C: Note that point C is in the longer segment. Data from above apply. Also, $x = 100$ mm from the right support to C.

$$y_c = \frac{-Pbx}{6EIL}(L^2 - b^2 - x^2)$$

$$y_c = \frac{-(3500)(150)(100)}{6(207 \times 10^3)(0.201 \times 10^6)(400)}(400^2 - 150^2 - 100^2)$$

$$y_c = -0.0670 \text{ mm}$$

Maximum deflection: Appendix A–22–b shows that the maximum deflection occurs in the longer segment of the beam at a distance x_1 from the support, where,

$$x_1 = \sqrt{a(L + b)/3} = \sqrt{(250)(400 + 150)/3} = 214 \text{ mm}$$

Then the deflection at that point is

$$y_{max} = \frac{-Pab(L + b)\sqrt{3a(L + b)}}{27(EIL)}$$

$$y_{max} = \frac{-(3500)(250)(150)(400 + 150)\sqrt{3(250)(400 + 150)}}{(27)(207 \times 10^3)(0.201 \times 10^6)(400)}$$

$$y_{max} = 0.103 \text{ mm}$$

Comment The results are shown in Figure 12–6.

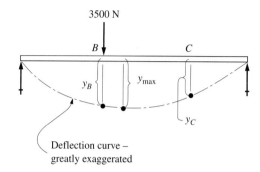

3500 N

B

C

y_B

y_{max}

y_C

Deflection curve – greatly exaggerated

FIGURE 12–6 Deflection curve for beam in Example Problem 12–2.

12–5 SUPERPOSITION USING DEFLECTION FORMULAS

Formulas such as those used in the preceding section are available for a large number of cases of loading and support conditions. Obviously, these cases would allow the solution of many practical beam deflection problems. An even larger number of situations can be handled by the use of the *principle of superposition*.

If a particular loading and support pattern can be broken into components such that each component is like one of the cases for which a formula is available, then the total deflection at a point on the beam is equal to the sum of the deflections caused by each component. The deflection due to one component load is *superposed* on deflections due to the other loads, thus giving the name *superposition*.

An example of where superposition can be applied is shown in Figure 12–7. Represented in the figure is a roof beam carrying a uniformly distributed roof load of

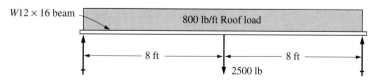

$W12 \times 16$ beam

800 lb/ft Roof load

8 ft

8 ft

2500 lb

FIGURE 12–7 Roof beam.

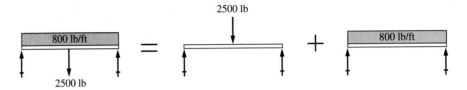

FIGURE 12-8 Illustration of the superposition principle.

800 lb/ft and also supporting a portion of a piece of process equipment which provides a concentrated load at the middle. Figure 12–8 shows how the loads are considered separately. Each component load produces a maximum deflection at the middle. Therefore, the maximum total deflection will occur there also. Let the subscript 1 refer to the concentrated load case and the subscript 2 refer to the distributed load case. Then

$$y_1 = \frac{-PL^3}{48EI}$$

$$y_2 = \frac{-5}{384} \frac{WL^3}{EI}$$

The total deflection will be

$$y_T = y_1 + y_2$$

The terms L, E, and I will be the same for both cases.

$$L = 16 \text{ ft} \times 12 \text{ in/ft} = 192 \text{ in}$$
$$E = 30 \times 10^6 \text{ psi} \quad \text{for steel}$$
$$I = 103 \text{ in}^4 \quad \text{for W12} \times 16 \text{ beam}$$

To compute y_1, let $P = 2500$ lb.

$$y_1 = \frac{-2500(192)^3}{48(30 \times 10^6)(103)} \text{ in} = -0.119 \text{ in}$$

To compute y_2, W is the total resultant of the distributed load.

$$W = (800 \text{ lb/ft})(16 \text{ ft}) = 12\,800 \text{ lb}$$

Then

$$y_2 = \frac{-5(12\,800)(192)^3}{384(30 \times 10^6)(103)} \text{ in} = -0.382 \text{ in}$$

and

$$y_T = y_1 + y_2 = -0.119 \text{ in} - 0.382 \text{ in} = -0.501 \text{ in}$$

Since this is the total deflection, we could check to see if it meets the recommendation that the maximum deflection should be less than $\frac{1}{360}$ times the span of the beam. The span is L.

$$\frac{L}{360} = \frac{192 \text{ in}}{360} = 0.533 \text{ in}$$

The computed deflection was 0.501 in, which is satisfactory.

The superposition principle is valid for any place on the beam, not just at the loads. The following example problem illustrates this.

Example Problem 12–3

Figure 12–9 shows a shaft carrying two gears and simply supported at its ends by bearings. The mating gears above exert downward forces which tend to separate the gears. Other forces acting horizontally are not considered in this analysis. Determine the total deflection at the gears B and C if the shaft is steel and has a diameter of 45 mm.

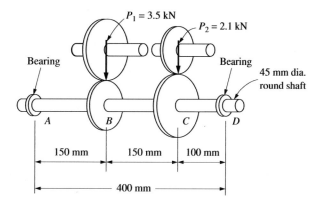

FIGURE 12–9   Shaft for Example Problem 12–3.

Solution   **Objective**   Compute the deflection at points B and C.

Given   Shaft $ABCD$ in Figure 12–9. Shaft is steel. $D = 45$ mm.
Load at $B = P_1 = 3.5$ kN $= 3500$ N. Load at $C = P_2 = 2.1$ kN $= 2100$ N.

Analysis   The two unsymmetrically placed, concentrated loads constitute a situation for which none of the given cases for beam deflection formulas is valid. However, it can be solved by using the formulas of Appendix A–22–b twice. Considering each load separately, the deflections at B and C can be found for each. The total, then, would be the sum of the component deflections. Figure 12–10 shows the logic from which we can say

$$y_B = y_{B1} + y_{B2}$$
$$y_C = y_{C1} + y_{C2}$$

where y_B = total deflection at B
y_C = total deflection at C

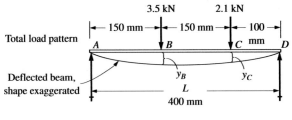

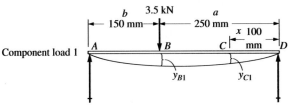

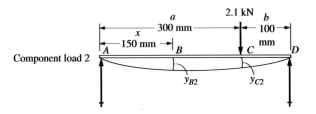

FIGURE 12–10 Superposition logic for deflection of shaft in Figure 12–9.

y_{B1} = deflection at B due to 3.5-kN load alone
y_{C1} = deflection at C due to 3.5-kN load alone
y_{B2} = deflection at B due to 2.1-kN load alone
y_{C2} = deflection at C due to 2.1-kN load alone

For all the calculations, the values of E, I, and L will be needed. These are the same as those in Example Problem 12–2.

$$E = 207 \times 10^3 \text{ N/mm}^2$$
$$I = 0.201 \times 10^6 \text{ mm}^4$$
$$L = 400 \text{ mm}$$

The product of EIL is present in all the formulas.

$$EIL = (207 \times 10^3)(0.201 \times 10^6)(400) = 16.64 \times 10^{12}$$

Now the individual component deflections will be computed. Note the values for the variables a, b, and x are different for each component load. See the data labeled in Figure 12–10.

Results   For component 1:

$$y_{B1} = \frac{-P_1 a^2 b^2}{3EIL} = \frac{-(3.5 \times 10^3)(250)^2(150)^2}{3(16.64 \times 10^{12})} = -0.0985 \text{ mm}$$

$$y_{C1} = \frac{-P_1 bx}{6EIL}(L^2 - b^2 - x^2)$$

$$y_{C1} = \frac{-(3.5 \times 10^3)(150)(100)}{6(16.64 \times 10^{12})}(400^2 - 150^2 - 100^2) = -0.0670 \text{ mm}$$

For component 2, the load will be 2.1 kN at point C. Then

$$y_{B2} = \frac{-P_2 bx}{6EIL}(L^2 - b^2 - x^2)$$

$$y_{B2} = \frac{-(2.1 \times 10^3)(100)(150)}{6(16.64 \times 10^{12})}(400^2 - 100^2 - 150^2) = -0.0402 \text{ mm}$$

$$y_{C2} = \frac{-P_2 a^2 b^2}{3EIL} = \frac{-(2.1 \times 10^3)(300)^2(100)^2}{3(16.64 \times 10^{12})} = -0.0378 \text{ mm}$$

Now, by superposition,

$$y_B = y_{B1} + y_{B2} = -0.0985 \text{ mm} - 0.0402 \text{ mm} = -0.1387 \text{ mm}$$
$$y_C = y_{C1} + y_{C2} = -0.0670 \text{ mm} - 0.0378 \text{ mm} = -0.1048 \text{ mm}$$

Comment   In Section 12–2 it was observed that a recommended limit for the movement of one gear relative to its mating gear is 0.13 mm. Thus this shaft is too flexible since the deflection at B exceeds 0.13 mm without even considering the deflection of the mating shaft.

12–6   BASIC PRINCIPLES FOR BEAM DEFLECTION BY SUCCESSIVE INTEGRATION METHOD

In this section we show the mathematical relationships among the moment, slope, and deflection curves from which you can solve for the actual equations for a given beam with a given loading and support condition.

Figure 12–11 shows a small segment of a beam in its initial straight shape and its deflected shape. The sides of the segment remain straight as the beam deflects, but they rotate with respect to a point at the neutral axis. This results in compression in the top portion of the segment and tension in the bottom portion, a fact used in the development of the flexure formula in Chapter 8.

The rotated sides of the segment intersect at the center of curvature and form the small angle $d\theta$. Note also the radius of curvature, R, measured from the center of curvature to the neutral axis. From the geometry shown in the figure,

$$\Delta s = R(d\theta) \tag{12–3}$$

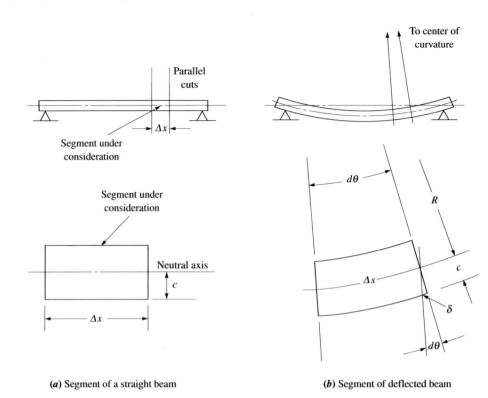

(a) Segment of a straight beam   **(b)** Segment of deflected beam

FIGURE 12–11   Relation between radius R and deformation δ.

and

$$\delta = c\,(d\theta) \qquad\qquad (12\text{--}4)$$

where Δs is the length of the segment at the neutral axis and δ is the elongation of the bottom line of the segment that occurs as the beam deflects. The term c has the same meaning as in the flexure formula, the distance from the neutral axis to the outer fiber of the section.

Recall that the definition of the neutral axis states that no strain occurs there. Then the length Δs in the deflected beam segment equals the length Δx in the undeflected segment, and Equation (12–3) can be written

$$\Delta x = R\,(d\theta) \qquad\qquad (12\text{--}5)$$

Now both Equations (12–4) and (12–5) can be solved for $d\theta$.

$$d\theta = \frac{\delta}{c}$$

$$d\theta = \frac{\Delta x}{R}$$

Equating these values of $d\theta$ to each other gives

$$\frac{\delta}{c} = \frac{\Delta x}{R}$$

Another form of this equation is

$$\frac{c}{R} = \frac{\delta}{\Delta x}$$

The right side of this equation conforms to the definition of strain, ϵ. Then

$$\epsilon = \frac{c}{R} \qquad (12-6)$$

Earlier it was shown that

$$\epsilon = \frac{\sigma}{E}$$

where σ, the stress due to bending, can be computed from the flexure formula,

$$\sigma = \frac{Mc}{I}$$

Then

$$\epsilon = \frac{\sigma}{E} = \frac{Mc}{EI}$$

Combining this with Equation (12-6) gives

$$\frac{c}{R} = \frac{Mc}{EI}$$

Dividing both sides by c gives

$$\frac{1}{R} = \frac{M}{EI} \qquad (12-7)$$

Equation (12-7) is useful in developing the moment-area method for finding beam deflections. See Section 12-8.

In analytic geometry, the reciprocal of the radius of curvature, $1/R$, is defined as the *curvature* and denoted by the κ, the lowercase Greek letter kappa. Then

$$\kappa = \frac{M}{EI} \qquad (12-8)$$

Equation (12–8) indicates that the curvature gets greater as the bending moment increases, which stands to reason. Similarly, the curvature decreases as the beam stiffness, EI, increases.

Another principle of analytic geometry states that if the equation of a curve is expressed as $y = f(x)$, that is, y is a function of x, then the curvature is

$$\kappa = \frac{d^2y}{dx^2} \tag{12–9}$$

Combining Equations (12–8) and (12–9) gives

$$\frac{M}{EI} = \frac{d^2y}{dx^2} \tag{12–10}$$

or

$$M = EI\frac{d^2y}{dx^2} \tag{12–11}$$

Equations (12–10) and (12–11) are useful in developing the successive integration method for finding beam deflections, described next.

12–7 BEAM DEFLECTIONS—SUCCESSIVE INTEGRATION METHOD—GENERAL APPROACH

A general approach will now be presented which allows the determination of deflection at any point on the beam. The advantages of this approach are listed below.

1. The result is a set of equations for deflection at all parts of the beam. Deflection at any point can then be found by substitution of the beam stiffness properties of E and I, and the position of the beam.

2. Data are easily obtained from which a plot of the shape of the deflection curve may be made.

3. The equations for the *slope* of the beam at any point are generated, as are deflections. This is important in some machinery applications such as shafts at bearings and shafts carrying gears. An excessive slope of the shaft would result in poor performance and reduced life of the bearings or gears.

4. The fundamental relationships among loads, manner of support, beam stiffness properties, slope, and deflections are emphasized in the solution procedure. The designer who understands these relationships can make more efficient designs.

5. The method requires the application of only simple mathematical concepts.

6. The point of maximum deflection can be found directly from the resulting equations.

The basis for the successive integration method has been developed in Sections 12–3 and 12–6. The five beam diagrams will be prepared, in a manner similar to that shown in Figure 12–1, to relate the loads, shearing forces, bending moments, slopes, and deflections over the entire length of the beam.

The load, shearing force, and bending moment diagrams can be drawn using the principles from Chapter 6. Then equations for the bending moment are derived for all segments of the bending moment diagram.

Equation (12–11) is then used to develop the slope and deflection equations from the moment equations by integrating twice with respect to the position on the beam, x, as follows.

$$M = EI\frac{d^2y}{dx^2} \tag{12–11}$$

Now, integrating once with respect to x gives

$$\int M\,dx = EI \int \frac{d^2x}{dx^2}dx = EI\frac{dy}{dx} \tag{12–12}$$

Earlier, in Section 12–3, Equation (12–2), we showed that $dy/dx = \theta$, the slope of the deflection curve. Then,

$$\int M\,dx = EI\theta = \theta EI \tag{12–13}$$

Equation (12–12) can be integrated again, giving

$$\int EI\theta\,dx = EI \int \frac{dy}{dx}dx = EIy = yEI \tag{12–14}$$

After the final values for $EI\theta$ and EIy have been determined, they will be divided by the beam stiffness, EI, to obtain the values for slope, θ, and deflection, y.

The steps indicated by Equations (12–11) through (12–14) are to be completed for each segment of the beam for which the moment diagram is continuous. Also, because our objective is to obtain discrete equations for slope and deflection for particular beam-loading patterns, we will need to evaluate a constant of integration for each integration performed.

The development of the equations for the bending moment versus position is often accomplished by integrating the equations for the shearing force versus x, as shown in Chapter 6. This follows from the rule that the change in bending moment between two points on a beam is equal to the area under the shearing force curve between the same two points.

The step-by-step method used to find the deflection of beams using the general approach is as follows.

Steps in the successive integration method for beam deflections

1. Determine the reactions at the supports for the beam.
2. Draw the shearing force and bending moment diagrams using the same procedures presented in Chapter 6, and identify the magnitudes at critical points.
3. Divide the beam into segments in which the shearing force diagram is continuous by naming points at the places where abrupt changes occur with the letters A, B, C, D, etc.
4. Write equations for the shearing force curve in each segment. In most cases, these will be equations of straight lines, that is, equations involving x to the first power. Sometimes, as for beams carrying concentrated loads, the equation will be simply of the form

$$V = \text{constant}$$

5. For each segment, perform the process,

$$M = \int V dx + C$$

To evaluate the constant of integration that ties the moment equation to the particular values already known for the moment diagram, insert known boundary conditions and solve for C.
6. For each segment, perform the process,

$$\theta EI = \int M dx + C$$

The constant of integration generated here cannot be evaluated directly right away. So each constant should be identified separately by a subscript such as C_1, C_2, C_3, etc. Then when they are evaluated (in step 9), they can be put in their proper places.
7. For each segment, perform the process,

$$yEI = \int \theta EI dx + C$$

Here again, the constants should be labeled with subscripts.
8. Establish *boundary conditions* for the slope and deflection diagrams. The same number of boundary conditions must be identified as there are unknown constants from steps 6 and 7. Boundary conditions express mathematically the special values of slope and deflection at cer-

tain points and the fact that both the slope curve and the deflection curve are continuous. Typical boundary conditions are:
   a. The deflection of the beam at each support is zero.
   b. The deflection of the beam at the end of one segment is equal to the deflection of the beam at the beginning of the next segment. This follows from the fact that the deflection curve is continuous; that is, it has no abrupt changes.
   c. The slope of the beam at the end of one segment is equal to the slope at the beginning of the next segment. The slope has no abrupt changes.
   d. For the special case of a cantilever, the slope of the beam at the support is also zero.
9. Combine all the boundary conditions to evaluate all the constants of integration. This typically involves the solution of a set of simultaneous equations in which the number of equations is equal to the number of unknown constants of integration. Equation-solving software or calculators are very helpful for this step.
10. Substitute the constants of integration back into the slope and deflection equations, thus completing them. The value of the slope or deflection at any point can then be evaluated by simply placing the proper value of the position on the beam in the equation. Points of maximum deflection in any segment can also be found.

The method will now be illustrated with an example problem.

Example Problem 12–4

Figure 12–12 shows a beam used as a part of a special structure for a machine. The 20 K (20 000-lb) load at *A* and the 30 K (30 000-lb) load at *C* represent places where heavy equipment is supported. Between the two supports at *B* and *D*, the uniformly distributed load of 2 K/ft (2000 lb/ft) is due to stored bulk materials which are in a bin supported by the beam. All loads are static. To maintain accuracy of the product produced by the machine, the maximum allowable deflection of the beam is 0.05 in. Specify an acceptable wide-flange steel beam, and also check the stress in the beam.

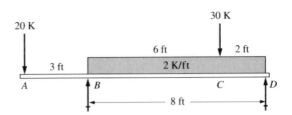

FIGURE 12–12 Beam for Example Problem 12–4.

Solution   **Objective**   Specify a steel wide-flange beam shape to limit the deflection to 0.05 in. Check the stress in the selected beam to ensure safety.

Given   Beam loading in Figure 12–12.

Analysis The beam will be analyzed to determine where the maximum deflection will occur. Then the required moment of inertia will be determined to limit the deflection to 0.05 in. A wide-flange beam which has the required moment of inertia will then be selected. The ten-step procedure discussed earlier will be used. The solution is shown in a programmed format. You should work through the problem yourself before looking at the given solution.

Results   Steps 1 and 2 call for drawing the shearing force and bending moment diagrams. Do this now before checking the result below.

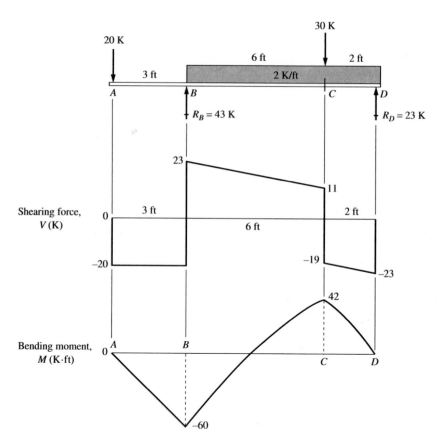

FIGURE 12–13 Load, shearing force, and bending moment diagrams for Example Problem 12–4.

Figure 12–13 shows the results. Now do step 3.

Three segments are required, *AB*, *BC*, and *CD*. These are the segments over which the shearing force diagram is continuous. Now do step 4 to get the shearing force curve equations.

The results are:

$$V_{AB} = -20 \tag{a}$$
$$V_{BC} = -2x + 29 \tag{b}$$
$$V_{CD} = -2x - 1 \tag{c}$$

In the segments BC and CD, the shearing force curve is a straight line with a slope of -2 kip/ft, the same as the load. Any method of writing the equation of a straight line can be used to derive these equations.

Now do step 5 of the procedure.

You should have the following for the moment equations. First,

$$M_{AB} = \int V_{AB}\,dx + C = \int -20\,dx + C = -20x + C$$

At $x = 0$, $M_{AB} = 0$. Therefore, $C = 0$ and

$$M_{AB} = -20x \tag{d}$$

Next.

$$M_{BC} = \int V_{BC}\,dx + C = \int (-2x + 29)\,dx + C = -x^2 + 29x + C$$

At $x = 3$, $M_{BC} = -60$. Therefore, $C = -138$ and

$$M_{BC} = -x^2 + 29x - 138 \tag{e}$$

Finally,

$$M_{CD} = \int V_{CD}\,dx + C = \int (-2x - 1)\,dx + C = -x^2 - x + C$$

At $x = 9$, $M_{CD} = 42$. Therefore, $C = 132$ and

$$M_{CD} = -x^2 - x + 132 \tag{f}$$

Now do step 6 to get equations for θEI.

By integrating the moment equations,

$$\theta_{AB} EI = \int M_{AB}\,dx + C = \int (-20x)\,dx + C$$
$$\theta_{AB} EI = -10x^2 + C_1 \tag{g}$$
$$\theta_{BC} EI = \int M_{BC}\,dx + C = \int (-x^2 + 29x - 138)\,dx + C$$
$$\theta_{BC} EI = x^3/3 + 14.5x^2 - 138x + C_2 \tag{h}$$

$$\theta_{CD} EI = \int M_{CD} \, dx + C = \int (-x^2 - x + 132) \, dx + C$$

$$\theta_{CD} EI = -x^3/3 - x^2/2 + 132x + C_3 \qquad\qquad \text{(i)}$$

Now in step 7, integrate Equations (g), (h), and (i) to get the yEI equations.

You should have

$$y_{AB} EI = \int \theta_{AB} EI \, dx + C$$

$$y_{AB} EI = -10x^3/3 + C_1 x + C_4 \qquad\qquad \text{(j)}$$

$$y_{BC} EI = \int \theta_{BC} EI \, dx + C$$

$$y_{BC} EI = -x^4/12 + 14.5x^3/3 - 69x^2 + C_2 x + C_5 \qquad\qquad \text{(k)}$$

$$y_{CD} EI = \int \theta_{CD} EI \, dx + C$$

$$y_{CD} EI = -x^4/12 - x^3/6 + 66x^2 + C_3 x + C_6 \qquad\qquad \text{(l)}$$

Step 8 calls for identifying boundary conditions. Six are required since there are six unknown constants of integration in Equations (g) through (l). Write them now.

Considering zero deflection points and the continuity of the slope and deflection curves, we can say

1. At $x = 3$, $y_{AB} EI = 0$ ⎫
2. At $x = 3$, $y_{BC} EI = 0$ ⎬ (zero deflection at supports).
3. At $x = 11$, $y_{CD} EI = 0$ ⎭
4. At $x = 9$, $y_{BC} EI = y_{CD} EI$ (continuous deflection curve at C).
5. At $x = 3$, $\theta_{AB} EI = \theta_{BC} EI$ ⎫
6. At $x = 9$, $\theta_{BC} EI = \theta_{CD} EI$ ⎬ (continuous slope curve at B and C).

We can now substitute the values of x above into the proper equations and solve for C_1 through C_6. First make the substitutions and reduce the resulting equations to the form involving the constants.

For the six conditions listed above, the following equations result:

1. $3C_1 + C_4$ $= 90$
2. $3C_2 + C_5$ $= 497.25$
3. $11C_3 + C_6$ $= -6544.08\overline{33} = -78\,529/12$
4. $9C_2 - 9C_3 + C_5 - C_6 = 7290$
5. $C_1 - C_2$ $= -202.5$
6. $C_2 - C_3$ $= 1215$

The value of the quantity on the right side of Equation 3 is expressed in excessively high precision. This is not often necessary but is being done for this example to eliminate the accumulation of round-off errors as the problem solution proceeds. There are many steps to the final solution, and inaccuracies at this stage can result in significant variation in the results that might frustrate you as you follow the solution. Note that writing the constant in Equation 3 as $-6544.08\overline{33}$ indicates that the 3's repeat to infinity. Thus this is an inherently inaccurate representation of the number. Entering the number as the exact fraction ($-78\,529/12$) into an equation solver would eliminate the error. Here is where the use of a computer-based equation solver such as MATHCAD, TK Solver, MATLAB, or MAPLE facilitates the laborious calculations involved in the balance of the procedure. Many high-level calculators having graphing capability also contain simultaneous equation solvers.

Now solve the six equations simultaneously for the values of C_1 through C_6.

The results are:

$$C_1 = 132.\overline{333} = 397/3 \qquad C_2 = 334.8\overline{33} = 4018/12$$
$$C_3 = -880.1\overline{66} = 5281/6 \qquad C_4 = -307 \text{ (exact)}$$
$$C_5 = -507.25 \text{ (exact)} \qquad C_6 = 3137.75 \text{ (exact)}$$

The final equations for θ and y can now be written by substituting the constants into Equations (g) through (l). The results are shown next.

$$\theta_{AB} EI = -10x^2 + 132.\overline{333}$$
$$\theta_{BC} EI = -x^3/3 + 14.5x^2 - 138x + 334.8\overline{33}$$
$$\theta_{CD} EI = -x^3/3 - x^2/2 + 132x - 880.1\overline{66}$$
$$y_{AB} EI = -10x^3/3 + 132.\overline{333}x - 307$$
$$y_{BC} EI = -x^4/12 + 14.5x^3/3 - 69x^2 + 334.8\overline{33}x - 507.25$$
$$y_{CD} EI = -x^4/12 - x^3/6 + 66x^2 - 880.1\overline{66}x + 3137.75$$

Having the completed equations, we can now determine the point of maximum deflection, which is the primary objective of the analysis. Based on the loading, the probable shape of the deflected beam would be like that shown in Figure 12–14. Therefore, the maximum deflection could occur at point A at the end of the overhang, at a point to the right of B (upward), or at a point near the load at C (downward). It is probable that there are two points of zero slope at the points E and F, as shown in Figure 12–14. We would need to know where the slope equation $\theta_{BC} EI$ equals zero to determine where the maximum deflections occur.

Notice that this is a third-degree equation. The use of a graphing calculator and an equation solver facilitates finding the points where $\theta_{BC} EI = 0$. Figure 12–15 shows an expanded plot of the BC segment of the beam showing that the zero points occur at $x = 3.836$ ft and at $x = 8.366$ ft.

We can now determine the values for yEI at points A, E, and F to find out which is larger.

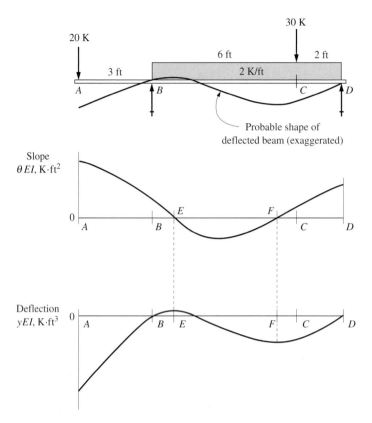

FIGURE 12–14 Slope and deflection curves for Example Problem 12–4.

Point A. At $x = 0$ in segment AB,

$$y_{AB}\,EI = -10x^3/3 + 132.\overline{333}x - 307$$
$$y_A\,EI = -10(0.00)^3/3 + 132.\overline{333}(0.00) - 307$$
$$y_A\,EI = -307\ \text{K·ft}^3$$

Point E. At $x = 3.836$ ft in segment BC,

$$y_{BC}\,EI = -x^4/12 + 14.5x^3/3 - 69x^2 + 334.8\overline{33}x - 507.25$$
$$y_E\,EI = -(3.836)^4/12 + 14.5(3.836)^3/3 - 69(3.836)^2$$
$$+\ 334.8\overline{33}(3.836) - 507.25$$
$$y_E\,EI = +16.62\ \text{K·ft}^3$$

Point F. At $x = 8.366$ ft in segment BC,

$$y_{BC}\,EI = -x^4/12 + 14.5x^3/3 - 69x^2 + 334.8\overline{33}x - 507.25$$
$$y_F\,EI = -(8.366)^4/12 + 14.5(8.366)^3/3 - 69(8.366)^2$$
$$+\ 334.8\overline{33}(8.366) - 507.25$$
$$y_F\,EI = +113.5\ \text{K·ft}^3$$

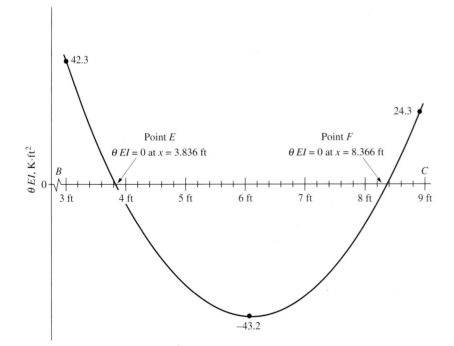

FIGURE 12–15 Graph showing points of zero slope.

The largest value occurs at point A, so that is the critical point. We must choose a beam that limits the deflection at A to 0.05 in or less.

$$y_A EI = -307 \text{ K·ft}^3$$

Let $y_A = -0.05$ in. Then the required I is

$$I = \frac{-307 \text{ K·ft}^3}{Ey_A} \times \frac{1000 \text{ lb}}{K} \times \frac{(12 \text{ in})^3}{\text{ft}^3}$$

$$= \frac{(-307)(1000)(1728) \text{ lb·in}^3}{(30 \times 10^6 \text{ lb/in}^2)(-0.05 \text{ in})} = 354 \text{ in}^4$$

Consult the table of wide-flange beams and select a suitable beam.

A W18 $\times$ 40 is the best choice from Appendix A–7 since it is the lightest beam that has a large enough value for I. For this beam, $I = 612 \text{ in}^4$, and the section modulus is $S = 68.4 \text{ in}^3$. Now compute the maximum bending stress in the beam.

In Figure 12–5, we find the maximum bending moment to be 60 K·ft. Then

$$\sigma = \frac{M}{S} = \frac{60 \text{ K·ft}}{68.4 \text{ in}^3} \times \frac{1000 \text{ lb}}{K} \times \frac{12 \text{ in}}{\text{ft}} = 10\,526 \text{ psi}$$

Since the allowable stress for structural steel under static loading is approximately 22 000 psi, the selected beam is safe.

12-8 BEAM DEFLECTIONS—MOMENT-AREA METHOD

The semigraphical procedure for finding beam deflections, called the *moment-area method,* is useful for problems in which a fairly complex loading pattern occurs or when the beam has a varying cross section along its length. Such cases are difficult to handle with the other methods presented in this chapter.

Shafts for mechanical drives are examples where the cross section varies throughout the length of the beam. Figure 12–16 shows a shaft designed to carry two gears where the changes in diameter provide shoulders against which to seat the gears and bearings to provide axial location. Note, also, that the bending moment decreases toward the ends of the shaft, allowing smaller sections to be safe with regard to bending stress.

In structural applications of beams, varying cross sections are often used to make more economical members. Larger sections having higher moments of inertia are used at sections where the bending moment is high while decreased section sizes are used in places where the bending moment is lower. Figure 12–17 shows an example.

The moment-area method uses the quantity M/EI, the bending moment divided by the stiffness of the beam, to determine the deflection of the beam at selected points. Then it is convenient to prepare such a diagram as a part of the beam analysis proce-

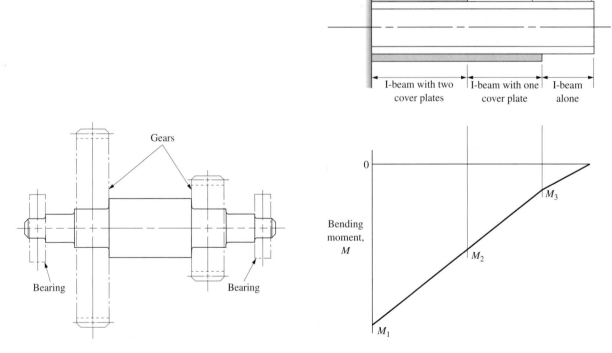

FIGURE 12–16 Shaft with varying cross sections.

FIGURE 12–17 Cantilever with varying cross sections.

dure. If the beam has the same cross section over its entire length, the M/EI diagram looks similar to the familiar bending moment diagram except that its values have been divided by the quantity EI. However, if the moment of inertia of the cross section varies along the length of the beam, the shape of the M/EI diagram will be different. This is shown in Figure 12–18.

Recall Equation (12–10) from Section 12–6,

$$\frac{M}{EI} = \frac{d^2y}{dx^2} \qquad (12\text{–}10)$$

This formula relates the deflection of the beam, y, as a function of position, x, the bending moment, M, and the beam stiffness, EI. The right side of Equation (12–10) can be rewritten as

$$\frac{d^2y}{dx^2} = \frac{d}{dx}\left(\frac{dy}{dx}\right)$$

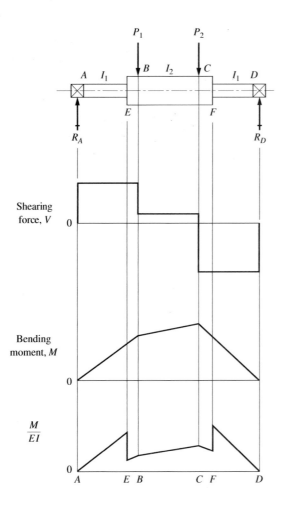

FIGURE 12–18 Illustration of M/EI diagrams for a beam with varying cross sections.

But note that dy/dx is defined as the slope of the deflection curve, θ; that is, $dy/dx = \theta$. Then

$$\frac{d^2y}{dx^2} = \frac{d\theta}{dx}$$

Equation (12–10) can then be written

$$\frac{M}{EI} = \frac{d\theta}{dx}$$

Solving for $d\theta$ gives

$$d\theta = \frac{M}{EI}\,dx \qquad\qquad (12\text{–}15)$$

The interpretation of Equation (12–15) can be seen in Figure 12–19 in which the right side, $(M/EI)\,dx$, is the area under the M/EI diagram over a small length dx. Then $d\theta$ is the change in the angle of the slope over the same distance dx. If tangent lines are drawn to the deflection curve of the beam at the two points marking the beginning and the end of the segment dx, the angle between them is $d\theta$.

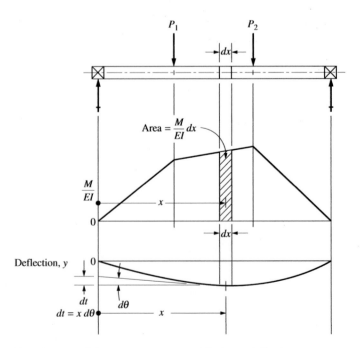

FIGURE 12–19   Principles of the moment-area method for beam deflections.

The change in angle $d\theta$ causes a change in the vertical position of a point at some distance x from the small element dx, as shown in Figure 12–19. Calling the change in vertical position dt, it can be found from

$$dt = x\,d\theta = \frac{M}{EI}x\,dx \tag{12–16}$$

To determine the effect of the change in angle over a larger segment of the beam, Equations (12–15) and (12–16) must be integrated over the length of the segment. For example, over the segment A–B shown in Figure 12–20 from Equation (12–15),

$$\int_A^B d\theta = \theta_B - \theta_A = \int_A^B \frac{M}{EI}\,dx \tag{12–17}$$

The last part of this equation is the area under the M/EI curve between A and B. This is equal to the change in the angle of the tangents at A and B, $\theta_B - \theta_A$.

From Equation (12–16),

$$\int_A^B dt = t_{AB} = \int_A^B \frac{M}{EI}x\,dx \tag{12–18}$$

Here the term t_{AB} represents the tangential deviation of the point A from the tangent to point B, as shown in Figure 12–20. Also, the right side of Equation (12–18) is the *moment of the area of the M/EI diagram between points A and B*. In practice, the moment of the area is computed by multiplying the area under the M/EI curve by the distance to the centroid of the area, also shown in Figure 12–20.

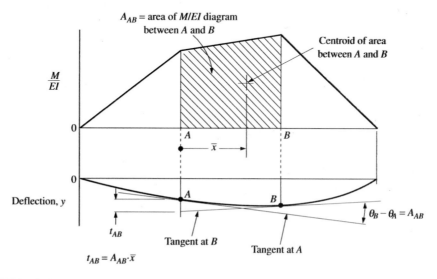

FIGURE 12–20 Illustrations of the two theorems of the moment-area method for beam deflections.

Equations (12–17) and (12–18) form the basis for the two *theorems of the moment-area method for finding beam deflections.* They are:

Theorem 1

> The change in angle, in radians, between tangents drawn at two points A and B on the deflection curve for a beam, is equal to the area under the M/EI diagram between A and B.

Theorem 2

> The vertical deviation of point A on the deflection curve for a beam from the tangent through another point B on the curve is equal to the moment of the area under the M/EI curve with respect to point A.

12–9 APPLICATIONS OF THE MOMENT-AREA METHOD

In this section we show several examples of the use of the moment-area method for finding the deflection of beams. Procedures are developed for each class of beam according to the manner of loading and support. Considered are:

1. Cantilevers with a variety of loads
2. Symmetrically loaded simply supported beams
3. Beams with varying cross section
4. Unsymmetrically loaded simply supported beams

Cantilevers. The definition of a cantilever includes the requirement that it is rigidly fixed to a support structure so that no rotation of the beam can occur at the support. Therefore, the tangent to the deflection curve at the support is always in line with the original position of the neutral axis of the beam in the unloaded state. If the beam is horizontal, as we usually picture it, this tangent is also horizontal.

The procedure for finding the deflection of any point on a cantilever, listed below, uses the two theorems developed in Section 12–8 along with the observation that the tangent to the deflection curve at the support is horizontal.

Procedure for finding the deflection of a cantilever— moment-area method

> 1. Draw the load, shearing force, and bending moment diagrams.
> 2. Divide the bending moment values by the beam stiffness, EI, and draw the M/EI diagram. The dimension for the quantity M/EI is $(\text{length})^{-1}$; for example, m^{-1}, ft^{-1}, or in^{-1}.
> 3. Compute the area of the M/EI diagram and locate its centroid. If the diagram is not a simple shape, break it into parts and find the area and centroid for each part separately. If the deflection at the end of the cantilever is desired, the area of the entire M/EI diagram is used. If the deflection of some other point is desired, only the area between the support and the point of interest is used.

> **4.** Use Theorem 2 to compute the vertical deviation of the point of interest from the tangent to the neutral axis of the beam at the support. Because the tangent is coincident with the original position of the neutral axis, the deviation thus found is the actual deflection of the beam at the point of interest. If all loads are in the same direction, the maximum deflection occurs at the end of the cantilever.

Example Problem 12–5 Use the moment-area method to determine the deflection at the end of the steel cantilever shown in Figure 12–21.

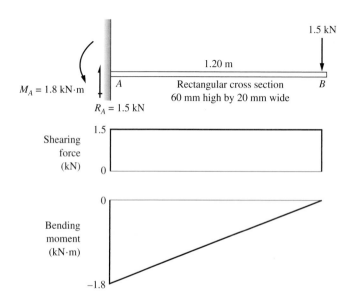

FIGURE 12–21 Load, shearing force, and bending moment diagrams for Example Problems 12–5 and 12–6.

Solution **Objective** Compute the deflection at the end of the cantilever.

Given Beam and loading in Figure 12–21.

Analysis Use the *Procedure for finding the deflection of a cantilever—moment-area method.*

Results *Step 1.* Figure 12–21 shows the load, shearing force, and bending moment diagrams.

Step 2. The stiffness is computed here:

$$E = 207 \text{ GPa} = 207 \times 10^9 \text{ N/m}^2$$

$$I = \frac{th^3}{12} = \frac{(0.02 \text{ m})(0.06 \text{ m})^3}{12} = 3.60 \times 10^{-7} \text{ m}^4$$

$$EI = (207 \times 10^9 \text{ N/m}^2)(3.60 \times 10^{-7} \text{ m}^4) = 7.45 \times 10^4 \text{ N·m}^2$$

The *M/EI* diagram is drawn in Figure 12–22. Note that the only changes from the bending moment diagram are the units and the values because the beam has a constant stiffness along its length.

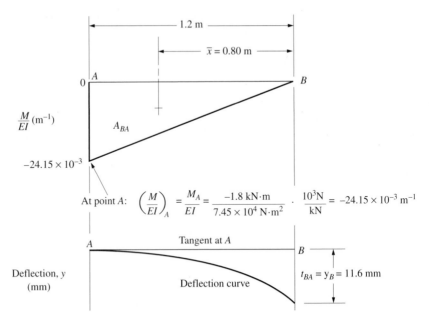

$$\left(\frac{M}{EI}\right)_A = \frac{M_A}{EI} = \frac{-1.8 \text{ kN} \cdot \text{m}}{7.45 \times 10^4 \text{ N} \cdot \text{m}^2} \cdot \frac{10^3 \text{N}}{\text{kN}} = -24.15 \times 10^{-3} \text{ m}^{-1}$$

FIGURE 12–22 *M/EI* curve and deflection curve for Example Problem 12–5.

Step 3. The desired area is that of the entire triangular shape of the *M/EI* diagram, called A_{BA} here to indicate that it is used to compute the deflection of point *B* relative to *A*.

$$A_{BA} = (0.5)(24.15 \times 10^{-3} \text{ m}^{-1})(1.20 \text{ m}) = 14.5 \times 10^{-3} \text{ rad}$$

The centroid of this area is two-thirds of the distance from *B* to *A*, 0.80 m.

Step 4. To implement Theorem 2, we need to compute the moment of the area found in step 3. This is equal to t_{BA}, the vertical deviation of point *B* from the tangent drawn to the deflection curve at point *A*.

$$t_{BA} = A_{BA} \times \bar{x} = (14.5 \times 10^{-3} \text{ rad})(0.80 \text{ m})$$
$$t_{BA} = y_B = 11.6 \times 10^{-3} \text{ m} = 11.6 \text{ mm}$$

Because the tangent to point *A* is horizontal, t_{BA} is equal to the deflection of the beam at its end, point *B*.

Comment This result is identical to that which would be found by applying the formula for Case *a* in Appendix A–23. The value of the moment-area method is much more evident when multiple loads are involved or if the cantilever has a varying cross section along its length.

Example Problem 12–6	For the same beam used in Example Problem 12–5 and shown in Figure 12–21, compute the deflection at a point 1.0 m from the support.

Solution

Objective Compute the deflection 1.0 m from the left end of the cantilever.

Given Beam and loading in Figure 12–21.

Analysis Use the *Procedure for finding the deflection of a cantilever—moment-area method*. Steps 1 and 2 from Example Problem 12–5 are identical, resulting in the load, shearing force, bending moment, and M/EI diagrams shown in Figures 12–21 and 12–22. The solution procedure is continued at step 3.

Results *Step 3.* This step changes because only the area of the M/EI diagram between the support and the point 1.0 m out on the beam is used, as shown in Figure 12–23. Calling the point of interest, point C, we

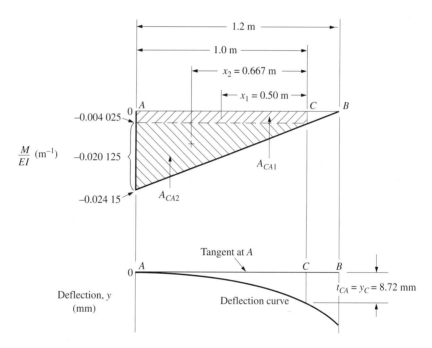

FIGURE 12–23 M/EI curve and deflection curve for Example Problem 12–6.

need to compute t_{CA}; that is, the vertical deviation of point C relative to the tangent drawn to point A at the support. The required area is a trapezoid and it is convenient to break it into a triangle and a rectangle and treat them as simple shapes. The calculation then takes the form

$$t_{CA} = A_{CA} \times \overline{x} = A_{CA1} \times x_1 + A_{CA2} \times x_2$$

where the areas and x-distances are shown in Figure 12–23. Note that the distances are measured *from the point C to the centroid of the component area.* Then

$$A_{CA1} \times x_1 = (0.004\,025 \text{ m}^{-1})(1.0 \text{ m})(0.50 \text{ m}) = 2.01 \times 10^{-3} \text{ m}$$
$$A_{CA2} \times x_2 = (0.5)(0.020\,125 \text{ m}^{-1})(1.0 \text{ m})(0.667 \text{ m}) = 6.71 \times 10^{-3} \text{ m}$$
$$t_{CA} = y_C = (2.01 + 6.71)(10^{-3}) \text{ m} = 8.72 \text{ mm}$$

As before, because the tangent to point A is horizontal, the vertical deviation, t_{CA}, is the true deflection of point C.

Symmetrically loaded simply supported beams. This class of problems has the advantage that it is known that the maximum deflection occurs at the middle of the span of the beam. An example is shown in Figure 12–24, in which the beam carries two identical loads equally spaced from the supports. Of course, any loading for which the point of maximum deflection can be predicted can be solved by the procedure illustrated here.

Procedure for finding the deflection of a symmetrically loaded simply supported beam— moment-area method

1. Draw the load, shearing force, and bending moment diagrams.
2. Divide the bending moment values by the beam stiffness, EI, and draw the M/EI diagram.
3. If the maximum deflection, at the middle of the span, is desired, use that part of the M/EI diagram between the middle and one of the supports; that is, half of the diagram.
4. Use Theorem 2 to compute the vertical deviation of the point at one of the supports from the tangent to the neutral axis of the beam at its middle. Because the tangent is horizontal and because the deflection at the support is actually zero, the deviation thus found is the actual deflection of the beam at its middle.
5. To determine the deflection at some other point on the same beam, use the area of the M/EI diagram between the middle and the point of interest. Use Theorem 2 to compute the vertical deviation of the point of interest from the point of maximum deflection at the middle of the beam. Then subtract this deviation from the maximum deflection found in step 4.

Example Problem 12–7

Determine the maximum deflection of the beam shown in Figure 12–24. The beam is an aluminum channel, C6 × 2.834, made from 6061-T6 aluminum, positioned with the legs downward.

Solution   **Objective**   Compute the maximum deflection of the beam.

Given   Beam and loading in Figure 12–24.

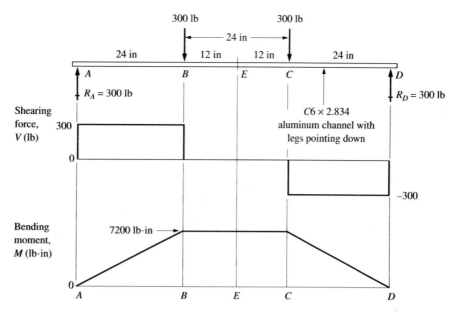

FIGURE 12–24 Load, shearing force, and bending moment diagrams for Example Problems 12–7 and 12–8.

Analysis   Use the *Procedure for finding the deflection of a symmetrically loaded simply supported beam—moment-area method,* steps 1–4. Because the loading pattern is symmetrical, the maximum deflection will occur at the middle of the beam.

Results   **Step 1.** The load, shearing force, and bending moment diagrams are shown in Figure 12–24, prepared in the traditional manner. The maximum bending moment is 7200 lb·in between *B* and *C*.

        Step 2. The stiffness of the beam, *EI*, is found using data from the Appendixes. From Appendix A–17, *E* for 6061-T6 aluminum is 10×10^6 psi. From Appendix A–10, the moment of inertia of the channel, with respect to the *Y-Y* axis, is 1.53 in⁴. Then

$$EI = (10 \times 10^6 \ \text{lb/in}^2)(1.53 \ \text{in}^4) = 1.53 \times 10^7 \ \text{lb·in}^2$$

        Because the stiffness of the beam is uniform over its entire length, the *M/EI* diagram has the same shape as the bending moment diagram, but the values are different, as shown in Figure 12–25. The maximum value of *M/EI* is $4.71 \times 10^{-4} \ \text{in}^{-1}$.

        Step 3. To determine the deflection at the middle of the beam, one-half of the *M/EI* diagram is used. For convenience, this is separated into a rectangle and a triangle and the centroid of each part is shown.

        Step 4. We need to find t_{AE}, the vertical deviation of point *A* from the tangent drawn to the deflection curve at point *E*, the middle of the beam. By Theorem 2,

$$t_{AE} = A_{AE1} \times x_{A1} + A_{AE2} \times x_{A2}$$

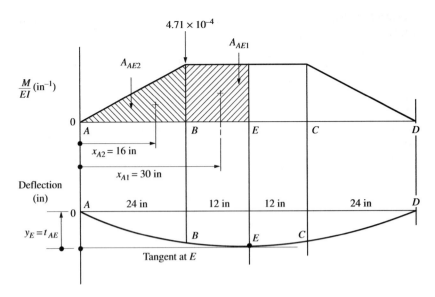

$$\frac{M}{EI} \text{ (in}^{-1})$$

4.71×10^{-4}

A_{AE1}

A_{AE2}

0

A   B   E   C   D

$x_{A2} = 16$ in

$x_{A1} = 30$ in

Deflection (in)

0

A   24 in   12 in   12 in   24 in   D

$y_E = t_{AE}$

B   E   C

Tangent at E

FIGURE 12–25 *M/EI* diagram and deflection curve for Example Problem 12–7.

The symbols x_{A1} and x_{A2} indicate that the distances to the centroids of the areas must be measured *from point A*.

$$A_{AE1} \times x_{A1} = (4.71 \times 10^{-4} \text{ in}^{-1})(12 \text{ in})(30 \text{ in}) = 0.170 \text{ in}$$
$$A_{AE2} \times x_{A2} = (0.5)(4.71 \times 10^{-4} \text{ in}^{-1})(24 \text{ in})(16 \text{ in}) = 0.090 \text{ in}$$
$$t_{AE} = y_E = 0.170 + 0.090 = 0.260 \text{ in}$$

This is the vertical deviation of point *A* from the tangent to point *E*. Because the tangent is horizontal and because the actual deflection of point *A* is zero, this represents the true deflection of point *E* relative to the original position of the neutral axis of the beam.

Comment   This result is identical to that which would be found by applying the formula for Case *c* in Appendix A–22.

Example Problem 12–8   For the same beam used for Example Problem 12–7, shown in Figure 12–24, determine the deflection at point *B* under one of the loads.

Solution   **Objective**   Compute the deflection at point *B* under one of the loads.

Given   Beam and loading in Figure 12–24.

Analysis   Use the *Procedure for finding the deflection of a symmetrically loaded simply supported beam—moment-area method,* steps 1–5. Steps 1–4 from Example Problem 12–7 are identical, resulting in the load, shearing force, bending moment, *M/EI*, and deflection diagrams shown in Figures 12–24 and 12–25. The solution procedure is continued at step 5.

Chapter 12 ▪ Deflection of Beams

Results We can use the moment-area method to determine the vertical deviation, t_{BE}, of point B from the tangent to point E at the middle of the beam. Then, subtracting that from the value of t_{AE} found in Example Problem 12–7 gives the true deflection of point B. Shown in Figure 12–26 are the data needed to compute t_{BE}.

$$t_{BE} = A_{BE1} \times x_{B1} = (4.71 \times 10^{-4} \text{ in}^{-1})(12 \text{ in})(6 \text{ in}) = 0.034 \text{ in}$$

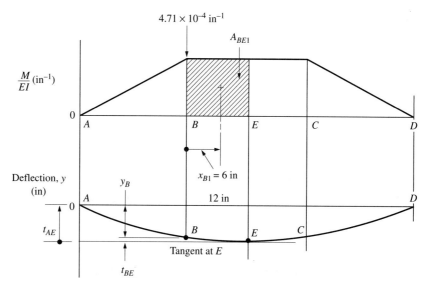

FIGURE 12–26 *M/EI diagram and deflection curve for Example Problem 12–8.*

Note that the distance x_{B1} must be measured from point B. Then the deflection of point B is

$$y_B = t_{AE} - t_{BE} = 0.260 - 0.034 = 0.226 \text{ in}$$

Beams with varying cross section. One of the major uses of the moment-area method is to find the deflection of a beam having a varying cross section along its length. Only one additional step is required as compared with beams having a uniform cross section, like those considered thus far.

An example of such a beam is illustrated in Figure 12–27. Note that it is a modification of the beam used in Example Problems 12–7 and 12–8 and shown in Figure 12–24. Here we have added a rectangular plate, 0.25 in by 6.0 in, to the underside of the original channel over the middle 48 in of the length of the beam. The box shape would provide a significant increase in stiffness, thereby reducing the deflection of the beam. The stress in the beam would also be reduced.

The change in the procedure for analyzing the deflection of the beam is in the preparation of the *M/EI* diagram. Figure 12–28 shows the load, shearing force, and bending moment diagrams as before. In the first and last 12 in of the *M/EI* diagram,

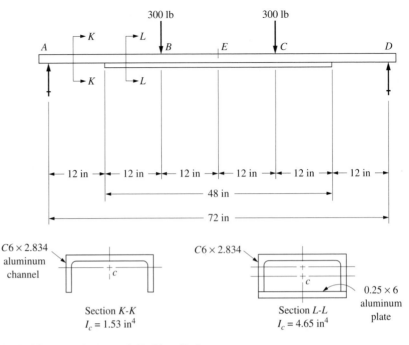

FIGURE 12–27 Beam for Example Problem 12–9.

the stiffness of the plain channel is used as before. For each segment over the middle 48 in, the stiffness of the box shape must be used. The M/EI diagram then includes the effect of the change in stiffness along the length of the beam.

Example Problem 12–9

Determine the deflection at the middle of the reinforced beam shown in Figure 12–27.

Solution

Step 1. The load, shearing force, and bending moment diagrams are prepared in the traditional manner, as shown in Figure 12–28.

Step 2. To prepare the M/EI diagram, two values for the stiffness, EI, are needed. The plain channel has the same value used in previous problems, 1.53×10^7 lb·in^2. For the box shape,

$$EI = (10 \times 10^6 \text{ lb/in}^2)(4.65 \text{ in}^4) = 4.65 \times 10^7 \text{ lb·in}^2$$

Then, at the point on the beam just before 12 in from A, where the bending moment is 3600 lb·in

$$\frac{M}{EI} = \frac{3600 \text{ lb·in}}{1.53 \times 10^7 \text{ lb·in}^2} = 2.35 \times 10^{-4} \text{ in}^{-1}$$

Just beyond 12 in from A,

$$\frac{M}{EI} = \frac{3600 \text{ lb·in}}{4.65 \times 10^7 \text{ lb·in}^2} = 7.74 \times 10^{-5} \text{ in}^{-1}$$

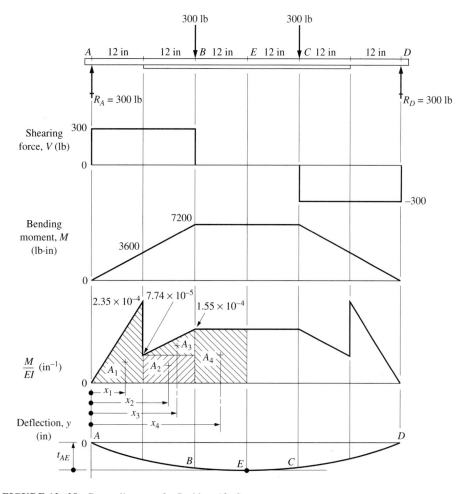

FIGURE 12–28 Beam diagrams for Problem 12–9.

At point *B*, where *M* = 7200 lb·in and *EI* = 4.65 × 10⁷ lb·in²,

$$\frac{M}{EI} = \frac{7200 \text{ lb·in}}{4.65 \times 10^7 \text{ lb·in}^2} = 1.55 \times 10^{-4} \text{ in}^{-1}$$

These values establish the critical points on the *M/EI* diagram.

Step 3. The moment area for the left half of the *M/EI* diagram will be used to determine the value of t_{AE}, as done in Example Problem 12–7. For convenience, the total area is divided into four parts, as shown in Figure 12–26, with the locations of the centroids indicated relative to point *A*. The distances are

$$x_1 = \left(\tfrac{2}{3}\right)(12 \text{ in}) = 8 \text{ in}$$
$$x_2 = \left(\tfrac{1}{2}\right)(12 \text{ in}) + 12 \text{ in} = 18 \text{ in}$$
$$x_3 = \left(\tfrac{2}{3}\right)(12 \text{ in}) + 12 \text{ in} = 20 \text{ in}$$
$$x_4 = \left(\tfrac{1}{2}\right)(12 \text{ in}) + 24 \text{ in} = 30 \text{ in}$$

Step 4. We can now use Theorem 2 to compute the value of t_{AE}, the deviation of point A from the tangent to point E, by computing the moment of each of the four areas shown crosshatched in the M/EI diagram of Figure 12–28.

$$t_{AE} = A_1 x_1 + A_2 x_2 + A_3 x_3 + A_4 x_4$$

$$A_1 x_1 = (0.5)(2.35 \times 10^{-4} \text{ in}^{-1})(12 \text{ in})(8 \text{ in}) = 1.128 \times 10^{-2} \text{ in}$$

$$A_2 x_2 = (7.74 \times 10^{-5} \text{ in}^{-1})(12 \text{ in})(18 \text{ in}) = 1.672 \times 10^{-2} \text{ in}$$

$$A_3 x_3 = (0.5)(7.74 \times 10^{-5} \text{ in}^{-1})(12 \text{ in})(20 \text{ in}) = 9.293 \times 10^{-3} \text{ in}$$

$$A_4 x_4 = (1.55 \times 10^{-4} \text{ in}^{-1})(12 \text{ in})(30 \text{ in}) = 5.580 \times 10^{-2} \text{ in}$$

Then

$$t_{AE} = y_E = \sum (A_i x_i) = 9.309 \times 10^{-2} \text{ in} = 0.093 \text{ in}$$

Comment   As before, this value is equal to the deflection of point E at the middle of the beam. Comparing it to the deflection of 0.260 in found in Example Problem 12–8, the addition of the cover plate reduced the maximum deflection by approximately 64%.

Unsymmetrically loaded simply supported beams.   The major difference between this type of beam and the ones considered earlier is that the point of maximum deflection is unknown. Special care is needed to describe the geometry of the M/EI diagram and the deflection curve for the beam.

The general procedure for finding the deflection at any point on the deflection curve for an unsymmetrically loaded simply supported beam is outlined below. Because of the myriad of different loading patterns, the specifics of applying this procedure may have to be adjusted for any given problem. You are advised to check the fundamental principles of the moment-area method as the solution of a problem is completed. The method will be demonstrated in an example problem.

Procedure for finding the deflection of an unsymmetrically loaded simply supported beam— moment-area method

1. Draw the load, shearing force, and bending moment diagrams.
2. Construct the M/EI diagram by dividing the bending moment at any point by the value of the beam stiffness, EI, at that point.
3. Sketch the probable shape of the deflection curve. Then draw the tangent to the deflection curve at one of the supports. Using Theorem 2, compute the vertical deviation of the other support from this tangent line. The moment of the entire M/EI diagram with respect to the second support is required.
4. Using proportions, compute the distance from the zero-axis to the tangent line from step 3 at the point for which the deflection is desired.
5. Using Theorem 2, compute the vertical deviation of the point of interest from the tangent line from step 3. The moment for that portion of the M/EI diagram between the first support and the point of interest will be used.
6. Subtract the deviation computed in step 5 from that found in step 4. The result is the deflection of the beam at the desired point.

Determine the deflection of the beam shown in Figure 12–29 at its middle, 1.0 m from the supports. The beam is an American Standard steel beam, S3 × 5.7.

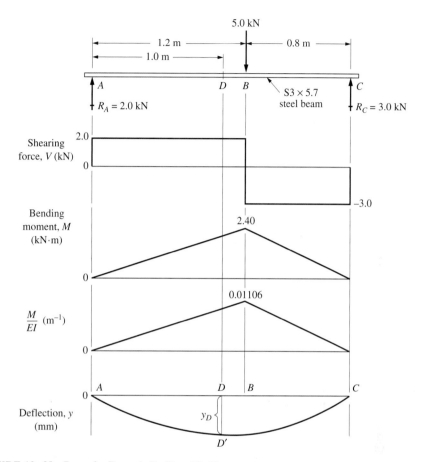

FIGURE 12–29 Beam for Example Problem 12–10.

Solution

Objective Compute the deflection at the middle of the beam.

Given Beam and loading shown in Figure 12–29. Beam is steel.
Beam shape is an American Standard S3 × 5.7.

Anaysis Use the *Procedure for finding the deflection of an unsymmetrically loaded simply supported beam—moment-area method.*

Results *Step 1.* The load, shearing force, and bending moment diagrams are shown in Figure 12–29.

Step 2. The beam has a uniform stiffness along its entire length resulting in the shape of the M/EI diagram being the same as the bending moment diagram. The value of M/EI at point B can be computed by dividing the bending moment there (2.40 kN·m or 2400 N·m) by EI.

We will use $E = 207$ GPa for steel. From the Appendix we find $I = 2.52$ in⁴, and this must be converted to metric units.

$$I = \frac{(2.52 \text{ in}^4)(0.0254 \text{ m})^4}{1.0 \text{ in}^4} = 1.049 \times 10^{-6} \text{ m}^4$$

Then the beam stiffness is

$$EI = (207 \times 10^9 \text{ N/m}^2)(1.049 \times 10^{-6} \text{ m}^4) = 2.17 \times 10^5 \text{ N·m}^2$$

The value of M/EI at point B on the beam can now be computed.

$$\left(\frac{M}{EI}\right)_B = \frac{2400 \text{ N·m}}{2.17 \times 10^5 \text{ N·m}^2} = 0.01106 \text{ m}^{-1}$$

The M/EI diagram is drawn in Figure 12–29. It is desired to compute the deflection of the beam at its middle, labeled point D.

Step 3. Figure 12–29 shows an exaggerated sketch of the deflection curve for the beam. It is likely that the maximum deflection will occur very close to the middle of the beam where we are to determine the deflection, point D. Figure 12–30 shows the tangent to the deflection curve at point A at the left support and the vertical deviation of point C from this line. Note that point C is a known point on the deflection curve because the deflection there is zero. Now we can use Theorem 2 to compute t_{CA}. The entire M/EI diagram is used, broken into two triangles.

$$t_{CA} = A_{CA1} x_{C1} + A_{CA2} x_{C2}$$
$$A_{CA1} x_{C1} = (0.5)(0.01106 \text{ m}^{-1})(0.8 \text{ m})(0.533 \text{ m}) = 0.002359 \text{ m}$$
$$A_{CA2} x_{C2} = (0.5)(0.01106 \text{ m}^{-1})(1.2 \text{ m})(1.2 \text{ m}) = 0.007963 \text{ m}$$

Then

$$t_{CA} = 0.002359 + 0.007963 = 0.010322 \text{ m} = 10.322 \text{ mm}$$

Step 4. Use the principle of proportion to determine the distance DD'' from D to the tangent line.

$$\frac{t_{CA}}{CA} = \frac{DD''}{AD}$$

or

$$DD'' = t_{CA} \times \frac{AD}{CA} = (10.322 \text{ mm}) \times \frac{1.0 \text{ m}}{2.0 \text{ m}} = 5.161 \text{ mm}$$

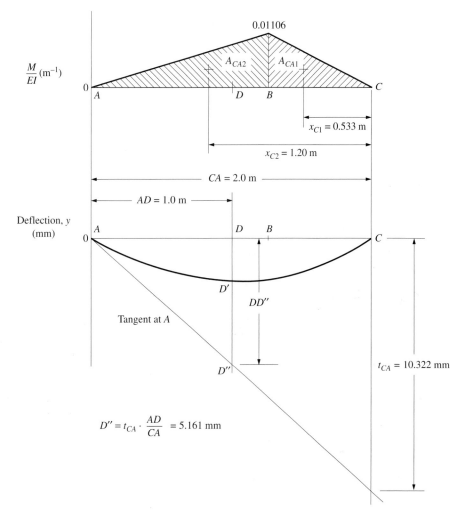

FIGURE 12–30 Moment-area diagrams for Example Problem 12–10.

Step 5. Compute the deviation, $t_{D'A}$, of point D' from the tangent line drawn to point A using Theorem 2. The part of the M/EI diagram between D and A is used as shown in Figure 12–31.

$$t_{D'A} = A_{D'A}x_{D1} = (0.5)(0.00922 \text{ m}^{-1})(1.0 \text{ m})(0.333 \text{ m}) = 0.001536 \text{ m}$$
$$t_{D'A} = 1.536 \text{ mm}$$

Step 6. From the geometry of the deflection diagram shown in Figure 12–31 the deflection at point D, y_D, is

$$y_D = DD' = DD'' - t_{D'A} = 5.161 - 1.536 = 3.625 \text{ mm}$$

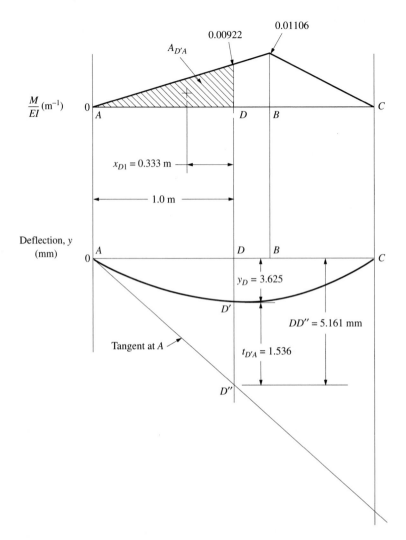

FIGURE 12–31 Moment-area diagrams for Example Problem 12–10.

12–10 BEAMS WITH DISTRIBUTED LOADS—MOMENT-AREA METHOD

The general procedure for determining the deflection of beams carrying distributed loads is the same as that shown for beams carrying concentrated loads. However, the shape of the bending moment and M/EI curves is different, requiring the use of other formulas for computing the area and centroid location for use in the moment-area method. The following example shows the kind of differences to be expected.

Example Problem 12-11

Determine the deflection at the end of the cantilever carrying a uniformly distributed load shown in Figure 12–32. The beam is a $6 \times 2 \times \frac{1}{4}$ hollow rectangular steel tube with the 6.0-in dimension horizontal.

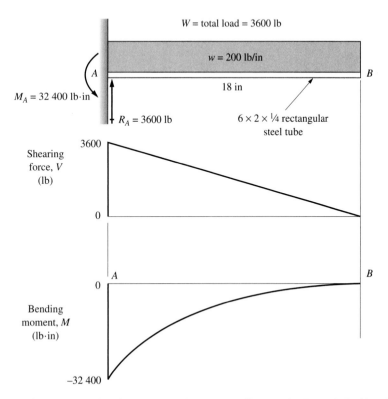

FIGURE 12-32 Load, shearing force, and bending moment diagrams for Example Problem 12–11.

Solution

Objective Compute the deflection at the end of the cantilever.

Given Beam and loading in Figure 12–32. Beam is a steel rectangular tube, $6 \times 2 \times 1/4$, with 6.0-in dimension horizontal.

Analysis The same basic procedure outlined for Example Problem 12–5 can be used.

Results The solution starts with the preparation of the load, shearing force, and bending moment diagrams, shown in Figure 12–32. Then the M/EI curve will have the same shape as the bending moment curve because the stiffness of the beam is uniform. From Appendix A–9 we find $I = 2.31$ in⁴. Then, using $E = 30 \times 10^6$ psi for steel,

$$EI = (30 \times 10^6 \text{ lb/in}^2)(2.31 \text{ in}^4) = 6.93 \times 10^7 \text{ lb·in}^2$$

Now Figure 12–33 shows the M/EI curve along with the deflection curve for the beam. The horizontal line on the deflection diagram is the tangent drawn to the beam shape at point A, where the beam is fixed to the support. Then, at the right end of the beam, the deviation of the beam deflection curve from this tangent, t_{BA}, is equal to the deflection of the beam itself.

Using Theorem 2, the deviation t_{BA} is equal to the product of the area of the M/EI curve between B and A times the distance from point B to the centroid of the area. That is,

$$t_{BA} = A_{BA} \cdot x_B$$

Recalling that the load, shearing force, and bending moment diagrams are related to each other in such a way that the higher curve is the derivative of the curve below it, the following conclusions can be drawn:

1. The shearing force curve is a first-degree curve (straight line with constant slope). Its equation is of the form

$$V = m \cdot x + b$$

where m is the slope of the line and b is its intercept with the vertical axis. The variable x is the position on the beam.

2. The bending moment curve is a second-degree curve, a parabola. The general equation of the curve is of the form,

$$M = a \cdot x^2 + b$$

Appendix A–1 shows the relationships for computing the area and the location of the centroid for areas bounded by second-degree curves. For an area having the shape of the bending moment or M/EI curves,

$$\text{area} = \frac{L \cdot h}{3}$$

$$x = \frac{L}{4}$$

where L = length of the base of the area
   h = height of the area
   x = distance from the side of the area to the centroid

Note that the corresponding distance from the vertex of the curve to the centroid is

$$x' = \frac{3L}{4}$$

Now, for the data shown in Figure 12–33,

$$A_{BA} = \frac{L \cdot h}{3} = \frac{(18\ \text{in})\,(-4.68 \times 10^{-4}\ \text{in}^{-1})}{3} = 2.808 \times 10^{-3}$$

$$x_B = \frac{3L}{4} = \frac{3(18\ \text{in})}{4} = 13.5\ \text{in}$$

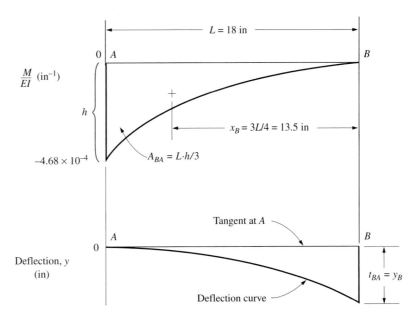

FIGURE 12–33 *M/EI* and deflection curves for Example Problem 12–11.

Theorem 2 can now be used:

$$t_{BA} = A_{BA}x_B = (2.808 \times 10^{-3})(13.5 \text{ in}) = 0.0379 \text{ in}$$

This is equal to the deflection of the beam at its end, y_B.

REFERENCES

1. Aluminum Association, *Engineering Data for Aluminum Structures,* 5th ed., Washington, DC, 1986.

2. Blodgett, Omer W., *Design of Weldments,* James F. Lincoln Arc Welding Foundation, Cleveland, OH, 1963.

3. Mott, Robert L., *Machine Elements in Mechanical Design,* 2nd ed., Macmillan Publishing Co., Columbus, OH, 1992.

4. Popov, E. P., *Engineering Mechanics of Solids,* Prentice-Hall, Englewood Cliffs, NJ, 1990.

5. Young, W. C., *Roark's Formulas for Stress and Strain,* 6th ed., McGraw-Hill Book Company, New York, 1989.

PROBLEMS

Formula Method

12–1.M A round shaft having a diameter of 32 mm is 700 mm long and carries a 3.0-kN load at its center. If the shaft is steel and simply supported at its ends, compute the deflection at the center.

12–2.M For the shaft in Problem 12–1, compute the deflection if the shaft is 6061-T6 aluminum instead of steel.

12–3.M For the shaft in Problem 12–1, compute the deflection if the ends are fixed against rotation instead of simply supported.

12–4.M For the shaft in Problem 12–1, compute the deflection if the shaft is 350 mm long rather than 700 mm.

12–5.M For the shaft in Problem 12–1, compute the deflection if the diameter is 25 mm instead of 32 mm.

12–6.M For the shaft in Problem 12–1, compute the deflection if the load is placed 175 mm from the left support rather than in the center. Compute the deflection both at the load and at the center of the shaft.

12–7.E A wide-flange steel beam, W12 × 16, carries the load shown in Figure P6–4. Compute the deflection at the loads and at the center of the beam.

12–8.E A standard steel $1\frac{1}{2}$ in schedule 40 pipe carries a 650-lb load at the center of a 28-in span, simply supported. Compute the deflection of the pipe at the load.

12–9.E An Aluminum Association Standard I-beam, I8 × 6.181, carries a uniformly distributed load of 1125 lb/ft over a 10-ft span. Compute the deflection at the center of the span.

12–10.E For the beam in Problem 12–9, compute the deflection at a point 3.5 ft from the left support of the beam.

12–11.E A wide-flange steel beam, W12 × 30, carries the load shown in Figure P6–12. Compute the deflection at the load.

12–12.E For the beam in Problem 12–11, compute the deflection at the load if the left support is moved 2.0 ft toward the load.

12–13.E For the beam in Problem 12–11, compute the maximum upward deflection and determine its location.

12–14.E A 1-in schedule 40 steel pipe is used as a cantilever beam 8 in long to support a load of 120 lb at its end. Compute the deflection of the pipe at the end.

12–15.M A round steel bar is to be used to carry a single concentrated load of 3.0 kN at the center of a 700-mm-long span on simple supports. Determine the required diameter of the bar if its deflection must not exceed 0.12 mm.

12–16.M For the bar designed in Problem 12–15, compute the stress in the bar and specify a suitable steel to provide a design factor of 8 based on ultimate strength.

12–17.E A flat strip of steel 0.100 in wide and 1.200 in long is clamped at one end and loaded at the other like a cantilever beam (as in Case *a* in Appendix A–23). What should be the thickness of the strip if it is to deflect 0.15 in under a load of 0.52 lb?

12–18.E A wood joist in a commercial building is 14 ft 6 in long and carries a uniformly distributed load of 50 lb/ft. It is 1.50 in wide and 9.25 in high. If it is made of southern pine, compute the maximum deflection of the joist. Also compute the stress in the joist due to bending and horizontal shear, and compare them with the allowable stresses for No. 2 grade southern pine.

Superposition

12–19.M The load shown in Figure P6–6 is being carried by an extruded aluminum (6061-T6) beam. Compute the deflection of the beam at each load. The beam shape is shown in Figure P7–11.

12–20.M The loads shown in Figure P6–5 represent the feet of a motor on a machine frame. The frame member has the cross section shown in Figure P7–12 which has a moment of inertia of 16 956 mm^4. Compute the deflection at each load. The aluminum alloy 2014-T4 is used for the frame.

12–21.C Compute the maximum deflection of a steel W18 × 55 beam when it carries the load shown in Figure P6–7.

12–22.E A 1-in schedule 40 steel pipe carries the two loads shown in Figure P6–18. Compute the deflection of the pipe at each load.

12–23.M A cantilever beam carries two loads as shown in Figure P6–21. If a rectangular steel bar 20 mm wide by 80 mm high is used for the beam, compute the deflection at the end of the beam.

12–24.M For the beam in Problem 12–23, compute the deflection if the bar is aluminum 2014-T4 rather than steel.

12–25.M For the beam in Problem 12–23, compute the deflection if the bar is magnesium, ASTM AZ 63A-T6, instead of steel.

12–26.E The load shown in Figure P6–55 is carried by a round steel bar having a diameter of 0.800 in.

Compute the deflection of the bar at the right end.

12–27.M Specify a standard wide-flange steel beam that would carry the loading shown in Figure P6–7 with a maximum deflection less than $\frac{1}{360}$ times the length of the beam.

12–28.E It is planned to use an Aluminum Association channel with its legs down to carry the loads shown in Figure P6–43 so that the flat face can contact the load. The maximum allowable deflection is 0.080 in. Specify a suitable channel.

Successive Integration Method

For the following problems, 12–29 through 12–37, use the general approach outlined in Section 12–7 to determine the equations for the shape of the deflection curves of the beams. Unless otherwise noted, compute the maximum deflection of the beam from the equations and tell where it occurs.

12–29.E The loading is shown in Figure P12–29. The beam is a rectangular steel bar, 1.0 in wide and 2.0 in high.

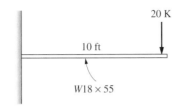

FIGURE P12–29

12–30.E The loading is shown in Figure P12–30. The beam is a steel wide-flange shape W18 × 55.

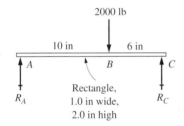

FIGURE P12–30

12–31.C The loading is shown in Figure P12–31. The beam is a $2\frac{1}{2}$-in schedule 40 steel pipe.

12–32.E The loading is shown in Figure P12–32. The beam is a steel wide-flange shape W24 × 76.

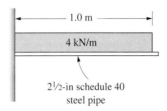

FIGURE P12–31

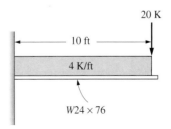

FIGURE P12–32

12–33.M The load is shown in Figure P12–33. Design a round steel bar that will limit the deflection at the end of the beam to 5.0 mm.

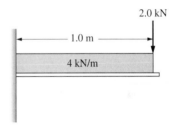

FIGURE P12–33

12–34.C The load is shown in Figure P12–34. Design a steel beam that will limit the maximum deflection to 1.0 mm. Use any shape, including those listed in the Appendix.

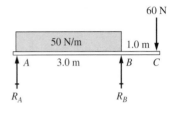

FIGURE P12–34

12–35.C The load is shown in Figure P12–35. Select an aluminum I-beam that will limit the stress to

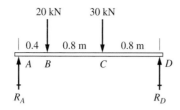

FIGURE P12–35

120 MPa; then compute the maximum deflection in the beam.

12–36.C A steel wide-flange beam W14 × 26 carries the loading shown in Figure P12–36. Compute the maximum deflection between the supports and at each end.

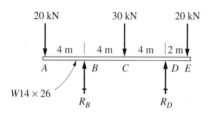

FIGURE P12–36

12–37.M The loading shown in Figure P12–37 represents a steel shaft for a large machine. The loads are due to gears mounted on the shaft. Assuming that the entire shaft will be the same diameter, determine the required diameter to limit the deflection at any gear to 0.13 mm.

Moment-Area Method

Use the moment-area method to solve the following problems.

12–38.E For the beam shown in Figure P12–29, compute the deflection at the load. The beam is a rectangular steel bar 1.0 in wide and 2.0 in high.

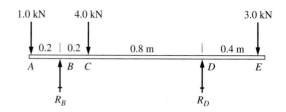

FIGURE P12–37

12–39.E For the beam shown in Figure P12–29, compute the deflection at the middle, 8.0 in from either support. The beam is a rectangular steel bar 1.0 in wide and 2.0 in high.

12–40.E For the beam shown in Figure P12–30, compute the deflection at the end. The beam is a steel wide-flange shape W18 × 55.

12–41.C For the beam shown in Figure P12–31, compute the deflection at the end. The beam is a $2\frac{1}{2}$-in schedule 40 steel pipe.

12–42.E For the beam shown in Figure P12–32, compute the deflection at the end. The beam is a steel wide-flange shape W24 × 76.

12–43.M For the beam shown in Figure P12–33, compute the deflection at the end. The beam is a round aluminum bar 6061-T6 with a diameter of 100 mm.

12–44.C For the beam shown in Figure P12–34, compute the deflection at the right end, point C. The beam is a square steel structural tube 2 × 2 × $\frac{1}{4}$.

12–45.C For the beam shown in Figure P12–35, compute the deflection at point C. The beam is an aluminum I-beam I7 × 5.800 made from 6061-T6.

12–46.C For the beam shown in Figure P12–36, compute the deflection at point A. The beam is a steel wide-flange shape W14 × 26.

12–47.E Figure P12–47 shows a stepped round steel shaft carrying a single concentrated load at its middle. Compute the deflection under the load.

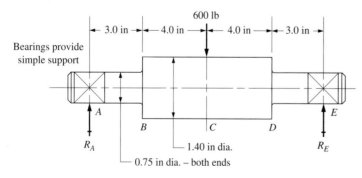

FIGURE P12–47

Chapter 12 ▪ Deflection of Beams

12–48.E Figure P12–48 shows a composite beam made from steel structural tubing. Compute the deflection at the load.

12–49.E Figure P12–49 shows a cantilever beam made from a steel wide-flange shape W18 × 55 with cover plates welded to the top and bottom of the

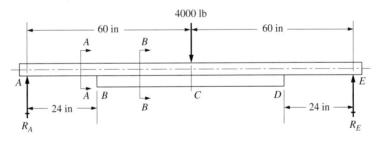

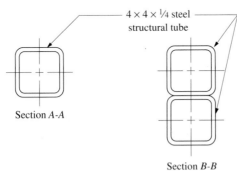

4 × 4 × ¼ steel
structural tube

Section A-A

Section B-B

FIGURE P12–48

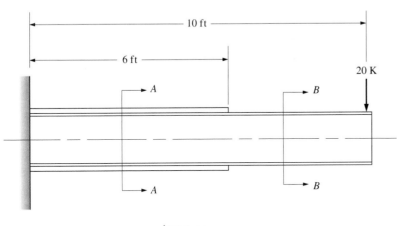

10 ft

6 ft

20 K

A

B

A

B

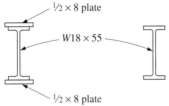

½ × 8 plate

W18 × 55

½ × 8 plate

Section A-A   Section B-B   **FIGURE P12–49**

beam over the first 6.0 ft. Compute the deflection at the end of the beam.

12–50.E The top member for a gantry crane is made as shown in Figure P12–50. The two end pieces are $3 \times 3 \times \frac{1}{4}$ steel structural tubing. The middle piece is $4 \times 4 \times \frac{1}{2}$. They are combined for a distance of 10.0 in with the 3-in tube fitted tightly inside the 4-in tube and brazed together. Compute the deflection at the middle of the beam.

12–51.M A crude diving board is made by nailing two wood planks together as shown in Figure P12–51. Compute the deflection at the end if the diver exerts a force of 1.80 kN at the end. The wood is No. 2 southern pine.

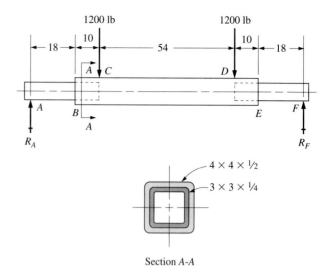

FIGURE P12–50 Gantry crane beam for Problem 12–50.

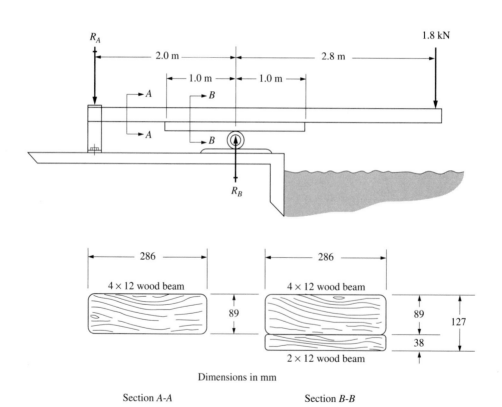

Dimensions in mm

Section *A-A*   Section *B-B*

FIGURE P12–51 Diving board for Problem 12–51.

COMPUTER PROGRAMMING ASSIGNMENTS

1. Write a program to evaluate the deflection of a simply supported beam carrying one concentrated load between the supports using the formulas given in Case *b* of Appendix A–22. The program should accept input for the length, position of the load, length of the overhang, if any, the beam stiffness values (*E* and *I*), and the point for which the deflection is to be computed.

Enhancements

(a) Design the program so that it computes the deflection at a series of points that could permit the plotting of the complete deflection curve.

(b) In addition to computing the deflection of the series of points, have the program plot the deflection curve on a plotter or printer.

(c) Have the program compute the maximum deflection and the point at which it occurs.

2. Repeat Assignment 1 for any of the loading and support patterns shown in Appendix A–22.

3. Write a program similar to that of Assignment 1 for Case *b* of Appendix A–22, but have it accept two or more concentrated loads at any point on the beam and compute the deflection at specified points using the principle of superposition.

4. Combine two or more programs that solve for the deflection of beams for a given loading pattern so that superposition can be used to compute the deflection at any point due to the combined loading. For example, combine Cases *b* and *d* of Appendix A–22 to handle any beam with a combination of concentrated loads and a complete uniformly distributed load. Or, add Case *g* to include a distributed load over only part of the length of the beam.

5. Repeat Assignments 1 through 4 for the cantilevers from Appendix A–23.

13

Statically Indeterminate Beams

13–1 OBJECTIVES OF THIS CHAPTER

The beams considered in earlier chapters include beams with two and only two simple supports or cantilevers that have one fixed end and the other end free. It was shown that all the unknown reaction forces and bending moments could be found using the classic equations of equilibrium:

$$\sum F = 0 \qquad \text{in any direction}$$

 Equilibrium Equations

$$\sum M = 0 \qquad \text{about any point}$$

Such beams are called *statically determinate*.

This chapter deals with beams that do not fall into the categories listed above and thus they are called *statically indeterminate*. Different methods of analyzing such beams are required and will be shown in this chapter. Also, a comparison will be made for the behavior of beams designed to perform a similar function but implemented with different support systems, some of which are statically determinate and some of which are statically indeterminate.

After completing this chapter, you should be able to:

1. Define *statically determinate* and *statically indeterminate*.

2. Recognize statically indeterminate beams from given descriptions of the loading and support conditions.

3. Define *continuous beam*.

4. Define *supported cantilever*.

5. Define *fixed-end beam*.

6. Use established formulas to analyze certain types of statically indeterminate beams.

7. Use superposition to combine simple cases for which formulas are available to solve more complex loading cases.

8. Use the *theorem of three moments* to analyze continuous beams having three or more supports and carrying any combination of concentrated and distributed loads.

9. Compare the relative strength and stiffness of beams having different support systems for giving loadings.

13–2 EXAMPLES OF STATICALLY INDETERMINATE BEAMS

Beams with more than two simple supports, cantilevers with a second support, or beams with two fixed ends are important examples of statically indeterminate beams. Figure 13–1 shows the conventional method of representing these kinds of beams. The typical shapes, although exaggerated, for the deflection curves of the beams are also shown. Significant differences should be noted between these and the curves for beams in the preceding chapter.

Figure 13–1(a) is called a *continuous beam,* the name coming from the fact that the beam is continuous over several supports. It is important to note that the shape of the deflection curve of the beam is also continuous through the supports. This fact will be useful in analyzing such beams. Continuous beams occur frequently in building structures and highway bridges. Many ranch-type homes with basements contain such a beam running the length of the house to carry loads from floor joists and partitions. Expressway bridges carrying local traffic over through-lanes are frequently supported at the ends on each side of the roadway and also at the middle in the median strip. Notice that the beams of such bridges are usually one piece or are connected to form a rigid *continuous* beam.

The *fixed-end* beam shown in Figure 13–1(b) is used in building structures and also in machine structures because of the high degree of rigidity provided. The creation of the fixed type of end condition requires that the connections at the ends restrain the beam from rotation as well as provide support for vertical loads. Figure 13–2 shows one way in which the fixed end condition can be achieved. Welding of the cross beam to the supporting columns would achieve the same result. Care must be used in evaluating fixed-ended beams to ensure that the connections can indeed restrain the beam

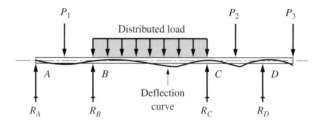

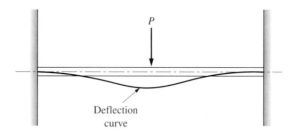

(**a**) Continuous beam - a beam on
three or more simple supports

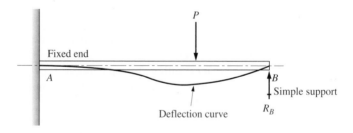

(**b**) Beam with two fixed ends

(**c**) Supported cantilever

FIGURE 13–1 Examples of statically indeterminate beams.

from rotating at the support and resist the moments created because of the restraint. Without due care, a condition *between* that of fixed ends and simple supports could result, making analysis difficult and subject to error.

The supported cantilever could be constructed as shown in Figure 13–3. The load on the flat roof is carried by a beam rigidly fixed to a column at one end and resting simply on another column at the other end.

Continuous beams and beams with fixed ends have certain advantages over simply supported beams: A greater rigidity is provided and generally lower stresses would be produced in beams of equivalent size; alternatively, smaller and lighter beams can be used to give the same degree of strength and rigidity. The disadvantages are that the analyses are more difficult and the effect of small variations in support conditions can be significant.

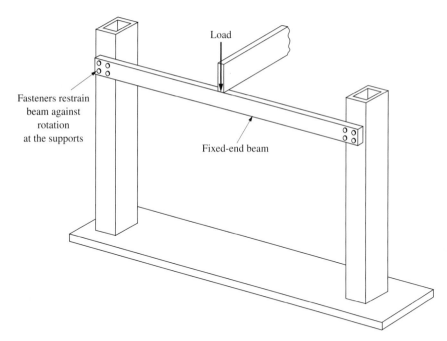

FIGURE 13–2 Fixed-end beam.

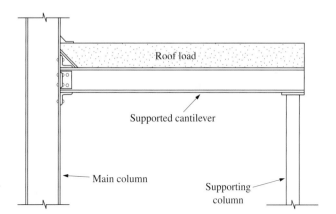

FIGURE 13–3 Supported cantilever.

13–3 FORMULAS FOR STATICALLY INDETERMINATE BEAMS

Appendix A–24 shows several examples of statically indeterminate beams along with formulas for computing the reactions at supports and the deflection at any point on the beams. These formulas can be applied directly in the manner demonstrated in Chapter 12 for statically determinate beams. Also shown are plots of the shearing force and bending moment diagrams along with the necessary formulas for computing the values at critical points.

The general characteristics of statically indeterminate beams are quite different from the statically determinate designs studied in earlier chapters. This is particularly true for the manner of computing the reaction moments and forces at supports, the distribution of the bending moment with respect to position on the beam, the magnitude of the deflection at various points on the beam, and the general shape of the deflection curve. The formulas given in Appendix A–24 were derived using the principles discussed in Chapters 6, 7, and 12 of this book along with special considerations to accommodate the statically indeterminate nature of the loading and support conditions. The techniques of *superposition* and the *theorem of three moments,* discussed later in this chapter, can be used to derive these formulas. As you review the formulas in Appendix A–24, note the following general characteristics.

General characteristics of statically indeterminate beams

Supported Cantilevers (Cases *a* to *d* in Appendix A–24)

1. The fixed end provides a rigid support that resists any tendency for the beam to rotate. Thus there is typically a significant bending moment there.
2. The second support is taken to be a simple support. If the simple support is at the free end of the beam, as in Cases *a*, *b*, and *c*, the bending moment there is zero.
3. If the supported cantilever has an overhang, as in Case *d*, the largest bending moment often occurs at the simple support. The shape of the bending moment curve is generally opposite that for the cases without an overhang.
4. There is a point of zero bending moment on a supported cantilever, generally near the fixed end.

Fixed-end beams (Cases *e*, *f*, and *g* in Appendix A–24)

1. The bending moments at the fixed ends are *not* zero and may be the maximum for the beam.
2. When loads are downward on a fixed-end beam, the bending moments at the ends are *negative,* indicating that the deflection curve near there is *concave downward.*
3. When loads are downward, the bending moments near the middle of fixed-end beams are *positive,* indicating that the deflection curve there is *concave upward*.
4. There are typically two points of zero bending moment in fixed-end beams.
5. The deflection curve for a fixed-end beam has a zero slope at the fixed ends because of the restraint provided there against rotation.
6. The means of fastening a fixed-end beam to its supports must be capable of resisting the bending moments and the shearing forces at these points. Chapter 16 on Connections should be consulted for techniques of designing and analyzing connections that must resist moments.

1. Points of maximum positive bending moment generally occur near the middle of spans between supports.

2. Points of maximum negative bending moment generally occur at the interior supports and these are often the maximum bending moments.

3. Particularly for long continuous beams, it is often economically desirable to tailor the cross-sectional shape and dimensions to provide reinforcement over the sections that have the larger bending moments to accommodate the locally high values. An example would be to design the main beam shape to withstand the maximum positive bending moment between the supports and then to add reinforcing plates to the top or bottom of the beam near the supports to increase the moment of inertia and section modulus in the regions of high bending moment. Another approach would be to increase the depth of the beam near the supports. Highway overpasses and bridges over rivers often have these design features.

4. Where long continuous beams must be made in sections that are fastened together on site, it may be desirable to place the connection near a point of zero bending moment to simplify the design of the connection.

Comparison of the manner of support for a beam using standard formulas. The application of the formulas for statically indeterminate beams is similar to the process used in Chapter 12 for statically determinate beams. In the following example problems we demonstrate the use of various formulas from Appendixes A–22, A–23, and A–24 and also generate data with which to compare the performance of four different types of beam support to accomplish the same objective; that is, to support a given load at a given distance from one or two supports. The comparison is based on the magnitude of the bending stress and deflection in the four beams where each has the same material, shape, and size. The best performing beam, then, is the one with the lower stress and the smaller deflection.

The parameters of the comparisons are the following:

1. The four types of beam support to be compared are:
   a. Cantilever
   b. Simply supported beam
   c. Supported cantilever
   d. Fixed-end beam

2. Each beam is to support a single, static, concentrated load of 1200 lb.

3. The load is to be placed at a distance of 30 in from any support.

4. Each beam will be made from ASTM A36 structural steel for which we will use the following properties: $s_y = 36\,000$ psi; $E = 30 \times 10^6$ psi.

5. The maximum allowable bending stress will be based on the AISC standard,

$$\sigma_d = 0.66 s_y = 0.66(36\,000 \text{ psi}) = 23\,760 \text{ psi}$$

6. The maximum allowable deflection will be based on a limit of $L/360$, where L is the length of the beam.

It can be shown that, of the four types of beams and support systems to be considered, the cantilever has the poorest performance with regard to stress and deflection. Therefore, we will start the process of comparison by specifying the beam cross section shape and size that will ensure that the cantilever meets the requirements stated in items 5 and 6. Then, the same shape and size will be used for the other three designs. Five example problems follow, generating the results for the desired comparison.

> Example Problem 13–1: Design and analysis of the cantilever.
>
> Example Problem 13–2: Analysis of the simply supported beam.
>
> Example Problem 13–3: Analysis of the supported cantilever.
>
> Example Problem 13–4: Analysis of the fixed-end beam.
>
> Example Problem 13–5: Comparison of the four beam designs.

Example Problem 13–1 For the cantilever shown in Figure 13–4, specify the lightest standard steel channel shape that will limit the bending stress to 23 760 psi and limit the maximum deflection to

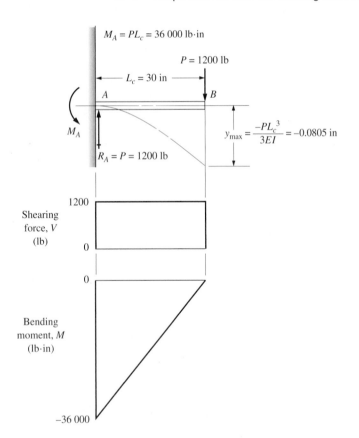

$M_A = PL_c = 36\,000$ lb·in

$P = 1200$ lb

$L_c = 30$ in

A B

M_A

$y_{max} = \dfrac{-PL_c^{\,3}}{3EI} = -0.0805$ in

$R_A = P = 1200$ lb

Shearing force, V (lb)

1200

0

Bending moment, M (lb·in)

0

−36 000

FIGURE 13–4 Cantilever.

$L/360$. The channel is to be positioned with the legs pointing downward and the flat side of the web on top to provide a surface for mounting the load. For the beam shape selected, compute the actual expected maximum stress and deflection.

Solution   **Objective**   Design the cantilever and compute the resulting stress and deflection.

Given   Beam dimensions and loading as shown in Figure 13–4. Beam shape to be a standard steel channel with the legs pointing downward.

$$\sigma_d = 23\,760 \text{ psi}; \qquad y_{max} = L/360 \quad \text{with } L = L_c = 30 \text{ in}$$

Analysis   Figure 13–4 contains the loading, shearing force, and bending moment diagrams from which we find that $M_{max} = 36\,000$ lb· in. Then,

$$\sigma = \sigma_d = M/S$$

Stress analysis: For a safe stress, the required section modulus is

$$S = M/\sigma_d = (36\,000 \text{ lb·in})/(23\,760 \text{ lb/in}^2) = 1.515 \text{ in}^3$$

From Appendix A–6, considering the properties with respect to the *Y-Y* axis, we select C12 × 25 as the lightest suitable section. Its properties are

$$S = 1.88 \text{ in}^3; \qquad I = 4.47 \text{ in}^4; \qquad w = 25 \text{ lb/ft}$$

Deflection: The maximum allowable deflection is

$$y_{max} = -L/360 = -(30 \text{ in})/360 = -0.0833 \text{ in}$$

From Appendix A–23, the formula for the maximum deflection is

$$y_{max} = -PL^3/3EI \qquad \text{at the end of the cantilever}$$

Then the required moment of inertia is

$$I = -PL^3/3Ey_{max}$$
$$I = \frac{-(1200 \text{ lb})(30 \text{ in})^3}{3(30 \times 10^6)(-0.0833 \text{ in})} = 4.32 \text{ in}^4$$

The previously selected beam shape is satisfactory for deflection.
Actual bending stress:

$$\sigma = M/S = (36\,000 \text{ lb·in})/(1.88 \text{ in}^3) = 19\,150 \text{ psi}$$

Actual deflection:

$$y_{max} = \frac{-PL^3}{3EI} = \frac{-(1200 \text{ lb})(30 \text{ in})^3}{3(30 \times 10^6 \text{ lb/in}^2)(4.47 \text{ in}^4)} = -0.0805 \text{ in}$$

Results *Summary of results:*
Beam shape: C12 × 25 standard steel channel

$$\sigma = 19\,150 \text{ psi}$$
$$y_{max} = -0.0805 \text{ in}$$

Comment These results will be compared with the other designs in the following example problems.

Example Problem 13–2

The simply supported beam shown in Figure 13–5 is to be made from a standard steel channel, C12 × 25, with the legs pointing downward. Compute the actual expected maximum stress and deflection and compare them with the results of Example Problem 13–1 for the cantilever.

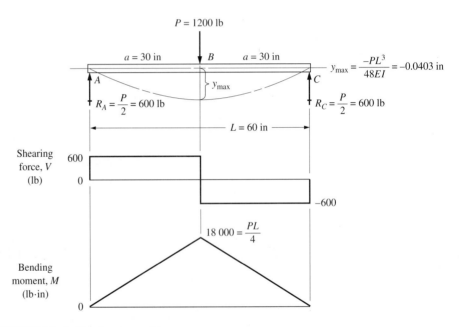

FIGURE 13–5 Simply supported beam.

Solution **Objective** Compute the maximum stress and deflection. Compare with cantilever.

Given Beam dimensions and loading as shown in Figure 13–5. Beam shape is a standard steel channel, C12 × 25, with the legs pointing downward.
Beam properties: $S = 1.88 \text{ in}^3$; $I = 4.47 \text{ in}^4$

Analysis From Figure 13–5 we find $M_{max} = 18\,000$ lb·in. Then,
Actual bending stress:

$$\sigma = M/S = (18\,000 \text{ lb·in})/(1.88 \text{ in}^3) = 9575 \text{ psi}$$

Actual deflection:

$$y_{max} = \frac{-PL^3}{48EI} = \frac{-(1200 \text{ lb})(60 \text{ in})^3}{48(30 \times 10^6 \text{ lb/in}^2)(4.47 \text{ in}^4)} = -0.0403 \text{ in}$$

Results *Comparison of results with cantilever:* Using the subscript *1* for the cantilever and *2* for the simply supported beam,

$$\sigma_2/\sigma_1 = (9575 \text{ psi})/(19\,150 \text{ psi}) = 0.50$$
$$y_2/y_1 = (-0.0403 \text{ in})/(-0.0805 \text{ in}) = 0.50$$

Comment These results will be compared with the other designs in Example Problem 13–5.

Example Problem 13–3 The supported cantilever shown in Figure 13–6 is to be made from a standard steel channel, C12 × 25, with the legs pointing downward. Compute the actual expected maximum stress and deflection and compare them with the results of Example Problem 13–1 for the cantilever.

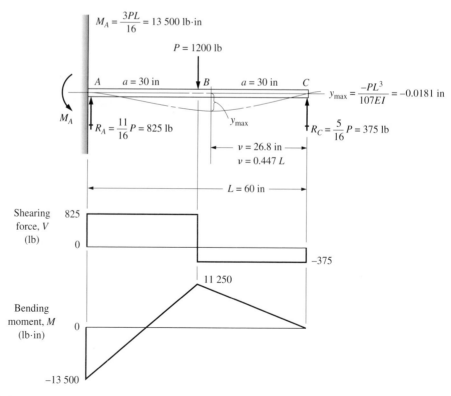

FIGURE 13–6 Supported cantilever.

Solution **Objective** Compute the maximum stress and deflection. Compare with cantilever.

Given Beam dimensions and loading as shown in Figure 13–6. Beam shape is a standard steel channel, C12 × 25, with the legs pointing downward. Beam properties: $S = 1.88$ in³; $I = 4.47$ in⁴

Analysis From Figure 13–6 we find $M_{max} = 13\,500$ lb·in. Then,
Actual bending stress:

$$\sigma = M/S = (13\,500 \text{ lb·in})/(1.88 \text{ in}^3) = 7180 \text{ psi}$$

Actual deflection: From Appendix A–24(a),

$$y_{max} = \frac{-PL^3}{107EI} = \frac{-(1200 \text{ lb})(60 \text{ in})^3}{107(30 \times 10^6 \text{ lb/in}^2)(4.47 \text{ in}^4)} = -0.0181 \text{ in}$$

Results *Comparison of results with cantilever.* Using the subscript *1* for the cantilever and *3* for the supported cantilever,

$$\sigma_3/\sigma_1 = (7180 \text{ psi})/(19\,150 \text{ psi}) = 0.375$$
$$y_3/y_1 = (-0.0181 \text{ in})/(-0.0805 \text{ in}) = 0.225$$

Comment These results will be compared with the other designs in Example Problem 13–5.

Example Problem 13–4 The fixed-end beam shown in Figure 13–7 is to be made from a standard steel channel, C12 × 25, with the legs pointing downward. Compute the actual expected maximum stress and deflection and compare them with the results of Example Problem 13–1 for the cantilever.

Solution **Objective** Compute the maximum stress and deflection. Compare with cantilever.

Given Beam dimensions and loading as shown in Figure 13–7. Beam shape is a standard steel channel, C12 × 25, with the legs pointing downward. Beam properties: $S = 1.88$ in³; $I = 4.47$ in⁴

Analysis From Figure 13–7 we find $M_{max} = 9000$ lb·in. Then,
Actual bending stress:

$$\sigma = M/S = (9000 \text{ lb·in})/(1.88 \text{ in}^3) = 4787 \text{ psi}$$

Actual deflection: From Appendix A–24(e),

$$y_{max} = \frac{-PL^3}{192EI} = \frac{-(1200 \text{ lb})(60 \text{ in})^3}{192(30 \times 10^6 \text{ lb/in}^2)(4.47 \text{ in}^4)} = -0.0101 \text{ in}$$

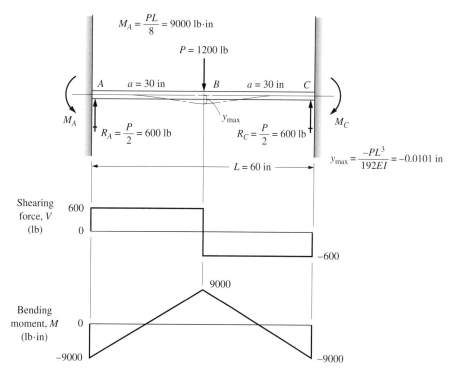

$$M_A = \frac{PL}{8} = 9000 \text{ lb·in}$$

$P = 1200 \text{ lb}$

A $a = 30 \text{ in}$ B $a = 30 \text{ in}$ C

M_A y_{max} M_C

$$R_A = \frac{P}{2} = 600 \text{ lb}$$ $$R_C = \frac{P}{2} = 600 \text{ lb}$$

$L = 60 \text{ in}$

$$y_{max} = \frac{-PL^3}{192EI} = -0.0101 \text{ in}$$

Shearing force, V (lb)   600   0   −600

Bending moment, M (lb·in)   9000   0   −9000   −9000

FIGURE 13–7   Fixed-end beam.

Results   *Comparison of results with cantilever:* Using the subscript *1* for the cantilever and *4* for the fixed-end beam,

$$\sigma_4/\sigma_1 = (4787 \text{ psi})/(19\,150 \text{ psi}) = 0.250$$
$$y_4/y_1 = (-0.0101 \text{ in})/(-0.0805 \text{ in}) = 0.125$$

Comment   These results will be compared with the other designs in Example Problem 13–5.

Example Problem 13–5   Compare the behavior of the four beams shown in Figures 13–4 through 13–7 with regard to shearing forces, bending moments, and maximum deflection. Use the results of Example Problems 13–1 through 13–4.

Solution   **Objective**   Compare the performance of the four beams.

Given   Beam designs shown in Figures 13–4 through 13–7.
Results of Example Problems 13–1 through 13–4.

Analysis   Figure 13–8 shows the comparison.

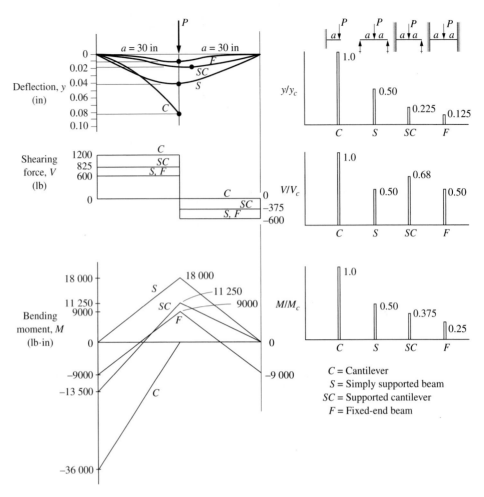

FIGURE 13-8 Comparison of performance of the four beam designs in Figures 13–4 through 13–7.

Results　In Figure 13–8, the graphs on the left show plots of the deflection, shearing force, and bending moment as a function of position on the beam for the four designs superimposed on each other. In each case, it is obvious that the cantilever produces the highest value by a significant margin. On the right are bar graphs of the relative performance where the values of deflection, shearing force, and bending moment for the cantilever are assigned a value of 1.0.

Comment　The results shown in Figure 13–8 indicate that it is desirable to provide two supports rather than one if that is practical. Also, providing two fixed ends for the beam is most desirable, producing only 1/8 of the deflection, 1/2 of the shearing force, and 1/4 of the bending moment as compared with the cantilever for a given load, distance from the load to the support, beam shape, and size. Of course, it must be practical to provide the second support.

13–4 SUPERPOSITION METHOD

Consider first the supported cantilever shown in Figure 13–9. Because of the restraint at A and the simple support at B, the unknown reactions include:

1. The vertical force R_B
2. The vertical force R_A
3. The restraining moment M_A

The conditions assumed for this beam are that the supports at A and B are absolutely rigid and at the same level, and that the connection at A does not allow any rotation of the beam at that point. Conversely, the support at B will allow rotation and cannot resist moments.

If the support at B is removed, the beam would deflect downward, as shown in Figure 13–10(a), by an amount y_{B1} due to the load P. Now if the load is removed and the reaction force R_B is applied upward at B, the beam would deflect upward by an amount y_{B2}, as shown in Figure 13–10(b). In reality, of course, both forces are applied, and the deflection at B is *zero*. The principle of superposition would then provide the conclusion that

$$y_{B1} + y_{B2} = 0 \qquad (13\text{--}1)$$

This equation, along with the normal equations of static equilibrium, will permit the evaluation of all three unknowns, as demonstrated in the example problem that follows.

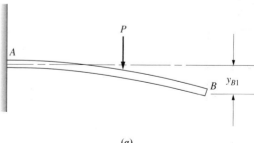

(a)

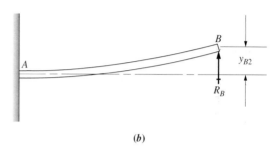

(b)

FIGURE 13–10 Superposition applied to the supported cantilever.

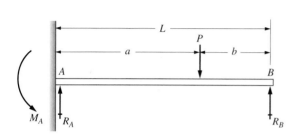

FIGURE 13–9 Supported cantilever.

It must be recognized that the principles of static equilibrium are still valid for statically indeterminate beams. However, they are not sufficient to allow a direct solution.

Example Problem 13–6

Determine the support reactions at A and B for the supported cantilever shown in Figure 13–9 if the load P is 2600 N and placed 1.20 m out from A. The total length of the beam is 1.80 m. Then draw the complete shearing force and bending moment diagrams, and design the beam by specifying a configuration, a material, and the required dimensions of the beam. Use a design factor of 8 based on ultimate strength since the load will be repeated often.

Solution **Objective** Determine the support reactions, draw the shearing force and bending moment diagrams, and design the beam.

Given Beam loading in Figure 13–9. $P = 2600$ N (repeated). $L = 1.80$ m. $a = 1.20$ m. Use a design factor of $N = 8$ based on s_u.

Analysis Use the superposition method.

Results *Reaction at B, R_B*

We must first determine the reaction at B using superposition. Equation (13–1) states that, for this situation,

$$y_{B1} + y_{B2} = 0 \qquad (13\text{–}1)$$

The equation for y_{B1} can be found from the beam deflection formulas in Appendix A–23. As suggested in Figure 13–10(a), the deflection at the end of a cantilever carrying an intermediate load is required. Then

$$y_{B1} = \frac{-Pa^2}{6EI}(3L - a)$$

where $P = 2600$ N
   $a = 1.20$ m
   $L = 1.80$ m

The values of E and I are still unknown, but we can express the deflection in terms of EI.

$$y_{B1} = \frac{(-2600 \text{ N})(1.20 \text{ m})^2}{6EI}[3(1.80 \text{ m}) - 1.20 \text{ m}]$$

$$= \frac{-2621 \text{ N·m}^3}{EI}$$

Now looking at Figure 13–10(b), we need the deflection at the end of the cantilever due to a concentrated load there. Then

$$y_{B2} = \frac{PL^3}{3EI} = \frac{R_B(1.8 \text{ m})^3}{3EI} = \frac{R_B(1.944 \text{ m}^3)}{EI}$$

Putting these values in Equation (13–1) gives

$$\frac{-2621 \text{ N·m}^3}{EI} + \frac{R_B(1.944 \text{ m}^3)}{EI} = 0$$

The term EI can be canceled out, allowing the solution for R_B.

$$R_B = \frac{2621 \text{ N·m}^3}{1.944 \text{ m}^3} = 1348 \text{ N}$$

The values of R_A and M_A can now be found using the equations of static equilibrium.
Reaction at A, R_A

$$\sum F = 0 \quad \text{(in the vertical direction)}$$

$$R_A + R_B - P = 0$$

$$R_A = P - R_B = 2600 \text{ N} - 1348 \text{ N} = 1252 \text{ N}$$

Bending moment at A, M_A
Summing moments about point *A* gives

$$0 = M_A - 2600 \text{ N} (1.2 \text{ m}) + 1348 \text{ N} (1.8 \text{ m})$$

$$M_A = 693 \text{ N·m}$$

The positive sign for the result indicates that the assumed sense of the reaction moment in Figure 13–9 is correct. However, this is a negative moment because it causes the beam to bend concave downward near the support *A*.
Shearing force and bending moment diagrams
     The shearing force and bending moment diagrams can now be drawn, as shown in Figure 13–11, using conventional techniques. The maximum bending moment occurs at the load where $M = 809$ N·m.
Beam design
     The beam can now be designed. Let's assume that the actual installation is similar to that sketched in Figure 13–12, with the beam welded at its left end and resting on another beam at the right end. A rectangular bar would work well in this arrangement, and a ratio of $h = 3t$ will be assumed. A carbon steel such as AISI 1040, hot rolled, provides an ultimate strength of 621 MPa. Its percent elongation, 25%, suggests good ductility, which will help it resist the repeated loads. The design should be based on bending stress.

$$\sigma = \frac{M}{S}$$

But let

$$\sigma = \sigma_d = \frac{s_u}{N} = \frac{621 \text{ MPa}}{8} = 77.6 \text{ MPa}$$

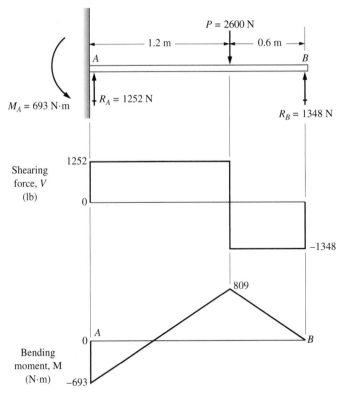

FIGURE 13–11 Shearing force and bending moment diagrams for the supported cantilever in Example Problem 13–6.

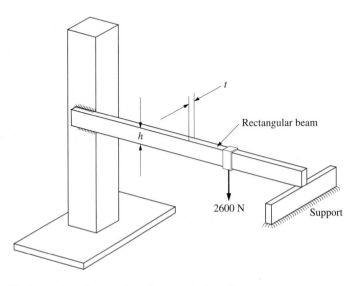

FIGURE 13–12 Physical implementation of a supported cantilever.

Then
$$S = \frac{M}{\sigma_d} = \frac{809 \text{ N·m}}{77.6 \text{ N/mm}^2} \times \frac{10^3 \text{ mm}}{\text{m}} = 10\,425 \text{ mm}^3$$

For a rectangular bar,

$$S = \frac{th^2}{6} = \frac{t(3t)^2}{6} = \frac{9t^3}{6} = 1.5t^3$$

Then
$$1.5t^3 = 10\,425 \text{ mm}^3$$
$$t = 19.1 \text{ mm}$$

Let's use the preferred size of 20 mm for t. Then

$$h = 3t = 3(20 \text{ mm}) = 60 \text{ mm}$$

Comment The final design can be summarized as a rectangular steel bar of AISI 1040, hot rolled, 20 mm thick and 60 mm high, welded to a rigid support at its left end and resting on a simple support at its right end. The maximum stress in the bar would be less than 77.6 MPa, providing a design factor of at least 8 based on ultimate strength.

The superposition method can be applied to any supported cantilever beam analysis for which the equation for the deflection due to the applied load can be found. Either the beam deflection formulas like those in the Appendix, the moment-area method, or the mathematical analysis method developed in Chapter 12 can be used.

Continuous beams can also be analyzed using superposition. Consider the beam on three supports shown in Figure 13–13. The three unknown support reactions make the beam statically indeterminate. The "extra" reaction R_C can be found using the technique suggested in Figure 13–14. Removing the support at C would cause the deflection y_{C1} downward due to the two 800-lb loads. Case c of Appendix A–22 can be used to

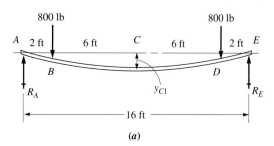

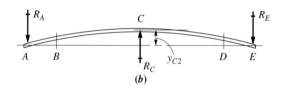

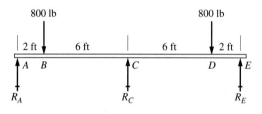

FIGURE 13–13 Continuous beam.

FIGURE 13–14 Superposition applied to a continuous beam.

find y_{C1}. Then if the loads are imagined to be removed and the reaction R_C is replaced, the upward deflection y_{C2} would result. The formulas of Case a in Appendix A–22 can be used.

Here again, of course, the actual deflection at C is zero because of the unyielding support. Therefore,

$$y_{C1} + y_{C2} = 0$$

From this relationship, the value of R_C can be computed. The remaining reactions R_A and R_E can then be found in the conventional manner, allowing the creation of the shearing force and bending moment diagrams.

13–5 CONTINUOUS BEAMS—THEOREM OF THREE MOMENTS

A continuous beam on any number of supports can be analyzed by using the *theorem of three moments*. The theorem actually relates the bending moments at three successive supports to each other and to the loads on the beam. For a beam with only three supports, the theorem allows the direct computation of the moment at the middle support. Known end conditions provide data for computing moments at the ends. Then the principles of statics can be used to find reactions.

For beams on more than three supports, the theorem is applied successively to sets of three adjacent supports (two spans), yielding a set of equations which can be solved simultaneously for the unknown moments.

The theorem of three moments can be used for any combination of loads. Special forms of the theorem have been developed for uniformly distributed loads and concentrated loads. These forms will be used in this chapter.

Uniformly distributed loads on adjacent spans. Figure 13–15 shows the arrangement of loads and the definition of terms applicable to Equation (13–2).

Three-Moment Equation— Distributed Loads

$$M_A L_1 + 2M_B(L_1 + L_2) + M_C L_2 = \frac{-w_1 L_1^3}{4} - \frac{w_2 L_2^3}{4} \qquad (13\text{–}2)$$

The values of w_1 and w_2 are expressed in units of force per unit length such as N/m, lb/ft, etc. The bending moments at the supports A, B, and C are M_A, M_B, and M_C. If M_A

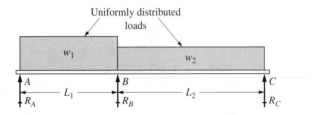

FIGURE 13–15 Uniformly distributed loads on a continuous beam of two spans.

and M_C at the ends of the beam are known, M_B can be found from Equation (13–2) directly. Example problems will demonstrate the application of this equation.

The special case in which two equal spans carry equal uniform loads allows the simplification of Equation (13–2). If $L_1 = L_2 = L$ and $w_1 = w_2 = w$, then

$$M_A + 4M_B + M_C = \frac{-wL^2}{2} \tag{13–3}$$

Concentrated loads on adjacent spans.
If adjacent spans carry only one concentrated load each, as shown in Figure 13–16, then Equation (13–4) applies.

Three Moment Equation— Concentrated Loads

$$M_A L_1 + 2M_B(L_1 + L_2) + M_C L_2 = \frac{-P_1 a}{L_1}(L_1^2 - a^2) - \frac{P_2 b}{L_2}(L_2^2 - b^2) \tag{13–4}$$

The special case of two equal spans carrying equal, centrally placed loads allows the simplification of Equation (13–4). If $P_1 = P_2 = P$ and $a = b = L/2$, then,

$$M_A + 4M_B + M_C = 3PL/4 \tag{13–5}$$

Combinations of uniformly distributed loads and several concentrated loads.
This is a somewhat general case, allowing each span to carry a uniformly distributed load and any number of concentrated loads, as suggested in Figure 13–17. The

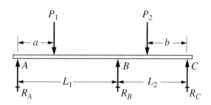

FIGURE 13–16 Continuous beam for two spans with one concentrated load on each span.

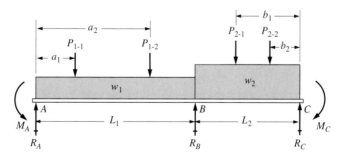

FIGURE 13–17 General notation for terms in Equation 13–6.

general equation for such a loading is a combination of Equations (13–2) and (13–4), given as Equation (13–6).

Three Moment
Equation—
General Form

$$M_A L_1 + 2M_B(L_1 + L_2) + M_C L_2 = -\sum \left[\frac{P_i a_i}{L_1}(L_1^2 - a_i^2) \right]_1 - \sum \left[\frac{P_i b_i}{L_2}(L_2^2 - b_i^2) \right]_2$$
$$- \frac{w_1 L_1^3}{4} - \frac{w_2 L_2^3}{4} \tag{13–6}$$

The term in the bracket with the subscript *1* is to be evaluated for *each* concentrated load on span 1 and then summed together. Similarly, the term having the subscript *2* is repeatedly applied for all loads on span 2. Notice that the distances a_i are measured from the reaction at *A* for each load on span 1, and the distances b_i are measured from the reaction at *C* for each load on span 2. The moments at the ends *A* and *C* can be due to concentrated moments applied there or to loads on overhangs beyond the supports. Any of the terms in Equation (13–6) may be left out of a problem solution if there is no appropriate load or moment existing at a particular section for which the equation is being written. Furthermore, other concentrated loads could be included in addition to those actually shown in Figure 13–17.

**Example Problem
13–7**

The loading composed of a combination of distributed loads and concentrated loads, shown in Figure 13–18, is to be analyzed to determine the reactions at the three supports and the complete shearing force and bending moment diagrams. The 17-m beam is to be used as a floor beam in an industrial building.

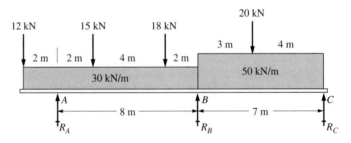

FIGURE 13–18   Beam for Example Problem 13–7.

Solution   **Objective**   Determine the support reactions and draw the shearing force and bending moment diagrams.

Given   Beam loading in Figure 13–18.

Analysis   Because the loading pattern contains both concentrated loads and uniformly distributed loads, we must use Equation (13–6). The subscript *1* refers to span 1 between supports *A* and *B*, and subscript *2* refers to span 2 between supports *B* and *C*. We must evaluate the actual magnitude of M_A and M_C to facilitate the solution of Equation (13–6). Because point *C* is at the end of a simply supported span, $M_C = 0$. At point *A* we can consider the overhanging part of the beam to the left of A to be a free body and then sum moments about A to find the internal moment in the beam at A. Along with the 12 kN load 2.0 m from A, a resultant of 60 kN of load due to the distributed load acts a distance of 1.0 m from A. Then,

$$M_A = -12 \text{ kN } (2 \text{ m}) - 60 \text{ kN } (1 \text{ m}) = -84 \text{ kN·m}$$

Each remaining term in Equation (13–6) will be evaluated.

$$M_A L_1 = -84 \text{ kN·m}(8 \text{ m}) = -672 \text{ kN·m}^2$$

$$2M_B(L_1 + L_2) = 2M_B(8 + 7) = M_B(30 \text{ m})$$

$$-\sum \left[\frac{P_i a_i}{L_1} (L_1^2 - a_i^2) \right]_1 = -\frac{15(2)}{8}(8^2 - 2^2) - \frac{18(6)}{8}(8^2 - 6^2)$$

$$= -603 \text{ kN·m}^2$$

$$-\sum \left[\frac{P_i b_i}{L_2} (L_2^2 - b_i^2) \right]_2 = -\frac{20(4)}{7}(7^2 - 4^2) = -377 \text{ kN·m}^2$$

$$-\frac{w_1 L_1^3}{4} = -\frac{30(8)^3}{4} = -3840 \text{ kN·m}^2$$

$$-\frac{w_2 L_2^3}{4} = -\frac{50(7)^3}{4} = -4288 \text{ kN·m}^2$$

Now putting these values into Equation (13–6) gives

$$-672 \text{ kN·m}^2 + M_B(30 \text{ m}) + 0 = (-603 - 377 - 3840 - 4288) \text{ kN·m}^2$$

Solving for M_B yields

$$M_B = 281 \text{ kN·m}$$

Support reactions. Having the three moments at the locations of the support points A, B, and C allows the computation of the reactions at those supports. The solution procedure starts by considering each of the two spans as separate free bodies, as shown in Figure 13–19. In each case, when the

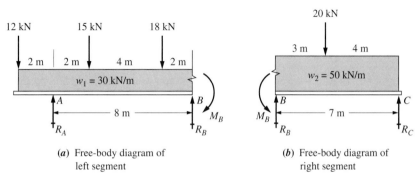

(a) Free-body diagram of left segment

(b) Free-body diagram of right segment

FIGURE 13–19 Free-body diagrams of beam segments used to find reactions R_A, R_B, and R_C.

beam is divided at B, the moment M_B is shown acting at the cut section to maintain equilibrium. Then, using the left segment, we can sum moments about point B and solve for the left reaction, R_A.

$$\sum M_B = 0 = 12 \text{ kN}(10 \text{ m}) + 15 \text{ kN}(6 \text{ m}) + 300 \text{ kN}(5 \text{ m}) - 281 \text{ kN·m}$$

$$- R_A(8 \text{ m})$$
$$R_A = 183 \text{ kN}$$

Similarly, using the right segment and summing moments about point B allows the determination of the right reaction, R_C.

$$\sum M_B = 0 = 20 \text{ kN}(3 \text{ m}) + 350 \text{ kN}(3.5 \text{ m}) - 281 \text{ kN·m} - R_C(7 \text{ m})$$

$$R_C = 143 \text{ kN}$$

Now we can use the $\sum F_v = 0$ to find the middle reaction, R_B.

$$\sum F_v = 0 = 12 \text{ kN} + 15 \text{ kN} + 18 \text{ kN} + 20 \text{ kN} + 300 \text{ kN} + 350 \text{ kN}$$

$$- 183 \text{ kN} - 143 \text{ kN} - R_B$$
$$R_B = 389 \text{ kN}$$

Shearing force and bending moment diagrams. We now have the necessary data to draw the complete diagrams, as shown in Figure 13–20.

Comment   In summary, the reaction forces are

$$R_A = 183 \text{ kN}$$
$$R_B = 389 \text{ kN}$$
$$R_C = 143 \text{ kN}$$

Figure 13–20 shows that the local maximum positive bending moments occur between the supports, and the local maximum negative bending moments occur at the supports. The overall maximum positive bending moment is 204 kN·m at a point 2.86 m from C where the shearing force curve crosses the zero axis. The actual maximum negative bending moment is −281 kN·m at support B. If a beam of uniform cross section is used, it would have to be designed to withstand a bending moment of 281 kN·m. But notice that this is a rather localized peak on the bending moment diagram. It may prove economical to design the beam to withstand the 204 kN·m bending moment and then adding reinforcing plates in the vicinity of support B to increase the section modulus there to a level that would be safe for the 281 kN·m moment. You might be able to notice many highway overpasses designed in this manner.

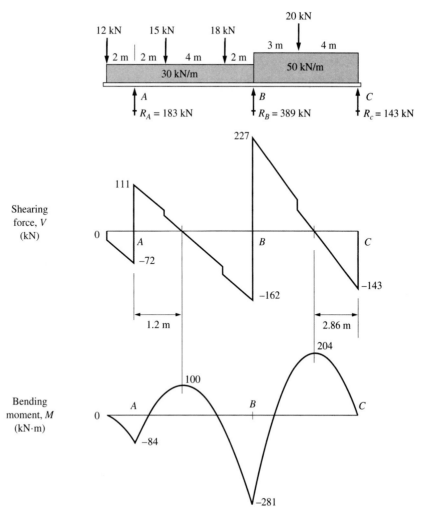

FIGURE 13-20 Shearing force and bending moment diagrams for Example Problem 13-7.

PROBLEMS

For Problems 13–1 through 13–27, use the formulas in Appendix A–24 to complete any of the following according to the instructions given for a particular assignment.

a. Determine the reactions and draw the complete shearing force and bending moment diagrams. Report the maximum shearing force and the maximum bending moment and indicate where they occur.

b. Where deflection formulas are available, also compute the maximum deflection of the beam expressed in the form,

$$y = C_d/EI$$

where EI is the beam stiffness, the product of the modulus of elasticity for the material of the beam and the moment of inertia of the cross section of the beam. The term C_d will then be the result of the computation of all other variables in the deflection equation for the particular beam support type, span length, and loading pattern.

c. Complete the design of the beam specifying a suitable material, and the shape and size for the cross section. The standard for design must include the specification that bending stresses and shearing stresses are safe for the given material. Unless otherwise specified by the assignment, consider all loads to be static.

d. Complete the design of the beam to limit the maximum deflection to some specified value as given by the assignment. In the absence of a specified limit, use $L/360$ as the maximum allowable deflection where L is the span between supports or the overall length of the beam. The design must specify a suitable material, and the shape and size for the cross section. This assignment may be linked to part b where the deflection was calculated in terms of the beam stiffness, EI. Then, for example, you may specify the material and its value for E, compute the limiting deflection, and solve for the required moment of inertia, I. The shape and size of the cross section can then be determined. Note that any design must also be shown to be safe with regard to bending stresses and shearing stresses as in part c.

13–1.M Use A–24(a) with $P = 35$ kN, $L = 4.0$ m.

13–2.M Use A–24(b) with $P = 35$ kN, $L = 4.0$ m, $a = 1.50$ m.

13–3.M Use A–24(b) with $P = 35$ kN, $L = 4.0$ m, $a = 2.50$ m.

13–4.E Use A–24(c) with $w = 400$ lb/ft, $L = 14.0$ ft.

13–5.E Use A–24(c) with $w = 50$ lb/in, $L = 16.0$ in.

13–6.E Use A–24(d) with $P = 350$ lb, $L = 10.8$ in, $a = 2.50$ in.

13–7.M Use A–24(e) with $P = 35$ kN, $L = 4.0$ m.

13–8.M Use A–24(f) with $P = 35$ kN, $L = 4.0$ m, $a = 1.50$ m.

13–9.M Use A–24(f) with $P = 35$ kN, $L = 4.0$ m, $a = 2.50$ m.

13–10.E Use A–24(g) with $w = 400$ lb/ft, $L = 14.0$ ft.

13–11.E Use A–24(g) with $w = 50$ lb/in, $L = 16.0$ in.

13–12.E Use A–24(h) with $w = 400$ lb/ft, $L = 7.0$ ft.

13–13.E Use A–24(h) with $w = 50$ lb/in, $L = 8.0$ in.

13–14.E Use A–24(i) with $w = 400$ lb/ft, $L = 56$ in.

13–15.E Use A–24(i) with $w = 50$ lb/in, $L = 5.333$ in.

13–16.E Use A–24(j) with $w = 400$ lb/ft, $L = 3.5$ ft.

13–17.E Use A–24(j) with $w = 50$ lb/in, $L = 4.0$ in.

13–18.M Use Figure P13–1.

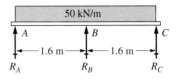

FIGURE P13–1

13–19.M Use Figure P13–2.

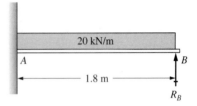

FIGURE P13–2

13–20.E Use Figure P13–3.

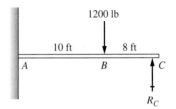

FIGURE P13–3

13–21.M Use A–24(d) with $P = 18$ kN, $L = 2.75$ m, $a = 1.40$ m.

13–22.E Use A–24(f) with $P = 8500$ lb, $L = 109$ in, $a = 75$ in.

13–23.E Use A–24(h) with $w = 4200$ lb/ft, $L = 16.0$ ft.

13–24.M Use A–24(i) with $w = 50$ kN/m, $L = 3.60$ m.

13–25.E Use A–24(j) with $w = 15$ lb/in, $L = 36$ in.

13–26.E Use A–24(e) with $P = 140$ lb, $L = 54$ in.

13–27.M Use A–24(b) with $P = 250$ N, $L = 55$ mm, $a = 15$ mm.

13–28.M Compare Problems 13–4, 13–10, 13–12, 13–14, and 13–16 with regard to the maximum values of shearing force, bending moment, and deflection.

13–29.M Compare Problems 13–5, 13–11, 13–13, 13–15, and 13–17 with regard to the maximum values of shearing force, bending moment, and deflection.

13–30.E Specify a suitable design for a wooden beam, simply supported at its ends, to carry a uniformly distributed load of 120 lb/ft for a span of 24 ft. The beam must be safe for both bending and shear stresses when made from No. 2 grade southern pine. Then compute the maximum deflection for the beam you have designed.

13–31.E Repeat Problem 13–30 except place an additional support at the middle of the beam, 12 ft from either end.

13–32.E Repeat Problem 13–30 except use four supports 8 ft apart.

13–33.E Compare the behavior of the three beams designed in Problems 13–30, 13–31, and 13–32 with regard to shearing force, bending moment, and maximum deflection. Complete the analysis in the manner used in Example Problems 13–1 through 13–5.

13–34.E A plank from a wooden deck carries a load like that shown in Case *j* in Appendix A–24 with $w = 100$ lb/ft and $L = 24$ in. The plank is a

standard 2 × 6 having a thickness of 1.50 in and a width of 5.50 in with the long dimension horizontal. Would the plank be safe if it is made from No. 2 grade southern pine?

13–35.M Two designs for a diving board are proposed, as shown in Figure P13–4. Compare the designs with regard to shearing force, bending moment, and deflection. Express the deflection in terms of the beam stiffness, *EI*. Note that cases from Appendixes A–22 and A–24 can be used.

13–36.M For each of the proposed diving board designs described in Problem 13–35 and shown in Figure P13–4, complete the design, specifying the material, the cross section, and the final dimensions. The board is to be 600 mm wide.

13–37.M Figure P13–5 shows a roof beam over a loading dock of a factory building. Compute the reactions at the supports and draw the complete shearing force and bending moment diagrams.

13–38.M For the roof beam described in Problem 13–37, complete its design, specifying the material, the cross section, and the final dimensions.

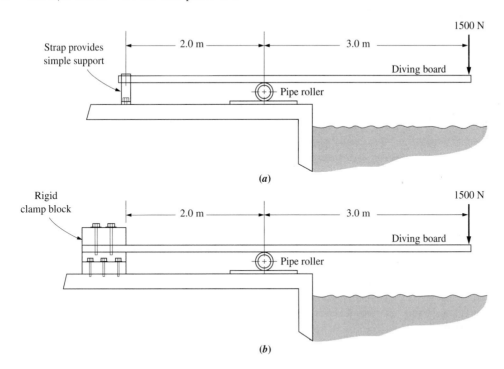

FIGURE P13–4 Proposals for diving boards for Problems 13–35 and 13–36. (a) Simple supports with overhang. (b) Supported cantilever with overhang.

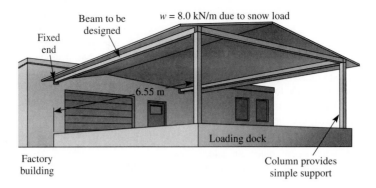

FIGURE P13–5 Roof beam for a loading dock for Problems 13–37 and 13–38.

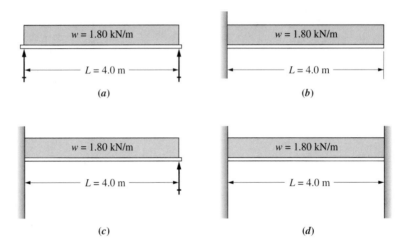

FIGURE P13–6 Beams for Problem 13–39. (a) Simply supported beam. (b) Cantilever. (c) Supported cantilever. (d) Fixed end beam.

13–39.M Compare the behavior of the four beams shown in Figure P13–6 with regard to shearing force, bending moment, and maximum deflection. In each case, the beam is designed to support a uniformly distributed load across a given span. Complete the analysis in a manner similar to that used in Example Problems 13–1 through 13–5.

For Problems 13–40 through 13–47, use the superposition method to determine the reactions at all supports and draw the complete shearing force and bending moment diagrams. Indicate the maximum shearing force and bending moment for each beam.

13–40.M Use Figure P13–1.

13–41.M Use Figure P13–2.

13–42.E Use Figure P13–3.

13–43.E Use Figure P13–7.

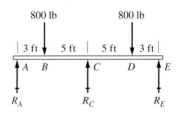

FIGURE P13–7

13–44.E Use Figure P13–8.

13–45.M Use Figure P13–9.

13–46.M Use Figure P13–10.

13–47.E Use Figure P13–11.

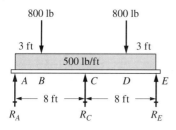

FIGURE P13–8

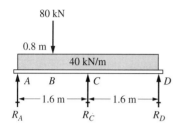

FIGURE P13–9

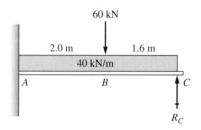

FIGURE P13–10

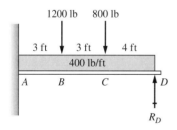

FIGURE P13–11

For Problems 13–48 through 13–55, use the theorem of three moments to determine the reactions at all supports and draw the complete shearing force and bending moment diagrams. Indicate the maximum shearing force and bending moment for each beam.

13–48.M Use Figure P13–1.

13–49.E Use Figure P13–7.

13–50.E Use Figure P13–8.

13–51.M Use Figure P13–9.

13–52.M Use Figure P13–12.

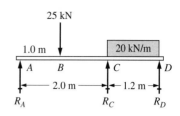

FIGURE P13–12

13–53.M Use Figure P13–13.

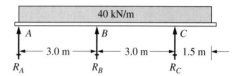

FIGURE P13–13

13–54.M Use Figure P13–14.

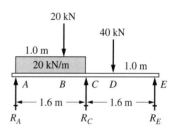

FIGURE P13–14

13–55.M Use Figure P13–15.

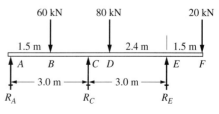

FIGURE P13–15

COMPUTER PROGRAMMING ASSIGNMENTS

1. Write a program to solve for the critical shearing forces and bending moments for any of the statically indeterminate beam types in Appendix A–24.

Enhancements to Assignment 1

(a) Use the graphics mode to plot the complete shearing force and bending moment diagrams for the beams.

(b) Compute the required section modulus for the beam cross section to limit the stress due to bending a specified amount.

(c) Include a table of section properties for steel wide-flange beams and search for suitable beam sizes to carry the load.

(d) Assuming that the beam cross section will be a solid circular section, compute the required diameter.

(e) Assuming that the beam cross section will be a rectangle with a given ratio of height to thickness, compute the required dimensions.

(f) Assuming that the beam cross section will be a rectangle with a given height *or* thickness, compute the other required dimension.

(g) Assuming that the beam is to be made from wood, in a rectangular shape, compute the required area of the beam cross section to limit the shearing stress to a specified value. Use the special shear formula for rectangular beams from Chapter 9.

(h) Add the computation of the deflection at specified points on the beam using the formulas of Appendix A–24.

2. Combine two or more of the beam formulas from Appendix A–22 in a program to use the superposition method to solve for the reactions, shearing forces, and bending moments in a statically indeterminate beam using the method outlined in Section 13–4.

3. Write a program to aid in the solution of Equation (13–6), the theorem of three moments applied to a continuous beam of two spans with combinations of uniformly distributed loads and several concentrated loads. Notice that this equation reduces to Equation (13–2), (13–3), (13–4), or (13–5) when certain terms are zero.

14

Columns

14–1 OBJECTIVES OF THIS CHAPTER

A column is a relatively long member loaded in compression.

The analysis of columns is different from what has been studied before because the mode of failure is different. In Chapter 3, when compression stress was discussed, it was assumed that the member failed by yielding of the material when a stress greater than the yield strength of the material was applied. This is true for short members.

A long, slender column fails by *buckling,* a common name given to *elastic instability.* Instead of crushing or tearing the material, the column deflects drastically at a certain critical load and then collapses suddenly. Any thin member can be used to illustrate the buckling phenomenon. Try it with a wood or plastic ruler, a thin rod or strip of metal, or a drinking straw. By gradually increasing the force, applied directly downward, the critical load is reached when the column begins to bend. Normally, the load can be removed without causing permanent damage since yielding does not occur. Thus a column fails by buckling at a stress lower than the yield strength of the material in the column. The objective of column analysis methods is to predict the load or stress level at which a column would become unstable and buckle.

After completing this chapter, you should be able to:

1. Define *column.*
2. Differentiate between a column and a short compression member.
3. Describe the phenomenon of *buckling,* also called *elastic instability.*

513

4. Define *radius of gyration* for the cross section of a column and be able to compute its magnitude.

5. Understand that a column is expected to buckle about the axis for which the radius of gyration is the minimum.

6. Define *end-fixity factor, K.*

7. Specify the appropriate value of the end-fixity factor, *K*, depending on the manner of supporting the ends of a column.

8. Define *effective length, L_e.*

9. Define *slenderness ratio* and compute its value.

10. Define *transition slenderness ratio,* also called the *column constant, C_c,* and compute its value.

11. Use the values for the slenderness ratio and the column constant to determine when a column is *long* or *short.*

12. Use the *Euler formula* for computing the critical buckling load for long columns.

13. Use the *J. B. Johnson formula* for computing the critical buckling load for short columns.

14. Apply a design factor to the critical buckling load to determine the *allowable load* on a column.

15. Recognize efficient shapes for column cross sections.

16. Design columns to safely carry given axial compression loads.

17. Apply the specifications of the American Institute of Steel Construction (AISC) code to the analysis of columns.

18. Apply the specifications of the Aluminum Association to the analysis of columns.

14–2 SLENDERNESS RATIO

A column has been described as a relatively long, slender member loaded in compression. This description is stated in relative terms and is not very useful for analysis.

The measure of the slenderness of a column must take into account the length, the cross-sectional shape and dimensions of the column, and the manner of attaching the ends of the column to the structures that supply loads and reactions to the column. The commonly used measure of slenderness is the *slenderness ratio,* defined as

 **Slenderness Ratio**

$$SR = \frac{KL}{r} = \frac{L_e}{r} \tag{14–1}$$

where L = *actual length* of the column between points of support or lateral restraint

K = *end-fixity factor*

L_e = *effective length,* taking into account the manner of attaching the ends (note that $L_e = KL$)

r = *smallest radius of gyration* of the cross section of the column

Each of these terms is discussed below.

Actual length, L. For a simple column having the load applied at one end and the reaction provided at the other, the actual length is, obviously, the length between its ends. But for components of structures loaded in compression where a means of restraining the member laterally to prevent buckling is provided, the actual length is taken between points of restraint. Each part is then considered to be a separate column.

End-fixity factor, K. The end-fixity factor is a measure of the degree to which each end of the column is restrained against rotation. Three classic types of end connections are typically considered: the pinned end, the fixed end, and the free end. Figure 14–1 shows these end types in several combinations with the corresponding values of K. Note that two values of K are given. One is the theoretical value and the other is the one typically used in practical situations, recognizing that it is difficult to achieve the truly fixed end, as discussed below.

Pinned ends of columns are essentially unrestrained against rotation. When a column with two pinned ends buckles, it assumes the shape of a smooth curve between its ends, as shown in Figure 14–1(a). This is the basic case of column buckling and the value of $K = 1.0$ is applied to columns having two pinned ends. An ideal implementation of the pinned end is the frictionless spherical ball joint that would permit the column to rotate in any direction about any axis. For a cylindrical pin joint, free rotation is permitted about the centerline of the pin, but some restraint is provided in the plane perpendicular to the centerline. Care must be exercised in applying end-fixity factors to cylindrical pins for this reason. It is assumed that the pinned end is guided in some way so that the line of action of the axial load remains unchanged.

Fixed ends provide theoretically perfect restraint against rotation of the column at its ends. As the column tends to buckle, the deflected shape of the axis of the

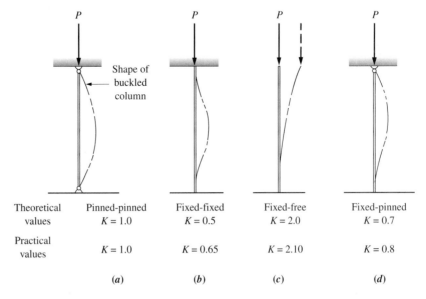

FIGURE 14–1 Values of K for effective length, $L_e = KL$, for different end connections.

column must approach the fixed end with a zero slope, as illustrated in Figure 14–1(b). The buckled shape bows out in the middle but exhibits two points of inflection to reverse the direction of curvature near the ends. The theoretical value of the end-fixity factor is $K = 0.5$, indicating that the column acts as if it were only one-half as long as it really is. Columns with fixed ends are much stiffer than pinned-end columns and can therefore take higher loads before buckling. It should be understood that it is very difficult to provide perfectly fixed ends for a column. It requires that the connection to the column is rigid and stiff and that the structure to which the loads are transferred is also rigid and stiff. For this reason, the higher value of $K = 0.65$ is recommended for practical use.

A free end for a column is unrestrained against rotation and also against translation. Because it can move in any direction, this is the worst case for column end-fixity. The only practical way of using a column with a free end is to have the opposite end fixed, as illustrated in Figure 14–1(c). Such a column is sometimes referred to as the *flagpole* case because the fixed end is similar to the flagpole inserted deeply into a tight-fitting socket while the other end is free to move in any direction. Called the fixed-free end condition, the theoretical value of K is 2.0. A practical value is $K = 2.10$.

The combination of one fixed end and one pinned end is shown in Figure 14–1(d). Notice that the buckled shape approaches the fixed end with a zero slope while the pinned end rotates freely. The theoretical value of $K = 0.7$ applies to such an end-fixity while $K = 0.80$ is recommended for practical use.

Effective length, L_e.   Effective length combines the actual length with the end-fixity factor; $L_e = KL$. In the problems in this book we use the recommended practical values for the end-fixity factor, as shown in Figure 14–1. In summary, the following relationships will be used to compute the effective length:

Effective Length

1. Pinned-end columns: $L_e = KL = 1.0(L) = L$
2. Fixed-end columns: $L_e = KL = 0.65(L)$
3. Fixed-free columns: $L_e = KL = 2.10(L)$
4. Fixed-pinned columns: $L_e = KL = 0.80(L)$

Radius of gyration, r.   The measure of slenderness of the cross section of the column is its radius of gyration, r, defined as

Radius of Gyration

$$r = \sqrt{\frac{I}{A}} \qquad (14\text{–}2)$$

where I = moment of inertia of the cross section of the column with respect to one of the principal axes
A = area of the cross section.

Because both I and A are geometrical properties of the cross section, the radius of gyration, r, is also. Formulas for computing r for several common shapes are given in Appendix A–1. Also, r is listed with other properties for some of the standard shapes in the Appendix. For those for which r is not listed, the values of I and A are available and Equation (14–2) can be used to compute r very simply.

Note that the value of the radius of gyration, r, is dependent on the axis about which it is to be computed. In most cases, it is required that you determine the axis for which the *radius of gyration is the smallest,* because that is the axis about which the column would likely buckle. Consider, for example, a column made from a rectangular section whose width is much greater than its thickness, as sketched in Figure 14–2. A simple ruler or a meter stick can be used to demonstrate that when loaded in axial compression with little or no restraint at the ends, the column will always buckle with respect to the axis through the thinnest dimension. For the rectangle, as shown in Figures 14–2(b) and (c),

$$r_{\min} = r_Y = 0.289t$$

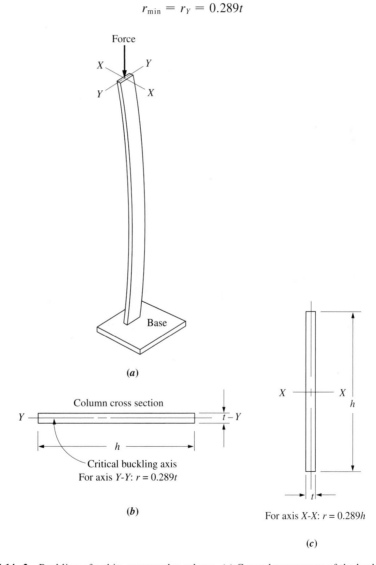

FIGURE 14–2 Buckling of a thin, rectangular column. (a) General appearance of the buckled column. (b) Radius of gyration for *Y-Y* axis. (c) Radius of gyration for *X-X* axis.

where t is the thickness of the rectangle. Note that

$$r_X = 0.289h$$

where h is the height of the rectangle and that $h > t$. Thus

$$r_X > r_Y$$

and then r_Y is the smallest radius of gyration for the section.

For the wide-flange beams (Appendix A–7) and American Standard beams (Appendix A–8), the minimum value of r is that computed with respect to the Y-Y axis; that is,

$$r_{min} = \sqrt{\frac{I_Y}{A}}$$

Similarly, for rectangular structural tubing (Appendix A–9), the minimum radius of gyration is that with respect to the Y-Y axis. Values for r are listed in the table.

For structural steel angles, called L-shapes, neither the X-X nor the Y-Y axis provides the minimun radius of gyration. As illustrated in Appendix A–5, r_{min} is computed with respect to the Z-Z axis, with the values listed in the table.

For symmetrical sections, the value of r is the same with respect to any principal axis. Such shapes are the solid or hollow circular section and the solid or hollow square section.

Summary of the method for computing the slenderness ratio

1. Determine the actual length of the column, L, between endpoints or between points of lateral restraint.
2. Determine the end-fixity factor from the manner of support of the ends and by reference to Figure 14–1.
3. Compute the effective length, $L_e = KL$.
4. Compute the *smallest* radius of gyration for the cross section of the column.
5. Compute the slenderness ratio from

$$SR = \frac{L_e}{r_{min}}$$

14–3 TRANSITION SLENDERNESS RATIO

When is a column considered long? The answer to this question requires the determination of the *transition slenderness ratio,* or column constant C_c.

Transition Slenderness Ratio

$$C_c = \sqrt{\frac{2\pi^2 E}{s_y}} \qquad (14\text{–}3)$$

The following rules are related to the value of C_c.

> If the actual effective slenderness ratio L_e/r is greater than C_c, then the column is long, and the Euler formula, defined in the next section, should be used to analyze the column.

> If the actual ratio L_e/r is less than C_c, then the column is short. In these cases, either the J. B. Johnson formula, special codes, or the direct compressive stress formula should be used, as discussed in later sections.

Where a given column is being analyzed to determine the load it will carry, the value of C_c and the actual ratio L_e/r should be computed first to determine which method of analysis should be used. Notice that C_c depends on the material properties of yield strength s_y and modulus of elasticity E. In working with steel, E is usually taken to be 207 GPa (30×10^6 psi). Using this value and assuming a range of values for yield strength, we obtain the values for C_c shown in Figure 14–3.

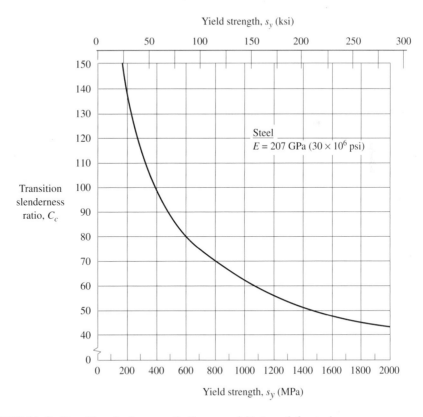

FIGURE 14–3 Transition slenderness ratio C_c versus yield strength for steel.

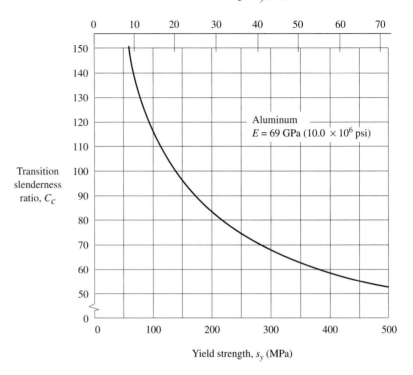

FIGURE 14–4 Transition slenderness ratio C_c versus yield strength for aluminum.

For aluminum, E is approximately 69 GPa (10×10^6 psi). The corresponding values for C_c are shown in Figure 14–4.

14–4 THE EULER FORMULA FOR LONG COLUMNS

For long columns having an effective slenderness ratio greater than the transition value C_c, the Euler formula can be used to predict the critical load at which the column would be expected to buckle. The formula is

> **Euler Formula for Long Columns**

$$P_{cr} = \frac{\pi^2 E A}{(L_e/r)^2} \qquad (14-4)$$

where A is the cross-sectional area of the column. An alternative form can be expressed in terms of the moment of inertia by noting that $r^2 = I/A$. Then the formula becomes

> **Euler Formula for Long Columns**

$$P_{cr} = \frac{\pi^2 E I}{L_e^2} \qquad (14-5)$$

14–5 THE J. B. JOHNSON FORMULA FOR SHORT COLUMNS

If a column has an actual effective slenderness ratio L_e/r less than the transition value C_c, the Euler formula predicts an unreasonably high critical load. One formula recommended for machine design applications in the range of L_e/r less than C_c is the J. B. Johnson formula.

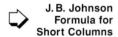
J. B. Johnson Formula for Short Columns

$$P_{cr} = As_y \left[1 - \frac{s_y(L_e/r)^2}{4\pi^2 E} \right] \qquad (14-6)$$

This is one form of a set of equations called parabolic formulas, and it agrees well with the performance of steel columns in typical machinery.

The Johnson formula gives the same result as the Euler formula for the critical load at the transition slenderness ratio C_c. Then for very short columns the critical load approaches that which would be predicted from the direct compressive stress equation, $\sigma = P/A$. Therefore, it could be said that the Johnson formula applies best to columns of intermediate length.

14–6 DESIGN FACTORS FOR COLUMNS AND ALLOWABLE LOAD

Because the mode of failure for a column is buckling rather than yielding or ultimate failure of the material, the methods used before for computing design stress do not apply to columns.

Instead, an *allowable load* is computed by dividing the critical buckling load from either the Euler formula [Equation (14–4)] or the Johnson formula [Equation (14–6)] by a design factor, N. That is,

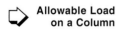
Allowable Load on a Column

$$P_a = \frac{P_{cr}}{N} \qquad (14-7)$$

where P_a = allowable, safe load
P_{cr} = critical buckling load
N = design factor

The selection of the design factor is the responsibility of the designer unless the project comes under a code. Factors to be considered in the selection of a design factor are similar to those used for design factors applied to stresses. A common factor used in mechanical design is $N = 3.0$, chosen because of the uncertainty of material properties, end-fixity, straightness of the column, or the possibility that the load will be applied with some eccentricity rather than along the axis of the column. Larger factors are sometimes used for critical situations and for very long columns.

In building construction, where the design is governed by the specifications of the American Institute of Steel Construction, AISC, a factor of 1.92 is recommended for long columns. The Aluminum Association calls for $N = 1.95$ for long columns. See Sections 14–9 and 14–10.

14–7 SUMMARY—METHOD OF ANALYZING COLUMNS

The purpose of this section is to summarize the concepts presented in Sections 14–2 through 14–6 into a procedure that can be used to analyze columns. It can be applied to a straight column having a uniform cross section throughout its length, and for which the compression load is applied in line with the centroidal axis of the column.

To start, it is assumed that the following factors are known:

Method of Analyzing Columns

1. The actual length, L
2. The manner of connecting the column to its supports
3. The shape of the cross section of the column and its dimensions
4. The material from which the column is made

Then the procedure is:

1. Determine the end-fixity factor, K, by comparing the manner of connection of the column to its supports with the information in Figure 14–1.
2. Compute the effective length, $L_e = KL$.
3. Compute the minimum value of the radius of gyration of the cross section from $r_{min} = \sqrt{I_{min}/A}$; or determine r_{min} from tables of data.
4. Compute the maximum slenderness ratio from

$$SR_{max} = \frac{L_e}{r_{min}}$$

5. Using the modulus of elasticity, E, and the yield strength, s_y, for the material, compute the column constant,

$$C_c = \sqrt{\frac{2\pi^2 E}{s_y}}$$

6. Compare the value of SR with C_c.
   a. If $SR > C_c$, the column is long. Use the Euler formula to compute the critical buckling load,

   $$P_{cr} = \frac{\pi^2 EA}{(SR)^2}$$

   b. If $SR < C_c$, the column is short. Use the Johnson formula to compute the critical buckling load,

   $$P_{cr} = A s_y \left[1 - \frac{s_y (SR)^2}{4\pi^2 E} \right]$$

7. Specify the design factor, N.
8. Compute the allowable load, P_a,

$$P_a = \frac{P_{cr}}{N}$$

Example Problem 14–1

A round compression member with both ends pinned and made of AISI 1020 cold-drawn steel is to be used in a machine. Its diameter is 25 mm, and its length is 950 mm. What maximum load can the member take before buckling would be expected? Also compute the allowable load on the column for a design factor of $N = 3$.

Solution

Objective Compute the critical buckling load for the column and the allowable load for a design factor of $N = 3$.

Given $L = 950$ mm. Cross section is circular; $D = 25$ mm. Pinned ends.
Column is steel; AISI 1020 cold-drawn.
From Appendix A–13: $s_y = 441$ MPa; $E = 207$ GPa $= 207 \times 10^9$ N/m^2

Analysis Use the *Guidelines for analyzing centrally loaded columns*.

Results **Step 1.** Determine the end-fixity factor. For the pinned-end column, $K = 1.0$.

Step 2. Compute the effective length.

$$L_e = KL = 1.0(L) = 950 \text{ mm}$$

Step 3. Compute the smallest value of the radius of gyration. From Appendix A–1, for any axis of a circular cross section, $r = D/4$. Then,

$$r = \frac{D}{4} = \frac{25 \text{ mm}}{4} = 6.25 \text{ mm}$$

Step 4. Compute the slenderness ratio, $SR = L_e/r$.

$$\frac{L_e}{r} = \frac{950 \text{ mm}}{6.25 \text{ mm}} = 152$$

Step 5. Compute the column constant, C_c.

$$C_c = \sqrt{\frac{2\pi^2 E}{s_y}} = \sqrt{\frac{2\pi^2 (207 \times 10^9 \text{ N/m}^2)}{441 \times 10^6 \text{ N/m}^2}} = 96.2$$

Step 6. Compare C_c with L_e/r and decide if column is long or short. Then use the appropriate column formula to compute the critical buckling load. Since L_e/r is greater than C_c, Euler's formula applies.

$$P_{cr} = \frac{\pi^2 EA}{(L_e/r)^2}$$

The area is

$$A = \frac{\pi D^2}{4} = \frac{\pi (25 \text{ mm})^2}{4} = 491 \text{ mm}^2$$

Then

$$P_{cr} = \frac{\pi^2 (207 \times 10^9 \text{ N/m}^2)(491 \text{ mm}^2)}{(152)^2} \times \frac{1 \text{ m}^2}{(10^3 \text{ mm})^2} = 43.4 \text{ kN}$$

Step 7. A design factor of $N = 3$ is specified.

Step 8. The allowable load, P_a, is

$$P_a = \frac{P_{cr}}{N} = \frac{43.4 \text{ kN}}{3} = 14.5 \text{ kN}$$

Example Problem 14–2

Determine the critical load on a steel column having a square cross section 12 mm on a side with a length of 300 mm. The column is to be made of AISI 1040, hot rolled. It will be rigidly welded to a firm support at one end and connected by a pin joint at the other. Also compute the allowable load on the column for a design factor of $N = 3$.

Solution

Objective Compute the critical buckling load for the column and the allowable load for a design factor of $N = 3$.

Given $L = 300$ mm. Cross section is square; each side is $b = 12$ mm. One pinned end; one fixed end. Column is steel; AISI 1040 hot rolled. From Appendix A–13: $s_y = 414$ MPa; $E = 207$ GPa $= 207 \times 10^9$ N/m^2

Analysis Use the *Guidelines for analyzing centrally loaded columns.*

Results **Step 1.** Determine the end-fixity factor. For the fixed-pinned-end column, $K = 0.80$ is a practical value. (Figure 14–1)

Step 2. Compute the effective length.

$$L_e = KL = 0.80(L) = 0.80(300 \text{ mm}) = 240 \text{ mm}$$

Step 3. Compute the smallest value of the radius of gyration. From Appendix A–1, for a square cross section, $r = b/\sqrt{12}$. Then,

$$r = \frac{b}{\sqrt{12}} = \frac{12 \text{ mm}}{\sqrt{12}} = 3.46 \text{ mm}$$

Step 4. Compute the slenderness ratio, $SR = L_e/r$.

$$\frac{L_e}{r} = \frac{KL}{r} = \frac{(0.8)(300 \text{ mm})}{3.46 \text{ mm}} = 69.4$$

Step 5. Normally, we would compute the value of the column constant, C_c. But, in this case, let's use Figure 14–3. For a steel having a yield strength of 414 MPa, $C_c = 96$, approximately.

Step 6. Compare C_c with L_e/r and decide if the column is long or short. Then use the appropriate column formula to compute the critical buckling load. Since L_e/r is less than C_c, the Johnson formula, Equation (14−6), should be used.

$$P_{cr} = As_y\left[1 - \frac{s_y(L_e/r)^2}{4\pi^2 E}\right]$$

The area of the square cross section is

$$A = b^2 = (12 \text{ mm})^2 = 144 \text{ mm}^2$$

Then,

$$P_{cr} = (144 \text{ mm}^2)\left(\frac{414 \text{ N}}{\text{mm}^2}\right)\left[1 - \frac{(414 \times 10^6 \text{ N/m}^2)(69.4)^2}{4\pi^2(207 \times 10^9 \text{ N/m}^2)}\right]$$

$$= 45.1 \text{ kN}$$

Step 7. A design factor of $N = 3$ is specified.

Step 8. The allowable load, P_a, is

$$P_a = \frac{P_{cr}}{N} = \frac{45.1 \text{ kN}}{3} = 15.0 \text{ kN}$$

14−8 EFFICIENT SHAPES FOR COLUMN CROSS SECTIONS

When designing a column to carry a specified load, the designer has the responsibility for selecting the general shape of the column cross section and then to determine the required dimensions. The following principles may aid in the initial selection of the cross-section shape.

An efficient shape is one that uses a small amount of material to perform a given function. For columns, the goal is to maximize the radius of gyration in order to reduce the slenderness ratio. Note also that because $r = \sqrt{I/A}$, maximizing the moment of inertia for a given area has the same effect.

When discussing moment of inertia in Chapters 7 and 8, it was noted that it is desirable to put as much of the area of the cross section as far away from the centroid as possible. For beams, discussed in Chapter 8, there was usually only one important axis, the axis about which the bending occurred. For columns, buckling can, in general, occur in any direction. Therefore, it is desirable to have uniform properties with respect to any axis. The hollow circular section, commonly called a pipe, then makes a very efficient shape for a column. Closely approximating that is the hollow square tube. Fabricated sections made from standard structural sections can also be used, as shown in Figure 14−5.

Building columns are often made from special wide-flange shapes called *column sections.* They have relatively wide, thick flanges as compared with the shapes typically selected for beams. This makes the moment of inertia with respect to the *Y-Y* axis more nearly equal to that for the *X-X* axis. The result is that the radii of gyration for the

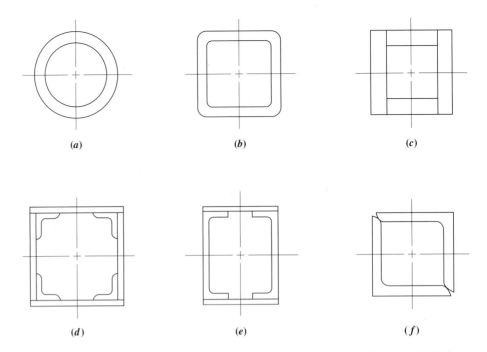

FIGURE 14–5 Examples of efficient column shapes. (a) Hollow circular section, pipe. (b) Hollow square tube. (c) Box section made from wood beams. (d) Equal-leg angles with plates. (e) Aluminum channels with plates. (f) Two equal-leg angles.

two axes are more nearly equal also. Figure 14–6 shows a comparison of two 12-in wide-flange shapes: one a column section and one a typical beam shape. Note that the smaller radius of gyration should be used in computing the slenderness ratio.

14–9 SPECIFICATIONS OF THE AISC

Columns are essential elements of many structures. The design and analysis of steel columns in construction applications are governed by the specifications of the AISC, the American Institute of Steel Construction (1). The specification defines an allowable unit load or stress for columns which is the allowable total load P_a divided by the area of the column cross section. The design formulas are expressed in terms of the transition slenderness ratio C_c, already defined in Equation (14–3), the yield strength of the column material, and the effective slenderness ratio L_e/r. For $L_e/r < C_c$,

$$\frac{P_a}{A} = \left[1 - \frac{(L_e/r)^2}{2C_c^2} \right] \frac{s_y}{FS} \tag{14–8}$$

where P_a = allowable or design load

$C_c = \sqrt{\dfrac{2\pi^2 E}{s_y}}$ (Use $E = 29 \times 10^6$ psi (200 GPa) for structural steel)

FS = factor of safety

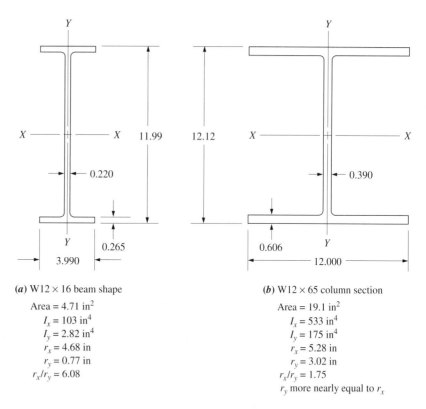

(a) W12 × 16 beam shape

Area = 4.71 in^2
I_x = 103 in^4
I_y = 2.82 in^4
r_x = 4.68 in
r_y = 0.77 in
r_x/r_y = 6.08

(b) W12 × 65 column section

Area = 19.1 in^2
I_x = 533 in^4
I_y = 175 in^4
r_x = 5.28 in
r_y = 3.02 in
r_x/r_y = 1.75
r_y more nearly equal to r_x

FIGURE 14–6 Comparison of a wide-flange beam shape with a column section.

Equation (14–8) was developed by the Column Research Council and is indentical to the Johnson formula. The factor of safety *FS* is a function of the ratio of the effective slenderness ratio to C_c in order to include the effect of accidental crookedness, a small eccentricity of the load, residual stresses, and any uncertainties in the evaluation of the effective length factor *K*. The equation for *FS* is

$$FS = \frac{5}{3} + \frac{3(L_e/r)}{8C_c} - \frac{(L_e/r)^3}{8C_c^3} \qquad (14-9)$$

The value of *FS* varies from 1.67 for the ratio $(L_e/r)/C_c = 0$ to 1.92 for $(L_e/r)/C_c = 1.0$.

For long columns, $L_e/r > C_c$, Euler's equation is used as defined before but with a factor of safety of 1.92.

$$\frac{P_a}{A} = \frac{\pi^2 E}{(L_e/r)^2(1.92)} \qquad (14-10)$$

For structural steel with $E = 29 \times 10^6$ psi,

$$\frac{P_a}{A} = \frac{149 \times 10^6}{(L_e/r)^2} \text{ psi} \qquad (14-11)$$

In the SI system, using $E = 200$ GPa for structural steel,

$$\frac{P_a}{A} = \frac{1028}{(L_e/r)^2} \text{ GPa} \tag{14-12}$$

14-10 SPECIFICATIONS OF THE ALUMINUM ASSOCIATION

The Aluminum Association publication, *Specifications for Aluminum Structures* (2), defines allowable stresses for columns for each of several aluminum alloys and their heat treatments. Three different equations are given for short, intermediate, and long columns defined in relation to slenderness limits. The equations are of the form

$$\frac{P_a}{A} = \frac{s_y}{FS} \qquad \text{(short columns)} \tag{14-13}$$

$$\frac{P_a}{A} = \frac{B_c - D_c(L/r)}{FS} \qquad \text{(intermediate columns)} \tag{14-14}$$

$$\frac{P_a}{A} = \frac{\pi^2 E}{FS(L/r)^2} \qquad \text{(long columns)} \tag{14-15}$$

In all three cases, it is recommended that $FS = 1.95$ for buildings and similar structures. The short column analysis assumes that buckling will not occur and that safety is dependent on the yield strength of the material. Equation (14-15) for long columns is the Euler formula with a factor of safety applied. The intermediate column formula [Equation (14-14)] depends on buckling constants B_c and D_c, which are functions of the yield strength of the aluminum alloy and the modulus of elasticity. The division between intermediate and long columns is similar to the C_c used previously in this chapter.

Following are the specific equations for the alloy 6061-T6 used in building structures in the forms of sheet, plate, extrusions, structural shapes, rod, bar, tube, and pipe. The slenderness ratio L/r should be evaluated using the actual L (pinned ends). Any end restraint is assumed to be allowed for in the factor of safety.

Short columns: $L/r < 9.5$

$$\frac{P_a}{A} = 19 \text{ ksi (131 MPa)} \tag{14-16}$$

Intermediate columns: $9.5 < L/r < 66$

$$\frac{P_a}{A} = \left(20.2 - 0.126\frac{L}{r}\right) \text{ksi} \tag{14-17a}$$

$$\frac{P_a}{A} = \left(139 - 0.869\frac{L}{r}\right) \text{MPa} \tag{14-17b}$$

Long columns: $L/r > 66$

$$\frac{P_a}{A} = \frac{51\,000}{(L/r)^2} \text{ ksi} \qquad (14\text{-}18a)$$

$$\frac{P_a}{A} = \frac{352\,000}{(L/r)^2} \text{ MPa} \qquad (14\text{-}18b)$$

See Reference 2 for column design stresses for other aluminum alloys.

14-11 NON-CENTRALLY LOADED COLUMNS

All of the analysis methods discussed in this chapter so far have been limited to loadings in which the compressive loads on the columns act in-line with the centroidal axis of the column cross section. Also, it is assumed that the column axis is perfectly straight prior to the application of the loads. We have used the term *straight centrally loaded column* to describe such a case.

Many real columns violate these assumptions to some degree. Figure 14–7 shows two such conditions. If a column is initially crooked, the applied compressive force on the column tends to cause bending in the column in addition to buckling, and failure would occur at a lower load than that predicted from the equations used in this chapter. An *eccentrically loaded column* is one in which there is some purposeful offsetting of the line of action of the compressive load from the centroidal axis of the column. Here, again, there is some bending stress produced in addition to the axial compressive stress that tends to cause buckling. References 3 and 4 provide additional methods

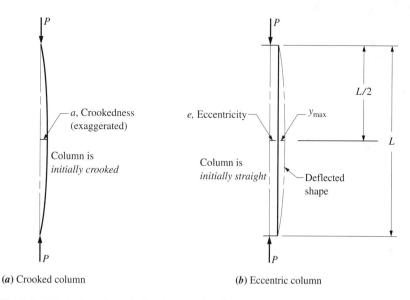

(**a**) Crooked column (**b**) Eccentric column

FIGURE 14–7 Illustration of crooked and eccentric columns.

for handling such non-centrally loaded columns. When there is a small amount of crookedness or eccentricity, the use of a larger design factor than normal would tend to compensate.

REFERENCES

1. American Institute of Steel Construction, *Manual of Steel Construction,* 9th ed., Chicago, 1989.

2. Aluminum Association, *Specifications for Aluminum Structures,* 5th ed., Washington, DC, 1986.

3. Mott, Robert L., *Machine Elements in Mechanical Design,* 2nd ed., Macmillan Publishing Co., New York, 1992.

4. Timoshenko, S., and Gere, J. M., *Theory of Elastic Stability,* 2nd ed., McGraw-Hill Book Company, 1961.

PROBLEMS

14–1.M Determine the critical load for a pinned-end column made of a circular bar of AISI 1020 hot-rolled steel. The diameter of the bar is 20 mm, and its length is 800 mm.

14–2.M Repeat Problem 14–1 with the length of 350 mm.

14–3.M Repeat Problem 14–1 with the bar made of 6061-T6 aluminum instead of steel.

14–4.M Repeat Problem 14–1 with the column ends fixed instead of pinned.

14–5.M Repeat Problem 14–1 with a square steel bar with the same cross-sectional area as the circular bar.

14–6.M For a 1-in schedule 40 steel pipe, used as a column, determine the critical load if it is 2.05 m long. The material is similar to AISI 1020 hot-rolled steel. Compute the critical load for each of the four end conditions described in Figure 14–1.

14–7.M A rectangular steel bar has cross-sectional dimensions of 12 mm by 25 mm and is 210 mm long. Assuming that the bar has pinned ends and is made from AISI 1141 OQT 1300 steel, compute the critical load when the bar is subjected to an axial compressive load.

14–8.M Compute the allowable load on a column with fixed ends if it is 5.45 m long and made from an S6 $\times$ 12.5 beam. The material is ASTM A36 steel. Use the AISC formula.

14–9.E A raised platform is 20 ft by 40 ft in area and is being designed for 75 pounds per square foot uniform loading. It is proposed that standard 3-in schedule 40 steel pipe be used as columns to support the platform 8 ft above the ground with the base fixed and the top free. How many columns would be required if a design factor of 3.0 is desired? Use $s_y = 30\,000$ psi.

14–10.M An aluminum I-beam, I10 $\times$ 8.646, is used as a column with two pinned ends. It is 2.80 m long and made of 6061-T6 aluminum. Using Equations (14–16) through (14–18b), compute the allowable load on the column.

14–11.M Compute the allowable load for the column described in Problem 14–10 if the length is only 1.40 m.

14–12.E A column is a W8 $\times$ 15 steel beam, 12.50 ft long, and made of ASTM A36 steel. Its ends are attached in such a way that L_e is approximately 0.80L. Using the AISC formulas, determine the allowable load on the column.

14–13.E A built-up column is made of four angles, as shown in Figure 14–8. The angles are held together with lacing bars, which can be neglected in the analysis of geometrical properties. Using the standard Johnson or Euler equations with $L_e = L$ and a design factor of 3.0, compute the allowable load on the column if it is 18.4 ft long. The angles are of ASTM A36 steel.

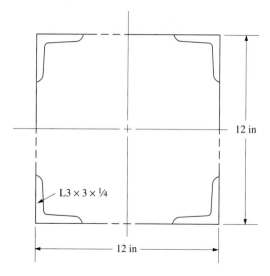

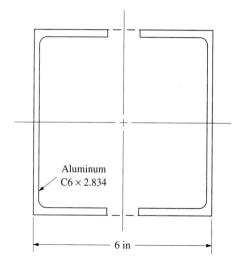

FIGURE 14–8 Built-up column cross section for Problem 14–13.

FIGURE 14–9 Built-up column cross section for Problem 14–14.

14–14.E Compute the allowable load on a built-up column having the cross section shown in Figure 14–9. Use $L_e = L$ and 6061-T6 aluminum. The column is 10.5 ft long. Use the Aluminum Association formulas.

14–15.C Figure 14–10 shows a beam supported at its ends by pin joints. The inclined bar at the top supports the right end of the beam, but also places an axial compressive force in the beam. Would a standard S6 × 12.5 beam be satisfac-

tory in this application if it carries 1320 kg at its end? The beam is made from ASTM A36 steel.

14–16.E A link in a mechanism which is 8.40 in long, has a rectangular cross section $\frac{1}{4}$ in × $\frac{1}{8}$ in, and is subjected to a compressive load of 50 lb. If the link has pinned ends, is it safe from buckling? Cold-drawn AISI 1040 steel is used in the link.

14–17.M A piston rod on a shock absorber is 12 mm in diameter and has a maximum length of 190 mm outside the shock absorber body. The rod is

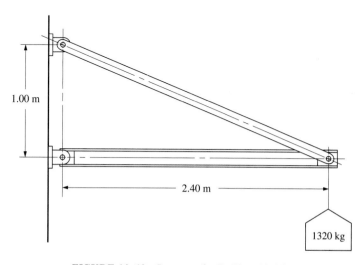

FIGURE 14–10 Structure for Problem 14–15.

made of AISI 1141 OQT 1300 steel. Consider one end to be pinned and the other to be fixed. What axial compressive load on the rod would be one-third of the critical buckling load?

14–18.E A stabilizing rod in an automobile suspension system is a round bar loaded in compression. It is subjected to 1375 lb of axial load and supported at its ends by pin-type connections, 28.5 in apart. Would a 0.800-in diameter bar of AISI 1020 hot-rolled steel be satisfactory for this application?

14–19.E A structure is being designed to support a large hopper over a plastic extruding machine, as sketched in Figure 14–11. The hopper is to be carried by four columns which share the load equally. The structure is cross-braced by rods. It is proposed that the columns be made from standard 2-in schedule 40 pipe. They will be fixed at the floor. Because of the cross-bracing, the top of each column is guided so that it behaves as if it were rounded or pinned. The material for the pipe is AISI 1020 steel, hot rolled. The hopper is designed to hold 20 000 lb of plastic powder. Are the proposed columns adequate for this load?

14–20.E Discuss how the column design in Problem 14–19 would be affected if a careless forklift driver runs into the cross braces and breaks them.

14–21.E The assembly shown in Figure 14–12 is used to test parts by pulling on them repeatedly with the hydraulic cylinder. A maximum force of 3000 lb can be exerted by the cylinder. The parts of the assembly of concern here are the columns. It is proposed that the two columns be made of $1\frac{1}{4}$-in square bars of aluminum alloy 6061-T6. The columns are fixed at the bottom and free at the top. Determine the acceptability of the proposal.

14–22.E Figure 14–13 shows the proposed design for a hydraulic press used to compact solid waste. A piston at the right is capable of exerting a force of 12 500 lb through the connecting rod to the ram. The rod is straight and centrally loaded. It is made from AISI 1040 OQT 1100 steel. Compute the resulting design factor for this design.

14–23.E For the conditions described in Problem 14–22, specify the required diameter of the connecting rod if it is made as a solid circular cross section. Use a design factor of 4.0.

14–24.E For the conditions described in Problem 14–22, specify a suitable standard steel pipe for use as the connecting rod. Use a design factor of 4.0. The pipe is to be made from ASTM A501 structural steel.

14–25.E For the conditions described in Problem 14–22, specify a suitable standard I-beam for use as the connecting rod. Use a design factor of 4.0. The I-beam is to be made from aluminum alloy 6061-T6. The connection at the piston is as shown in Figure 14–14.

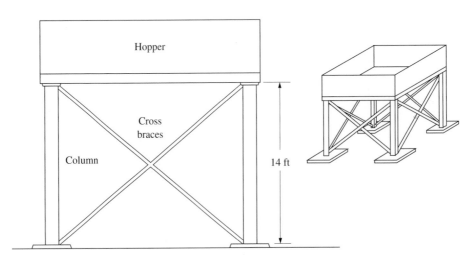

FIGURE 14–11 Hopper for Problems 14–19 and 14–20.

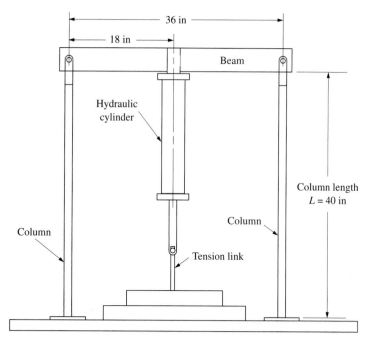

Note: Cylinder pulls up on tension link and down
on beam with a force of 3000 lb.

FIGURE 14–12 Test fixture for Problem 14–21.

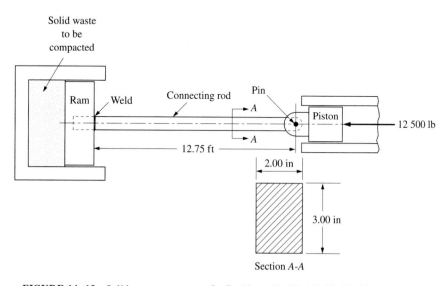

FIGURE 14–13 Solid waste compactor for Problems 14–22, 14–23, 14–24, and 14–25.

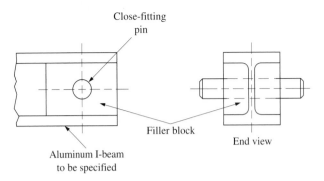

Close-fitting
pin

Filler block

End view

Aluminum I-beam
to be specified

FIGURE 14–14 End connection for I-beam for Problem 14–25.

14–26.E A hollow square tube, $3 \times 3 \times \frac{1}{4}$, made from ASTM A500 steel, grade B is used as a building column having a length of 16.5 ft. Using $L_e = 0.80L$, compute the allowable load on the column for a design factor of 3.0.

14–27.E A hollow rectangular tube, $4 \times 2 \times \frac{1}{4}$, made from ASTM A500 steel, grade B is used as a building column having a length of 16.5 ft. Using $L_e = 0.80L$, compute the allowable load on the column for a design factor of 3.0.

14–28.E A column is made by welding two standard steel angles, $3 \times 3 \times \frac{1}{4}$, into the form shown in Figure 14–5(f). The angles are made from ASTM A36 structural steel. If the column has a length of 16.5 ft and $L_e = 0.8L$, compute the allowable load on the column for a design factor of 3.0.

14–29.M A rectangular steel bar, made of AISI 1020 hot-rolled steel, is used as a safety brace to hold the ram of a large punch press while dies are installed in the press. The bar has cross-sectional dimensions of 60 mm by 40 mm. Its length is 750 mm and its ends are welded to heavy flat plates which rest on the flat bed of the press and the flat underside of the ram. Specify a safe load that could be applied to the brace.

14–30.M It is planned to use an aluminum channel, C4 × 1.738, as a column having a length of 4.25 m. The ends can be considered to be pinned. The aluminum is alloy 6061-T4. Compute the allowable load on the column for a design factor of 4.0.

14–31.M In an attempt to improve the load-carrying capacity of the column described in Problem 14–30, alloy 6061-T6 is proposed in place of the 6061-T4 to take advantage of its higher strength. Evaluate the effect of this proposed change on the allowable load.

14–32.E Compute the allowable load on the steel W12 × 65 column section shown in Figure 14–6(b) if it is 22.5 ft long, made from ASTM A36 steel, and installed such that $L_e = 0.8L$. Use the AISC code.

COMPUTER PROGRAMMING ASSIGNMENTS

1. Write a computer program to analyze proposed column designs using the procedure outlined in Section 14–7. Have all the essential design data for material, end-fixity, length, and cross-section properties input by the user. Have the program output the critical load and the allowable load for a given design factor.

Enhancements to Assignment 1

  (a) Include a table of data for standard schedule 40 steel pipe for use by the program to determine the cross section properties for a specified pipe size.

  (b) Design the program to handle columns made from solid circular cross sections and compute the cross-section properties for a given diameter.

  (c) Add a table of data for standard structural steel square tubing for use by the program to determine the cross-section properties for a specified size.

  (d) Have the program use the specifications of the AISC as stated in Section 14–9 for computing the allowable load and factor of safety for steel columns.

  (e) Have the program use the specifications of the Aluminum Association as stated in Section 14–10 for computing the allowable load for columns made from 6061-T6.

2. Write a program to design a column with a solid circular cross section to carry a given load with a given design factor. Note that the program will have to check to see that the correct method of analysis is used, either the Euler formula for long columns or the Johnson formula for short columns, after an initial assumption is made.

3. Write a program to design a column with a solid square cross section to carry a given load with a given design factor.

4. Write a program to select a suitable schedule 40 steel pipe to carry a given load with a given design factor. The program could be designed to search through a table of data for standard pipe sections from the smallest to the largest until a suitable pipe was found. For each trial section, the allowable load could be computed using the Euler or Johnson formula, as required, and compared with the design load.

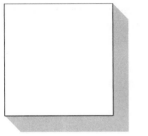

15

Pressure Vessels

15–1 OBJECTIVES OF THIS CHAPTER

The most common forms of pressure vessels designed to hold liquids or gases under internal pressure are spheres and closed-end cylinders. The internal pressure tends to burst the vessel because there are tensile stresses developed in its walls. The overall objective of this chapter is to describe the manner in which these stresses are developed and to present formulas that can be used to compute the magnitude of the stresses.

After completing this chapter, you should be able to:

1. Determine whether a pressure vessel should be classified as *thin-walled* or *thick-walled*.
2. Draw the free-body diagram for a part of a sphere when subjected to internal pressure to identify the force that must be resisted by the wall of the sphere.
3. Describe *hoop stress* as it is applied to spheres carrying an internal pressure.
4. State the formula for computing the hoop stress developed in the wall of a thin-walled sphere due to internal pressure.
5. Use the hoop stress formula to compute the maximum stress in the wall of a thin-walled sphere.
6. Determine the required wall thickness of the sphere to resist a given internal pressure.

7. Draw the free-body diagram for a part of a cylinder when subjected to internal pressure to identify the force that must be resisted by the wall of the cylinder.

8. Describe *hoop stress* as it is applied to cylinders carrying an internal pressure.

9. State the formula for computing the hoop stress developed in the wall of a thin-walled cylinder due to internal pressure.

10. Use the hoop stress formula to compute the maximum stress in the wall of a thin-walled cylinder.

11. Determine the required wall thickness of the cylinder to resist a given internal pressure safely.

12. Describe *longitudinal stress* as it is applied to cylinders carrying internal pressure.

13. State the formula for computing the longitudinal stress in the wall of a thin-walled cylinder due to an internal pressure.

14. Use the longitudinal stress formula to compute the stress in the wall of a thin-walled cylinder that acts in the direction parallel to the axis of the cylinder.

15. Determine the required wall thickness of a thin-walled cylinder to resist a given internal pressure safely.

16. Identify the hoop stress, longitudinal stress, and radial stress developed in the wall of a thick-walled sphere or cylinder due to internal pressure.

17. Apply the formulas for computing the maximum values of the hoop stress, longitudinal stress, and radial stress in the wall of a thick-walled sphere or cylinder.

18. Apply the formulas for computing the magnitudes of the hoop stress, longitudinal stress, and radial stress at any radius within the wall of a thick-walled cylinder or sphere.

15–2 DISTINCTION BETWEEN THIN-WALLED AND THICK-WALLED PRESSURE VESSELS

In general, the magnitude of the stress in the wall of a pressure vessel varies as a function of position within the wall. A precise analysis should enable the computation of the stress at any point. The formulas for making such a computation will be shown in a later section.

However, when the wall thickness of the pressure vessel is small, the assumption that the stress is uniform throughout the wall results in very little error. Also, this assumption permits the development of relatively simple formulas for stress. Figure 15–1 shows the definition of key diameters, radii, and the wall thickness for cylinders and spheres.

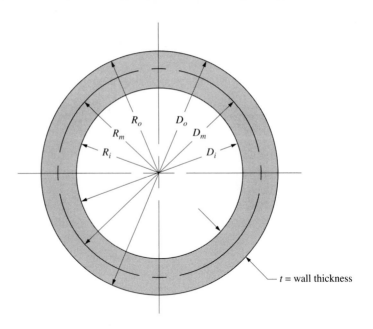

t = wall thickness

FIGURE 15–1 Definition of key diameters, radii, and wall thickness for cylinders and spheres.

The criterion for determining when a pressure vessel can be considered thin-walled is as follows:

> If the ratio of the mean radius of the vessel to its wall thickness is 10 or greater, the stress is very nearly uniform and it can be assumed that all the material of the wall shares equally to resist the applied forces. Such pressure vessels are called thin-walled vessels.

The mean radius is defined as the average of the outside radius and the inside radius. That is,

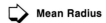 **Mean Radius**

$$R_m = \frac{R_o + R_i}{2} \tag{15-1}$$

Then a pressure vessel is considered to be thin if

$$\frac{R_m}{t} \geq 10 \tag{15-2}$$

where t is the wall thickness of the vessel. We can also define the *mean diameter* as

 **Mean Diameter**

$$D_m = \frac{D_o + D_i}{2} \tag{15-3}$$

Because the diameter is twice the radius, the criterion for a vessel to be considered thin-walled is

$$\frac{D_m}{t} \geq 20 \qquad (15-4)$$

Obviously, if the vessel does not satisfy the criteria listed in Equations (15–2) and (15–4), it is considered to be thick-walled.

In addition to Equations (15–1) and (15–3) for the mean radius and mean diameter, the following forms can be useful:

$$R_i = R_o - t$$
$$R_m = R_o - \frac{t}{2}$$
$$R_m = R_i + \frac{t}{2}$$
$$D_i = D_o - 2t$$
$$D_m = D_o - t$$
$$D_m = D_i + t$$

The next two sections are devoted to the analysis of *thin-walled* spheres and cylinders. Then Section 15–5 is concerned with *thick-walled* spheres and cylinders.

15–3 THIN-WALLED SPHERES

In analyzing a spherical pressure vessel, the objective is to determine the stress in the wall of the vessel to ensure safety. Because of the symmetry of a sphere, a convenient free body for use in the analysis is one-half of the sphere, as shown in Figure 15–2. The internal pressure of the liquid or gas contained in the sphere acts perpendicular to the walls, uniformly over all the interior surface. Because the sphere was cut through a diameter, the forces in the walls all act horizontally. Therefore, only the horizontal component of the forces due to the fluid pressure needs to be considered in determining the magnitude of the force in the walls. If a pressure P acts on an area A, the force exerted on the area is

$$F = pA \qquad (15-5)$$

Taking all of the force acting on the entire inside of the sphere and finding the horizontal component, we find the resultant force in the horizontal direction to be

$$F_R = pA_p \qquad (15-6)$$

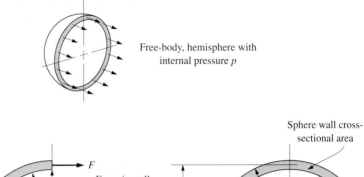

Free-body, hemisphere with
internal pressure p

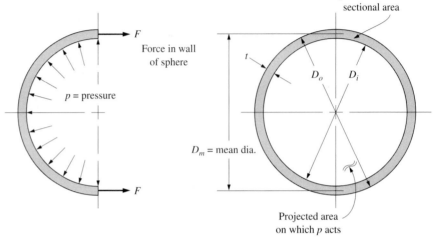

FIGURE 15–2 Free-body diagram for a sphere carrying internal pressure.

where A_p is the *projected area* of the sphere on the plane through the diameter. Therefore,

$$A_p = \frac{\pi D_m^2}{4} \tag{15–7}$$

Because of the equilibrium of the horizontal forces on the free body, the forces in the walls must also equal F_R, as computed in Equation (15–6). These tensile forces acting on the cross-sectional area of the walls of the sphere cause tensile stresses to be developed. That is,

$$\sigma = \frac{F_R}{A_w} \tag{15–8}$$

where A_w is the area of the annular ring cut to create the free body, as shown in Figure 15–2. The actual area is

$$A_w = \frac{\pi}{4}(D_o^2 - D_i^2) \tag{15–9}$$

However, for thin-walled spheres with a wall thickness t, less than about $\frac{1}{10}$ of the radius of the sphere, the wall area can be closely approximated as

$$A_w = \pi D_m t \qquad (15-10)$$

This is the area of a rectangular strip having a thickness t and a length equal to the mean circumference of the sphere, πD_m.

Equations (15–6) and (15–8) can be combined to yield an equation for stress,

$$\sigma = \frac{F_R}{A_w} = \frac{pA_p}{A_w} \qquad (15-11)$$

Expressing A_p and A_w in terms of D_m and t from Equations (15–7) and (15–10) gives

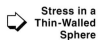
Stress in a Thin-Walled Sphere

$$\sigma = \frac{p(\pi D_m^2/4)}{\pi D_m t} = \frac{pD_m}{4t} \qquad (15-12)$$

This is the expression for the stress in the wall of a thin-walled sphere subjected to internal pressure. Very little error (less than 5%) will result from using either the outside or inside diameter in place of the mean diameter.

Example Problem 15–1 Compute the stress in the wall of a sphere having an inside diameter of 300 mm and a wall thickness of 1.50 mm when carrying nitrogen gas at 3500 kPa internal pressure.

Solution **Objective** Compute the stress in the wall of the sphere.

Given $p = 3500$ kPa; $D_i = 300$ mm; $t = 1.50$ mm.

Analysis We must first determine if the sphere can be considered to be thin-walled by computing the ratio of the mean diameter to the wall thickness.

$$D_m = D_i + t = 300 \text{ mm} + 1.50 \text{ mm} = 301.5 \text{ mm}$$
$$D_m/t = 301.5 \text{ mm}/1.50 \text{ mm} = 201$$

Because this is far greater than the lower limit of 20, the sphere is thin. Then Equation (15–12) should be used to compute the stress.

Results $$\sigma = \frac{pD_m}{4t} = \frac{(3500 \times 10^3 \text{ Pa})(301.5 \text{ mm})}{4(1.50 \text{ mm})}$$
$$\sigma = 175.9 \times 10^6 \text{ Pa} = 175.9 \text{ MPa}$$

15–4 THIN-WALLED CYLINDERS

Cylinders are frequently used for pressure vessels, for example, as storage tanks, hydraulic and pneumatic actuators, and for piping of fluids under pressure. The stresses in the walls of cylinders are similar to those found for spheres, although the maximum value is greater.

Two separate analyses are shown here. In one case, the tendency for the internal pressure to pull the cylinder apart in a direction parallel to its axis is found. This is called *longitudinal stress.* Next, a ring around the cylinder is analyzed to determine the stress tending to pull the ring apart. This is called *hoop stress,* or *tangential stress.*

Longitudinal stress. Figure 15–3 shows a part of a cylinder, which is subjected to an internal pressure, cut perpendicular to its axis to create a free body. Assuming that the end of the cylinder is closed, the pressure acting on the circular area of the end would produce a resultant force of

$$F_R = pA = p\left(\frac{\pi D_m^2}{4}\right) \tag{15–14}$$

This force must be resisted by the force in the walls of the cylinder, which, in turn, creates a tensile stress in the walls. The stress is

$$\sigma = \frac{F_R}{A_w} \tag{15–15}$$

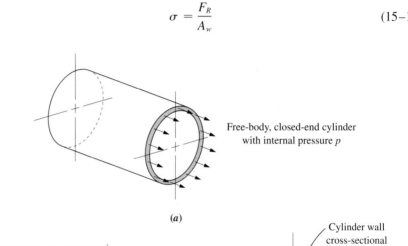

Free-body, closed-end cylinder with internal pressure p

(a)

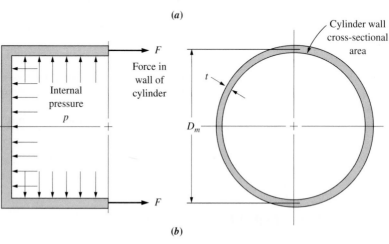

(b)

FIGURE 15–3 Free-body diagram of a cylinder carrying internal pressure showing longitudinal stress.

Assuming that the walls are thin, as we did for spheres,

$$A_w = \pi D_m t$$

where t is the wall thickness.

Now combining Equations (15–14) and (15–15),

**Longitudinal Stress in a Thin-Walled Cylinder**

$$\sigma = \frac{F_R}{A_w} = \frac{p(\pi D_m^2/4)}{\pi D_m t} = \frac{pD_m}{4t} \qquad (15\text{–}16)$$

This is the stress in the wall of the cylinder in a direction parallel to the axis, called the longitudinal stress. Notice that it is of the same magnitude as that found for the wall of a sphere. But this is not the maximum stress, as shown next.

Hoop stress.   The presence of the tangential or hoop stress can be visualized by isolating a ring from the cylinder, as shown in Figure 15–4. The internal pressure pushes outward evenly all around the ring. The ring must develop a tensile stress in a direction tangential to the circumference of the ring to resist the tendency of the pressure to burst the ring. The magnitude of the stress can be determined by using half of the ring as a free body, as shown in Figure 15–4(b).

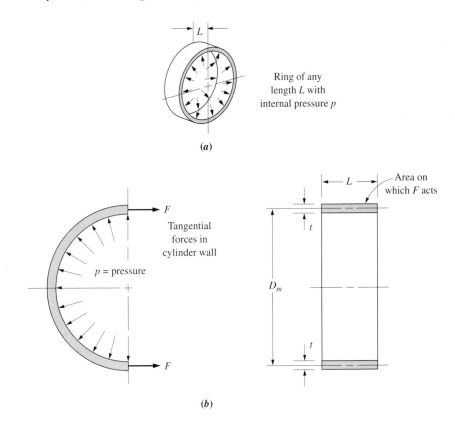

FIGURE 15–4   Free-body diagram of a cylinder carrying internal pressure showing hoop stress.

The resultant of the forces due to the internal pressure must be determined in the horizontal direction and balanced with the forces in the walls of the ring. Using the same reasoning as we did for the analysis of the sphere, we find that the resultant force is the product of the pressure and the *projected area* of the ring. For a ring with a diameter D and a length L,

$$F_R = pA_p = p(D_m L) \tag{15–17}$$

The tensile stress in the wall of the cylinder is equal to the resisting force divided by the cross-sectional area of the wall. Again assuming that the wall is thin, the wall area is

$$A_w = 2tL \tag{15–18}$$

Then the stress is

$$\sigma = \frac{F_R}{A_w} = \frac{F_R}{2tL} \tag{15–19}$$

Combining Equations (15–17) and (15–19) gives

Hoop Stress in a Thin-Walled Cylinder

$$\sigma = \frac{F_R}{A_w} = \frac{pD_m L}{2tL} = \frac{pD_m}{2t} \tag{15–20}$$

This is the equation for the hoop stress in a thin cylinder subjected to internal pressure. Notice that the magnitude of the hoop stress is *twice* that of the longitudinal stress. Also, the hoop stress is twice that of the stress in a spherical container of the same diameter carrying the same pressure.

Example Problem 15–2

A cylindrical tank holding oxygen at 2000 kPa pressure has an outside diameter of 450 mm and a wall thickness of 10 mm. Compute the hoop stress and the longitudinal stress in the wall of the cylinder.

Solution   **Objective**   Compute the hoop stress and the longitudinal stress in the wall of the cylinder.

Given $p = 2000$ kPa; $D_o = 450$ mm; $t = 10$ mm.

Analysis   We must first determine if the cylinder can be considered to be thin-walled by computing the ratio of the mean diameter to the wall thickness.

$$D_m = D_o - t = 450 \text{ mm} - 10 \text{ mm} = 440 \text{ mm}$$
$$D_m/t = 440 \text{ mm}/10 \text{ mm} = 44$$

Because this is far greater than the lower limit of 20, the cylinder is thin. Then Equation (15–20) should be used to compute the hoop stress and Equation (15–16) should be used to compute the longitudinal stress.

Results
$$\sigma = \frac{pD_m}{2t} = \frac{(2000 \times 10^3 \text{ Pa})(440 \text{ mm})}{2(10 \text{ mm})} = 44.0 \text{ MPa}$$

The longitudinal stress, from Equation (15–12), is

$$\sigma = \frac{pD_m}{4t} = 22.0 \text{ MPa}$$

Example Problem 15–3 Determine the pressure required to burst a standard 8-in schedule 40 steel pipe if the ultimate tensile strength of the steel is 40 000 psi.

Solution   **Objective**   Compute the pressure required to burst the steel pipe.

Given   Ultimate tensile strength of steel = s_u = 40 000 psi
Pipe is a standard 8-in schedule 40 steel pipe.
The dimensions of the pipe are found in Appendix A–12 to be

$$\text{outside diameter} = 8.625 \text{ in} = D_o$$
$$\text{inside diameter} \ \ = 7.981 \text{ in} = D_i$$
$$\text{wall thickness} \ \ \ = 0.322 \text{ in} = t$$

Analysis   We should first check to determine if the pipe should be called a thin-walled cylinder by computing the ratio of the mean diameter to the wall thickness.

$$D_m = \text{mean diameter} = \frac{D_o + D_i}{2} = 8.303 \text{ in}$$

$$\frac{D_m}{t} = \frac{8.303 \text{ in}}{0.322 \text{ in}} = 25.8$$

Since this ratio is greater than 20, the thin-wall equations can be used. The hoop stress is the maximum stress and should be used to compute the bursting pressure.

Results   Use Equation (15–20).

$$\sigma = \frac{pD_m}{2t} \tag{15–20}$$

Letting σ = 40 000 psi and using the mean diameter gives the bursting pressure to be

$$p = \frac{2t\sigma}{D_m} = \frac{(2)(0.322 \text{ in})(40\,000 \text{ lb/in}^2)}{8.303 \text{ in}} = 3102 \text{ psi}$$

Comment   A design factor of 6 or greater is usually applied to the bursting pressure to get an allowable operating pressure. This cylinder would be limited to approximately 500 psi of internal pressure.

15–5 THICK-WALLED CYLINDERS AND SPHERES

The formulas in the preceding sections for thin-walled cylinders and spheres were derived under the assumption that the stress is uniform throughout the wall of the container. As stated, if the ratio of the diameter of the container to the wall thickness is greater than 20, this assumption is reasonably correct. Conversely, if the ratio is less than 20, the walls are considered to be thick, and a different analysis technique is required.

The detailed derivation of the thick-wall formulas will not be given here because of their complexity. See References 1 and 2. But the application of the formulas will be shown.

For a thick-walled cylinder, Figure 15–5 shows the notation to be used. The geometry is characterized by the inner radius a, the outer radius b, and any radial position between a and b, called r. The *longitudinal stress* is called σ_1; the *hoop stress* is σ_2. These have the same meaning as they did for thin-walled vessels, except now they will have varying magnitudes at different positions in the wall. In addition to hoop and longitudinal stresses, a *radial stress* σ_3 is created in a thick-walled vessel. As the name implies, the radial stress acts along a radius of the cylinder or sphere. It is a compressive stress and varies from a magnitude of zero at the outer surface to a maximum at the inner surface, where it is equal to the internal pressure.

Table 15–1 shows a summary of the formulas needed to compute the three stresses in the walls of thick-walled cylinders and spheres subjected to internal pressure. The terms *longitudinal stress* and *hoop stress* do not apply to spheres. Instead, we refer to the *tangential stress,* which is the same in all directions around the sphere. Then

$$\text{tangential stress} = \sigma_1 = \sigma_2$$

15–6 PROCEDURE FOR ANALYZING AND DESIGNING SPHERICAL AND CYLINDRICAL PRESSURE VESSELS

Presented here is a summary of the principles discussed in this chapter related to the stress analysis of both thin-walled and thick-walled spheres and cylinders. The summary is given in the form of general procedures for analyzing and designing pressure vessels. For design stresses, it is advised that Section 3–3 be reviewed. It will be assumed here that the failure of a pressure vessel subjected to internal pressure is due to

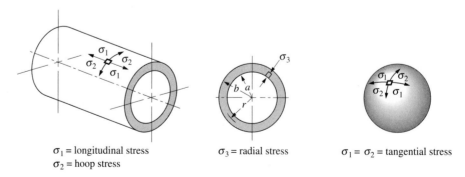

σ_1 = longitudinal stress
σ_2 = hoop stress

σ_3 = radial stress

$\sigma_1 = \sigma_2$ = tangential stress

FIGURE 15–5 Notation for stresses in thick-walled cylinders and spheres.

TABLE 15–1 Stresses in thick-walled cylinders and spheres*

	Stress at position r	Maximum stress
Thick-walled cylinder		
Longitudinal	$\sigma_1 = \dfrac{pa^2}{b^2 - a^2}$	$\sigma_1 = \dfrac{pa^2}{b^2 - a^2}$ (uniform throughout wall)
Hoop (tangential)	$\sigma_2 = \dfrac{pa^2(b^2 + r^2)}{r^2(b^2 - a^2)}$	$\sigma_2 = \dfrac{p(b^2 + a^2)}{b^2 - a^2}$ (at inner surface)
Radial	$\sigma_3 = \dfrac{-pa^2(b^2 - r^2)}{r^2(b^2 - a^2)}$	$\sigma_3 = -p$ (at inner surface)
Thick-walled sphere		
Tangential	$\sigma_1 = \sigma_2 = \dfrac{pa^3(b^3 + 2r^3)}{2r^3(b^3 - a^3)}$	$\sigma_1 = \sigma_2 = \dfrac{p(b^3 + 2a^3)}{2(b^3 - a^3)}$ (at inner surface)
Radial	$\sigma_3 = \dfrac{-pa^3(b^3 - r^3)}{r^3(b^3 - a^3)}$	$\sigma_3 = -p$ (at inner surface)

*Symbols used here are as follows: a = inner radius; b = outer radius; r = any radius between a and b; p = internal pressure, uniform in all directions. Stresses are tensile when positive, compressive when negative.

the tensile stresses occurring tangentially in the walls of the vessel. Design stresses should take into account the material from which the vessel is made, the operating environment, and whether the pressure is constant or varying in a cyclic fashion.

See also Section 15–7 for a discussion of other modes of failure in vessels having penetrations, structural supports, reinforcing rings, and other features that differ from the simple cylindrical or spherical shape.

Design stresses. *For steady pressure,* the design stress can be based on the yield strength of the material,

$$\sigma_d = s_y/N$$

The choice of the design factor, N, is often dictated by code because of the danger created when a pressure vessel fails. This is particularly true for vessels containing gases or steam under pressure because failures produce violent expulsion of the gas as the high level of stored energy is released. In the absence of a code, we will use $N = 4$ as a minimum and larger values should be used for critical applications or where uncertainty exists in the operating conditions or material properties. Another suggested guideline is to limit the pressure in a vessel to no more than 1/6 of the pressure that would be predicted to burst it. This effectively calls for a design stress related to the ultimate tensile strength of the material of,

$$\sigma_d = s_u/N = s_u/6$$

For cycling pressure, base the design stress on the ultimate strength,

$$\sigma_d = s_u/N$$

Use $N = 8$ as a minimum to produce a design stress related to the fatigue strength of the material.

A. Procedure for Analyzing Pressure Vessels

Given   Internal pressure in the vessel, p

Material from which the vessel is made. Ductile metal assumed.

Outside diameter, D_o, inside diameter, D_i, and wall thickness, t, for the vessel.

Objective   Determine the maximum stress in the vessel and check the safety of that stress level with regard to the design stress in the material from which the vessel is made.

1. Compute the mean diameter, D_m, for the vessel from Equation (15–3): $D_m = (D_o + D_i)/2$.
2. Compute the ratio of the mean diameter to the wall thickness for the vessel, D_m/t.
3. If $D_m/t > 20$, the vessel can be considered to be thin-walled. Use Equation (15–12) for spheres or Equation (15–20) for cylinders to compute the maximum tangential stress in the walls of the vessel.

$$\sigma = pD_m/4t \qquad \text{for spheres} \qquad\qquad (15\text{–}12)$$

$$\sigma = pD_m/2t \qquad \text{for cylinders} \qquad\qquad (15\text{–}20)$$

4. If $D_m/t < 20$, the vessel must be considered to be thick-walled. Use Equations from Table 15–1 to compute the maximum tangential or hoop stress in the walls of the vessel.

$$\sigma = \frac{p(b^3 + 2a^3)}{2(b^3 - a^3)} \qquad \text{for spheres}$$

$$\sigma = \frac{p(b^2 + a^2)}{b^2 - a^2} \qquad \text{for cylinders}$$

5. Compute the design stress for the material from which the vessel is made.
6. The actual maximum stress must be less than the design stress for safety.

B. Procedure for Designing Pressure Vessels for a Given Material

Given   Internal pressure in the vessel, p.

Material from which the vessel is made. Ductile metal assumed.

Nominal internal diameter of the vessel based on the desired volumetric capacity.

Objective   Specify the outside diameter, D_o, inside diameter, D_i, and wall thickness, t, for the vessel to ensure the safety of the vessel with regard to the design stress in the material from which the vessel is made.

1. Use the given diameter as an estimate of the mean diameter, D_m, for the vessel.
2. Assume at first that the vessel will be thin-walled and that the maximum stress can be computed from Equation (15–12) for a sphere or Equation (15–20) for a cylinder. This assumption will be checked later.
3. Compute the design stress for the material from which the vessel is to be made.
4. In the appropriate stress equation, substitute the design stress for the maximum stress and solve for the minimum required wall thickness, t.
5. Specify convenient values for t, D_i, and D_o, based on available material thicknesses. Appendix Table A–2 may also be used to specify preferred basic sizes.
6. Compute the actual mean diameter for the vessel using the specified dimensions.
7. Compute the ratio of the mean diameter to the wall thickness for the vessel, D_m/t.
8. If $D_m/t > 20$, the vessel is thin-walled as assumed and the design is finished.
9. If $D_m/t < 20$, the vessel must be considered to be thick-walled. Use Equations from Table 15–1 to compute the maximum tangential or hoop stress in the walls of the vessel and compare that stress with the design stress. If the actual stress is less than the design stress, the design is satisfactory. If the actual maximum stress is greater than the design stress, increase the wall thickness and
recompute the resulting stress. Continue this process until a satisfactory stress level and convenient dimensions for the vessel are obtained. Equation solving computer software or a graphing calculator with equation-solving capability may facilitate this process.

C. Procedure for Specifying a Ductile Metal for a Pressure Vessel of a Given Size

Given   Internal pressure in the vessel, p
Outside diameter, D_o, inside diameter, D_i, and wall thickness, t, for the vessel.

Objective   Specify a suitable ductile metal from which the vessel is to be made.

1. Compute the mean diameter, D_m, for the vessel from Equation (15–3): $D_m = (D_o + D_i)/2$.
2. Compute the ratio of the mean diameter to the wall thickness for the vessel, D_m/t.
3. If $D_m/t > 20$, the vessel can be considered to be thin-walled. Use Equation (15–12) for spheres or Equation (15–20) for

cylinders to compute the maximum tangential stress in the walls of the vessel.

$$\sigma = pD_m/4t \qquad \text{for spheres} \qquad\qquad (15–12)$$
$$\sigma = pD_m/2t \qquad \text{for cylinders} \qquad\qquad (15–20)$$

4. If $D_m/t < 20$, the vessel must be considered to be thick-walled. Use Equations from Table 15–1 to compute the maximum tangential or hoop stress in the walls of the vessel.

$$\sigma = \frac{p(b^3 + 2a^3)}{2(b^3 - a^3)} \qquad \text{for spheres}$$

$$\sigma = \frac{p(b^2 + a^2)}{b^2 - a^2} \qquad \text{for cylinders}$$

5. Specify a suitable equation for the design stress from the discussion earlier in this section.
6. Let the design stress equal the computed maximum stress from step 3 or 4. Then solve for the appropriate material strength, either s_y or s_u from the design stress equation.
7. Specify a suitable material that has a strength greater than the minimum required value.

Example Problem 15–4

Compute the magnitude of the maximum longitudinal, hoop, and radial stresses in a cylinder carrying helium at a steady pressure of 10 000 psi. The outside diameter is 8.00 in and the inside diameter is 6.40 in. Specify a suitable material for the cylinder.

Solution

Objective Compute the maximum stresses and specify a material.

Given Pressure = p = 10 000 psi. D_o = 8.00 in. D_i = 6.40 in.

Analysis Use Procedure C from this section.

Results *Step 1.* $D_m = (D_o + D_i)/2 = (8.00 + 6.40)/2 = 7.20$ in

Step 2. $t = (D_o - D_i)/2 = (8.00 - 6.40)/2 = 0.80$ in
$D_m/t = 7.20/0.80 = 9.00$

Step 3. This step does not apply. Cylinder is thick.

Step 4. Use equations from Table 15–1.

$$a = D_i/2 = 6.40/2 = 3.20 \text{ in}$$
$$b = D_o/2 = 8.00/2 = 4.00 \text{ in}$$
$$\sigma_1 = \frac{pa^2}{b^2 - a^2} = \frac{(10\,000 \text{ psi})(3.20 \text{ in})^2}{(4.00^2 - 3.20^2) \text{ in}^2} = 17\,780 \text{ psi longitudinal} \, \blacksquare$$

$$\sigma_2 = \frac{p(b^2 + a^2)}{b^2 - a^2} = \frac{(10\,000 \text{ psi})\,(4.00^2 + 3.20^2) \text{ in}^2}{(4.00^2 - 3.20^2) \text{ in}^2} = 45\,560 \text{ psi}$$
hoop

$$\sigma_3 = -p = -10\,000 \text{ psi radial}$$

All three stresses are a maximum at the inner surface of the cylinder.

Step 5. Let the design stress $= \sigma_d = s_y/4$.

Step 6. The maximum stress is the hoop stress, $\sigma_2 = 45\,560$ psi. Then the required yield strength for the material is

$$s_y = N(\sigma_2) = 4(45\,560 \text{ psi}) = 182\,200 \text{ psi} = 182 \text{ ksi}$$

Step 7. From Appendix A–13, we can specify AISI 4140 OQT 700 steel that has a yield strength of 212 ksi.

Example Problem 15–5

Compute the magnitude of the maximum tangential and radial stresses in a sphere carrying helium at a steady pressure of 10 000 psi. The outside diameter is 8.00 in and the inside diameter is 6.40 in. Specify a suitable material for the cylinder.

Solution

Objective Compute the maximum stresses and specify a material.

Given Pressure $= p = 10\,000$ psi. $D_o = 8.00$ in. $D_i = 6.40$ in.

Analysis Use Procedure C from this section. These data are the same as those used in Example Problem 15–4. Some values will be carried forward.

Results **Steps 1, 2, 3.** Sphere is thick-walled.

Step 4. Use equations from Table 15–1. $a = 3.20$ in. $b = 4.00$ in.

$$\sigma_1 = \sigma_2 = \frac{p(b^3 + 2a^3)}{2(b^3 - a^3)} = \frac{(10\,000 \text{ psi})\,[4.00^3 + 2(3.20)^3] \text{ in}^3}{2(4.00^3 - 3.20^3) \text{ in}^3}$$

$$\sigma_1 = \sigma_2 = 20\,740 \text{ psi tangential}$$

$$\sigma_3 = -p = -10\,000 \text{ psi radial}$$

Each of these stresses is a maximum at the inner surface.

Steps 5, 6. For a maximum stress of 20 740 psi, the required yield strength for the material is

$$s_y = N(\sigma_2) = 4(20\,740 \text{ psi}) = 82\,960 \text{ psi} = 83 \text{ ksi}$$

Step 7. From Appendix A–13, we can specify AISI 4140 OQT 1300 steel that has a yield strength of 101 ksi. Others could be used.

Comment The maximum stress in the sphere is less than half that in the cylinder of the same size, allowing a material with a much lower strength to be used. Alternatively, it would be possible to design the sphere with the same material but with a smaller wall thickness.

A cylindrical vessel has an outside diameter of 400 mm and an inside diameter of 300 mm. For an internal pressure of 20.1 MPa, compute the hoop stress σ_2 at the inner and outer surfaces and at points within the wall at intervals of 10 mm. Plot a graph of σ_2 versus the radial position in the wall.

Solution   **Objective**   Compute the hoop stress at specified positions in the wall of the cylinder.

Given   Pressure = p = 20.1 MPa. D_o = 400 mm. D_i = 300 mm.
Use 10 mm increments for radius within the wall from the outside surface to the inside surface.

Analysis   Use Steps 1–4 from Procedure A from this section.

Results   **Step 1.** $D_m = (D_o + D_i)/2 = (400 + 300)/2 = 350$ mm

Step 2. $t = (D_o - D_i)/2 = (400 - 300)/2 = 50$ mm
$D_m/t = 350/50 = 7.00 < 20$; thick-walled cylinder

Step 3. This step does not apply.

Step 4. Use equation for tangential stress from Table 15–1.

$$\sigma_2 = \frac{pa^2(b^2 + r^2)}{r^2(b^2 - a^2)}$$

$$a = D_i/2 = 300/2 = 150 \text{ mm}$$
$$b = D_o/2 = 400/2 = 200 \text{ mm}$$

The results are shown in tabular form below.

r (mm)	σ_2 (MPa)	
200	51.7	(Minimum at outer surface)
190	54.5	
180	57.7	
170	61.6	
160	66.2	
150	71.8	(Maximum at inner surface)

Comment   Figure 15–6 shows the graph of tangential stress versus position in the wall. The graph illustrates clearly that the assumption of uniform stress in the wall of a thick-walled cylinder would *not* be valid.

Design a cylinder to be made from aged titanium Ti-6Al-4V to carry compressed natural gas at 7500 psi. The internal diameter must be 24.00 in to provide the necessary volume. The design stress is to be 1/6 of the ultimate strength of the titanium.

Solution   **Objective**   Design the cylinder.

Given   Pressure = p = 7500 psi. D_i = 24.0 in.
Titanium Ti-6Al-4V; s_u = 170 ksi (Appendix A–14)

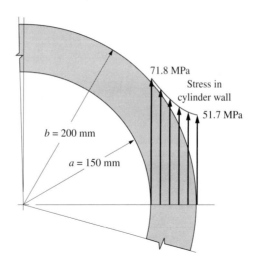

71.8 MPa
Stress in
cylinder wall
51.7 MPa

$b = 200$ mm

$a = 150$ mm

FIGURE 15–6 Variation of tangential stress in the wall of
the thick-walled cylinder in Example Problem 15–6.

Analysis   Use Procedure B from this section.

Results   **Step 1.** Let $D_m = 24.00$ in

Step 2. Assume thin-walled cylinder.

Step 3. Design stress,

$$\sigma_d = s_u/6 = (170\,000 \text{ psi})/6 = 28\,333 \text{ psi}$$

Step 4. Use Equation (15–20) to compute nominal value for t.

$$t = \frac{pD_m}{2\sigma_d} = \frac{(7500 \text{ psi})(24.0 \text{ in})}{2(28\,333 \text{ psi})} = 3.18 \text{ in}$$

Step 5. Trial #1: $D_i = 24.00$; $t = 3.50$ in; $D_o = D_i + 2t = 31.00$ in

Step 6. $D_m = D_i + t = 24.00 + 3.50 = 27.50$ in

Step 7. $D_m/t = 27.50/3.50 = 7.86 < 20$; thick-walled

Step 8. This step does not apply.

Step 9. Use equation for σ_2 from Table 15–1.

$$a = D_i/2 = 24.00/2 = 12.00 \text{ in}$$
$$b = D_o/2 = 31.00/2 = 15.50 \text{ in}$$
$$\sigma_2 = \frac{p(b^2 + a^2)}{(b^2 - a^2)} = \frac{(7500 \text{ psi})(15.50^2 + 12.00^2)}{(15.50^2 - 12.00^2)}$$
$$\sigma_2 = 29\,940 \text{ psi} \quad \text{slightly high. Repeat steps 5 and 9}$$

Step 5. Increase $t = 3.75$ in; $D_o = D_i + 2t = 31.50$ in; thick-walled

Step 9. Use equation for σ_2 from Table 15–1.

$$a = D_i/2 = 24.00/2 = 12.00 \text{ in}$$
$$b = D_o/2 = 31.50/2 = 15.75 \text{ in}$$

Then $\sigma_2 = 28\,250$ psi. This is less than the design stress. OK

Comment The wall thickness is quite large which would result in a heavy cylinder. Consider using a sphere and higher strength material for the vessel.

15–7 OTHER DESIGN CONSIDERATIONS FOR PRESSURE VESSELS

Pressure vessel design and analysis techniques presented thus far have related only to the basic stress analysis of ideal cylinders and spheres without consideration of penetrations or other changes in geometry. Of course, most practical pressure vessels incorporate several types of features that cause the vessel to differ from the ideal shape. Also, external loads are often applied that create stresses that combine with the stress due to internal pressure. For example,

- A spherical or cylindrical pressure vessel would typically have one or more ports used to fill or empty the vessel. The ports would often be welded into the vessel causing a discontinuity in the geometry as well as modifying the material properties in the vicinity of the weld.

- Some pressure vessels used for chemical reactions or other material processing applications contain view ports for observation of the process. The view ports may contain flanges to hold the transparent window.

- Cylindrical vessels often are made with domed or hemispherical ends to provide a more optimum design to resist the internal pressure. But, because the tangential stress in the spherical end is less than that in the cylinder, special attention should be paid to the design at the intersection of the ends with the straight cylindrical portion.

- Large cylinders may have reinforcing bands or ribs applied to the inside or outside to stiffen the vessel structurally.

- Large cylinders and spheres may experience large stresses due to the actual weight of the vessel and its contents that combine with the stresses produced by the internal pressure. For example: a relatively long cylindrical tank laid horizontally and supported near its ends is subjected to bending stresses; a cylindrical tank positioned with its axis vertical is subjected to axial compressive stress.

- Large cylinders and spheres must be fitted with supports that transmit the weight of the vessel and its contents to a floor or the earth. Special stress conditions exist in the vicinity of such supports.

- Pressure vessels used in ground transportation equipment often experience dynamic loads caused by stopping, starting, product movement within the vessel, and vibrations caused by uneven roadways.

- Pressure vessels on aircraft and spacecraft are subjected to high acceleration forces during landings, takeoffs, launches, and rapid maneuvers.
- Joints between sections of pressure vessels made from two or more pieces often contain geometric discontinuities requiring special analysis techniques and careful fabrication.

Analysis techniques for these conditions are not covered in this book. Some are discussed in References 2, 3, and 4. The *Boiler and Pressure Vessel Code,* published by the American Society of Mechanical Engineers (ASME), contains numerous standards and techniques of analysis governing the design, fabrication, and inspection of boilers and pressure vessels to provide reasonably certain protection of life and property. Many commercial vendors offer computer software packages that perform the complex computations required to design and analyze pressure vessels and their attachments.

The applications and examples presented in this chapter have emphasized the use of metals for the structural walls of pressure vessels. Other materials, particularly composite materials and reinforced plastics, are often used as well. The special characteristics of these materials must be understood when applying them to pressure vessels.

Composite pressure vessels. High-strength composite materials are well suited for use in fabricating pressure vessels. The fact that the primary stresses are either tangential (hoop) or longitudinal leads the designer of composite vessels to call for alignment of the composite fibers in the direction of the maximum stresses. Circumferential wrapping of a prepreg tape around a liner made from metal or plastic offers significant weight savings as compared with a design using only metal or plastic. To resist longitudinal stresses due to internal pressure along with other external forces, some tanks are wrapped in a helical fashion in addition to the circumferential wrapping. The thickness and direction of plies can be tailored to the specific set of loads expected in a particular application.

Materials selected for composite pressure vessels include E-glass/epoxy, structural glass/epoxy, and carbon/epoxy. Cost is a major factor in material specification.

The primary applications for composite pressure vessels include those in which light weight is a major design goal. The air supply tank for self-contained breathing apparatus (SCBA) used by firefighters is a good example because the lighter tank allows more mobility and less fatigue. Weight reductions in space and aeronautical applications allow greater payloads or higher performance of aerospace vehicles. The development of ground vehicles using compressed natural gas (CNG) calls for the production of lightweight cylinders to carry the CNG fuel. Demonstrations are being carried out in buses, commercial fleet vehicles, and some passenger cars. Examples of practical weight savings are reported in Reference 1. A composite compressed air reservoir for a transportation application weighing 27 pounds replaced one of steel, saving almost 100 pounds. A structural glass/epoxy SCBA tank weighs 18 pounds compared with an aluminum tank that weighs over 36 pounds.

Care must be exercised to ensure that the composite material bonds well and fits the geometry of any liner used in the vessel. Particular attention is needed in the domed ends of pressure cylinders and in the locations of ports. Ports are typically placed at the top or bottom at the poles of the domed ends in such a manner that the composite fibers are continuous. Placing ports in the sides of a tank would interrupt the integrity of the filament windings. Also, the geometry of the shape of the tank is often tailored to produce gradually varying stresses at joints between the cylindrical part and the domed ends. The thickness of composite plies is also varied to match the expected stresses.

REFERENCES

1. Advanstar Communications, Inc., *Design Guide for Advanced Composites Applications,* Duluth, MN, 1993.

2. American Society of Mechanical Engineers, *ASME Boiler & Pressure Vessel Code,* Fairfield, NJ, 1992.

3. Muvdi, B. B., and J. W. McNabb, *Engineering Mechanics of Materials,* 3rd ed., Springer-Verlag, New York, 1990.

4. Young, Warren C., *Roark's Formulas for Stress and Strain,* 6th ed., McGraw-Hill, New York, 1989.

PROBLEMS

15–1.M Compute the stress in a sphere having an outside diameter of 200 mm and an inside diameter of 184 mm if an internal pressure of 19.2 MPa is applied.

15–2.M A large, spherical storage tank for a compressed gas in a chemical plant is 10.5 m in diameter and is made of AISI 1040 hot-rolled steel plate, 12 mm thick. What internal pressure could the tank withstand if a design factor of 4.0 based on yield strength is desired?

15–3.M Titanium 6Al-4V is to be used to make a spherical tank having an outside diameter of 1200 mm. The working pressure in the tank is to be 4.20 MPa. Determine the required thickness of the tank wall if a design factor of 4.0 based on yield strength is desired.

15–4.M If the tank of Problem 15–3 was made of aluminum 2014-T6 sheet instead of titanium, compute the required wall thickness. Which design would weigh less?

15–5.E Compute the hoop stress in the walls of a 10-in schedule 40 steel pipe if it carries water at 150 psi.

15–6.M A pneumatic cylinder has a bore of 80 mm and a wall thickness of 3.5 mm. Compute the hoop stress in the cylinder wall if an internal pressure of 2.85 MPa is applied.

15–7.M A cylinder for carrying acetylene has a diameter of 300 mm and will hold the acetylene at 1.7 MPa. If a design factor of 4 is desired based on yield strength, compute the required wall

thickness for the tank. Use AISI 1040 cold-drawn steel.

15–8.M The companion oxygen cylinder for the acetylene discussed in Problem 15–7 carries oxygen at 15.2 MPa. Its diameter is 250 mm. Compute the required wall thickness using the same design criteria.

15–9.M A propane tank for a recreation vehicle is made of AISI 1040 hot-rolled steel, 2.20 mm thick. The tank diameter is 450 mm. Determine what design factor would result based on yield strength if propane at 750 kPa is put into the tank.

15–10.M The supply tank for propane at the distributor is a cylinder having a diameter of 1800 mm. If it is desired to have a design factor of 4 based on yield strength using AISI 1040 hot-rolled steel, compute the required thickness of the tank walls when the internal pressure is 750 kPa.

15–11.M Oxygen on a spacecraft is carried at a pressure of 70.0 MPa in order to minimize the volume required. The spherical vessel has an outside diameter of 250 mm and a wall thickness of 18 mm. Compute the maximum tangential and radial stresses in the sphere.

15–12.M Compute the maximum longitudinal, hoop, and radial stresses in the wall of a standard $\frac{1}{2}$-in schedule 40 steel pipe when carrying an internal pressure of 1.72 MPa (250 psi).

15–13.M The barrel of a large field-artillery piece has a bore of 220 mm and an outside diameter of 300 mm. Compute the magnitude of the hoop

15–14.M A $1\frac{1}{2}$-in schedule 40 steel pipe has a mean radius less than 10 times the wall thickness and thus should be classified as a thick-walled cylinder. Compute what maximum stresses would result from both the thin-wall and the thick-wall formulas due to an internal pressure of 10.0 MPa.

15–15.M A cylinder has an outside diameter of 50 mm and an inside diameter of 30 mm. Compute the maximum tangential stress in the wall of the cylinder due to an internal pressure of 7.0 MPa.

15–16.M For the cylinder of Problem 15–15, compute the tangential stress in the wall at increments of 2.0 mm from the inside to the outside. Then plot the results for stress versus radius.

15–17.M For the cylinder of Problem 15–15, compute the radial stress in the wall at increments of 2.0 mm from the inside to the outside. Then plot the results for stress versus radius.

15–18.M For the cylinder of Problem 15–15, compute the tangential stress that would have been predicted if thin-walled theory were used instead of thick-walled theory. Compare the result with the stress found in Problem 15–15.

15–19.M A sphere is made from stainless steel, AISI 501 OQT 1000. Its outside diameter is 500 mm and the wall thickness is 40 mm. Compute the maximum pressure that could be placed in the sphere if the maximum stress is to be one-fourth of the yield strength of the steel.

15–20.M A sphere has an outside diameter of 500 mm and an inside diameter of 420 mm. Compute the tangential stress in the wall at increments of 5.0 mm from the inside to the outside. Then plot the results. Use a pressure of 100 MPa.

15–21.M A sphere has an outside diameter of 500 mm and an inside diameter of 420 mm. Compute the radial stress in the wall at increments of 5.0 mm from the inside to the outside. Then plot the results. Use a pressure of 100 MPa.

15–22.M To visualize the importance of using the thick-walled formulas for computing stresses in the walls of a cylinder, compute the maximum predicted tangential stress in the wall of a cylinder from both the thin-walled and thick-walled formulas for the following conditions. The outside diameter for all designs is to be 400 mm.

The wall thickness is to vary from 5.0 mm to 85.0 mm in 10.0 mm increments. Use a pressure of 10.0 MPa. Then compute the ratio of D_m/t and plot percent difference between the stress from the thick-walled and thin-walled theory versus that ratio. Note the increase in the percent difference as the value of D_m/t decreases, that is, as t increases.

15–23.M A sphere has an outside diameter of 400 mm and an inside diameter of 325 mm. Compute the variation of the tangential stress from the inside to the outside in increments of 7.5 mm. Use a pressure of 10.0 MPa.

15–24.M A sphere has an outside diameter of 400 mm and an inside diameter of 325 mm. Compute the variation of the radial stress from the inside to the outside in increments of 7.5 mm. Use a pressure of 10.0 MPa.

15–25.E Appendix A–12 lists the dimensions of American National Standard schedule 40 steel pipe. Which of these pipe sizes should be classified as thick-walled and which can be considered thin-walled?

15–26.E Design a cylindrical pressure vessel to carry compressed air for a self-contained breathing apparatus for use by firefighters when operating in smoke-filled buildings. The minimum inside diameter is to be 6.00 in and the length of the cylindrical portion of the tank is to be 15.0 in. It must withstand a service pressure of 4500 psi. Use a design stress of $s_u/8$ to account for a large number of pressurization cycles. Also, check the final design for its ability to withstand a maximum pressure of 13 500 psi by computing the design factor based on yield strength. The tank is to be made from aluminum alloy 6061-T6. Compute the weight of just the cylindrical portion.

15–27.E Repeat Problem 15–26 but use titanium Ti-6Al-4V.

15–28.E Repeat Problem 15–26 but use stainless steel 17-4PH H900.

15–29.E For any of the designs for the SCBA air cylinder for Problems 15–26, 15–27, or 15–28, make a sketch of the complete tank by placing hemispherical heads on each end. Show a port at one end for attaching the pressure regulator and discharge device. Assuming that the wall thickness of the heads is the same as the wall thickness of the cylindrical portion, compute the approximate weight of the complete tank.

15–30.E Repeat Problem 15–26 but use the graphite/epoxy composite material listed in Table 2–6 in Chapter 2 that has a tensile strength of 278 ksi. Check the final design by computing the design factor with respect to tensile strength against the maximum pressure of 13 500 psi. The tank will be lined with a thin polymeric shell and wrapped fully with the unidirectional composite in a circumferential pattern to resist the hoop stress in the cylinder. Neglect the contribution of the liner in the stress analysis and in the weight calculation. (Note that the cylindrical shell would also likely require some plies of the composite to be placed in a helical fashion to resist the longitudinal stress and to permit the formation of the dome-shaped ends. Therefore, the final weight will be somewhat higher than computed for just the circumferentially wrapped part.)

15–31.E Design a spherical tank to carry oxygen at a pressure of 3000 psi with an internal diameter of 18.0 in. Use stainless steel AISI 501 OQT 1000 and a design factor of 6 bsaed on the ultimate strength. Compute the weight of the tank.

15–32.E Repeat Problem 15–31 but use aluminum alloy 7075-T6.

15–33.E Repeat Problem 15–31 but use titanium alloy Ti-6A1-4V.

15–34.M Design a cylindrical tank for compressed natural gas at a pressure of 4.20 MPa. The minimum inside diameter is to be 450 mm. Use aluminum alloy 6061-T6 and a design factor of 8 based on the ultimate strength.

15–35.E Design a cylindrical tank for compressed air that will be used to provide remote service for truck tire repair. The air pressure will be 300 psi. The minimum internal diameter for the tank is to be 24 in. Use AISI 1040 cold-drawn steel and a design factor of 8 based on the ultimate strength. Check the final design for a maximum pressure of 900 psi by computing the design factor based on yield strength.

COMPUTER PROGRAMMING ASSIGNMENTS

1. Write a program to compute the tangential stress in the wall of a thin-walled sphere. Include the computation of the mean diameter and the ratio of mean diameter to thickness to verify that it is thin-walled.

2. Write a program to compute the tangential stress in the wall of a thin-walled cylinder. Include the computation of the mean diameter and the ratio of mean diameter to thickness to verify that it is thin-walled.

3. Write a program to compute the longitudinal stress in the wall of a thin-walled cylinder. Include the computation of the mean diameter and the ratio of mean diameter to thickness to verify that it is thin-walled.

4. Combine the programs of Assignments 2 and 3.

5. Combine the programs of Assignments 1, 2, and 3, and let the user specify whether the vessel is a cylinder or a sphere.

6. Rewrite the programs of Assignments 1, 2, and 5 so that the objective is to compute the required wall thickness for the pressure vessel to produce a given maximum stress for a given internal pressure.

7. Write a program to compute the maximum longitudinal, hoop, and radial stress in the wall of a thick-walled cylinder using the formulas from Table 15–1.

8. Write a program to compute the tangential stress at any radius within the wall of a thick-walled cylinder using the formulas from Table 15–1.

9. Write a program to compute the radial stress at any radius within the wall of a thick-walled cylinder using the formulas from Table 15–1.

10. Write a program to compute the tangential stress at any radius within the wall of a thick-walled sphere using the formulas from Table 15–1.

11. Write a program to compute the radial stress at any radius within the wall of a thick-walled sphere using the formulas from Table 15–1.

12. Combine the programs of Assignments 8 through 11.

13. Write a program to compute the tangential stress distribution within the wall of a thick-walled cylinder using the formulas from Table 15–1. Start at the inside radius and specify a number of increments between the inside and the outside.

14. Write a program to compute the radial stress distribution within the wall of a thick-walled cylinder using the formulas from Table 15–1. Start at the inside radius and specify a number of increments between the inside and the outside.

15. Write a program to compute the tangential stress distribution within the wall of a thick-walled sphere using the formulas from Table 15–1. Start at the inside radius and specify a number of increments between the inside and the outside.

16. Write a program to compute the radial stress distribution within the wall of a thick-walled sphere using the formulas from Table 15–1. Start at the inside radius and specify a number of increments between the inside and the outside.

17. Write a program to perform the computations of the type called for in Problem 15–22.

18. Write a program to perform the computations of the type called for in Problem 15–22, except do it for a sphere.

19. Write a program to compute the maximum tangential stress in any standard schedule 40 pipe for a given internal pressure. Include a table of data for the dimensions of the pipe sizes listed in Appendix A–12. Include a check to see if the pipe is thick-walled or thin-walled.

16

Connections

16-1 OBJECTIVES OF THIS CHAPTER

Load-carrying members that make up structures and machines must act *together* to perform their desired functions. After completing the design or analysis of the primary members, it is necessary to specify suitable connections between them. As their name implies, connections provide the linkage between members.

The primary objective of this chapter is to provide data and methods of analysis for the safe design of riveted joints, bolted joints, and welded joints. After completing this chapter, you should be able to:

1. Describe the typical geometry of riveted and bolted joints.
2. Identify the probable modes of failure for a joint.
3. Recognize typical styles of rivets.
4. Identify when a fastener is in single shear or double shear.
5. Analyze a riveted or bolted joint for shearing force capacity.
6. Analyze a riveted or bolted joint for tensile force capacity.
7. Analyze a riveted or bolted joint for bearing capacity.
8. Use the allowable stresses for steel structural connections as published by the American Institute of Steel Construction (AISC).
9. Describe the difference between a friction-type connection and a bearing-type connection and complete the appropriate analysis.

10. Use the allowable stresses for aluminum structural connections as published by the Aluminum Association.
11. Analyze both symmetrically loaded joints and eccentrically loaded joints.
12. Analyze welded joints with concentric loads.

16–2 TYPES OF CONNECTIONS

Structures and mechanical devices rely on the connections between load-carrying elements to maintain the integrity on the assemblies. The connections provide the path by which loads are transferred from one element to another.

Three common types of connections are riveting, welding, and bolting. Figure 16–1 shows a bulk storage hopper supported by rectangular straps from a tee beam. During fabrication of the hopper, the support tabs were welded to the outside of the side walls. The tabs contain a pattern of holes, allowing the straps to be bolted on at the assembly site. Prior to installation of the tee beam, the straps were riveted to the web.

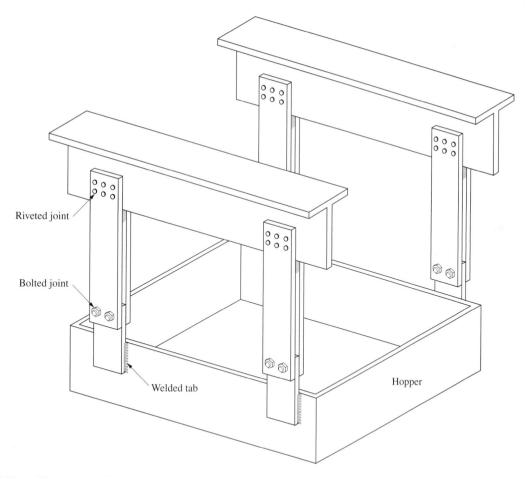

FIGURE 16–1 Illustration of three types of joints—riveted, bolted, and welded.

The load due to the weight of the hopper and the materials in it must be transferred from the hopper walls into the tabs through the welds. Then the four bolts transfer the load to the straps, which act as tension members. Finally, the six rivets transfer the load into the tee.

16–3 MODES OF FAILURE

For riveted and bolted connections, four possible modes of failure exist related to four different types of stresses present in the vicinity of the joint. Figure 16–2 illustrates these kinds of stresses for a simple lap joint in which two flat plates are joined together by two rivets. The joint is prepared by drilling or punching matching holes in each plate. Then the rivets, originally having one of the shapes shown in Figure 16–3, are

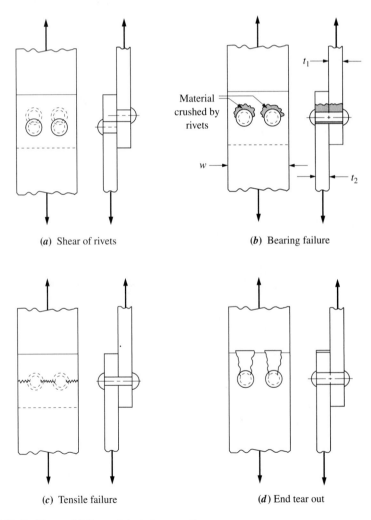

(a) Shear of rivets

(b) Bearing failure

(c) Tensile failure

(d) End tear out

FIGURE 16–2 Types of failure of riveted connections.

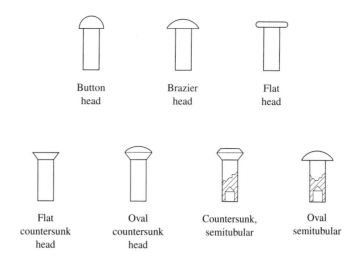

FIGURE 16-3 Examples of rivet styles.

inserted into the holes, and the heads are upset, gripping the two plates together. In a good riveted joint, the body of the rivet is also upset somewhat, causing the rivet to completely fill the hole. This makes a tight joint which does not allow relative motion of the members joined.

As the joint is loaded with a tensile force, a shearing force is transmitted across the cross section of the rivets between the two plates. Thus *shear failure* is a mode of joint failure. The body of the rivet must bear against the material of the plates being joined, with the possibility of *failure in bearing.* This would result in a crushing of the material, normally in the plates. *Tensile failure* of the plates being joined must be investigated because the presence of the holes for the rivets causes the cross section of material at the joint to be less than in the main part of the tension member. The fourth possible failure mode is *end tear out,* in which the rivet causes the material between the edge of the plate and the hole to tear out.

Properly designed riveted and bolted joints should have an edge distance from the center of the rivet or bolt to the edge of the plate being joined of at least two times the diameter of the bolt or rivet. The edge distance is measured in the direction toward which the bearing pressure is directed. If this recommendation is heeded, then end tear out should not occur. This will be assumed in example problems for this chapter. Thus shear, bearing, and tensile failure modes only will be considered in evaluating joint strength.

Welded joints fail by shear in the weld material or by fracture of the base metal of the parts joined by the welds. A properly designed and fabricated welded joint will always fail in the base metal. Thus the design of welded connections has the objective of determining the required size of weld and the length of weld in the joint.

16-4 RIVETED CONNECTIONS

In riveted connections, it is assumed that the joined plates are *not* clamped together tightly enough to cause frictional forces between the plates to transmit loads. Therefore, the rivets do bear on the holes, and bearing failure must be investigated. Both shear failure and tensile failure could also occur. The method of analysis of these three modes of failure is outlined below.

Shear failure. The rivet body is assumed to be in direct shear when a tensile load is applied to a joint, provided that the line of action of the load passes through the centroid of the pattern of rivets. It is also assumed that the total applied load is shared equally among all the rivets. The capacity of a joint with regard to shear of the rivets is

$$F_s = \tau_a A_s \tag{16-1}$$

where F_s = capacity of the joint in shear
$\quad\quad \tau_a$ = allowable shear stress in rivets
$\quad\quad A_s$ = area in shear

The area in shear is dependent on the number of cross sections of rivets available to resist shear. Calling this number N_s,

$$A_s = \frac{N_s \pi D^2}{4} \tag{16-2}$$

where D is the rivet diameter. In some cases, particularly for rivets driven hot, the body expands to fill the hole, and a larger area is available for resisting shear. However, the increase is small, and only the nominal diameter will be used here.

To determine N_s, it must be observed whether *single shear* or *double shear* exists in the joint. Figure 16–2 shows an example of single shear. Only one cross section of each rivet resists the applied load. Then N_s is equal to the number of rivets in the joint. The straps used to support the bin in Figure 16–1 place the rivets and bolts in double shear. Two cross sections of each rivet resist the applied load. Then N_s is twice the number of rivets in the joint.

Bearing failure. When a cylindrical rivet bears against the wall of a hole in the plate, a nonuniform pressure exists between them. As a simplification of the actual stress distribution, it is assumed that the area in bearing, A_b, is the rectangular area found by multiplying the plate thickness t by the diameter of the rivet D. This can be considered to be the *projected area* of the rivet hole. Then the bearing capacity of a joint is

$$F_b = \sigma_{ba} A_b \tag{16-3}$$

where F_b = capacity of the joint in bearing
$\quad\quad \sigma_{ba}$ = allowable bearing stress
$\quad\quad A_b$ = bearing area = $N_b D t$ $\tag{16-4}$
$\quad\quad N_b$ = number of bearing surfaces
$\quad\quad t$ = plate thickness

Tensile failure. A direct tensile force applied through the centroid of the rivet pattern would produce a tensile stress. Then the capacity of the joint in tension would be

$$F_t = \sigma_{ta} A_t \qquad (16-5)$$

where F_t = capacity of the joint in tension
  σ_{ta} = allowable stress in tension
  A_t = net tensile area

The evaluation of A_t requires the subtraction of the diameter of all the holes from the width of the plates being joined. Then

$$A_t = (w - ND_H)t \qquad (16-6)$$

where w = width of plate
  D_H = hole diameter (in structures use $D_H = D + \frac{1}{16}$ in or $D = 2$ mm)
  N = number of holes at the section of interest
  t = thickness of plate

16–5 ALLOWABLE STRESSES

For members not covered by codes and specifications, the allowable stresses can be determined using the design factors presented in Appendix A–20. For the design of steel building structures, the specifications of the American Institute of Steel Construction (AISC) are usually used (1). For aluminum structures, the Aluminum Association has published its *Specifications for Aluminum Structures* (2). Table 16–1 shows

TABLE 16–1 Allowable stresses for steel structural connections*

Rivets	Allowable shear stress		Allowable tensile stress	
	ksi	MPa	ksi	MPa
ASTM A502				
Grade 1	17.5	121	23	159
Grade 2	22	152	29	200

Bolts	Allowable shear stress†		Allowable tensile stress	
	ksi	MPa	ksi	MPa
ASTM A325	17.5	121	44	303
ASTM A490	22	152	54	372

Connected members	Allowable bearing stress‡§	Allowable tensile stress‡
All alloys	$1.20s_u$	$0.6s_y$

*AISC specifications.

†For friction-type connection. For bearing-type connection with no threads in the shear plane, use 30 ksi (207 MPa) for A325 and 40 ksi (276 MPa) for A490.

‡See Appendix A–15 for structural steels.

§Bearing stress not considered in friction-type bolted joint.

TABLE 16–2 Allowable stresses for aluminum structural connections

Rivets			

Alloy and temper		Allowable shear stress	
Before driving	After driving*	ksi	MPa
1100-H14	1100-F	4	27
2017-T4	2017-T3	14.5	100
6053-T61	6053-T61	8.5	58
6061-T6	6061-T6	11	76

Bolts				

Alloy and temper	Allowable shear stress[†]		Allowable tensile stress[†]	
	ksi	MPa	ksi	MPa
2024-T4	16	110	26	179
6061-T6	12	83	18	124
7075-T73	17	117	28	193

Connected members		

Alloy and temper	Allowable bearing stress	
	ksi	MPa
1100-H14	12.5	86
2014-T6	49	338
3003-H14	15	103
6061-T6	34	234
6063-T6	24	165

Source: Aluminum Association, *Specifications for Aluminum Structures,* 5th ed., Washington, DC, 1986.

*All cold-driven.

[†]Stresses are based on the area corresponding to the nominal diameter of the bolt unless the threads are in the shear plane. Then the shear area is based on the root diameter.

allowable stresses for steel structures. Table 16–2 summarizes some allowable stresses for aluminum.

16–6 BOLTED CONNECTIONS

The analysis of bolted connections is the same as for riveted connections if the bolt is allowed to bear on the hole, a bearing-type connection. This would occur in joints where the clamping force provided by the bolts is small. However, most bolted connections are made with high-strength bolts, such as A325 and A490, and tightened to a high tension level. The resulting large clamping forces make a friction-type joint in which the friction forces between the two mating surfaces transmit much of the joint load. The bolts are still designed for shear, using the strengths listed in Table 16–1. But bearing stress is not considered for a friction-type joint.

16–7 EXAMPLE PROBLEMS—RIVETED AND BOLTED JOINTS

Example Problem 16–1 For the single lap joint shown in Figure 16–2, determine the allowable load on the joint if the two plates are $\frac{1}{4}$ in thick by 2 in wide and joined by two steel rivets, $\frac{1}{4}$ in diameter, ASTM A502, grade 1. The plates are ASTM A36 structural steel.

Solution **Objective** Compute the allowable load on the joint.

Given Plate thickness $= t = 0.25$ in; plate width $= w = 2.00$ in
Plates are ASTM A36 structural steel.
Rivets: Diameter $= D = 0.25$ in; ASTM A502, grade 1.

Analysis Shear, bearing, and tensile failure will be investigated to determine the capacity of the joint relative to all three modes. The lowest of the three values is then the limiting load on the joint.

Results **Shear failure**

$$F_s = \tau_a A_s \qquad (16\text{--}1)$$

From Table 16–1, $\tau_a = 17.5$ ksi. For the two rivets in a single shear, there are two shear planes. Then

$$A_s = \frac{N_s \pi D^2}{4} = \frac{2\pi(0.25 \text{ in})^2}{4} = 0.098 \text{ in}^2 \qquad (16\text{--}2)$$

The capacity of the joint in shear is

$$F_s = (17\,500 \text{ lb/in}^2)\,(0.098 \text{ in}^2) = 1715 \text{ lb}$$

Bearing failure

$$F_b = \sigma_{ba} A_b \qquad (16\text{--}3)$$

In bearing, $\sigma_{ba} = 1.20\ s_u$. For ASTM A36 steel, $s_u = 58$ ksi. Then

$$\sigma_{ba} = 1.20(58\,000 \text{ psi}) = 69\,600 \text{ psi}$$

The bearing area is

$$A_b = N_b D t \qquad (16\text{--}4)$$

The force on either plate is resisted by two bearing surfaces. Then

$$A_b = 2(0.25 \text{ in})\,(0.25 \text{ in}) = 0.125 \text{ in}^2$$

The capacity of the joint in bearing is

$$F_b = (69\,600 \text{ lb/in}^2)\,(0.125 \text{ in}^2) = 8700 \text{ lb}$$

Tensile failure. The plates would fail in tension across a section through the rivet holes, as indicated in Figure 16–2(c).

$$F_t = \sigma_{ta} A_t$$

From Table 16–1,

$$\sigma_{ta} = 0.6 s_y = 0.6(36\,000 \text{ psi}) = 21\,600 \text{ psi}$$

The net tensile area, assuming that $D_H = D + \frac{1}{16}$ in, is

$$A_t = (w - ND_H)t = [2.0 \text{ in} - 2(0.25 + 0.063) \text{ in}] (0.25 \text{ in})$$
$$= 0.344 \text{ in}^2$$

The capacity of the joint in tension is

$$F_t = (21\,600 \text{ lb/in}^2)(0.344 \text{ in}^2) = 7425 \text{ lb}$$

Comment Since shear failure would occur at a load of 1715 lb, that is the capacity of the joint.

Example Problem 16–2 Determine the allowable load for a joint of the same dimensions as used in Example Problem 16–1, except use two $\frac{3}{8}$-in-diameter ASTM A490 bolts in a bearing-type connection with no threads in the shear plane.

Solution **Objective** Compute the allowable load on the joint.

Given Plate thickness $= t = 0.25$ in; plate width $= w = 2.00$ in
Plates are ASTM A36 structural steel.
Bolts: Diameter $= D = 0.375$ in; ASTM A490
Bearing type connection; no threads in the shear plane.

Analysis As in Example Problem 16–1, possible failure in shear, bearing, and tension will be investigated. The lowest of the three values is the limiting load on the joint.

Results **Shear failure**

$$F_s = \tau_a A_s$$
$$\tau_a = 40\,000 \text{ psi} \qquad \text{(Table 16–1, footnote)}$$
$$A_s = \frac{2\pi(0.375 \text{ in})^2}{4} = 0.221 \text{ in}^2$$

Then

$$F_s = (40\,000 \text{ lb/in}^2)(0.221 \text{ in}^2) = 8840 \text{ lb}$$

Bearing failure

$$F_b = \sigma_{ba} A_b$$
$$\sigma_{ba} = 1.20(58\,000 \text{ psi}) = 69\,600 \text{ psi}$$
$$A_b = N_b Dt = (2)(0.375 \text{ in})(0.25 \text{ in}) = 0.188 \text{ in}^2$$

Then

$$F_b = (69\,600 \text{ lb/in}^2)(0.188 \text{ in}^2) = 13\,050 \text{ lb}$$

Tensile failure

$$F_t = \sigma_{ta} A_t$$
$$\sigma_{ta} = 0.6(36\,000 \text{ psi}) = 21\,600 \text{ psi}$$
$$A_t = [2.0 \text{ in} - 2(0.375 + 0.063) \text{ in}](0.25 \text{ in}) = 0.281 \text{ in}^2$$

Then

$$F_t = (21\,600 \text{ lb/in}^2)(0.281 \text{ in}^2) = 6070 \text{ lb}$$

Comment Now the capacity in tension is the lowest, so the capacity of the joint is 6070 lb.

16–8 ECCENTRICALLY LOADED RIVETED AND BOLTED JOINTS

Previously considered joints were restricted to cases in which the line of action of the load on the joint passed through the centroid of the pattern of rivets or bolts. In such cases, the applied load is divided equally among all the fasteners. When the load does not pass through the centroid of the fastener pattern, it is called an *eccentrically loaded joint,* and a nonuniform distribution of forces occurs in the fasteners.

In eccentrically loaded joints, the effect of the moment or couple on the fastener must be considered. Figure 16–4 shows a bracket attached to the side of a column and used to support an electric motor. The net downward force exerted by the weight of the motor and the belt tension acts at a distance a from the center of the column flange. Then the total force system acting on the bolts of the bracket consists of the direct shearing force P plus the moment $P \times a$. Each of these components can be considered separately and then added together by using the principle of superposition.

Figure 16–5(a) shows that for the direct shearing force P, each bolt is assumed to carry an equal share of the load, just as in concentrically loaded joints. But in part (b) of the figure, because of the moment, each bolt is subjected to a force acting perpendicular to a radial line from the centroid of the bolt pattern. It is assumed that the

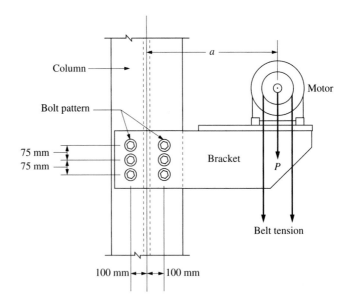

FIGURE 16–4 Eccentric load on a bolted joint.

magnitude of the force in a bolt due to the moment load is proportional to its distance r from the centroid. This magnitude is

$$R_i = \frac{Mr_i}{\sum r^2} \qquad (16\text{–}7)$$

where R_i = shearing force in bolt i due to the moment M
r_i = radial distance to bolt i from the centroid of the bolt pattern
$\sum r^2$ = sum of the radial distances to *all* bolts in the pattern squared

If it is more convenient to work with horizontal and vertical components of forces, they can be found from

$$R_{ix} = \frac{My_i}{\sum r^2} = \frac{My_i}{\sum (x^2 + y^2)} \qquad (16\text{–}8)$$

$$R_{iy} = \frac{Mx_i}{\sum r^2} = \frac{Mx_i}{\sum (x^2 + y^2)} \qquad (16\text{–}9)$$

where y_i = vertical distance to bolt i from centroid
x_i = horizontal distance to bolt i from centroid
$\sum (x^2 + y^2)$ = sum of horizontal and vertical distances squared for all bolts in the pattern

Finally, all horizontal forces are summed and all vertical forces are summed for any particular bolt. Then the resultant of the horizontal and vertical forces is determined.

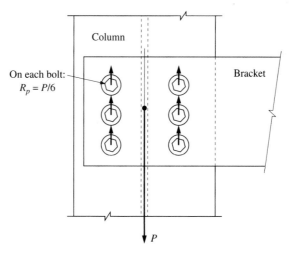

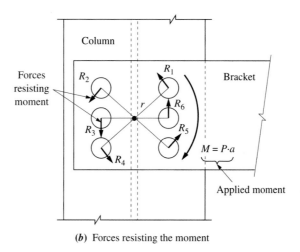

(a) Forces resisting P, the shearing force

(b) Forces resisting the moment

FIGURE 16–5 Loads on bolts that are eccentrically loaded.

Example Problem 16–3

In Figure 16–4 the net downward force P is 26.4 kN on each side plate of the bracket. The distance a is 0.75 m. Determine the required size of ASTM A325 bolts to secure the bracket. Consider the connection to be of the friction type.

Solution

Objective Specify the size of the bolts in the joint.

Given Load $= P = 26.4$ kN downward. Moment arm $= a = 0.75$ m.
Bolt pattern in Figure 16–5. Bolts: ASTM A325
Friction-type connection.

Analysis   To determine the force in each bolt to carry the direct vertical shearing force of $P = 26.4$ kN, each of the six bolts will be assumed to carry an equal share of the load. Then Equations (16−8) and (16−9) will be used to compute the forces in the most highly stressed bolt to resist the moment load, where

$$M = P \times a$$

The resulting forces will be combined vectorially to determine the resultant load on the most highly stressed bolt. Then the required size of that bolt will be computed based on the allowable shearing stress for the ASTM A325 bolts.

Results   ***Direct shearing force.*** The total downward shearing force is divided among six bolts. Then the load on each bolt, called R_p, is

$$R_p = \frac{P}{6} = \frac{26.4 \text{ kN}}{6} = 4.4 \text{ kN}$$

Figure 16−5(a) shows this to be an upward reaction force on each bolt.
Forces resisting the moment.   In Equations (16−8) and (16−9), the following term is required:

$$\sum (x^2 + y^2) = 6(100 \text{ mm})^2 + 4(75 \text{ mm})^2 = 82\,500 \text{ mm}^2$$

The moment on the joint is

$$M = P \times a = 26.4 \text{ kN} (0.75 \text{ m}) = 19.8 \text{ kN·m}$$

Starting first with bolt 1 at the upper right (see Figure 16−6),

$$R_{1x} = \frac{My_1}{\sum (x^2 + y^2)} = \frac{19.8 \text{ kN·m}(75 \text{ mm})}{82\,500 \text{ mm}^2} \times \frac{10^3 \text{ mm}}{\text{m}}$$

$$R_{1x} = 18.0 \text{ kN} \quad \leftarrow \quad \text{(acts toward the left)}$$

$$R_{1y} = \frac{Mx_1}{\sum (x^2 + y^2)} = \frac{(19.8 \text{ kN·m}) (100 \text{ mm})}{82\,500 \text{ mm}^2} \times \frac{10^3 \text{ mm}}{\text{m}}$$

$$R_{1y} = 24.0 \text{ kN} \quad \uparrow \quad \text{(acts upward)}$$

Now the resultant of these forces can be found. In the vertical direction, R_p and R_{1y} both act upward.

$$R_p + R_{1y} = 4.4 \text{ kN} + 24.0 \text{ kN} = 28.4 \text{ kN}$$

Only R_{1x} acts in the horizontal direction. Calling the total resultant force on bolt 1, R_{t1},

$$R_{t1} = \sqrt{28.4^2 + 18.0^2} = 33.6 \text{ kN}$$

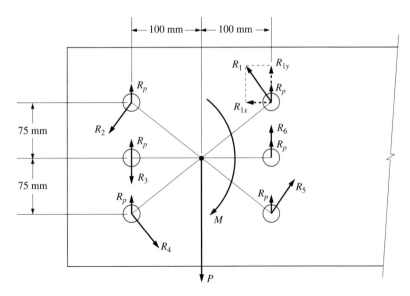

FIGURE 16–6 Forces on each bolt.

Investigating the other five bolts in a similar manner would show that bolt 1 is the most highly stressed. Then its diameter will be determined to limit the shear stress to 121 MPa (17.5 ksi) for the ASTM A325 bolts, as listed in Table 16–1.

$$\tau = \frac{R_{t1}}{A}$$

$$A = \frac{R_{t1}}{\tau_a} = \frac{33.6\ \text{kN}}{121\ \text{N/mm}^2} = 280\ \text{mm}^2 = \frac{\pi D^2}{4}$$

$$D = \sqrt{\frac{4A}{\pi}} = \sqrt{\frac{4(280)\ \text{mm}^2}{\pi}} = 18.9\ \text{mm}$$

The nearest preferred metric size is 20 mm. If conventional inch units are to be specified,

$$D = 18.9\ \text{mm} \times \frac{1\ \text{in}}{25.4\ \text{mm}} = 0.74\ \text{in}$$

The nearest standard size is $\frac{3}{4}$ in (0.750 in).

Comment Specify either $D = 20$ mm or $D = \frac{3}{4}$ in bolts.

16–9 WELDED JOINTS WITH CONCENTRIC LOADS

Welding is a joining process in which heat is applied to cause two pieces of metal to become metallurgically bonded. The heat may be applied by a gas flame, an electric arc, or by a combination of electric resistance heating and pressure.

Types of welds include groove, fillet, and spot welds (as shown in Figure 16–7), and others. Groove and fillet welds are most frequently used in structural connections since they are readily adaptable to the shapes and plates that make up the structures. Spot welds are used for joining relatively light gauge steel sheets and cold-formed shapes.

The variables involved in designing welded joints are the shape and size of the weld, the choice of filler metal, the length of weld, and the position of the weld relative to the applied load.

Fillet welds are assumed to have a slope of 45 deg between the two surfaces joined, as shown in Figure 16–7(b). The size of the weld is denoted as the height of one side of the triangular-shaped fillet. Typical sizes range from $\frac{1}{8}$ in to $\frac{1}{2}$ in in steps of $\frac{1}{16}$ in. The stress developed in fillet welds is assumed to be *shear stress* regardless of the direction of application of the load. The maximum shear stress would occur at the throat

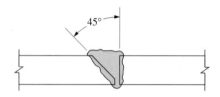

(*a*) Butt weld with single bevel groove

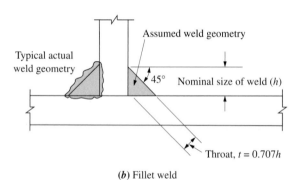

(*b*) Fillet weld

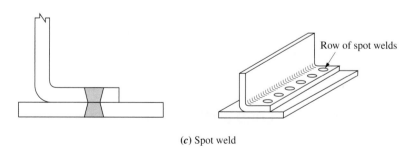

(*c*) Spot weld

FIGURE 16–7   Types of welds.

of the fillet (see Figure 16–7), where the thickness is 0.707 times the nominal size of the weld. Then the shearing stress in the weld due to a load P is

$$\tau = \frac{P}{Lt} \qquad (16-10)$$

where L is the length of weld and t is the thickness at the throat. Equation (16–10) is used *only* for concentrically loaded members. This requires that the line of action of the force on the welds passes through the centroid of the weld pattern. Eccentricity of the load produces a moment, in addition to the direct shearing force, which must be resisted by the weld metal. References 1, 3, and 4 at the end of this chapter contain pertinent information with regard to eccentrically loaded welded joints.

In electric arc welding, used mostly for structural connections, a filler rod is normally used to add metal to the welded zone. As the two parts to be joined are heated to a molten state, filler metal is added, which combines with the base metal. On cooling, the resulting weld metal is normally stronger than the original base metal. Therefore, a properly designed and made welded joint should fail in the base metal rather than in the weld. In structural welding, the electrodes are given a code beginning with an E and followed by two or three digits, such as E60, E80, or E100. The number denotes the ultimate tensile strength in ksi of the weld metal in the rod. Thus an E80 rod would have a tensile strength of 80 000 psi. Other digits may be added to the code number to denote special properties. Complete specifications can be found in the standards of ASTM A233 and A316. The allowable shear stress for fillet welds using electrodes is 0.3 times the tensile strength of the electrode according to AISC. Table 16–3 lists some common electrodes and their allowable stresses.

Aluminum products are welded using either the inert gas-shielded arc process or the resistance welding process. For the inert gas-shielded arc process, filler alloys are specified by the Aluminum Association for joining particular base metal alloys, as indicated in Table 16–4. The allowable shear stresses for such welds are also listed. It should be noted that the heat of welding lowers the properties of most aluminum alloys within 1.0 in of the weld, and allowance for this must be made in the design of welded assemblies.

TABLE 16–3 Properties of welding electrodes for steel

Electrode type	Minimum tensile strength		Allowable shear stress		Typical metals joined
	ksi	MPa	ksi	MPa	
E60	60	414	18	124	A36, A500
E70	70	483	21	145	A242, A441
E80	80	552	24	165	A572, Grade 65
E90	90	621	27	186	—
E100	100	690	30	207	—
E110	110	758	33	228	A514

TABLE 16-4 Allowable shear stresses in fillet welds in aluminum building-type structures

| | Filler alloy | | | | | | | |
| | 1100 | | 4043 | | 5356 | | 5556 | |
Base metal	ksi	MPa	ksi	MPa	ksi	MPa	ksi	MPa
1100	3.2	22	4.8	33	—	—	—	—
3003	3.2	22	5.0	34	—	—	—	—
6061	—	—	5.0	34	7.0	48	8.5	59
6063	—	—	5.0	34	6.5	45	6.5	45

**Example Problem
16-4**

A lap joint is made by placing two $\frac{3}{8}$-in fillet welds across the full width of two $\frac{1}{2}$-in ASTM A36 steel plates, as shown in Figure 16-8. The shielded metal-arc method is used, using an E60 electrode. Compute the allowable load P that can be applied to the joint.

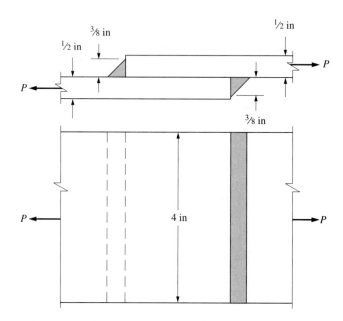

FIGURE 16-8 Welded lap joint.

Solution **Objective** Compute the allowable load, P, on the joint.

Given Joint design in Figure 16-8. Plates are ASTM A36 steel.
E60 electrode used in shielded metal-arc method.

Analysis The load is assumed to be equally distributed on all parts of the weld, so that Equation (16-10) can be used with $L = 8.0$ in.

$$\tau = \frac{P}{Lt}$$

Let τ equal the allowable stress of 18 ksi, listed in Table 16–3. The thickness t is

$$t = 0.707 \left(\tfrac{3}{8} \text{ in}\right) = 0.265 \text{ in}$$

Now we can solve for P.

$$P = \tau_a Lt = (18\,000 \text{ lb/in}^2)(8.0 \text{ in})(0.265 \text{ in}) = 38\,200 \text{ lb}$$

REFERENCES

1. American Institute of Steel Construction, *Manual of Steel Construction,* 9th ed., Chicago, IL, 1989.

2. Aluminum Association, *Specifications for Aluminum Structures,* 5th ed., Washington, DC, 1986.

3. Blodgett, O.W., *Design of Weldments,* James F. Lincoln Arc Welding Foundation, Cleveland, OH, 1963.

4. Mott, R. L., *Machine Elements in Mechanical Design,* 2nd ed., Merrill, an imprint of Macmillan Publishing Co., New York, 1992.

PROBLEMS

16–1.E Determine the allowable loads on the joints shown in Figure 16–9. The fasteners are all steel rivets, ASTM A502, grade 1. The plates are all ASTM A36 steel.

16–2.E Determine the allowable loads on the joints shown in Figure 16–10. The fasteners are all steel rivets, ASTM A502, grade 2. The plates are all ASTM A242 steel.

16–3.E Determine the allowable loads on the joints shown in Figure 16–9 if the fasteners are all ASTM A325 steel bolts providing a bearing-type connection and the plates are all ASTM A242 steel.

16–4.E Determine the allowable loads on the joints shown in Figure 16–10. The fasteners are all ASTM A490 steel bolts providing a friction-type joint. The plates are all ASTM A514 steel.

16–5.E Determine the required diameter of the bolts used to attach the cantilever beam to the column, as shown in Figure 16–11. Use ASTM A325 steel bolts.

16–6.M Design the connection of the channel to the column for the hanger shown in Figure 16–12. Both members are ASTM A36 steel. Specify the type of fastener (rivet, bolt), the pattern and spacing, the number of fasteners, and the material for the fasteners. Use AISC specifications.

16–7.E For the connection shown in Figure 16–9(a), assume that, instead of the two rivets, the two plates were welded across the ends of the 3-in-wide plates using $\tfrac{5}{16}$-in welds. The plates are ASTM A36 steel and the electric arc welding technique is used with E60 electrodes. Determine the allowable load on the connection.

16–8.E Determine the allowable load on the joint shown in Figure 16–10(c) if $\tfrac{1}{4}$-in welds using E70 electrodes were placed along both ends of both cover plates. The plates are ASTM A242 steel.

16–9.M Design the joint at the top of the straps in Figure 16–1 if the total load in the hopper is 54.4 megagrams (Mg). The beam is a WT12 × 34 made of ASTM A36 steel and has a web thickness of 10.6 mm. The clear vertical height of the web is about 250 mm. Use steel rivets and specify the pattern, number of rivets, rivet diameter, rivet material, strap material, and strap dimensions. Specify dimensions, using the preferred metric sizes shown in Appendix A–2.

16–10.M Design the joint at the bottom of the straps in Figure 16–1 if the total load in the hopper is

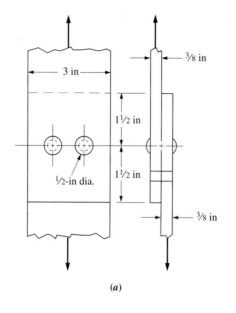

(a)

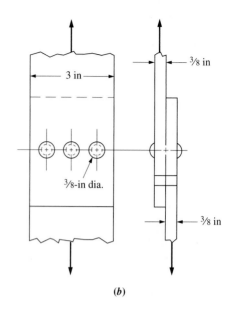

(b)

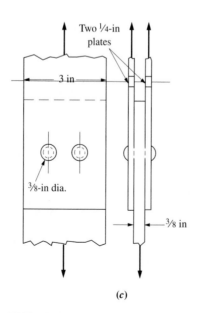

(c)

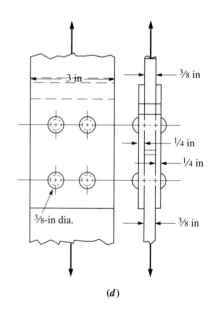

(d)

FIGURE 16–9 Joints for Problems 16–1 and 16–3.

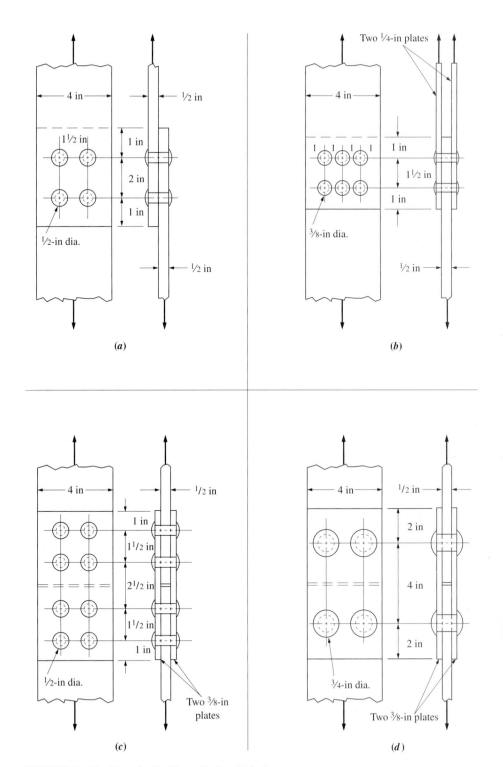

FIGURE 16–10 Joints for Problems 16–2 and 16–4.

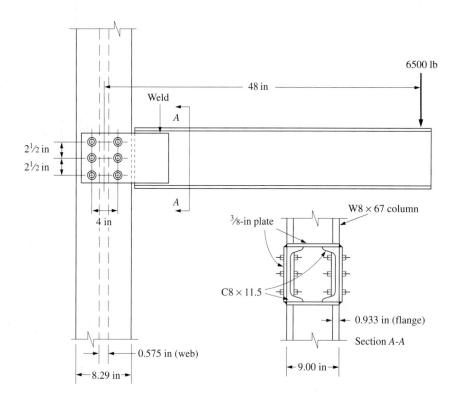

FIGURE 16–11 Connection for Problem 16–5.

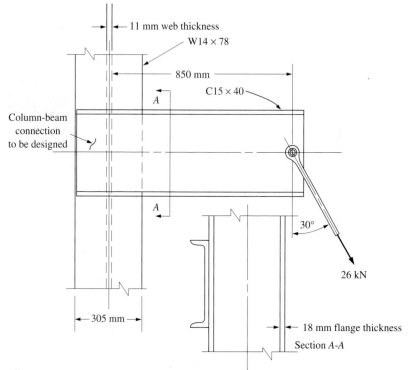

FIGURE 16–12 Connection for Problem 16–6.

54.4 Mg. Use steel bolts and a bearing-type connection. Specify the pattern, number of bolts, bolt diameter, bolt material, strap material, and strap dimensions. You may want to coordinate the strap design with the results of Problem 16–9. The design of the tab in Problem 16–11 is also affected by the design of the bolted joint.

16–11.M Design the tab to be welded to the hopper for connection to the support straps, as shown in Figure 16–1. The hopper load is 54.4 Mg. The material from which the hopper is made is ASTM A36 steel. Specify the width and thickness of the tab and the design of the welded joint. You may want to coordinate the tab design with the bolted connection called for in Problem 16–10.

Appendix

A–1 Properties of areas*

Circle

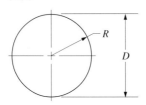

$$A = \frac{\pi D^2}{4} = \pi R^2 \qquad r = \frac{D}{4} = \frac{R}{2}$$

$$I = \frac{\pi D^4}{64} \qquad J = \frac{\pi D^4}{32}$$

$$S = \frac{\pi D^3}{32} \qquad Z_p = \frac{\pi D^3}{16}$$

Circumference $= \pi D = 2\pi R$

Hollow circle (tube)

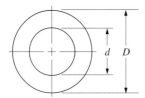

$$A = \frac{\pi(D^2 - d^2)}{4} \qquad r = \frac{\sqrt{D^2 + d^2}}{4}$$

$$I = \frac{\pi(D^4 - d^4)}{64} \qquad J = \frac{\pi(D^4 - d^4)}{32}$$

$$S = \frac{\pi(D^4 - d^4)}{32D} \qquad Z_p = \frac{\pi(D^4 - d^4)}{16D}$$

*Symbols used are:

A = area r = radius of gyration = $\sqrt{I/A}$

I = moment of inertia J = polar moment of inertia

S = section modulus Z_p = polar section modulus

Square

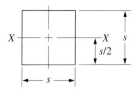

$$A = s^2$$

$$I_x = \frac{s^4}{12} \qquad r_x = \frac{s}{\sqrt{12}}$$

$$S_x = \frac{s^3}{6}$$

Rectangle

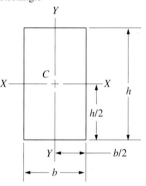

$$A = bh$$

$$I_x = \frac{bh^3}{12} \qquad S_x = \frac{bh^2}{6} \qquad r_x = \frac{h}{\sqrt{12}}$$

$$I_y = \frac{hb^3}{12} \qquad S_y = \frac{hb^2}{6} \qquad r_y = \frac{b}{\sqrt{12}}$$

Triangle

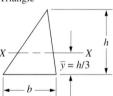

$$A = \frac{bh}{2}$$

$$I_x = \frac{bh^3}{36} \qquad r_x = \frac{h}{\sqrt{18}}$$

$$S_x = \frac{bh^2}{24}$$

Semicircle

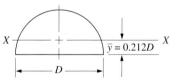

$$A = \frac{\pi D^2}{8}$$

$$I_x = 0.007D^4 \qquad r_x = 0.132D$$

$$S_x = 0.024D^3$$

Regular hexagon

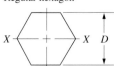

$$A = 0.866D^2$$

$$I_x = 0.06D^4 \qquad r_x = 0.264D$$

$$S_x = 0.12D^3$$

583

Area under a second-degree curve

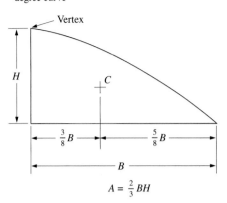

$$A = \tfrac{2}{3} BH$$

Area over a second-degree curve

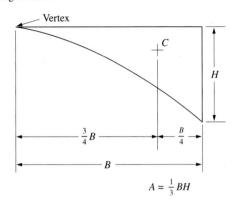

$$A = \tfrac{1}{3} BH$$

Area under a third-degree curve

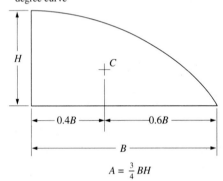

$$A = \tfrac{3}{4} BH$$

Area over a third-degree curve

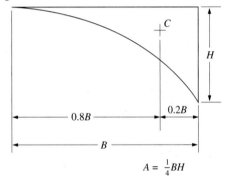

$$A = \tfrac{1}{4} BH$$

A–2 Preferred basic sizes

Fractional (in)				Decimal (in)			Metric (mm) First	Second	First	Second	First	Second
$\frac{1}{64}$	0.015 625	5	5.000	0.010	2.00	8.50	1		10		100	
$\frac{1}{32}$	0.031 25	$5\frac{1}{4}$	5.250	0.012	2.20	9.00		1.1		11		110
$\frac{1}{16}$	0.062 5	$5\frac{1}{2}$	5.500	0.016	2.40	9.50	1.2		12		120	
$\frac{3}{32}$	0.093 75	$5\frac{3}{4}$	5.750	0.020	2.60	10.00		1.4		14		140
$\frac{1}{8}$	0.125 0	6	6.000	0.025	2.80	10.50	1.6		16		160	
$\frac{5}{32}$	0.156 25	$6\frac{1}{2}$	6.500	0.032	3.00	11.00		1.8		18		180
$\frac{3}{16}$	0.187 5	7	7.000	0.040	3.20	11.50	2		20		200	
$\frac{1}{4}$	0.250 0	$7\frac{1}{2}$	7.500	0.05	3.40	12.00		2.2		22		220
$\frac{5}{16}$	0.312 5	8	8.000	0.06	3.60	12.50	2.5		25		250	
$\frac{3}{8}$	0.375 0	$8\frac{1}{2}$	8.500	0.08	3.80	13.00		2.8		28		280
$\frac{7}{16}$	0.437 5	9	9.000	0.10	4.00	13.50	3		30		300	
$\frac{1}{2}$	0.500 0	$9\frac{1}{2}$	9.500	0.12	4.20	14.00		3.5		35		350
$\frac{9}{16}$	0.562 5	10	10.000	0.16	4.40	14.50	4		40		400	
$\frac{5}{8}$	0.625 0	$10\frac{1}{2}$	10.500	0.20	4.60	15.00		4.5		45		450
$\frac{11}{16}$	0.687 5	11	11.000	0.24	4.80	15.50	5		50		500	
$\frac{3}{4}$	0.750 0	$11\frac{1}{2}$	11.500	0.30	5.00	16.00		5.5		55		550
$\frac{7}{8}$	0.875 0	12	12.000	0.40	5.20	16.50	6		60		600	
1	1.000	$12\frac{1}{2}$	12.500	0.50	5.40	17.00		7		70		700
$1\frac{1}{4}$	1.250	13	13.000	0.60	5.60	17.50	8		80		800	
$1\frac{1}{2}$	1.500	$13\frac{1}{2}$	13.500	0.80	5.80	18.00		9		90		900
$1\frac{3}{4}$	1.750	14	14.000	1.00	6.00	18.50					1000	
2	2.000	$14\frac{1}{2}$	14.500	1.20	6.50	19.00						
$2\frac{1}{4}$	2.250	15	15.000	1.40	7.00	19.50						
$2\frac{1}{2}$	2.500	$15\frac{1}{2}$	15.500	1.60	7.50	20.00						
$2\frac{3}{4}$	2.750	16	16.000	1.80	8.00							
3	3.000	$16\frac{1}{2}$	16.500									
$3\frac{1}{4}$	3.250	17	17.000									
$3\frac{1}{2}$	3.500	$17\frac{1}{2}$	17.500									
$3\frac{3}{4}$	3.750	18	18.000									
4	4.000	$18\frac{1}{2}$	18.500									
$4\frac{1}{4}$	4.250	19	19.000									
$4\frac{1}{2}$	4.500	$19\frac{1}{2}$	19.500									
$4\frac{3}{4}$	4.750	20	20.000									

(a) American Standard thread dimensions, numbered sizes

Size	Basic major diameter, D (in)	Coarse threads: UNC		Fine threads: UNF	
		Threads per inch, n	Tensile stress area (in^2)	Threads per inch, n	Tensile stress area (in^2)
0	0.060 0	—	—	80	0.001 80
1	0.073 0	64	0.002 63	72	0.002 78
2	0.086 0	56	0.003 70	64	0.003 94
3	0.099 0	48	0.004 87	56	0.005 23
4	0.112 0	40	0.006 04	48	0.006 61
5	0.125 0	40	0.007 96	44	0.008 30
6	0.138 0	32	0.009 09	40	0.010 15
8	0.164 0	32	0.014 0	36	0.014 74
10	0.190 0	24	0.017 5	32	0.020 0
12	0.216 0	24	0.024 2	28	0.025 8

(b) American Standard thread dimensions, fractional sizes

Size	Basic major diameter, D (in)	Coarse threads: UNC		Fine threads: UNF	
		Threads per inch, n	Tensile stress area (in^2)	Threads per inch, n	Tensile stress area (in^2)
$\frac{1}{4}$	0.250 0	20	0.031 8	28	0.036 4
$\frac{5}{16}$	0.312 5	18	0.052 4	24	0.058 0
$\frac{3}{8}$	0.375 0	16	0.077 5	24	0.087 8
$\frac{7}{16}$	0.437 5	14	0.106 3	20	0.118 7
$\frac{1}{2}$	0.500 0	13	0.141 9	20	0.159 9
$\frac{9}{16}$	0.562 5	12	0.182	18	0.203
$\frac{5}{8}$	0.625 0	11	0.226	18	0.256
$\frac{3}{4}$	0.750 0	10	0.334	16	0.373
$\frac{7}{8}$	0.875 0	9	0.462	14	0.509
1	1.000	8	0.606	12	0.663
$1\frac{1}{8}$	1.125	7	0.763	12	0.856
$1\frac{1}{4}$	1.250	7	0.969	12	1.073
$1\frac{3}{8}$	1.375	6	1.155	12	1.315
$1\frac{1}{2}$	1.500	6	1.405	12	1.581
$1\frac{3}{4}$	1.750	5	1.90	—	—
2	2.000	$4\frac{1}{2}$	2.50	—	—

Basic major diameter, D (mm)	Coarse threads		Fine threads	
	Pitch (mm)	Tensile stress area (mm^2)	Pitch (mm)	Tensile stress area (mm^2)
1	0.25	0.460	—	—
1.6	0.35	1.27	0.20	1.57
2	0.4	2.07	0.25	2.45
2.5	0.45	3.39	0.35	3.70
3	0.5	5.03	0.35	5.61
4	0.7	8.78	0.5	9.79
5	0.8	14.2	0.5	16.1
6	1	20.1	0.75	22.0
8	1.25	36.6	1	39.2
10	1.5	58.0	1.25	61.2
12	1.75	84.3	1.25	92.1
16	2	157	1.5	167
20	2.5	245	1.5	272
24	3	353	2	384
30	3.5	561	2	621
36	4	817	3	865
42	4.5	1 121	—	—
48	5	1 473	—	—

A–4 Properties of standard wood beams

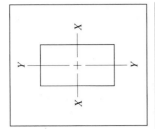

Nominal size	Actual size		Area of section		Moment of inertia, I_x		Section modulus, S_x	
	in	mm	in^2	mm$^2 \times 10^{-3}$	in^4	mm$^4 \times 10^{-6}$	in^3	mm$^3 \times 10^{-3}$
2 × 4	1.5 × 3.5	38 × 89	5.25	3.39	5.36	2.23	3.06	50.1
2 × 6	1.5 × 5.5	38 × 140	8.25	5.32	20.8	8.66	7.56	124
2 × 8	1.5 × 7.25	38 × 184	10.87	7.01	47.6	19.8	13.14	215
2 × 10	1.5 × 9.25	38 × 235	13.87	8.95	98.9	41.2	21.4	351
2 × 12	1.5 × 11.25	38 × 286	16.87	10.88	178	74.1	31.6	518
4 × 4	3.5 × 3.5	89 × 89	12.25	7.90	12.51	5.21	7.15	117
4 × 6	3.5 × 5.5	89 × 140	19.25	12.42	48.5	20.2	17.65	289
4 × 8	3.5 × 7.25	89 × 184	25.4	16.39	111.1	46.2	30.7	503
4 × 10	3.5 × 9.25	89 × 235	32.4	20.90	231	96.1	49.9	818
4 × 12	3.5 × 11.25	89 × 286	39.4	25.42	415	172	73.9	1211
6 × 6	5.5 × 5.5	140 × 140	30.3	19.55	76.3	31.8	27.7	454
6 × 8	5.5 × 7.5	140 × 191	41.3	26.65	193	80.3	51.6	846
6 × 10	5.5 × 9.5	140 × 241	52.3	33.74	393	164	82.7	1355
6 × 12	5.5 × 11.5	140 × 292	63.3	40.84	697	290	121	1983
8 × 8	7.5 × 7.5	191 × 191	56.3	36.32	264	110	70.3	1152
8 × 10	7.5 × 9.5	191 × 241	71.3	46.00	536	223	113	1852
8 × 12	7.5 × 11.5	191 × 292	86.3	55.68	951	396	165	2704
10 × 10	9.5 × 9.5	241 × 241	90.3	58.26	679	283	143	2343
10 × 12	9.5 × 11.5	241 × 292	109.3	70.52	1204	501	209	3425
12 × 12	11.5 × 11.5	292 × 292	132.3	85.35	1458	607	253	4146

A–5 Properties of steel angles
Equal legs and unequal legs
L-Shapes*

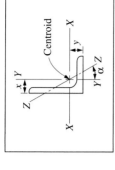

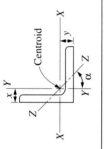

Designation	Area (in²)	Weight per foot (lb)	Axis X-X			Axis Y-Y			Axis Z-Z	
			I (in⁴)	S (in³)	y (in)	I (in⁴)	S (in³)	x (in)	r (in)	α (deg)
L8 × 8 × 1	15.0	51.0	89.0	15.8	2.37	89.0	15.8	2.37	1.56	45.0
L8 × 8 × ½	7.75	26.4	48.6	8.36	2.19	48.6	8.36	2.19	1.59	45.0
L8 × 4 × 1	11.0	37.4	69.6	14.1	3.05	11.6	3.94	1.05	0.846	13.9
L8 × 4 × ½	5.75	19.6	38.5	7.49	2.86	6.74	2.15	0.859	0.865	14.9
L6 × 6 × ¾	8.44	28.7	28.2	6.66	1.78	28.2	6.66	1.78	1.17	45.0
L6 × 6 × ⅜	4.36	14.9	15.4	3.53	1.64	15.4	3.53	1.64	1.19	45.0
L6 × 4 × ¾	6.94	23.6	24.5	6.25	2.08	8.68	2.97	1.08	0.860	23.2
L6 × 4 × ⅜	3.61	12.3	13.5	3.32	1.94	4.90	1.60	0.941	0.877	24.0
L4 × 4 × ½	3.75	12.8	5.56	1.97	1.18	5.56	1.97	1.18	0.782	45.0
L4 × 4 × ¼	1.94	6.6	3.04	1.05	1.09	3.04	1.05	1.09	0.795	45.0
L4 × 3 × ½	3.25	11.1	5.05	1.89	1.33	2.42	1.12	0.827	0.639	28.5
L4 × 3 × ¼	1.69	5.8	2.77	1.00	1.24	1.36	0.599	0.896	0.651	29.2
L3 × 3 × ½	2.75	9.4	2.22	1.07	0.932	2.22	1.07	0.932	0.584	45.0
L3 × 3 × ¼	1.44	4.9	1.24	0.577	0.842	1.24	0.577	0.842	0.592	45.0
L2 × 2 × ⅜	1.36	4.7	0.479	0.351	0.636	0.479	0.351	0.636	0.389	45.0
L2 × 2 × ¼	0.938	3.19	0.348	0.247	0.592	0.348	0.247	0.592	0.391	45.0
L2 × 2 × ⅛	0.484	1.65	0.190	0.131	0.546	0.190	0.131	0.546	0.398	45.0

*Data taken from a variety of sources. Sizes listed represent a small sample of the sizes available.

Example designation: L4 × 3 × ½

4 = length of longer leg (in), 3 = length of shorter leg (in), ½ = thickness of legs (in)

Z-Z is axis of minimum moment of inertia (I) and radius of gyration (r)

I = moment of inertia, S = section modulus, r = radius of gyration

A–6 Properties of American Standard steel channels
C-Shapes*

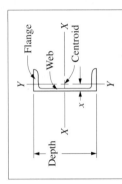

Designation	Area (in²)	Depth (in)	Web thickness (in)	Flange Width (in)	Flange Average thickness (in)	Axis X-X I (in⁴)	Axis X-X S (in³)	Axis Y-Y I (in⁴)	Axis Y-Y S (in³)	Axis Y-Y x (in)
C15 × 50	14.7	15.00	0.716	3.716	0.650	404	53.8	11.0	3.78	0.798
C15 × 40	11.8	15.00	0.520	3.520	0.650	349	46.5	9.23	3.37	0.777
C12 × 30	8.82	12.00	0.510	3.170	0.501	162	27.0	5.14	2.06	0.674
C12 × 25	7.35	12.00	0.387	3.047	0.501	144	24.1	4.47	1.88	0.674
C10 × 30	8.82	10.00	0.673	3.033	0.436	103	20.7	3.94	1.65	0.649
C10 × 20	5.88	10.00	0.379	2.739	0.436	78.9	15.8	2.81	1.32	0.606
C9 × 20	5.88	9.00	0.448	2.648	0.413	60.9	13.5	2.42	1.17	0.583
C9 × 15	4.41	9.00	0.285	2.485	0.413	51.0	11.3	1.93	1.01	0.586
C8 × 18.75	5.51	8.00	0.487	2.527	0.390	44.0	11.0	1.98	1.01	0.565
C8 × 11.5	3.38	8.00	0.220	2.260	0.390	32.6	8.14	1.32	0.781	0.571
C6 × 13	3.83	6.00	0.437	2.157	0.343	17.4	5.80	1.05	0.642	0.514
C6 × 8.2	2.40	6.00	0.200	1.920	0.343	13.1	4.38	0.693	0.492	0.511
C5 × 9	2.64	5.00	0.325	1.885	0.320	8.90	3.56	0.632	0.450	0.478
C5 × 6.7	1.97	5.00	0.190	1.750	0.320	7.49	3.00	0.479	0.378	0.484
C4 × 7.25	2.13	4.00	0.321	1.721	0.296	4.59	2.29	0.433	0.343	0.459
C4 × 5.4	1.59	4.00	0.184	1.584	0.296	3.85	1.93	0.319	0.283	0.457
C3 × 6	1.76	3.00	0.356	1.596	0.273	2.07	1.38	0.305	0.268	0.455
C3 × 4.1	1.21	3.00	0.170	1.410	0.273	1.66	1.10	0.197	0.202	0.436

*Data taken from a variety of sources. Sizes listed represent a small sample of the sizes available.

Example designation: C15 × 50

15 = depth (in), 50 = weight per unit length (lb/ft)

I = moment of inertia, S = section modulus

A–7 Properties of steel wide-flange shapes
W-Shapes*

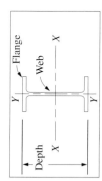

Designation	Area (in^2)	Depth (in)	Web thickness (in)	Flange Width (in)	Flange Thickness (in)	Axis X-X I (in^4)	Axis X-X S (in^3)	Axis Y-Y I (in^4)	Axis Y-Y S (in^3)
W24 × 76	22.4	23.92	0.440	8.990	0.680	2100	176	82.5	18.4
W24 × 68	20.1	23.73	0.415	8.965	0.585	1830	154	70.4	15.7
W21 × 73	21.5	21.24	0.455	8.295	0.740	1600	151	70.6	17.0
W21 × 57	16.7	21.06	0.405	6.555	0.650	1170	111	30.6	9.35
W18 × 55	16.2	18.11	0.390	7.530	0.630	890	98.3	44.9	11.9
W18 × 40	11.8	17.90	0.315	6.015	0.525	612	68.4	19.1	6.35
W14 × 43	12.6	13.66	0.305	7.995	0.530	428	62.7	45.2	11.3
W14 × 26	7.69	13.91	0.255	5.025	0.420	245	35.3	8.91	3.54
W12 × 30	8.79	12.34	0.260	6.520	0.440	238	38.6	20.3	6.24
W12 × 16	4.71	11.99	0.220	3.990	0.265	103	17.1	2.82	1.41
W10 × 15	4.41	9.99	0.230	4.000	0.270	69.8	13.8	2.89	1.45
W10 × 12	3.54	9.87	0.190	3.960	0.210	53.8	10.9	2.18	1.10
W8 × 15	4.44	8.11	0.245	4.015	0.315	48.0	11.8	3.41	1.70
W8 × 10	2.96	7.89	0.170	3.940	0.205	30.8	7.81	2.09	1.06
W6 × 15	4.43	5.99	0.230	5.990	0.260	29.1	9.72	9.32	3.11
W6 × 12	3.55	6.03	0.230	4.000	0.280	22.1	7.31	2.99	1.50
W5 × 19	5.54	5.15	0.270	5.030	0.430	26.2	10.2	9.13	3.63
W5 × 16	4.68	5.01	0.240	5.000	0.360	21.3	8.51	7.51	3.00
W4 × 13	3.83	4.16	0.280	4.060	0.345	11.3	5.46	3.86	1.90

*Data taken from a variety of sources. Sizes listed represent a small sample of the sizes available.

Example designation: W14 × 43

14 = nominal depth (in), 43 = weight per unit length (lb/ft)

I = moment of inertia, S = section modulus

A–8 Properties of American Standard steel beams
S-Shapes*

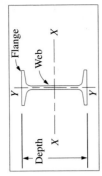

Designation	Area (in²)	Depth (in)	Web thickness (in)	Flange Width (in)	Flange Average thickness (in)	Axis X-X I (in⁴)	Axis X-X S (in³)	Axis Y-Y I (in⁴)	Axis Y-Y S (in³)
S24 × 90	26.5	24.00	0.625	7.125	0.870	2250	187	44.9	12.6
S20 × 96	28.2	20.30	0.800	7.200	0.920	1670	165	50.2	13.9
S20 × 75	22.0	20.00	0.635	6.385	0.795	1280	128	29.8	9.32
S20 × 66	19.4	20.00	0.505	6.255	0.795	1190	119	27.7	8.85
S18 × 70	20.6	18.00	0.711	6.251	0.691	926	103	24.1	7.72
S15 × 50	14.7	15.00	0.550	5.640	0.622	486	64.8	15.7	5.57
S12 × 50	14.7	12.00	0.687	5.477	0.659	305	50.8	15.7	5.74
S12 × 35	10.3	12.00	0.428	5.078	0.544	229	38.2	9.87	3.89
S10 × 35	10.3	10.00	0.594	4.944	0.491	147	29.4	8.36	3.38
S10 × 25.4	7.46	10.00	0.311	4.661	0.491	124	24.7	6.79	2.91
S8 × 23	6.77	8.00	0.441	4.171	0.426	64.9	16.2	4.31	2.07
S8 × 18.4	5.41	8.00	0.271	4.001	0.426	57.6	14.4	3.73	1.86
S7 × 20	5.88	7.00	0.450	3.860	0.392	42.4	12.1	3.17	1.64
S6 × 12.5	3.67	6.00	0.232	3.332	0.359	22.1	7.37	1.82	1.09
S5 × 10	2.94	5.00	0.214	3.004	0.326	12.3	4.92	1.22	0.809
S4 × 7.7	2.26	4.00	0.193	2.663	0.293	6.08	3.04	0.764	0.574
S3 × 5.7	1.67	3.00	0.170	2.330	0.260	2.52	1.68	0.455	0.390

*Data taken from a variety of sources. Sizes listed represent a small sample of the sizes available.
Example designation: S10 × 35

10 = nominal depth (in), 35 = weight per unit length (lb/ft)

I = moment of inertia, *S* = section modulus

A–9 Properties of steel structural tubing
Square and rectangular*

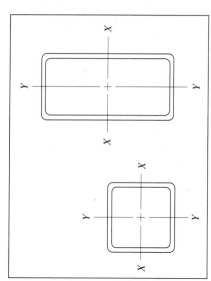

Size	Area (in²)	Weight per foot (lb)	Axis X-X			Axis Y-Y		
			I (in⁴)	S (in³)	r (in)	I (in⁴)	S (in³)	r (in)
8 × 8 × ½	14.4	48.9	131	32.9	3.03	131	32.9	3.03
8 × 8 × ¼	7.59	25.8	75.1	18.8	3.15	75.1	18.8	3.15
8 × 4 × ½	10.4	35.2	75.1	18.8	2.69	24.6	12.3	1.54
8 × 4 × ¼	5.59	19.0	45.1	11.3	2.84	15.3	7.63	1.65
8 × 2 × ¼	4.59	15.6	30.1	7.52	2.56	3.08	3.08	0.819
6 × 6 × ½	10.4	35.2	50.5	16.8	2.21	50.5	16.8	2.21
6 × 6 × ¼	5.59	19.0	30.3	10.1	2.33	30.3	10.1	2.33
6 × 4 × ¼	4.59	15.6	22.1	7.36	2.19	11.7	5.87	1.60
6 × 2 × ¼	3.59	12.2	13.8	4.60	1.96	2.31	2.31	0.802
4 × 4 × ½	6.36	21.6	12.3	6.13	1.39	12.3	6.13	1.39
4 × 4 × ¼	3.59	12.2	8.22	4.11	1.51	8.22	4.11	1.51
4 × 2 × ¼	2.59	8.81	4.69	2.35	1.35	1.54	1.54	0.770
3 × 3 × ¼	2.59	8.81	3.16	2.10	1.10	3.16	2.10	1.10
3 × 2 × ¼	2.09	7.11	2.21	1.47	1.03	1.15	1.15	0.742
2 × 2 × ¼	1.59	5.41	0.766	0.766	0.694	0.766	0.766	0.694

*Data taken from a variety of sources. Sizes listed represent a small sample of the sizes available.

Example size: 6 × 4 × ¼

6 = vertical depth (in), 4 = width (in), ¼ = wall thickness (in)

I = moment of inertia, S = section modulus, r = radius of gyration

A–10 · Aluminum Association standard channels: dimensions, areas, weights, and section properties

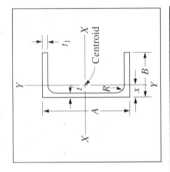

Section properties

Size		Area	Weight	Flange thickness,	Web thickness,	Fillet radius,	Axis X-X			Axis Y-Y			
Depth, A (in)	Width B (in)	(in²)	(lb/ft)	t_1 (in)	t (in)	R (in)	I (in⁴)	S (in³)	r (in)	I (in⁴)	S (in³)	r (in)	x (in)
2.00	1.00	0.491	0.577	0.13	0.13	0.10	0.288	0.288	0.766	0.045	0.064	0.303	0.298
2.00	1.25	0.911	1.071	0.26	0.17	0.15	0.546	0.546	0.774	0.139	0.178	0.391	0.471
3.00	1.50	0.965	1.135	0.20	0.13	0.25	1.41	0.94	1.21	0.22	0.22	0.47	0.49
3.00	1.75	1.358	1.597	0.26	0.17	0.25	1.97	1.31	1.20	0.42	0.37	0.55	0.62
4.00	2.00	1.478	1.738	0.23	0.15	0.25	3.91	1.95	1.63	0.60	0.45	0.64	0.65
4.00	2.25	1.982	2.331	0.29	0.19	0.25	5.21	2.60	1.62	1.02	0.69	0.72	0.78
5.00	2.25	1.881	2.212	0.26	0.15	0.30	7.88	3.15	2.05	0.98	0.64	0.72	0.73
5.00	2.75	2.627	3.089	0.32	0.19	0.30	11.14	4.45	2.06	2.05	1.14	0.88	0.95
6.00	2.50	2.410	2.834	0.29	0.17	0.30	14.35	4.78	2.44	1.53	0.90	0.80	0.79
6.00	3.25	3.427	4.030	0.35	0.21	0.30	21.04	7.01	2.48	3.76	1.76	1.05	1.12
7.00	2.75	2.725	3.205	0.29	0.17	0.30	22.09	6.31	2.85	2.10	1.10	0.88	0.84
7.00	3.50	4.009	4.715	0.38	0.21	0.30	33.79	9.65	2.90	5.13	2.23	1.13	1.20
8.00	3.00	3.526	4.147	0.35	0.19	0.30	37.40	9.35	3.26	3.25	1.57	0.96	0.93
8.00	3.75	4.923	5.789	0.41	0.25	0.35	52.69	13.17	3.27	7.13	2.82	1.20	1.22
9.00	3.25	4.237	4.983	0.35	0.23	0.35	54.41	12.09	3.58	4.40	1.89	1.02	0.93
9.00	4.00	5.927	6.970	0.44	0.29	0.35	78.31	17.40	3.63	9.61	3.49	1.27	1.25
10.00	3.50	5.218	6.136	0.41	0.25	0.35	83.22	16.64	3.99	6.33	2.56	1.10	1.02
10.00	4.25	7.109	8.360	0.50	0.31	0.40	116.15	23.23	4.04	13.02	4.47	1.35	1.34
12.00	4.00	7.036	8.274	0.47	0.29	0.40	159.76	26.63	4.77	11.03	3.86	1.25	1.14
12.00	5.00	10.053	11.822	0.62	0.35	0.45	239.69	39.95	4.88	25.74	7.60	1.60	1.61

Source: Aluminum Association, *Aluminum Standards and Data*, 11th ed., Washington, DC, © 1993, p. 187.

See Footnotes for Table A–11.

A–11 Aluminum Association standard I-beams: dimensions, areas, weights, and section properties

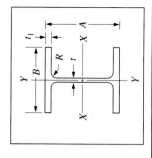

| Size | | | | | | | Section properties[‡] | | | | | |
Depth, A (in)	Width B (in)	Area* (in²)	Weight[†] (lb/ft)	Flange thickness, t_1 (in)	Web thickness, t (in)	Fillet radius, R (in)	Axis X-X I (in⁴)	Axis X-X S (in³)	Axis X-X r (in)	Axis Y-Y I (in⁴)	Axis Y-Y S (in³)	Axis Y-Y r (in)
3.00	2.50	1.392	1.637	0.20	0.13	0.25	2.24	1.49	1.27	0.52	0.42	0.61
3.00	2.50	1.726	2.030	0.26	0.15	0.25	2.71	1.81	1.25	0.68	0.54	0.63
4.00	3.00	1.965	2.311	0.23	0.15	0.25	5.62	2.81	1.69	1.04	0.69	0.73
4.00	3.00	2.375	2.793	0.29	0.17	0.25	6.71	3.36	1.68	1.31	0.87	0.74
5.00	3.50	3.146	3.700	0.32	0.19	0.30	13.94	5.58	2.11	2.29	1.31	0.85
6.00	4.00	3.427	4.030	0.29	0.19	0.30	21.99	7.33	2.53	3.10	1.55	0.95
6.00	4.00	3.990	4.692	0.35	0.21	0.30	25.50	8.50	2.53	3.74	1.87	0.97
7.00	4.50	4.932	5.800	0.38	0.23	0.30	42.89	12.25	2.95	5.78	2.57	1.08
8.00	5.00	5.256	6.181	0.35	0.23	0.30	59.69	14.92	3.37	7.30	2.92	1.18
8.00	5.00	5.972	7.023	0.41	0.25	0.30	67.78	16.94	3.37	8.55	3.42	1.20
9.00	5.50	7.110	8.361	0.44	0.27	0.30	102.02	22.67	3.79	12.22	4.44	1.31
10.00	6.00	7.352	8.646	0.41	0.25	0.40	132.09	26.42	4.24	14.78	4.93	1.42
10.00	6.00	8.747	10.286	0.50	0.29	0.40	155.79	31.16	4.22	18.03	6.01	1.44
12.00	7.00	9.925	11.672	0.47	0.29	0.40	255.57	42.60	5.07	26.90	7.69	1.65
12.00	7.00	12.153	14.292	0.62	0.31	0.40	317.33	52.89	5.11	35.48	10.14	1.71

Source: Aluminum Association, *Aluminum Standards and Data*, 11th ed., Washington, DC, © 1993, p. 187.

*Areas listed are based on nominal dimensions.

[†]Weights per foot are based on nominal dimensions and a density of 0.098 pound per cubic inch which is the density of alloy 6061.

[‡]I = moment of inertia; S = section modulus; r = radius of gyration.

A–12 Properties of American national standard schedule 40 welded and seamless wrought steel pipe

	Diameter (in)						Properties of sections		
Nominal	Actual inside	Actual outside	Wall thickness (in)	Cross-sectional area of metal (in²)	Moment of inertia, I (in⁴)	Radius of gyration (in)	Section modulus, S (in³)	Polar section modulus, Z_p (in³)
$\frac{1}{8}$	0.269	0.405	0.068	0.072	0.00106	0.122	0.00525	0.01050
$\frac{1}{4}$	0.364	0.540	0.088	0.125	0.00331	0.163	0.01227	0.02454
$\frac{3}{8}$	0.493	0.675	0.091	0.167	0.00729	0.209	0.02160	0.04320
$\frac{1}{2}$	0.622	0.840	0.109	0.250	0.01709	0.261	0.04070	0.08140
$\frac{3}{4}$	0.824	1.050	0.113	0.333	0.03704	0.334	0.07055	0.1411
1	1.049	1.315	0.133	0.494	0.08734	0.421	0.1328	0.2656
$1\frac{1}{4}$	1.380	1.660	0.140	0.669	0.1947	0.539	0.2346	0.4692
$1\frac{1}{2}$	1.610	1.900	0.145	0.799	0.3099	0.623	0.3262	0.6524
2	2.067	2.375	0.154	1.075	0.6658	0.787	0.5607	1.121
$2\frac{1}{2}$	2.469	2.875	0.203	1.704	1.530	0.947	1.064	2.128
3	3.068	3.500	0.216	2.228	3.017	1.163	1.724	3.448
$3\frac{1}{2}$	3.548	4.000	0.226	2.680	4.788	1.337	2.394	4.788
4	4.026	4.500	0.237	3.174	7.233	1.510	3.215	6.430
5	5.047	5.563	0.258	4.300	15.16	1.878	5.451	10.90
6	6.065	6.625	0.280	5.581	28.14	2.245	8.496	16.99
8	7.981	8.625	0.322	8.399	72.49	2.938	16.81	33.62
10	10.020	10.750	0.365	11.91	160.7	3.674	29.91	59.82
12	11.938	12.750	0.406	15.74	300.2	4.364	47.09	94.18
16	15.000	16.000	0.500	24.35	732.0	5.484	91.50	183.0
18	16.876	18.000	0.562	30.79	1172	6.168	130.2	260.4

A–13 Typical properties of carbon and alloy steels*

Material AISI no.	Condition†	Ultimate strength, s_u		Yield strength, s_y		Percent elongation
		ksi	MPa	ksi	MPa	
1020	Annealed	57	393	43	296	36
1020	Hot-rolled	65	448	48	331	36
1020	Cold-drawn	75	517	64	441	20
1040	Annealed	75	517	51	352	30
1040	Hot-rolled	90	621	60	414	25
1040	Cold-drawn	97	669	82	565	16
1040	WQT 700	127	876	93	641	19
1040	WQT 900	118	814	90	621	22
1040	WQT 1100	107	738	80	552	24
1040	WQT 1300	87	600	63	434	32
1080	Annealed	89	614	54	372	25
1080	OQT 700	189	1303	141	972	12
1080	OQT 900	179	1234	129	889	13
1080	OQT 1100	145	1000	103	710	17
1080	OQT 1300	117	807	70	483	23
1141	Annealed	87	600	51	352	26
1141	Cold-drawn	112	772	95	655	14
1141	OQT 700	193	1331	172	1186	9
1141	OQT 900	146	1007	129	889	15
1141	OQT 1100	116	800	97	669	20
1141	OQT 1300	94	648	68	469	28
4140	Annealed	95	655	60	414	26
4140	OQT 700	231	1593	212	1462	12
4140	OQT 900	187	1289	173	1193	15
4140	OQT 1100	147	1014	131	903	18
4140	OQT 1300	118	814	101	696	23
5160	Annealed	105	724	40	276	17
5160	OQT 700	263	1813	238	1641	9
5160	OQT 900	196	1351	179	1234	12
5160	OQT 1100	149	1027	132	910	17
5160	OQT 1300	115	793	103	710	23

*Other properties approximately the same for all carbon and alloy steels:
  Modulus of elasticity in tension = 30 000 000 psi (207 GPa)
  Modulus of elasticity in shear = 11 500 000 psi (80 GPa)
  Density = 0.283 lb_m/in^3 (7680 kg/m^3)
†OQT means oil-quenched and tempered. WQT means water-quenched and tempered.

A–14 Typical properties of stainless steels and nonferrous metals

Material and condition	Ultimate strength, s_u		Yield strength, s_y		Percent elongation	Density		Modulus of elasticity, E	
	ksi	MPa	ksi	MPa		lb/in$^{3\,\dagger}$	kg/m^3	psi $\times$ 10^{-6}	GPa
Stainless steels									
AISI 301 annealed	110	758	40	276	60	0.290	8030	28	193
AISI 301 full hard	185	1280	140	965	8	0.290	8030	28	193
AISI 430 annealed	75	517	40	276	30	0.280	7750	29	200
AISI 430 full hard	90	621	80	552	15	0.280	7750	29	200
AISI 501 annealed	70	483	30	207	28	0.280	7750	29	200
AISI 501 OQT 1000	175	1210	135	931	15	0.280	7750	29	200
17-4PH H900	210	1450	185	1280	14	0.281	7780	28.5	197
PH 13-8 Mo H1000	215	1480	205	1410	13	0.279	7720	29.4	203
Copper and its alloys									
C14500 copper, soft	32	221	10	69	50	0.323	8940	17	117
hard	48	331	44	303	20				
C17000 copper, soft	175	1210	150	1030	5	0.298	8250	19	131
hard	215	1482	200	1379	2				
C54400 bronze, soft	68	469	57	393	20	0.318	8800	17	117
hard	75	517	63	434	15				
C36000 brass, soft	49	338	18	124	53	0.308	8530	16	110
hard	68	469	45	310	18				
Magnesium—cast									
ASTM AZ 63A-T6	40	276	19	131	5	0.066	1830	6.5	45
Zinc—cast-ZA 12	58	400	47	324	5	0.218	6030	12	83
Titanium alloy									
Ti-6Al-4V, aged	170	1170	155	1070	8	0.160	4430	16.5	114

†This can be used as specific weight or mass density in lb$_m$/in^3.

A–15 Properties of structural steels

Material ASTM no. and products	Ultimate strength, s_u*		Yield strength, s_y*		Percent elongation in 2 in
	ksi	MPa	ksi	MPa	
A36—Carbon steel shapes, plates, and bars	58	400	36	248	21
A242—High-strength low-alloy shapes, plates, and bars					
$\leq\frac{3}{4}$ in thick	70	483	50	345	21
$\frac{3}{4}$ to $1\frac{1}{2}$ in thick	67	462	46	317	21
$1\frac{1}{2}$ to 4 in thick	63	434	42	290	21
A500—Cold-formed structural tubing					
Round, grade A	45	310	33	228	25
Round, grade B	58	400	42	290	23
Round, grade C	62	427	46	317	21
Shaped, grade A	45	310	39	269	25
Shaped, grade B	58	400	46	317	23
Shaped, grade C	62	427	50	345	21
A501—Hot-formed structural tubing, round or shaped	58	400	36	248	23
A514—High-yield strength, quenched and tempered alloy steel plate					
$\leq 2\frac{1}{2}$ in thick	110	758	100	690	18
$2\frac{1}{2}$ to 6 in thick	100	690	90	620	16
A572—High-strength low-alloy columbium-vanadium steel shapes, plates, and bars					
Grade 42	60	414	42	290	24
Grade 50	65	448	50	345	21
Grade 60	75	517	60	414	18
Grade 65	80	552	65	448	17

*Minimum values; may range higher.

The American Institute of Steel Construction specifies $E = 29 \times 10^6$ psi (200 GPa) for structural steel.

A–16 Typical properties of cast iron*

| Material type and grade | Ultimate strength | | | | | | Yield strength | | Modulus of elasticity, $E^\ddagger$ | | Percent elongation |
| | $s_u^\dagger$ | | $s_{uc}^\ddagger$ | | s_{us} | | $s_{yt}^\ddagger$ | | | | |
	ksi	MPa	ksi	MPa	ksi	MPa	ksi	MPa	psi $\times 10^{-6}$	GPa	
Gray iron ASTM A48											
Grade 20	20	138	80	552	32	221	—	—	12.2	84	<1
Grade 40	40	276	140	965	57	393	—	—	19.4	134	<0.8
Grade 60	55	379	170	1170	72	496	—	—	21.5	148	<0.5
Ductile iron ASTM A536											
60-40-18	60	414	—	—	57	393	40	276	24	165	18
80-55- 6	80	552	—	—	73	503	55	379	24	165	6
100-70- 3	100	690	—	—	—	—	70	483	24	165	3
120-90- 2	120	827	180	1240	—	—	90	621	23	159	2
Austempered ductile iron (ADI)											
Grade 1	125	862	—	—	—	—	85	586	24	165	10
Grade 2	150	1034	—	—	—	—	100	690	24	165	7
Grade 3	175	1207	—	—	—	—	120	827	24	165	4
Grade 4	200	1379	—	—	—	—	140	965	24	165	2
Malleable iron ASTM A220											
45008	65	448	240	1650	49	338	45	310	26	170	8
60004	80	552	240	1650	65	448	60	414	27	186	4
80002	95	655	240	1650	75	517	80	552	27	186	2

*The density of cast iron ranges from 0.25 to 0.27 lb_m/in^3 (6920 to 7480 kg/m^3).

†Minimum values; may range higher.

‡Approximate values; may range higher or lower by about 15%.

A-17 Typical properties of aluminum alloys*

Alloy and temper	Ultimate strength, s_u		Yield strength, s_y		Percent elongation	Shear strength, s_{us}	
	ksi	MPa	ksi	MPa		ksi	MPa
1100-H12	16	110	15	103	25	10	69
1100-H18	24	165	22	152	15	13	90
2014-0	27	186	14	97	18	18	124
2014-T4	62	427	42	290	20	38	262
2014-T6	70	483	60	414	13	42	290
3003-0	16	110	6	41	40	11	76
3003-H12	19	131	18	124	20	12	83
3003-H18	29	200	27	186	10	16	110
5154-0	35	241	17	117	27	22	152
5154-H32	39	269	30	207	15	22	152
5154-H38	48	331	39	269	10	28	193
6061-0	18	124	8	55	30	12	83
6061-T4	35	241	21	145	25	24	165
6061-T6	45	310	40	276	17	30	207
7075-0	33	228	15	103	16	22	152
7075-T6	83	572	73	503	11	48	331
Casting alloys (permanent mold castings)							
204.0-T4	48	331	29	200	8	—	—
356.0-T6	33	228	22	152	3	—	—

*Modulus of elasticity E for most aluminum alloys, including 1100, 3003, 6061, and 6063, is 10×10^6 psi (69 GPa). For 2014, $E = 10.6 \times 10^6$ psi (73 GPa). For 5154, $E = 10.2 \times 10^6$ psi (70 GPa). For 7075, $E = 10.4 \times 10^6$ psi (72 GPa). Density of most aluminum alloys is approximately 0.10 lb_m/in^3 (2770 kg/m^3).

A–18 Typical properties of wood

	Allowable stress											Modulus of elasticity	
	Bending		Tension parallel to grain		Horizontal shear		Compression						
							Perpendicular to grain		Parallel to grain				
Type and grade	psi	MPa	psi	MPa	psi	MPa	psi	MPa	psi	MPa	ksi	GPa
Douglas fir—2 to 4 in thick, 6 in and wider												
No. 1	1750	12.1	1050	7.2	95	0.66	385	2.65	1250	8.62	1800	12.4
No. 2	1450	10.0	850	5.9	95	0.66	385	2.65	1000	6.90	1700	11.7
No. 3	800	5.5	475	3.3	95	0.66	385	2.65	600	4.14	1500	10.3
Hemlock—2 to 4 in thick, 6 in and wider												
No. 1	1400	9.6	825	5.7	75	0.52	245	1.69	1000	6.90	1500	10.3
No. 2	1150	7.9	675	4.7	75	0.52	245	1.69	800	5.52	1400	9.7
No. 3	625	4.3	375	2.6	75	0.52	245	1.69	500	3.45	1200	8.3
Southern pine—2½ to 4 in thick, 6 in and wider												
No. 1	1400	9.6	825	5.7	80	0.55	270	1.86	850	5.86	1600	11.0
No. 2	1000	6.9	575	4.0	70	0.48	230	1.59	550	3.79	1300	9.0
No. 3	650	4.5	375	2.6	70	0.48	230	1.59	400	2.76	1300	9.0

A–19 Typical properties of selected plastics

Type*	Tensile strength		Flexural strength		Tensile modulus		Density	
	ksi	MPa	ksi	MPa	ksi	GPa	lb_m/in^3	kg/m^3
ABS	7	48	11	76	360	2.5	0.036	995
Acetal copolymer	9	62	13	90	400	2.8	0.051	1410
Epoxy, molded	15	103	30	207	3000	20.7	0.069	1910
Nylon 6/6	26	179	35	241	1300	9.0	0.041	1135
Phenolic	9	62	18	124	2500	17.2	0.066	1825
Polycarbonate	16	110	19	131	860	5.9	0.049	1355
Polyester PET	22	152	31	214	1700	11.7	0.059	1630
Polyimide	27	186	50	345	3200	22.1	0.069	1910
Polypropylene	10	69	14	97	800	5.5	0.041	1135
Polystyrene	12	83	17	117	800	5.5	0.042	1165

*Reinforced with glass or other fibers.

A–20 Design stress guidelines—direct normal stresses

Manner of loading	Ductile material	Brittle material
Static	$\sigma_d = s_y/2$	$\sigma_d = s_u/6$
Repeated	$\sigma_d = s_u/8$	$\sigma_d = s_u/10$
Impact or shock	$\sigma_d = s_u/12$	$\sigma_d = s_u/15$

A–21–1   Axially loaded grooved round bar in tension

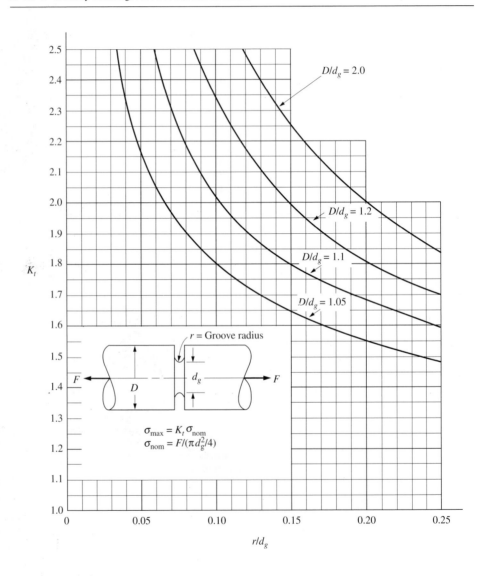

A–21–2 Axially loaded stepped round bar in tension

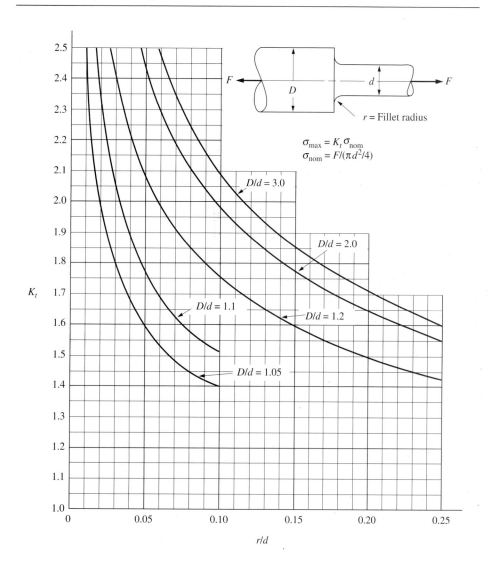

$\sigma_{max} = K_t \sigma_{nom}$
$\sigma_{nom} = F/(\pi d^2/4)$

A–21–3 Axially loaded stepped flat plate in tension

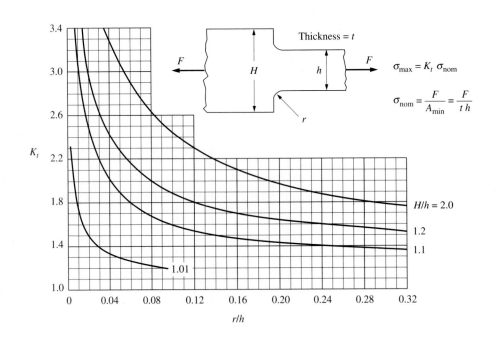

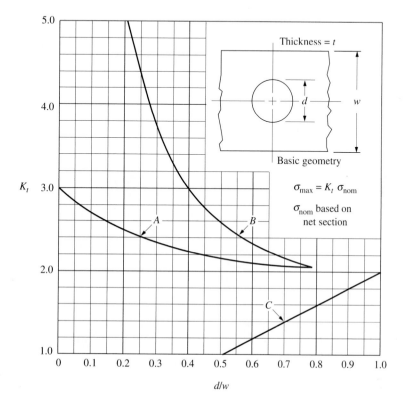

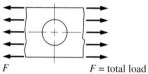

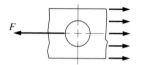

Curve A

Direct tension
on plate

$$\sigma_{nom} = \frac{F}{A_{net}} = \frac{F}{(w-d)t}$$

F F = total load

Curve B

Tension load
applied through
a pin in the hole

$$\sigma_{nom} = \frac{F}{A_{net}} = \frac{F}{(w-d)t}$$

Curve C

Bending in
the plane
of the plate

$$\sigma_{nom} = \frac{Mc}{I_{net}} = \frac{6Mw}{(w^3-d^3)t}$$

Note: $K_t = 1.0$ for $d/w < 0.5$

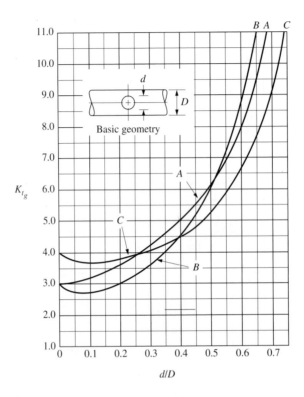

Note: K_{t_g} is based on the nominal stress in a round
bar without a hole (gross section).

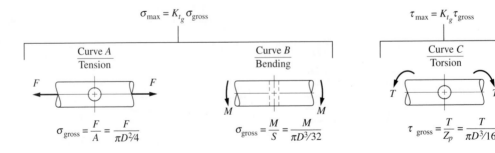

$$\sigma_{max} = K_{t_g}\, \sigma_{gross}$$

$$\tau_{max} = K_{t_g}\, \tau_{gross}$$

Curve *A*	Curve *B*	Curve *C*
Tension	Bending	Torsion

$$\sigma_{gross} = \frac{F}{A} = \frac{F}{\pi D^2/4}$$

$$\sigma_{gross} = \frac{M}{S} = \frac{M}{\pi D^3/32}$$

$$\tau_{gross} = \frac{T}{Z_p} = \frac{T}{\pi D^3/16}$$

Grooved round bar in torsion

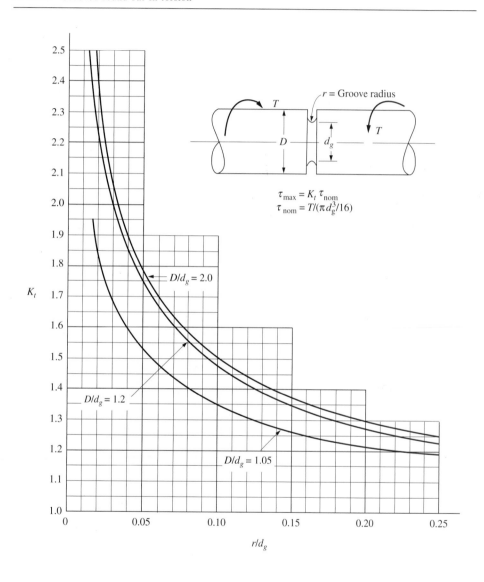

$$\tau_{\max} = K_t\,\tau_{\text{nom}}$$
$$\tau_{\text{nom}} = T/(\pi d_g^3/16)$$

A–21–7   Stepped round bar in torsion

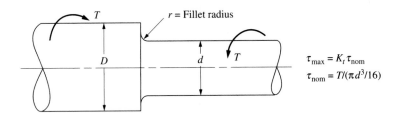

$\tau_{max} = K_t \tau_{nom}$

$\tau_{nom} = T/(\pi d^3/16)$

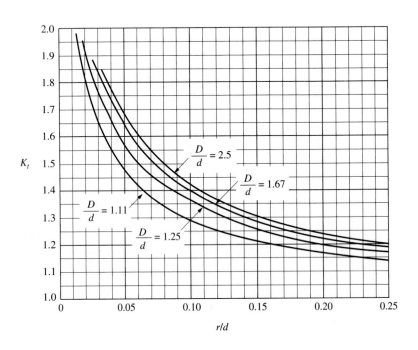

A–21–8   Grooved round bar in bending

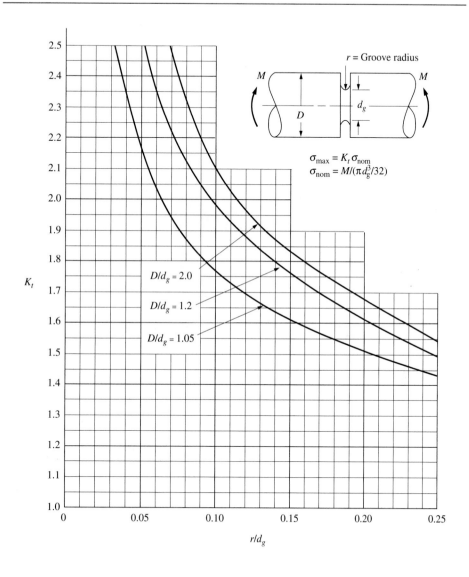

$\sigma_{max} = K_t \sigma_{nom}$
$\sigma_{nom} = M/(\pi d_g^3/32)$

r = Groove radius

$D/d_g = 2.0$
$D/d_g = 1.2$
$D/d_g = 1.05$

K_t

r/d_g

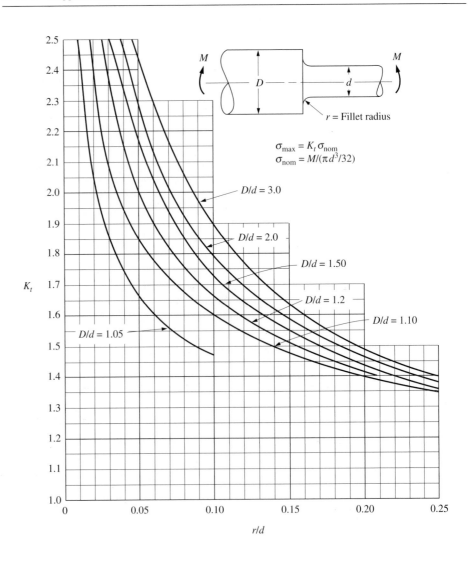

$\sigma_{max} = K_t \sigma_{nom}$

$\sigma_{nom} = M/(\pi d^3/32)$

$D/d = 3.0$

$D/d = 2.0$

$D/d = 1.50$

$D/d = 1.2$

$D/d = 1.10$

$D/d = 1.05$

K_t

r/d

r = Fillet radius

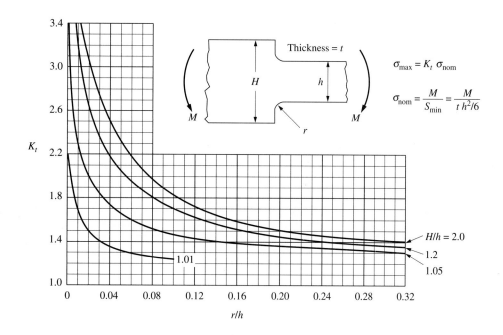

Type of keyseat	K_t*
Sled-runner	1.6
Profile	2.0

*K_t is to be applied to the stress computed for the full nominal diameter of the shaft where the keyseat is located.

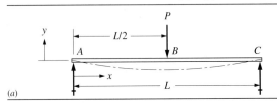

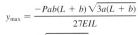

$$y_B = y_{max} = \frac{-PL^3}{48EI} \quad \text{at center}$$

Between A and B:

$$y = \frac{-Px}{48EI}(3L^2 - 4x^2)$$

(a)

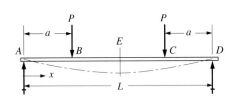

$$y_{max} = \frac{-Pab(L + b)\sqrt{3a(L + b)}}{27EIL}$$

at $x_1 = \sqrt{a(L + b)/3}$

$$y_B = \frac{-Pa^2b^2}{3EIL} \quad \text{at load}$$

Between A and B (the longer segment):

$$y = \frac{-Pbx}{6EIL}(L^2 - b^2 - x^2)$$

Between B and C (the shorter segment):

$$y = \frac{-Pav}{6EIL}(L^2 - v^2 - a^2)$$

At end of overhang at D:

$$y_D = \frac{Pabc}{6EIL}(L + a)$$

(b)

$$y_E = y_{max} = \frac{-Pa}{24EI}(3L^2 - 4a^2) \quad \text{at center}$$

$$y_B = y_C = \frac{-Pa^2}{6EI}(3L - 4a) \quad \text{at loads}$$

Between A and B:

$$y = \frac{-Px}{6EI}(3aL - 3a^2 - x^2)$$

Between B and C:

$$y = \frac{-Pa}{6EI}(3Lx - 3x^2 - a^2)$$

(c)

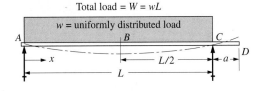

$$y_B = y_{max} = \frac{-5wL^4}{384EI} = \frac{-5WL^3}{384EI} \quad \text{at center}$$

Between A and B:

$$y = \frac{-wx}{24EI}(L^3 - 2Lx^2 + x^3)$$

At D at end:

$$y_D = \frac{wL^3a}{24EI}$$

(d)

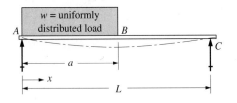

Between A and B:

$$y = \frac{-wx}{24EIL}[a^2(2L - a)^2 - 2ax^2(2L - a) + Lx^3]$$

Between B and C:

$$y = \frac{-wa^2(L - x)}{24EI}(4Lx - 2x^2 - a^2)$$

(e)

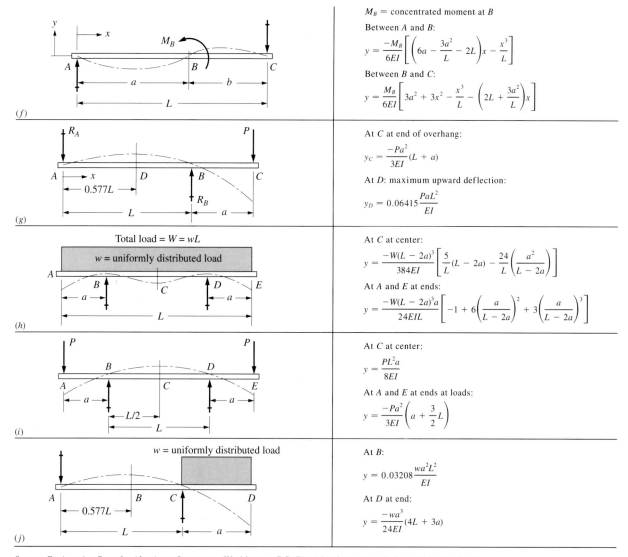

(f)

M_B = concentrated moment at B

Between A and B:

$$y = \frac{-M_B}{6EI}\left[\left(6a - \frac{3a^2}{L} - 2L\right)x - \frac{x^3}{L}\right]$$

Between B and C:

$$y = \frac{M_B}{6EI}\left[3a^2 + 3x^2 - \frac{x^3}{L} - \left(2L + \frac{3a^2}{L}\right)x\right]$$

(g)

At C at end of overhang:

$$y_C = \frac{-Pa^2}{3EI}(L + a)$$

At D: maximum upward deflection:

$$y_D = 0.06415\frac{PaL^2}{EI}$$

(h)

Total load = $W = wL$

w = uniformly distributed load

At C at center:

$$y = \frac{-W(L - 2a)^3}{384EI}\left[\frac{5}{L}(L - 2a) - \frac{24}{L}\left(\frac{a^2}{L - 2a}\right)\right]$$

At A and E at ends:

$$y = \frac{-W(L - 2a)^3 a}{24EIL}\left[-1 + 6\left(\frac{a}{L - 2a}\right)^2 + 3\left(\frac{a}{L - 2a}\right)^3\right]$$

(i)

At C at center:

$$y = \frac{PL^2a}{8EI}$$

At A and E at ends at loads:

$$y = \frac{-Pa^2}{3EI}\left(a + \frac{3}{2}L\right)$$

(j)

w = uniformly distributed load

At B:

$$y = 0.03208\frac{wa^2L^2}{EI}$$

At D at end:

$$y = \frac{-wa^3}{24EI}(4L + 3a)$$

Source: Engineering Data for Aluminum Structures (Washington, DC: The Aluminum Association, 1986), pp. 63–77.

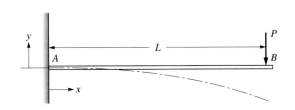

(a)

At B at end:

$$y_B = y_{max} = \frac{-PL^3}{3EI}$$

Between A and B:

$$y = \frac{-Px^2}{6EI}(3L - x)$$

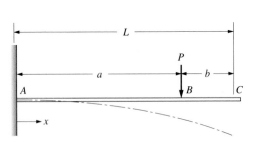

(b)

At B at load:

$$y_B = \frac{-Pa^3}{3EI}$$

At C at end:

$$y_C = y_{max} = \frac{-Pa^2}{6EI}(3L - a)$$

Between A and B:

$$y = \frac{-Px^2}{6EI}(3a - x)$$

Between B and C:

$$y = \frac{-Pa^2}{6EI}(3x - a)$$

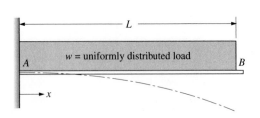

(c)

W = total load = wL

At B at end:

$$y_B = y_{max} = \frac{-WL^3}{8EI}$$

Between A and B:

$$y = \frac{-Wx^2}{24EIL}[2L^2 + (2L - x)^2]$$

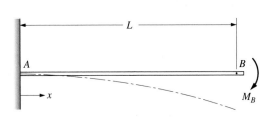

(d)

M_B = concentrated moment at end

At B at end:

$$y_B = y_{max} = \frac{-M_B L^2}{2EI}$$

Between A and B:

$$y = \frac{-M_B x^2}{2EI}$$

Source: Engineering Data for Aluminum Structures (Washington, DC: The Aluminum Association, 1986), pp. 63–77.

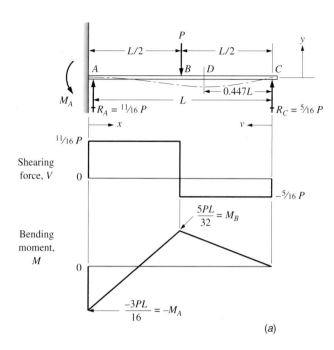

(a)

Deflections

At B at load:

$$y_B = \frac{-7}{768}\frac{PL^3}{EI}$$

y_{max} is at $v = 0.447L$ at D:

$$y_D = y_{max} = \frac{-PL^3}{107EI}$$

Between A and B:

$$y = \frac{-Px^2}{96EI}(9L - 11x)$$

Between B and C:

$$y = \frac{-Pv}{96EI}(3L^2 - 5v^2)$$

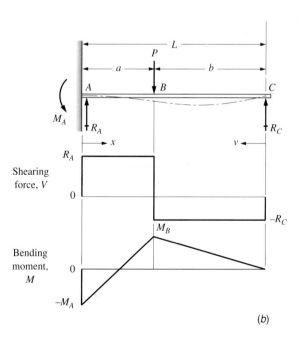

(b)

Reactions

$$R_A = \frac{Pb}{2L^3}(3L^2 - b^2)$$

$$R_C = \frac{Pa^2}{2L^3}(b + 2L)$$

Moments

$$M_A = \frac{-Pab}{2L^2}(b + L)$$

$$M_B = \frac{Pa^2b}{2L^3}(b + 2L)$$

Deflections

At B at load:

$$y_B = \frac{-Pa^3b^2}{12EIL^3}(3L + b)$$

Between A and B:

$$y = \frac{-Px^2b}{12EIL^3}(3C_1 - C_2x)$$

$$C_1 = aL(L + b); \; C_2 = (L + a)(L + b) + aL$$

Between B and C:

$$y = \frac{-Pa^2v}{12EIL^3}[3L^2b - v^2(3L - a)]$$

617

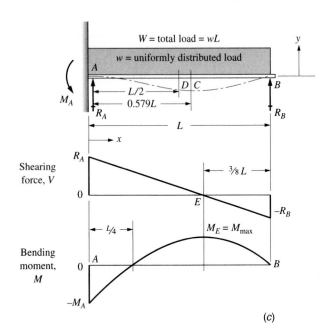

Reactions

$$R_A = \tfrac{5}{8}W$$

$$R_B = \tfrac{3}{8}W$$

Moments

$$M_A = -0.125WL$$

$$M_E = 0.0703WL$$

Deflections

At C at $x = 0.579L$:

$$y_C = y_{max} = \frac{-WL^3}{185EI}$$

At D at center:

$$y_D = \frac{-WL^3}{192EI}$$

Between A and B:

$$y = \frac{-Wx^2(L - x)}{48EIL}(3L - 2x)$$

(c)

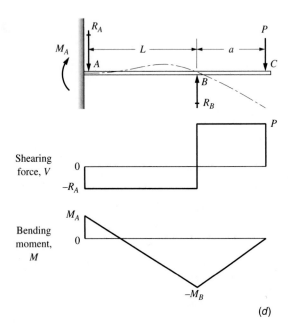

Reactions

$$R_A = \frac{-3Pa}{2L}$$

$$R_B = P\left(1 + \frac{3a}{2L}\right)$$

Moments

$$M_A = \frac{Pa}{2}$$

$$M_B = -Pa$$

Deflection

At C at end:

$$y_C = \frac{-PL^3}{EI}\left(\frac{a^2}{4L^2} + \frac{a^3}{3L^3}\right)$$

(d)

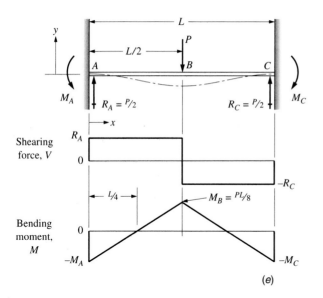

Moments

$$M_A = M_B = M_C = \frac{PL}{8}$$

Deflections

At B at center:

$$y_B = y_{max} = \frac{-PL^3}{192EI}$$

Between A and B:

$$y = \frac{-Px^2}{48EI}(3L - 4x)$$

Shearing force, V

Bending moment, M

$M_B = {}^{PL}\!/_8$

(e)

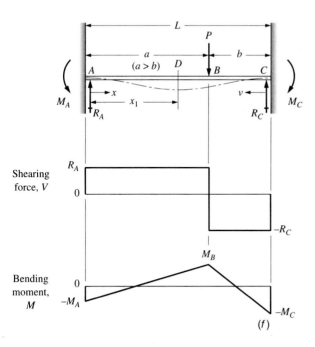

Reactions

$$R_A = \frac{Pb^2}{L^3}(3a + b)$$

$$R_C = \frac{Pa^2}{L^3}(3b + a)$$

Moments

$$M_A = \frac{-Pab^2}{L^2}$$

$$M_B = \frac{2Pa^2b^2}{L^3}$$

$$M_C = \frac{-Pa^2b}{L^2}$$

Deflections

At B at load:

$$y_B = \frac{-Pa^3b^3}{3EIL^3}$$

At D at $x_1 = \dfrac{2aL}{3a + b}$

$$y_D = y_{max} = \frac{-2Pa^3b^2}{3EI(3a + b)^2}$$

Between A and B (longer segment):

$$y = \frac{-Px^2b^2}{6EIL^3}[2a(L - x) + L(a - x)]$$

Between B and C (shorter segment):

$$y = \frac{-Pv^2a^2}{6EIL^3}[2b(L - v) + L(b - v)]$$

Shearing force, V

Bending moment, M

M_B

(f)

619

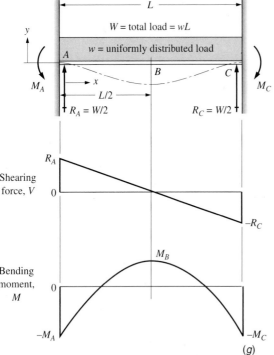

Moments

$$M_A = M_C = \frac{-WL}{12}$$

$$M_B = \frac{WL}{24}$$

Deflections

At B at center:

$$y_B = y_{max} = \frac{-WL^3}{384EI}$$

Between A and C:

$$y = \frac{-wx^2}{24EI}(L - x)^2$$

(g)

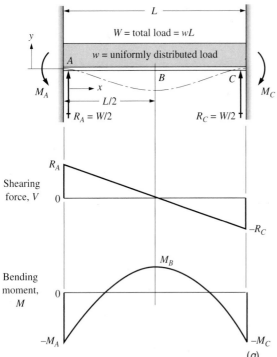

Reactions

$$R_A = R_C = \frac{3wL}{8}$$

$$R_B = 1.25wL$$

Shearing forces

$$V_A = V_C = R_A = R_C = \frac{3wL}{8}$$

$$V_B = \frac{5wL}{8}$$

Moments

$$M_D = M_E = 0.0703wL^2$$
$$M_B = -0.125wL^2$$

Deflections

At $x_1 = 0.4215L$ from A or C:

$$y_{max} = \frac{-wL^4}{185EI}$$

Between A and B:

$$y = \frac{-w}{48EI}(L^3x - 3Lx^3 + 2x^4)$$

(h)

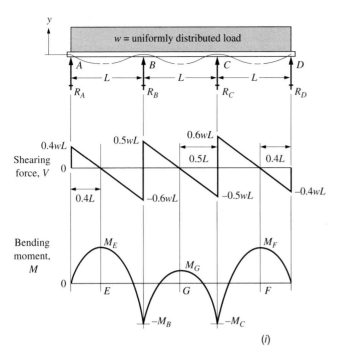

Reactions

$R_A = R_D = 0.4wL$

$R_B = R_C = 1.10wL$

Moments

$M_E = M_F = 0.08wL^2$

$M_B = M_C = -0.10wL^2 = M_{max}$

$M_G = 0.025wL^2$

(i)

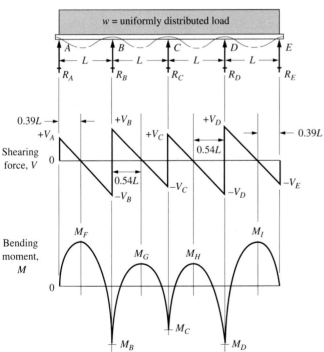

Reactions

$R_A = R_E = 0.393wL$

$R_B = R_D = 1.143wL$

$R_C = 0.928wL$

Shearing forces

$V_A = +0.393wL$

$-V_B = -0.607wL$

$+V_B = +0.536wL$

$-V_C = -0.464wL$

$+V_C = +0.464wL$

$-V_D = -0.536wL$

$+V_D = +0.607wL$

$-V_E = -0.393wL$

Moments

$M_B = M_D = -0.1071wL^2 = M_{max}$

$M_F = M_I = 0.0772wL^2$

$M_C = -0.0714wL^2$

$M_G = M_H = 0.0364wL^2$

(j)

Source: Engineering Data for Aluminum Structures (Washington, DC: The Aluminum Association, 1986), pp. 63–77.

A–25 Conversion factors

Quantity	U.S. Customary unit	SI unit	U.S. Customary to SI	SI to U.S. Customary
			Multiply given value by factor to convert from:	
Area	in^2	mm^2	645.16	1.550×10^{-3}
	ft^2	m^2	0.0929	10.76
Loading	lb/in^2	kPa	6.895	0.1450
	lb/ft^2	kPa	0.0479	20.89
Bending moment	lb·in	N·m	0.1130	8.851
	lb·ft	N·m	1.356	0.7376
Density	lb_m/in^3	kg/m^3	2.768×10^4	3.613×10^{-5}
	lb_m/ft^3	kg/m^3	16.02	0.0624
Force	lb	N	4.448	0.2248
	kip	kN	4.448	0.2248
Length	in	mm	25.4	0.03937
	ft	m	0.3048	3.281
Mass	lb_m	kg	0.454	2.205
Torque	lb·in	N·m	0.1130	8.851
Power	hp	kW	0.7457	1.341
Stress or pressure	psi	kPa	6.895	0.1450
	ksi	MPa	6.895	0.1450
	psi	GPa	6.895×10^{-6}	1.450×10^5
	ksi	GPa	6.895×10^{-3}	145.0
Section modulus	in^3	mm^3	1.639×10^4	6.102×10^{-5}
Moment of inertia	in^4	mm^4	4.162×10^5	2.403×10^{-6}

Answers to Selected Problems

Chapter 1

1–17. 7.85 kN front
11.77 kN rear

1–19. 54.5 mm

1–23. 1765 lb front
2646 lb rear

1–25. 55.1 lb
25.7 lb/in
2.14 in

1–27. 398 slugs

1–29. 8274 kPa

1–31. 96.5 to 524 MPa

1–33. 9097 mm^2

1–35. Area = 324 in^2
Area = 2.09 × 10^5 mm^2
Vol. = 3888 in^3
Vol. = 6.37 × 10^7 mm^3
Vol. = 6.37 × 10^{-2} m^3

1–37. 40.7 MPa

1–39. 5375 psi

1–41. 79.8 MPa

1–43. 803 psi

1–45. σ_{AB} = 107.4 MPa
σ_{BC} = 75.2 MPa
σ_{BD} = 131.1 MPa

1–47. σ_{AB} = 167 MPa tension
σ_{BC} = 77.8 MPa tension
σ_{CD} = 122 MPa tension

1–49. σ_{AB} = 20 471 psi tension
σ_{BC} = 3 129 psi tension

1–51. Forces: AD = CD = 10.5 kN
AB = BC = 9.09 kN
Stresses: σ_{AB} = σ_{BC} = 25.3 MPa tension
σ_{BD} = 17.5 MPa tension
σ_{AD} = σ_{CD} = 21.0 MPa compression

1–53. 50.0 MPa

1–55. 11 791 psi

1–57. 146 MPa

1–59. 151 MPa

1–61. 81.1 MPa

1–63. 24.7 MPa

1–65. Pin: τ = 50 930 psi
Collar: τ = 38 800 psi

1–67. 183 MPa

1–69. 73.9 MPa

1–71. 22.6 MPa

1–73. (a) 2187 psi
(b) 530 psi

1–75. 18 963 psi

1–77. 28.3 MPa

1–79. 5.39 MPa

Chapter 2

2–15. 1020 HR

2–19. 16.4 lb

2–21. Magnesium

2–29. $s_{ut} = 40$ ksi; $s_{uc} = 140$ Ksi

2–31. Bending: $\sigma_d = 1450$ psi
Tension: $\sigma_d = 850$ psi
Compression: $\sigma_d = 1000$ psi parallel to grain
Compression: $\sigma_d = 385$ psi perpendicular to grain
Shear: $\tau_d = 95$ psi

Chapter 3

3–1. Required $s_y = 216$ MPa

3–3. Required $s_u = 86\,000$ psi

3–5. No. Tensile stress too high.

3–7. $d_{min} = 0.824$ in

3–9. $d_{min} = 12.4$ mm

3–11. Required $\sigma_d > 803$ psi

3–13. 16.7 kN

3–15. $d_{min} = 0.412$ in

3–17. For sides B and H: $B_{min} = 22.2$ mm;
$H_{min} = 44.4$ mm

3–19. Required $s_y = 360$ MPa

3–21. Required $s_u = 400$ MPa

3–23. 20 260 lb

3–25. 0.577 m on a side

3–26. 36.4 kip

3–27. 52.3 kip

3–28. (a) 30.4 kip
(b) 43.6 kip

3–29. (a) $\sigma_b = 1572$ psi
(b) One possible design: Square plate on bottom of each leg; 2.50 in on a side.

3–31. $L_{min} = 0.556$ in based on shear.

3–33. $d_{min} = 0.808$ in based on shear
For $d = 1.00$ in, $L_{min} = 0.285$ in for bearing

3–35. $\tau = 151$ MPa; required
$s_u = 1473$ MPa

3–37. 83 kN

3–39. 256 kN

3–41. 18 300 lb

3–43. 62 650 lb

3–45. 119 500 lb

3–47. 119 700 lb

3–49. Pin A: $d_{min} = 13.8$ mm
Pins B and C: $d_{min} = 17.7$ mm

3–51. Lifting force = 2184 lb
$\tau = 6275$ psi if pin is in double shear

3–53. $d_{min} = 0.199$ in

3–55. N at hole = 3.91
N at fillets = 7.20

3–57. 19.15 MPa at circular grooves

Chapter 4

4–1. 0.041 in

4–3. Force = 2357 lb
$\sigma = 3655$ psi

4–5. (a) and (b) $d_{min} = 10.63$ mm; mass = 0.430 kg
(c) $d_{min} = 18.4$ mm; mass = 0.465 kg

4–7. (a) 0.857 mm
(b) 0.488 mm

4–9. Elongation = 0.0040 in
Compression = 0.00045 in

4–11. Force = 3214 lb; unsafe

4–13. $\delta = 0.016$ mm
$\sigma = 27.9$ MPa

4–15. 0.804 mm

4–17. 2.22 mm shorter

4–19. (a) $\delta = 0.276$ in; $\sigma = 37\,300$ psi (close to s_y)
(b) $\sigma = 62\,200$ psi—greater than s_u. Wire will break.

4–21. Force = 6737 lb
$\delta = 0.055$ in

4–23. Mass = 132 kg
$\sigma = 183$ MPa

4–25. 0.806 in

4–27. 180 MPa

4–29. (a) 0.459 mm
(b) 213 MPa

4–31. 693 psi

4–33. 234.8°C

4–35. Brass: $\delta = 6.46$ mm
Stainless steel: $\delta = 3.51$ mm

4–37. 38.7 MPa

4–39. $\sigma = 37500$ psi compression. Bar should fail in compression and/or buckle.

4–41. 0.157 in

4–43. 154 MPa

4–45. $\sigma_c = 17.1$ MPa; $\sigma_s = 109$ MPa

4–47. 13.8 in

4–49. $d_{min} = 6.20$ mm

4–51. $\sigma_s = 427$ MPa; $\sigma_a = 49.4$ MPa

Chapter 5

5–1. 178 MPa

5–3. 4042 psi

5–5. 83.8 MPa

5–7. $\tau = 6716$ psi; safe

5–9. $\tau = 5190$ psi;
required $s_y = 62300$ psi

5–11. $\tau = 65.5$ MPa; $\theta = 0.0378$ rad;
required $S_y = 262$ MPa

5–13. $D_i = 46.5$ mm; $D_o = 58.1$ mm

5–15. $d_{min} = 0.512$ in

5–17. Power $= 0.0686$ hp; $\tau = 8488$ psi;
required $s_y = 67900$ psi

5–19. $D_i = 12.09$ in; $D_o = 15.11$ in

5–21. 1.96 N·m

5–23. 0.1509 rad

5–25. 0.267 rad

5–27. 0.0756 rad

5–29. 0.278 rad

5–31. $\theta_{AB} = 0.0636$ rad; $\theta_{AC} = 0.0976$ rad

5–33. $\tau = 9.06$ MPa; $\theta = 0.0046$ rad

5–35. 49.0 MPa

5–37. 1370 lb·in

5–39. 2902 lb·in

5–41. 0.083 rad

5–43. 0.112 rad

5–45. 0.0667 rad

5–47. 1.82 MPa

5–49. 0.00363 rad

5–51. 0.0042 rad

5–53. 82 750 lb·in

5–55. 153 600 lb·in

5–57. $\tau_{tube}/\tau_{pipe} = 1.028$
$\theta_{tube}/\theta_{pipe} = 1.186$

Chapter 6

NOTE: The following answers refer to Figures P6–1 through P6–84. For reactions, R_1 is the left; R_2 is the right. V and M refer to the maximum absolute values for shearing force and bending moment, respectively. The complete solutions require the construction of the complete shearing force and bending moment diagrams.

P6–1. $R_1 = R_2 = 325$ lb
$V = 325$ lb
$M = 4550$ lb·in

P6–3. $R_1 = 11.43$ K; $R_2 = 4.57$ K
$V = 11.43$ K
$M = 45.7$ K·ft

P6–5. $R_1 = 575$ N; $R_2 = 325$ N
$V = 575$ N
$M = 195$ N·m

P6–7. $R_1 = 46.36$ kN; $R_2 = 23.64$ kN
$V = 46.36$ kN
$M = 71.54$ kN·m

P6–9. $R_1 = 1557$ lb; $R_2 = 1743$ lb
$V = 1557$ lb
$M = 6228$ lb·in

P6–11. $R_1 = 7.5$ K; $R_2 = 37.5$ K
$V = 20$ K
$M = 60$ K·ft

P6–13. $R_1 = R_2 = 250$ N
$V = 850$ N
$M = 362.5$ N·m

P6–15. $R_1 = 37.4$ kN (down); $R_2 = 38.3$ kN (up)
$V = 24.9$ kN
$M = 50$ kN·m

P6–17. $R = 120$ lb
$V = 120$ lb
$M = 960$ lb·in

P6–19. $R = 24$ K
$V = 24$ K
$M = 168$ K·ft

P6–21. $R = 1800$ N
$V = 1800$ N
$M = 1020$ N·m

P6–23. $R = 120$ kN
$V = 120$ kN
$M = 240$ kN·m

P6–25. $R_1 = R_2 = 180$ lb
$V = 180$ lb
$M = 810$ lb·in

P6–27. $R_1 = 240$ lb; $R_2 = 120$ lb
$V = 240$ lb
$M = 640$ lb·in

P6–29. $R_1 = 99.2$ N; $R_2 = 65.8$ N
$V = 99.2$ N
$M = 9.9$ N·m

P6–31. $R_1 = 42$ kN; $R_2 = 50$ kN
$V = 50$ kN
$M = 152.2$ kN·m

P6–33. $R_1 = R_2 = 440$ lb
$V = 240$ lb
$M = 360$ lb·in

P6–35. $R_1 = 1456$ N; $R_2 = 644$ N
$V = 956$ N
$M = 125$ N·m

P6–37. $R_1 = 35.3$ N; $R_2 = 92.3$ N
$V = 52.2$ N
$M = 4.0$ N·m

P6–39. $R = 360$ lb
$V = 360$ lb
$M = 1620$ lb·in

P6–41. $R = 600$ N
$V = 600$ N
$M = 200$ N·m

P6–43. $R_1 = R_2 = 330$ lb
$V = 330$ lb
$M = 4200$ lb·in

P6–45. $R_1 = 36.6$ K; $R_2 = 30.4$ K
$V = 36.6$ K
$M = 183.2$ K·ft

P6–47. $R_1 = R_2 = 450$ N
$V = 450$ N
$M = 172.5$ N·m

P6–49. $R_1 = 180$ kN; $R_2 = 190$ kN
$V = 190$ kN
$M = 630$ kN·m

P6–51. $R_1 = 636$ lb; $R_2 = 1344$ lb
$V = 804$ lb
$M = 2528$ lb·in

P6–53. $R_1 = 4950$ N; $R_2 = 3100$ N
$V = 2950$ N
$M = 3350$ N·m

P6–55. $R = 236$ lb
$V = 236$ lb
$M = 1504$ lb·in

P6–57. $R = 1130$ N
$V = 1130$ N
$M = 709$ N·m

P6–59. $R = 230$ kN
$V = 230$ kN
$M = 430$ kN·m

P6–61. $R = 1400$ lb
$V = 1500$ lb
$M = 99\,000$ lb·in

P6–63. $R = 1250$ N
$V = 1250$ N
$M = 1450$ N·m

P6–65. $R_1 = 1333$ lb; $R_2 = 2667$ lb
$V = 2667$ lb
$M = 5132$ lb·ft

P6–67. $R_1 = R_2 = 75$ N
$V = 75$ N
$M = 15$ N·m

P6–69. $R_1 = 8.60$ kN; $R_2 = 12.2$ kN
$V = 12.2$ kN
$M = 9.30$ kN·m

P6–71. $R_1 = R_2 = 5400$ lb
$V = 5400$ lb
$M = 19\,800$ lb·ft

P6–73. $R = 10.08$ kN
$V = 10.08$ kN
$M = 8.064$ kN·m

P6–75. $R = 7875$ lb
$V = 7875$ lb
$M = 21\,063$ lb·ft

For Problems P6–77 through P6–83, results are shown for the main horizontal section only.

P6–77. $R_1 = R_2 = 282$ N
$V = 282$ N
$M = 120$ N·m

P6–79. $R_1 = R_2 = 162$ N
$V = 162$ N
$M = 42.2$ N·m

P6–81. $R_1 = 165.4$ N; $R_2 = 18.4$ N
$V = 165.4$ N
$M = 16.54$ N·m

P6–83. $R_1 = 4.35$ N; $R_2 = 131.35$ N
$V = 127$ N
$M = 6.35$ N·m

Chapter 7

NOTE: The following answers refer to Figures P7–1 through P7–39. The first number is the distance to the centroid from the bottom of the section unless noted otherwise. The second number is the moment of inertia with respect to the horizontal centroidal axis.

P7–1. 0.663 in; 0.3156 in^4

P7–3. 4.00 in; 184 in^4

P7–5. 35.0 mm; 2.66×10^5 mm^4

P7–7. 20.0 mm; 7.29×10^4 mm^4

P7–9. 20.0 mm; 1.35×10^5 mm^4

P7–11. 21.81 mm; 1.86×10^5 mm^4

P7–13. 23.33 mm; 1.41×10^5 mm^4

P7–15. 1.068 in; 0.3572 in^4

P7–17. 125 mm; 6.73×10^7 mm^4

P7–19. 0.9305 in; 1.2506 in^4

P7–21. 4.25 in; 151.4 in^4

P7–23. 2.25 in; 107.2 in^4

P7–25. 7.33 in; 1155 in^4

P7–27. 7.40 in; 423.5 in^4

P7–29. 3.50 in; 88.94 in^4

P7–31. 3.00 in from center of either pipe; 22.91 in^4

P7–33. 2.723 in; 46.64 in^4

P7–35. 2.612 in; 58.17 in^4

P7–37. 3.50 in; 17.89 in^4

P7–39. 3.355 in; 45.07 in^4

Chapter 8

8–1. 94.4 MPa

8–3. (a) 20 620 psi
(b) 41 240 psi

8–5. 21 050 psi

8–7. $\sigma_\tau = 6882$ psi
$\sigma_c = 12\,970$ psi

8–9. 5794 psi (39.9 MPa)

8–11. 13 963 psi

8–13. Required $S = 2636$ mm^3; for $h/b = 3.0$;
$b = 12.1$ mm; $h = 36.3$ mm
Convenient dimensions from Appendix A–2:
$b = 12$ mm; $h = 40$ mm; $S = 3200$ mm^3;

$A = 480$ mm^2; $h/b = 3.33$
$b = 14$ mm; $h = 35$ mm; $S = 2858$ mm^3;
$A = 490$ mm^2; $h/b = 2.5$

8–15. Required $s_u = 290$ MPa; possible material—6061-T6

8–17. Required $S = 88.2$ in^3; W20 × 66 steel beam

8–19. Required $s_u = 10.6$ ksi; OK for 6061-T4

8–21. Required $s_y = 284$ MPa; OK for 2014-T4

8–23. Required $S = 1.35$ in^3; 3-in schedule 40 steel pipe

8–24. Required $S = 6.89$ in^3; 6 × 4 × $\frac{1}{4}$ or 8 × 2 × $\frac{1}{4}$ steel tube

8–25. Required $S = 6.72$ in^3; 6I × 4.030 aluminum beam

8–26. Required $S = 7.47$ in^3; W8 × 10 steel beam

8–27. Required $S = 7.47$ in^3; no suitable channel

8–28. Required $S = 7.47$ in^3; 6-in schedule 40 steel pipe

8–31. Required $S = 9.76$ in^3; W10 × 12 steel beam

8–33. Required $S = 23.1$ in^3; W14 × 26 steel beam

8–35. Required $S = 16.1$ in^3; W12 × 16 steel beam

8–37. Required $S = 63.3$ in^3; W18 × 40 steel beam

8–39. Required $S = 20.2$ in^3; W14 × 26 steel beam

8–41. Required $S = 0.540$ in^3; W8 × 10 steel beam

8–43. Required $S = 23.1$ in^3; S10 × 25.4 steel beam

8–45. Required $S = 16.1$ in^3; S8 × 23 steel beam

8–47. Required $S = 63.3$ in^3; S15 × 50 steel beam

8–49. Required $S = 20.2$ in^3; S10 × 25.4 steel beam

8–51. Required $S = 0.540$ in^3; S3 × 5.7 steel beam

8–53. Required $S = 13.85$ in^3; W12 × 16 steel beam

8–55. Required $S = 9.66$ in^3; W10 × 12 steel beam

8–57. Required $S = 38.0$ in^3; W12 × 30 steel beam

8–59. Required $S = 12.12$ in^3; W10 × 15 steel beam

8–61. Required $S = 0.32$ in^3; W8 × 10 steel beam

8–63. Required $S = 18.8$ in^3; 2 × 10 wood beam

8–65. 2 × 8 wood beam

8–68. Required $S = 11.1$ in^3; 2 × 8 wood beam

8–69. Required $S = 25.0$ in^3; 2 × 12 wood beam

8–70. Required $S = 2.79$ in^3; 2 × 4 wood beam

8–71. Required $S = 12.5$ in^3; 2 × 8 wood beam

8–75. 10 × 12 wood beam; No. 2 southern pine
W10 × 12 steel beam; ASTM A36 steel

8–77. $\sigma_d = 4.3$ MPa
At A: $\sigma = 3.81$ MPa; OK
At B: $\sigma = 5.16$ MPa; unsafe
At C: $\sigma = 4.62$ MPa; unsafe

8–79. 5.31 N/mm

8–81. 1180 N

8–83. 1045 lb

8–85. 102 lb

8–87. 6.77 lb/in

8–89. 3.20 ft from wall to joint
4-in pipe is safe at wall

8–91. 398 MPa at C

8–93. At fulcrum, $\sigma = 8000$ psi
At bottom hole, $\sigma = 5067$ psi
At next hole, $\sigma = 3800$ psi
At next hole, $\sigma = 2534$ psi
At next hole, $\sigma = 1267$ psi

8–94. At fulcrum, $\sigma = 8000$ psi
At bottom hole, $\sigma = 10\,000$ psi
At next hole, $\sigma = 7506$ psi
At next hole, $\sigma = 5004$ psi
At next hole, $\sigma = 2500$ psi

8–95. (a) With pivot in end hole as shown:
At fulcrum, $\sigma = 8000$ psi
At bottom hole, $\sigma = 8064$ psi
For pivot in any other hole, maximum stress is
at fulcrum. For pivot in:
Hole 2: $\sigma = 6800$ psi
Hole 3: $\sigma = 5600$ psi
Hole 4: $\sigma = 4400$ psi
Hole 5: $\sigma = 3200$ psi

8–97. 109 MPa

8–98. 149 MPa

8–99. Required $s_u = 1195$ MPa
AISI 4140 OQT 900 (others possible)

8–100. 2513 N

8–101. 1622 N

8–103. Impossible

8–105. Yes. $d_{max} = 37.2$ mm

8–106. 118 MPa at first step (L_3)

8–107. Required $s_u = 946$ MPa
AISI 1141 OQT 900 (others possible)

8–109. $L_{1max} = 206$ mm
$L_{2max} = 83.4$ mm
$L_{3max} = 24.7$ mm

8–111.

x (mm)	σ (MPa)
0	0
40	52.1
80	76.5
120	87.9
160	92.6
200	93.8
240	112.5

8–113. $h_1 = 22$ mm; $h_2 = 22$ mm

8–115. Required $S = 22.73$ in^3
W14 $\times$ 26

8–117. For composite beam, $w = 4.18$ K/ft
For S-beam alone: $w = 3.58$ K/ft

8–118. 11.5 mm

8–120. 0.805 in

8–122. 25.5 mm

8–124. 46 mm from center

8–126. $e = 12.9$ mm

8–127. 4.94 lb/in

8–129. 822 N

8–131. 625 N

8–133. 21.0 kN

8–134. 6.94 kN/m

8–135. 9.69 kN/m

8–136. 48.0 kN

8–137. 126 kN

Chapter 9

9–1. 1.125 MPa

9–3. 1724 psi

9–5. 3.05 MPa

9–7. 3180 psi

9–9. 1661 psi

9–11. 69.3 psi

9–13. 7.46 MPa

9–15. 2.79 MPa

9–17. 10.3 MPa

9–19. 2342 psi

9–21. 245 lb

9–23. 788 lb

9–25. 5098 lb

9–27. 661 lb

9–29. 787 lb

9–31. Let y = distance from bottom of the I-shape:

y (in)	τ (psi)
0.0	0.0
0.5	5.65
1.0(−)	10.54
1.0(+)	63.25
1.5	67.39
2.0	70.78
2.5	73.42
3.0	75.30
3.5	76.43
4.0	76.81

9–33. 8698 psi

9–35. From web shear formula:
τ = 8041 psi
Approximately 8% low compared with τ_{max} from 9–33.

9–37. Let y = distance from bottom of the I-shape:

y (in)	τ (psi)
0.0	0
0.175	155
0.35(−)	303
0.35(+)	6582
1.0	7073
2.0	7638
3.0	7976
4.0	8089

9–39. τ = 4352 psi;
τ_d = 14 400 psi; OK
σ = 26 100 psi;
σ_d = 23 760 psi; unsafe

9–41. W18 × 55; τ = 6300 psi;
τ_d = 14 400 psi; OK

9–43. $1\frac{1}{4}$ in schedule 40 pipe;
τ = 2404 psi;
τ_d = 8000 psi; OK

9–45. 4 × 10 beam

9–47. 1256 lb

9–49. 116.3 MPa

9–51. (a) τ = 1125 psi
(b) σ = 6750 psi
(c) s_y = 20 250 psi; any steel

9–53. 35.4 mm; τ = 1.59 MPa;
N = 86.7 for shear

9–55. 24.5 mm; τ = 1.28 MPa;
σ_d = 120 MPa

9–57. Required S = 0.45 in^3; 2-in schedule 40 pipe

9–59. q = 736 lb/in; τ = 526 psi

9–61. 431 lb/ft based on strength of rivets

9–63. 1722 lb based on bending

9–65. Maximum spacing = 4.36 in

9–67. Maximum spacing = 3.03 in

Chapter 10

NOTE: The complete solutions for Problems 10–1 through 10–27 require the construction of the complete Mohr's circle and the drawing of the principal stress element and the maximum shear stress element. Listed below are the significant numerical results.

Prob. No.	σ_1	σ_2	ϕ (deg)	τ_{max}	σ_{avg}	ϕ' (deg)
1	315.4 MPa	−115.4 MPa	10.9 cw	215.4 MPa	100.0 MPa	34.1 ccw
3	110.0 MPa	−40.0 MPa	26.6 cw	75.0 MPa	35.0 MPa	18.4 ccw
5	23.5 ksi	−8.5 ksi	19.3 ccw	16.0 ksi	7.5 ksi	64.3 ccw
7	79.7 ksi	−9.7 ksi	31.7 ccw	44.7 ksi	35.0 ksi	76.7 ccw
9	677.6 kPa	−977.6 kPa	77.5 ccw	827.6 kPa	−150.0 kPa	57.5 cw
11	327.0 kPa	−1202.0 kPa	60.9 ccw	764.5 kPa	−437.5 kPa	74.1 cw
13	570.0 psi	−2070.0 psi	71.3 cw	1320.0 psi	−750.0 psi	26.3 cw
15	4180.0 psi	−5180.0 psi	71.6 cw	4680.0 psi	−500.0 psi	26.6 cw
17	360.2 MPa	−100.2 MPa	27.8 ccw	230.2 MPa	130.0 MPa	72.8 ccw
19	23.9 ksi	−1.9 ksi	15.9 cw	12.9 ksi	11.0 ksi	29.1 ccw
21	4.4 ksi	−32.4 ksi	20.3 cw	18.4 ksi	−14.0 ksi	24.7 ccw
23	321.0 MPa	−61.0 MPa	66.4 ccw	191.0 MPa	130.0 MPa	68.6 cw
25	225.0 MPa	−85.0 MPa	0.0	155.0 MPa	70.0 MPa	45.0 ccw
27	775.0 kPa	−145.0 kPa	0.0	460.0 kPa	315.0 kPa	45.0 ccw

For Problems 10–29 through 10–39, Mohr's circle from the given data results in both principal stresses having the same sign. For this class of problems, the supplementary circle is drawn using the procedures discussed in Section 10–11 of the text. The results include three principal stresses where $\sigma_1 > \sigma_2 > \sigma_3$. Also, the maximum shear stress is found from the radius of the circle containing σ_1 and σ_3, and is equal to $\frac{1}{2}\sigma_1$ or $\frac{1}{2}\sigma_3$, whichever has the greatest magnitude. Angles of rotation of the resulting elements are not requested.

Prob. No.	σ_1	σ_2	σ_3	τ_{max}
29	328.1 MPa	71.9 MPa	0.0 MPa	164.0 MPa
31	214.5 MPa	75.5 MPa	0.0 MPa	107.2 MPa
33	35.0 ksi	10.0 ksi	0.0 ksi	17.5 ksi
35	55.6 ksi	14.4 ksi	0.0 ksi	27.8 ksi
37	0.0 kPa	−307.9 kPa	−867.1 kPa	433.5 kPa
39	0.0 psi	−295.7 psi	−1804.3 psi	902.1 psi

For Problems 10–41 through 10–49, Mohr's circles from earlier problems are used to find the stress condition on the element at some specified angle of rotation. The listed results include the two normal stresses and the shear stress on the specified element.

Prob. No.	σ_A	$\sigma_{A'}$	τ_A
41	130.7 MPa	69.3 MPa	213.2 MPa cw
43	−37.9 MPa	197.9 MPa	31.6 MPa ccw
45	3.6 ksi	−21.6 ksi	43.9 ksi cw
47	−2010.3 psi	510.3 psi	392.6 psi cw
49	8363.5 psi	86.5 psi	1421.2 psi cw

10–51. $\tau_{max} = 230.2$ MPa

10–53. $\tau_{max} = 12.9$ ksi

Chapter 11

11–1. −10 510 psi

11–3. $\sigma_N = 9480$ psi;
$\sigma_M = -7530$ psi

11–5. $\sigma_N = 13\,980$ psi;
$\sigma_M = -15\,931$ psi

11–7. −64.3 MPa

11–9. 415 N

11–11. $\sigma = 724$ MPa;
required $s_y = 1448$ MPa;
AISI 4140 OQT 700

11–13. At B: $\sigma = 24\,183$ psi tension on top of beam
$\sigma = -18\,674$ psi compression on bottom of beam

11–15. Load = 9650 N;
Mass = 983 kg

11–17. 26 mm

11–19. 18.7 mm

11–21. 71.6 MPa

11–23. 2548 psi

11–25. 7189 psi

11–27. 51.6 MPa

11–29. 982 MPa

11–31. $\tau = 9923$ psi, $s_y = 119$ ksi

11–33. 51.5 MPa

11–35. 7548 psi

11–37. 7149 psi

11–39. 61.5 MPa

11–41. (a) 3438 psi
(b) 3438 psi
(c) 344 psi
(d) 2300 psi
(e) 1186 psi

11–43.

	Middle of beam	Near supports	15 in from A
(a)	1875 psi	0 psi	1406 psi
(b)	1875 psi	0 psi	1406 psi
(c)	0 psi	375 psi	188 psi
(d)	1250 psi	208 psi	943 psi
(e)	625 psi	333 psi	498 psi

11–45. 75.3 MPa

11–47. (a) $P = 43\,167$ lb;
$\tau_{max} = 8333$ psi
(b) $\tau_{max} = 8862$ psi; $N = 2.82$

Chapter 12

12–1. -2.01 mm

12–3. -0.503 mm

12–5. -5.40 mm

12–7. At loads: -0.251 in
At center: -0.385 in

12–9. -0.424 in

12–11. -0.271 in

12–13. $+0.093$-in deflection at $x = 69.2$ in

12–15. $D = 64.8$ mm

12–17. $t = 0.020$ in

12–19. $y_B = -1.291$ mm at 840 N load
$y_C = -3.055$ mm at 600 N load
$y_D = -1.353$ mm at 1200 N load

12–22. $y_B = -0.0140$ in at 85 lb load
$y_C = -0.0262$ in at 75 lb load at end

12–23. -0.869 mm

12–25. -3.997 mm

12–26. -0.0498 in

12–29. $y = -0.0078$ in at $x = 8.56$ in

12–31. $y = -3.79$ mm

12–33. $D = 69.2$ mm

12–35. I7 $\times$ 5.800 aluminum beam
$y = -5.37$ mm at $x = 1.01$ m

12–37. $D = 109$ mm

12–39. -0.0078 in

12–41. -3.79 mm

12–43. -3.445 mm

12–45. -5.13 mm

12–47. -0.01138 in

12–49. -0.257 in

12–51. -71.7 mm

Chapter 13

For Problems 13–1 through 13–27, the following values are reported:

> Reactions at all supports
> Shearing forces at critical points
> Bending moments at critical points
> Maximum deflection or deflection at selected points in the form, $y = C_d/EI$

When you specify the beam material, shape, and dimensions, you can compute the beam stiffness, EI, and use

that to calculate the deflection. Deflections will be in the given unit of length when E and I are in the same units of force and length given in the answers. For example, in Problem 13–1, length is in m, force is in N. Then deflection is in m when E is in N/m^2 (Pa) and I is in m^4.

13–1. $R_A = V_A = 24\,063$ N, $R_C = -V_C = 10\,938$ N
$M_A = -26\,250$ N·m, $M_B = 21\,875$ N·m,
$M_C = 0$ N·m
$y_{max} = (-20\,934)/EI$ at 1.79 m from C.

13–3. $R_A = V_A = 18\,760$ N, $R_C = -V_C = 16\,235$ N
$M_A = -22\,560$ N·m, $M_B = 24\,353$ N·m,
$M_C = 0$ N·m
$y_B = (-21\,629)/EI$ at load

13–5. $R_A = V_A = 500$ lb, $R_B = -V_B = 300$ lb
$M_A = -1600$ lb·in, M_E (max) $= 900$ lb·in at
$x = 10.0$ in, $M_B = 0$ lb·in
$y_{max} = (-17\,712)/EI$ at $x = 9.264$ in

13–7. $R_A = V_A = 17\,500$ N, $R_C = -V_C = 17\,500$ N
$M_A = -17\,500$ N·m, $M_B = 17\,500$ N·m,
$M_C = -17\,500$ N·m
$y_{max} = (-11\,667)/EI$ at B at center

13–9. $R_A = V_A = 11\,074$ N, $R_C = -V_C = 23\,926$ N
$M_A = -12\,305$ N·m, $M_B = 15\,381$ N·m,
$M_C = -20\,508$ N·m
$y_{max} = (-10\,127)/EI$ at B at center

13–11. $R_A = V_A = 400$ lb, $R_B = -V_B = 400$ lb
$M_A = -1067$ lb·in, $M_B = 533$ lb·in,
$M_C = -1067$ lb·in
$y_{max} = (-8533)/EI$ at center

13–13. $R_A = V_A = 150$ lb, $R_B = 500$ lb,
$R_C = -V_C = 150$ lb
$M_A = 0$ lb·in, $M_B = -400$ lb·in, $M_C = 0$ lb·in
$M_D = M_E = 225$ lb·in at $x = 3.00$ in from A and C
$y_{max} = (-1107)/EI$ at $x = 3.372$ in from A or C

13–15. $R_A = R_D = 106.7$ lb, $R_B = R_C = 293.3$ lb
$V_A = -V_D = 106.7$ lb, $-V_B = V_C = 160$ lb
$M_A = M_D = 0$ lb·in, $M_B = M_C = -142.2$ lb·in,
$M_G = 35.6$ lb·in
$M_E = M_F = 113.8$ lb·in at $x = 2.13$ in from A and D

13–17. $R_A = R_E = 78.6$ lb, $R_B = R_D = 228.6$ lb,
$R_C = 185.6$ lb
$V_A = -V_E = 78.6$ lb, $-V_B = V_D = 121.4$ lb,
$V_C = 92.8$ lb
$M_A = M_E = 0$ lb·in, $M_B = M_D = -85.7$ lb·in,
$M_C = -57.1$ lb·in
$M_F = M_I = 61.8$ lb·in at $x = 1.56$ in from A and E
$M_G = M_H = 29.1$ lb·in at $x = 2.16$ in from B and D

13–19. $R_A = V_A = 22\,500$ N, $R_B = V_B = 13\,500$ N
$M_A = -8100$ N·m, $M_E = 4556$ N·m at 0.675 in
from B
$y_{max} = -1135/EI$ at 1.042 in from A

13–21. $R_A = V_A = -13\,750$ N, $R_B = 31\,750$ N
$V_B = -18\,000$ N
$M_A = 12\,600$ N·m, $M_B = -25\,200$ N·m, $M_C = 0$
$Y_C = -40\,719/EI$ at overhang at right end

13–23. $R_A = R_C = V_A = -V_C = 25\,200$ lb, $R_B = 8400$ lb
$V_B = 42\,000$ lb
$M_A = M_C = 0$, $M_B = -134\,400$ lb·ft
$M_D = M_E = 75\,587$ lb·ft at $x = 6.0$ ft from A or C

$y_{max} = (-1.488 \times 10^6 \text{ lb·ft}^3)/EI$ at 6.74 ft from A
or C

13–25. $R_A = R_E = V_A = -V_E = 212$ lb,
$R_B = R_D = 617$ lb, $R_C = 501$ lb
$V_B = V_D = 328$ lb, $V_C = 251$ lb
$M_A = M_E = 0$, $M_C = -1388$ lb·in
$M_B = M_D = -2082$ lb·in
$M_F = M_I = 1501$ lb·in at 14.04 in from A or E
$M_G = M_H = 708$ lb·in at 19.44 in from B or D

13–27. $R_A = V_A = 224.6$ N, $R_C = -V_C = 25.4$ N
$M_A = -2.355$ N·m, $M_B = 1.014$ N·m, $M_C = 0$
$y_B = (-1.386 \times 10^{-4})/EI$ at load

13–29. Comparison of results of five problems:

	V_{max}	V/V_1	M_{max}		M/M_1	y_{max}	y/y_1
1. 13–5	500 lb	1.0	−1600	lb·in	1.0	−17712/EI	1.0
2. 13–11	400 lb	0.80	−1067	lb·in	0.667	−8533/EI	0.482
3. 13–13	250 lb	0.50	−400	lb·in	0.250	−1107/EI	0.0625
4. 13–15	160 lb	0.320	−142	lb·in	0.089	—	—
5. 13–17	121 lb	0.243	−85.7	lb·in	0.054	—	—

13–33. Comparison of results of three problems:

	V_{max}	V/V_1	M_{max}	M/M_1	y_{max}	y/y_1	A	A/A_1
1. 13–30	1440 lb	1.0	103 680 lb·in	1.0	$-896 \times 10^6/EI$	1.0	63.3 in²	1.0
2. 13–31	900 lb	0.625	25 920 lb·in	0.25	$-372 \times 10^6/EI$	0.415	25.4 in²	0.401
3. 13–33	576 lb	0.40	9 216 lb·in	0.089	—	—	10.87 in²	0.172

13–35. Comparison of two designs in Figure 13–4:
(a) $V = 2250$ N, $M = -4500$ N·m,
   $y = -22\,500/EI$
(b) $V = 3375$ N, $M = -4500$ N·m,
   $y = -20\,250/EI = 0.90\,y_a$
Little difference in the performance of the two
designs.

13–37. $R_A = V_A = 32.75$ kN, $R_B = -V_B = 19.65$ kN
$M_A = -42.9$ kN·m, $M_E = 24.1$ kN·m

13–39. Comparison of four beam designs to carry a uniformly distributed load:

	V_{max}	V/V_1	M_{max}	M/M_1	y_{max}	y/y_1
(a)	3600 N	1.0	3600 N·m	1.0	−6000/EI	1.0
(b)	7200 N	2.0	−14 400 N·m	4.0	−57 600/EI	9.6
(c)	4500 N	1.25	−3600 N·m	1.0	−2490/EI	0.415
(a)	3600 N	1.0	−2400 N·m	0.67	−1200/EI	0.20

13–41. $R_A = V_A = 22\,500$ N, $R_B = V_B = 13\,500$ N
$M_A = -8100$ N·m, $M_E = 4556$ N·m at 0.675 in
from B
$y_{max} = -1135/EI$ at 1.042 in from A

13–43. $R_A = R_E = 371$ lb at ends of beam
$R_C = 858$ lb at middle support
$V_{max} = 429$ lb from B to D between two 800 lb loads
$M_{max} = 1113$ lb·ft under each load

13–45. $R_A = 56.5$ kN, $R_C = 135$ kN at middle support
$R_D = 16.5$ kN at right end
$V_{max} = -87.5$ kN at C
$M_{max} = 32.4$ kN·m at 80 kN load

13–47. $R_A = 4009$ lb at fixed end
$R_D = 1991$ lb at right support
$V_{max} = 4009$ lb at A
$M_{max} = -8490$ lb·ft at A

13–49. $R_A = R_E = 371$ lb at ends of beam
$R_C = 858$ lb at middle support
$V_{max} = 429$ lb from B to D between two 800 lb loads
$M_{max} = 1113$ lb·ft under each load

13–51. $R_A = 56.5$ kN, $R_C = 135$ kN at middle support
$R_D = 16.5$ kN at right end
$V_{max} = -87.5$ kN at C
$M_{max} = 32.4$ kN·m at 80 kN load

13–53. $R_A = 8.90$ kN, $R_C = 34.1$ kN at middle support
$R_D = 6.00$ kN at right end
$V_{max} = 18.0$ kN at C
$M_{max} = 8.9$ kN·m at B at 25 kN load

13–55. $R_A = 21.1$ kN, $R_C = 101.8$ kN at middle support
$R_E = 37.1$ kN
$V_{max} = 62.9$ kN at C
$M_{max} = 31.7$ kN·m at B under 60 kN load

Chapter 14

14–1. 25.1 kN

14–3. 8.35 kN

14–5. 26.2 kN

14–7. 111 kN

14–9. $P_a = 7318$ lb/column; use 9 columns

14–11. 499 kN

14–13. 65 300 lb

14–15. Axial force = 31.1 kN; critical load = 260 kN; $N = 8.37$ OK

14–17. 15.1 kN

14–19. Critical load = 10 914 lb; actual load = 5 000 lb; $N = 2.18$; low

14–21. Critical load = 2 849 lb; actual load = 1 500 lb; $N = 1.90$; low

14–23. 2.68 in

14–25. Required I = 5.02 in⁴; 17 × 5.800

14–27. 5869 lb

14–29. 245 kN

14–31. No improvement

Chapter 15

15–1. 115 MPa

15–3. 4.70 mm minimum

15–5. 2134 psi

15–7. 1.80 mm

15–9. $N = 3.80$

15–11. $\sigma_2 = 212$ MPa; $\sigma_3 = -70.0$ MPa

15–13.

Radius (mm)	σ_2 (MPa)
110	166 inner surface
120	149
130	136
140	125
150	116 outer surface

15–15. 14.88 MPa

15–17.

Radius (mm)	σ_3 (MPa)
15	−7.00 inner surface
17	−4.58
19	−2.88
21	−1.64
23	−0.71
25	0.00 outer surface

15–19. 86.8 MPa

15–21.

Radius (mm)	σ_3 (MPa)
210	−100.0 inner surface
215	−83.3
220	−68.0
225	−54.1
230	−41.4
235	−29.7
240	−19.0
245	−9.1
250	0.00 outer surface

15–23.

Radius (mm)	σ_2 (MPa)
162.5	22.35 inner surface
170	20.99
177.5	19.84
185	18.88
192.5	18.06
200	17.35 outer surface

15–25. Sizes 1/8 through 4 are thick-walled.
Sizes 5 through 18 are thin-walled.

Chapter 16

16–1. (a) $F_s = 6872$ lb (allowable); $F_b = 26\,100$ lb;
$F_t = 15\,179$ lb
(c) $F_s = 7732$ lb (allowable); $F_b = 19\,595$ lb;
$F_t = 17\,204$ lb

16–2. (a) $F_s = 17\,279$ lb (allowable); $F_b = 84\,000$ lb;
$F_t = 43\,110$ lb
(c) $F_s = 34\,558$ lb (allowable); $F_b = 84\,000$ lb;
$F_t = 43\,110$ lb

16–3. (a) $F_s = 11\,780$ lb (allowable); $F_b = 31\,500$ lb;
$F_t = 21\,083$ lb
(c) $F_s = 13\,253$ lb (allowable); $F_b = 23\,625$ lb;
$F_t = 23\,895$ lb

16–4. (a) $F_s = 17\,279$ lb (allowable); $F_t = 86\,220$ lb
(c) $F_s = 34\,558$ lb (allowable); $F_t = 86\,220$ lb

16–5. 1.25 in

16–7. 23 860 lb on welds

Index

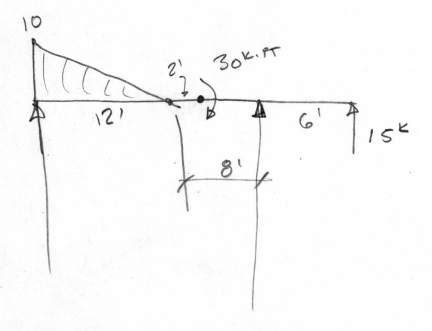

10

2'

30k.ft

12'

6'

15^K

8'